Textbook of

ANALYTICAL GEOMETRY

Textbook of
ANALYTICAL GEOMETRY

THIRD EDITION

P K Jain
Former
Professor and Head
Department of Mathematics
University of Delhi
Delhi

Khalil Ahmad
Former
Professor and Head
Department of Mathematics
Jamia Millia Islamia University
New Delhi

Committed to Educate the Nation

Publishing Globally

IN INDIA
NEW AGE INTERNATIONAL (P) LIMITED, PUBLISHERS
LONDON • NEW DELHI
Bangalore • Chennai • Cochin • Guwahati • Hyderabad • Kolkata • Mumbai
Visit us at **www.newagepublishers.com**

Published by New Age International (P) Ltd., Publishers
First Edition: 1985
Third Edition: 2014
Reprint: 2022

GLOBAL OFFICES

- **New Delhi** — **NEW AGE INTERNATIONAL (P) LIMITED, PUBLISHERS**
  7/30 A, Daryaganj, New Delhi-110002, (INDIA), **Tel.:** (011) 23253472, 23253771
  E-mail: contactus@newagepublishers.com • Visit us at www.newagepublishers.com
- **London** — **NEW AGE INTERNATIONAL (UK) LTD.**
  27 Old Gloucester Street, London, WC1N 3AX, UK
  E-mail: info@newacademicscience.co.uk • Visit us at www.newacademicscience.co.uk

BRANCHES

- **Bangalore** — 37/10, 8th Cross (Near Hanuman Temple), Azad Nagar, Chamarajpet, Bangalore- 560 018
  Tel.: (080) 26756823, **E-mail: bangalore@newagepublishers.com**
- **Chennai** — 26, Damodaran Street, T. Nagar, Chennai-600 017, **Tel.:** (044) 24353401
  E-mail: chennai@newagepublishers.com
- **Cochin** — CC-39/1016, Carrier Station Road, Ernakulam South, Cochin-682 016
  Tel.: (0484) 4051304, **E-mail: cochin@newagepublishers.com**
- **Guwahati** — Hemsen Complex, Mohd. Shah Road, Paltan Bazar, Near Starline Hotel
  Guwahati-781 008, **Tel.:** (0361) 2513881, **E-mail: guwahati@newagepublishers.com**
- **Hyderabad** — 105, 1st Floor, Madhiray Kaveri Tower, 3-2-19, Azam Jahi Road, Near Kumar Theater
  Nimboliadda Kachiguda, Hyderabad-500 027, **Tel.:** (040) 24652456
  E-mail: hyderabad@newagepublishers.com
- **Kolkata** — RDB Chambers (Formerly Lotus Cinema) 106A, 1st Floor, S N Banerjee Road
  Kolkata-700 014, **Tel.:** (033) 22273773, **E-mail: kolkata@newagepublishers.com**
- **Mumbai** — 142C, Victor House, Ground Floor, N.M. Joshi Marg, Lower Parel, Mumbai-400 013
  Tel.: (022) 24927869, 24915415, **E-mail: mumbai@newagepublishers.com**
- **New Delhi** — 22, Golden House, Daryaganj, New Delhi-110 002, **Tel.:** (011) 23262368, 23262370
  E-mail: sales@newagepublishers.com

ISBN: 978-81-224-3450-7
C-21-07-12788

Printed in India at Ajit Printing Press, New Delhi.
Typeset at In-house, New Delhi.

NEW AGE INTERNATIONAL (P) LIMITED, PUBLISHERS
7/30 A, Daryaganj, New Delhi - 110002
Visit us at **www.newagepublishers.com**
(CIN: U74899DL1966PTC004618)

Preface to the Third Edition

Till the seventeenth century, Algebra and Geometry had been considered as different Sciences, each having developed independent of the other. However, in 1637 René Descartes, a French mathematician and a philosopher published his book 'La Géométré' which introduced a device for unifying these two branches of mathematics. The basic feature of this new process, now called 'analytical geometry', concerns the use of a coordinate system. By means of coordinate systems algebraic methods can be applied powerfully to the study of geometry. Perhaps algebra can gain more through the pictorial representation of algebraic equations. In other words analytical geometry deals with a geometric problem by changing it into an algebraic problem so that it obeys certain rules of algebra and the proof of the problem becomes simpler. Since the time of Descartes analytical geometry has had a tremendous impact on the development of mathematical knowledge, and today analytical methods enter vitally into the diverse theoretical and practical applications of mathematics.

The present book on 'Analytical Geometry' is prepared by combining the following two earlier books by the authors:

1. *A textbook of Analytical Geometry of Two Dimensions. Wiley Eastern Limited, New Delhi, 1983. Revised 2nd Ed. July 1986.*
2. *A textbook of Analytical Geometry of Three Dimensions. Wiley Eastern Limited, New Delhi, 1986. Reprint August 1987. Revised 2nd Ed. February 1991.*

Both the above books were initially published by Wiley Eastern Limited which later on renamed as New Age International Publishers, New Delhi. After the second edition, each of the books has been reprinted several times till recent.

The contents of the book have been drawn from various streams of courses given to students by authors at various institutions during the last over forty years. This book is intended to serve as a textbook for a course on analytical geometry, of two dimensions and three dimensions, for the undergraduate students at various universities in India and abroad. It can be equally useful for the engineering students and also those taking competitive examinations such as IAS, IFS, PCS etc. In addition, this book will be of much help to the students at the Senior Secondary School level, particularly, those who are above average.

The book has been presented in two parts – Part A on Two dimensions and Part B on Three Dimensions. There are many examples, exercises, remarks and notes at different places which would help the readers' understanding and insight into the subject. The exercises cover a wide range from simple numerical problems to the development of the theory. In additions, exercises have been provided at the end of each section and each chapter ends with a miscellaneous set of exercises. Further, answers to the problems in the set of exercises are supplied at the end of the book. Care has been taken to grade the exercises on the difficulty scale.

Without claiming originality, we do claim simplicity and lucidity of presentation. Our debt to a large number of authors who have contributed to the discipline in the past. Although we have not tried to attribute the various results and proofs to their original discovers, we take it as our duty to place on record our gratitude to all of them. We are thankful to the generation of students who have made valuable contributions to injecting simplicity in presentation of the material so as to be intelligible to the student community in general. The authors are grateful to the readers who have given their observations and pointed out typos for bringing out this revised and combined edition of the earlier two books. In future also the authors will be happy to receive the suggestions and comments for further improvement.

Finally, our thanks go to Mr. Saumya Gupta, Managing Director, New Age International Pvt. Ltd., who has shown interest in bringing out this book in the present form, in fact, it was his idea to combine the earlier two books into one.

P. K. Jain

Khalil Ahmad

Contents

PART A Two Dimensions

PART A

Two Dimensions

Preliminary

1

CHAPTER

1. Introduction

The purpose of this chapter is to explain the basic concepts of the subject like coordinate systems, ratio formula and the idea of locus which are necessary for the development of the subject in the subsequent chapters. Assuming that the students have already studied an elementary course at the school level. The details are omitted, it is only for the sake of completeness that these concepts have been discussed here.

2. Cartesian Coordinates

The lines $X'OX$ and $Y'OY$ are called the x-axis and y-axis, respectively. These two lines are also referred to coordinate axes or simply the axes. The axes divide the plane into four parts XOY, YOX', $X'OY'$ and $Y'OX$ which are called respectively the first, second, third and fourth quadrants. The point O is called the origin. Here too we have the same rule for the sign convention as in trigonometry i.e., the lengths measured along OX and OY are positive while the lengths measured along OX' and OY' are negative.

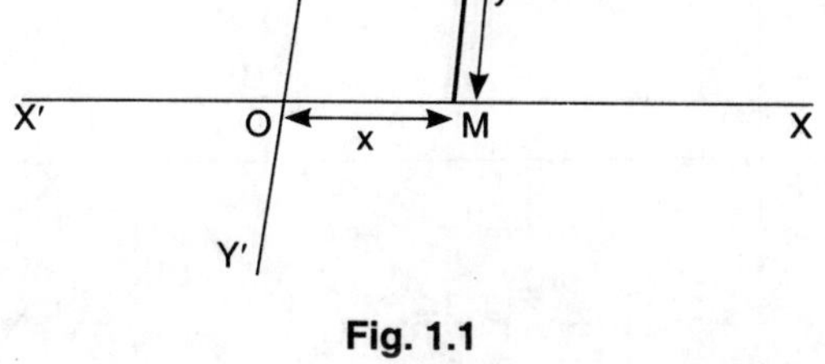

Fig. 1.1

Let P be any arbitrary point in the plane. Draw a line from P parallel to y-axis meeting the x-axis in the point M. Let $OM = x$ and $MP = y$. Then x and y are, respectively, called the x-coordinate and y-coordinate of the point P. They are also called the **abscissa** and **ordinate** of P. The pair (x, y) is referred to the cartesian coordinates or simply coordinates of the point P with reference to the coordinate axes $X'OX$ and $Y'OY$ and is usually denoted by $P(x, y)$.

Note. When the axes $X'OX$ and $Y'OY$ are perpendicular to each other, the coordinate axes are called rectangular, otherwise they are called oblique axes. Throughout we shall be dealing with rectangular coordinate axes unless specified otherwise and therefore by coordinate axes or simply axes we shall mean rectangular coordinate axes. However, for those students who are interested, we are giving some details about the oblique axes in the Appendix.

Remark. With reference to an arbitrary but fixed system of coordinate axes, for a given point P in the plane, there correspond unique coordinates (x, y), and conversely for a given ordered pair of real numbers (x, y), there corresponds a unique point in the plane with (x, y) as its coordinates. Thus, there is one-to-one correspondence between the set of all points in the plane and the set of all ordered pairs of real numbers, of course, with reference to a fixed system of coordinate axes.

2.1 Distance between two points. *To express the distance between two points in terms of their coordinates.*

Let $P(x_1, y_1)$ and $Q(x_2, y_2)$ be any two given points.

Draw PM, QN parallel to y-axis and PL parallel to x-axis as shown in the Fig. 1.2.

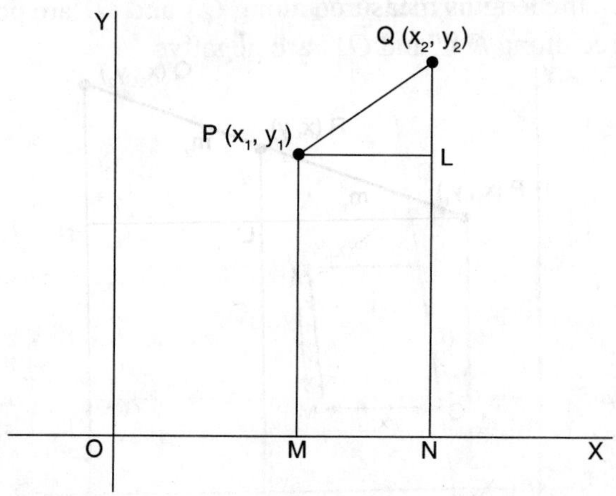

Fig. 1.2

Then $OM = x_1,\ MP = y_1,$

$ON = x_2,\ NQ = y_2.$

Now $PL = MN = ON - OM = x_2 - x_1,$

and $LQ = NQ - NL = NQ - MP = y_2 - y_1.$

Therefore, from right angled triangle PLQ, we have

$$PQ^2 = PL^2 + LQ^2 = (x_2 - x_1)^2 + (y_2 - y_1)^2$$

i.e. $$PQ = \sqrt{(x_2 - x_1)^2 + (y_2 - y_1)^2}.$$

Thus, we have obtained the distance formula between two points P and Q when both of these are lying in the first quadrant. But this formula will be found to be true for all positions of P and Q in the plane when due regard is paid to the signs of the coordinates (x_1, y_1) and (x_2, y_2).

2.2 The ratio formula. *To find the coordinates of a point which divides the straight line joining two given points in a given ratio.*

Let $P\ (x_1, y_1)$ and $Q\ (x_2, y_2)$ be any two given points. Let the point $R\ (x, y)$ divide PQ internally in the ratio $m_1 : m_2$.

Draw PM, RL, QN parallel to y-axis and $PL'N'$ parallel to x-axis. Then

$PL' = ML = OL - OM = x - x_1,$

and $L'N' = LN = ON - OL = x_2 - x.$

By geometry, we have

$$\frac{PL'}{L'N'} = \frac{PR}{RQ} = \frac{m_1}{m_2}$$

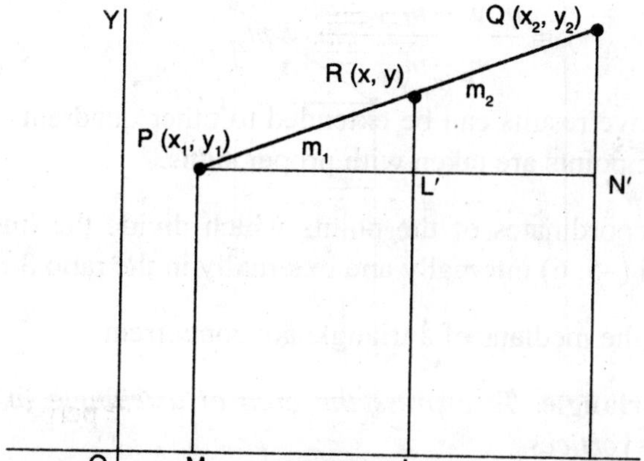

Fig. 1.3

$$\Rightarrow \qquad \frac{x - x_1}{x_2 - x} = \frac{m_1}{m_2}$$

$$\Rightarrow \qquad x = \frac{m_1 x_2 + m_2 x_1}{m_1 + m_2}.$$

Similarly $$y = \frac{m_1 y_2 + m_2 y_1}{m_1 + m_2}.$$

Corollary 1. The coordinates of the middle point of the line joining the points (x_1, y_1) and (x_2, y_2) are

$$\left(\frac{x_1 + x_2}{2}, \frac{y_1 + y_2}{2}\right)$$

Corollary 2. If R divides PQ externally in the ratio $m_1 : m_2$, then the coordinates of R are

$$\left(\frac{m_1 x_2 - m_2 x_1}{m_1 - m_2}, \frac{m_1 y_2 - m_2 y_1}{m_1 - m_2}\right)$$

Since, in this case the distances PL' and $L'N'$ are measured in the opposite sense, we have

$$\frac{PL'}{L'N'} = -\frac{m_1}{m_2}$$

which gives $$x = \frac{m_1 x_2 - m_2 x_1}{m_1 - m_2}, m_1 \neq m_2.$$

Similarly $$y = \frac{m_1 y_2 - m_2 y_1}{m_1 - m_2}, m_1 \neq m_2.$$

Remark. The above results can be extended to other quadrants provided the coordinates of the points are taken with proper signs.

Ex. 1. Find the coordinates of the points which divide the line joining the points (2, –8) and (–5, 6) internally and externally in the ratio 3 : 4.

Ex. 2. Show that the medians of a triangle are concurrent.

2.3. Area of a triangle. *To express the area of a triangle in terms of the coordinates of its vertices.*

Let ABC be the triangle. Let the coordinates of A, B and C be (x_1, y_1), (x_2, y_2) and (x_3, y_3), respectively.

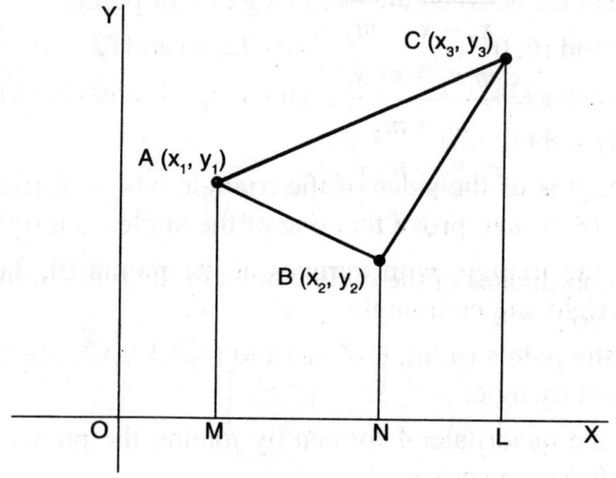

Fig. 1.4

Draw the lines AM, BN and CL parallel to y-axis. Then

$$\begin{aligned}\Delta ABC &= \text{trapezium } AMLC - \text{trapezium } AMNB - \text{trapezium } BNLC\\ &= \tfrac{1}{2}\,(MA + LC)\cdot ML - \tfrac{1}{2}\,(MA + NB)\cdot MN - \tfrac{1}{2}(NB + LC)\cdot NL\\ &= \tfrac{1}{2}\,(y_1 + y_3)(x_3 - x_1) - \tfrac{1}{2}(y_1 + y_2)(x_2 - x_1) - \tfrac{1}{2}(y_2 + y_3)(x_3 - x_2)\\ &= \tfrac{1}{2}\,\{(x_1y_2 + x_2y_3 + x_3y_1) - (y_1x_2 + y_2x_3 + y_3x_1)\}.\end{aligned}$$

Thus, *the area of the triangle whose vertices are* (x_1, y_1), (x_2, y_2) *and* (x_3, y_3) *is*

$$\tfrac{1}{2}\{(x_1y_2 + x_2y_3 + x_3y_1) - (y_1x_2 + y_2x_3 + y_3x_1)\}.$$

Note. The area of the triangle can also be expressed in the determinant form

$$\frac{1}{2}\begin{vmatrix} x_1 & y_1 & 1\\ x_2 & y_2 & 1\\ x_3 & y_3 & 1\end{vmatrix}.$$

Remark. The area of the triangle ABC is positive if the vertices A, B and C are taken in the counter-clockwise order but negative if the vertices are taken in clockwise order. However, in most of the cases only the absolute area is required.

Corollary. Three points are collinear if and only if the area of the triangle so formed by taking the points as vertices is zero.

Ex. Find the value of λ so that the points $(\lambda, 2)$, $(-2, 1)$ and $(-3, -2)$ are collinear.

Exercises

1. Find the distance between the following pair of points:
   (*i*) (5, 1) and (6, 0). (*ii*) (2, 8) and (2, –3).
   (*iii*) (1, 2) and (3, 4). (*iv*) (–1, –3) and (5, –6).
   (*v*) (2, –2) and (1, –5).
2. Find the lengths of the sides of the triangle whose vertices are (5, 1), (–3, 7) and (8, 5) and prove that one of the angles is a right angle.
3. Prove that the triangle with vertices at the points (0, 3), (–2, 1) and (–1, 4) is a right angled triangle.
4. Show that the points (a, a), $(-a, -a)$ and $(-a\sqrt{3}, a\sqrt{3})$ are the vertices of an equilateral triangle.
5. Show that the quadrilateral formed by joining the points (1, 0), (6, 1), (5, 6) and (0, 5) is a square.
6. Show that the four points (1, –4), (1, 0), (3, –2) and (–1, –2) form the vertices of a square and calculate the length of a diagonal.
7. Find the coordinates of the point which divides the straight line joining the given points in a given ratio:
   (*i*) (–7, 3), (2, –4); ratio 4 : 5. (*ii*) (0, 0), (7, 5); ratio 5 : 2.
   (*iii*) (5, 0), (–1, 2); ratio –4 : 3. (*iv*) (5, 2), (–1, 1); ratio –1 : 2.
   (*v*) (–3, 9), (5, –7); ratio 5 : 3.
8. Find the coordinates of the point which divides the line joining the points (–3, 4) and (5, 6) internally in the ratio 3 : 2.
9. Find the coordinates of the point which divides the line joining the points (–1, –5) and (1, –2) externally in the ratio 4 : 3.
10. Find the centroid of the triangle whose vertices are (–4, 6), (2, –2) and (2, 5).
11. In what ratio does the point (–1, –1) divide the join of (–5, –3) and (5, 2)?
12. Find the areas of the triangles with the following vertices:
    (*i*) (0, 0), (12, 0), (0, 5). (*ii*) (–2, 3), (4, 3), (1, 1).
    (*iii*) $(0, 0), (a, b), (b, a)$. (*iv*) $(0, 0), (2a \sin \alpha, 2a \cos \alpha), (\sin \beta, \cos \beta)$.
13. Find the area of the quadrilateral whose vertices taken in order are (1, 2), (6, 2), (5, 3) and (3, 4).
14. If *A, B, C* and *D* are the points (3, 1), (7, –3), (8, –1) and (19, –3), respectively, prove that the areas of the triangles *ABC* and *ADC* are equal in magnitude but opposite in sign.
15. Find the coordinates of the middle points of the sides of the triangle whose vertices are (x_1, y_1), (x_2, y_2), (x_3, y_3) and show that the area of the triangle formed by joining these points is one-fourth of that of the original triangle.

16. Show that the points (0, 4), (3, 2) and (6, 0) are collinear.

17. For what value of λ the points $(0, \lambda)$, $(-2, 1)$ and $(-3, -2)$ are collinear?

3. Polar Coordinates

There are various types of coordinate systems. The cartesian system, in particular, the rectangular cartesian system with which we have been dealing with is probably the most important. In that system, a point is located by its distances from two perpendicular lines. We shall introduce in this section a coordinate system in which the coordinates of a point in a plane are its distance from a fixed point and its direction from a fixed line. The coordinates given in this way are called polar coordinates. The proper choice of a coordinate system depends on the nature of the problem at hand. For some problems either the rectangular cartesian or the polar system may be satisfactory; usually, however, one of the two is preferable. In some situations, it is advantageous to use both systems, shifting from one to the other.

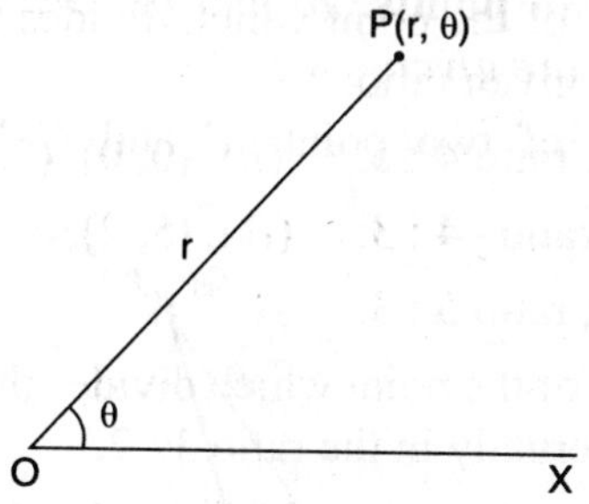

Fig. 1.5

In polar system of coordinates, we consider a fixed point O, called the **origin** (or **pole**) and a fixed straight line OX, called the **polar axis (or initial line)**. Consider a point P in the plane and let $OP = r$ and $\angle XOP = \theta$. The length r is called the **radius vector** of the point P and the angle XOP the **vectorial angle**. The ordered pair (r, θ) is called the **polar coordinates** of the point P.

Note. The radius vector is positive if it is measured from the pole along the line bounding the vectorial angle and is negative in the opposite direction.

If PO is produced to P' so that $OP' = OP$ in magnitude, and if (r, θ) be the coordinates of P, then coordinates of P' will be either $(-r, \theta)$ or $(r, \theta + \pi)$.

It is easy to see that

$$(r, \theta), (r, \theta + 2\pi), (r, \theta + 4\pi), \ldots;$$

$$(r, \theta - 2\pi), (r, \theta - 4\pi), \ldots;$$

$$(-r, \theta + \pi), (-r, \theta + 3\pi), \ldots;$$

and

$$(-r, \theta - \pi), (-r, \theta - 3\pi), \ldots;$$

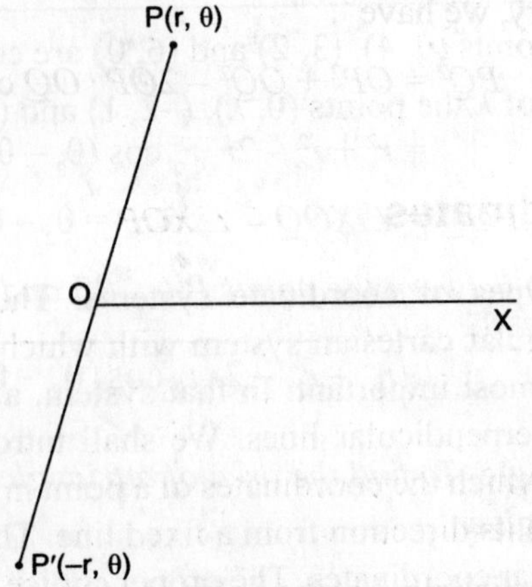

Fig. 1.6

all represent the coordinates of the same point P. Thus, in polar coordinates, unlike in cartesian system; the representation of a point is not unique.

3.1 Distance between two points. *To find the distance between two points whose polar coordinates are given.*

Let the coordinates of two points P and Q be (r_1, θ_1) and (r_2, θ_2), respectively.

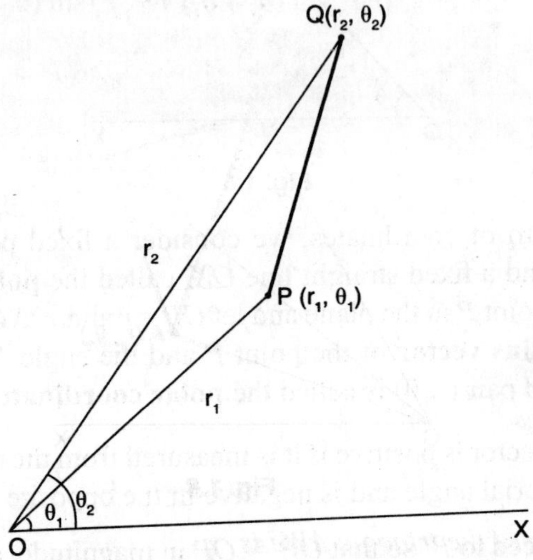

Fig. 1.7

Then
$$OP = r_1, \quad OQ = r_2,$$
$$\angle XOP = \theta_1, \quad \angle XOQ = \theta_2.$$

By trigonometry, we have

$$PQ^2 = OP^2 + OQ^2 - 2OP \cdot OQ \cos POQ$$
$$= r_1^2 + r_2^2 - 2r_1 r_2 \cos(\theta_2 - \theta_1)$$

since $\qquad \angle POQ = \angle XOQ - \angle XOP = \theta_2 - \theta_1.$

Hence, the distance between the points $P(r_1, \theta_1)$ *and* $Q(r_2, \theta_2)$ *is*

$$\sqrt{r_1^2 + r_2^2 - 2r_1r_2 \cos(\theta_2 - \theta_1)}.$$

3.2 Area of a triangle. *To find the area of a triangle the polar coordinates of whose vertices are given.*

Let ABC be the triangle. Let the polar coordinates of A, B and C be (r_1, θ_1), (r_2, θ_2) and (r_3, θ_3), respectively. Then

$$\Delta ABC = \Delta OAB + \Delta OBC - \Delta OAC$$
$$= \tfrac{1}{2} OA \cdot OB \sin AOB + \tfrac{1}{2} OB \cdot OC \sin BOC - \tfrac{1}{2} OA \cdot OC \sin AOC$$
$$= \tfrac{1}{2} r_1 r_2 \sin(\theta_2 - \theta_1) + \tfrac{1}{2} r_2 r_3 \sin(\theta_3 - \theta_2) - \tfrac{1}{2} r_3 r_1 \sin(\theta_3 - \theta_1)$$
$$= \tfrac{1}{2} \{r_1 r_2 \sin(\theta_2 - \theta_1) + r_2 r_3 \sin(\theta_3 - \theta_2) + r_3 r_1 \sin(\theta_1 - \theta_3)\}.$$

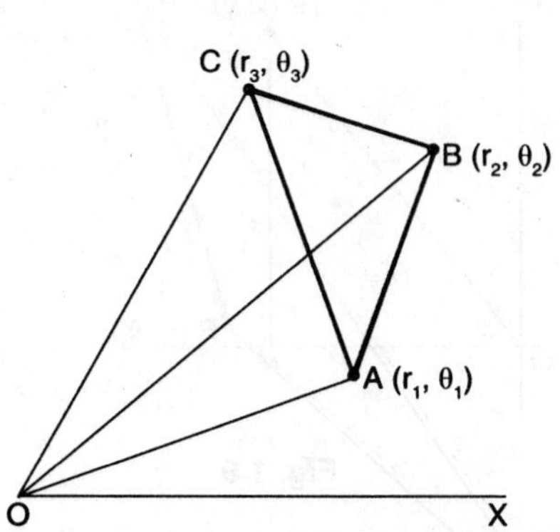

Fig. 1.8

Hence, *the area of the triangle ABC is*

$$\tfrac{1}{2}\{r_1 r_2 \sin(\theta_2 - \theta_1) + r_2 r_3 \sin(\theta_3 - \theta_2) + r_3 r_1 \sin(\theta_1 - \theta_3)\}.$$

4. Transformation

To change the cartesian coordinates of a point into polar coordinates and vice versa.

Let P be any point with cartesian coordinates (x, y) and polar coordinates (r, θ).

Draw PM parallel to y-axis. Then

$$OM = x, MP = y,$$

$$OP = r, \angle XOP = \theta.$$

Therefore

$$x = OP \cos\theta = r\cos\theta$$

and $$y = OP \sin\theta = r\sin\theta.$$

Also $$r = OP = \sqrt{x^2 + y^2}$$

and $$\tan\theta = \frac{MP}{OP} = \frac{y}{x}.$$

From these transforms it is easy to change any equation from cartesian form to polar form and vice versa.

Ex. 1. Transform $x^2 + y^2 - 2x + 2y = 0$ into polar form.

Ex. 2. Transform $r = 4a \cos\theta$ into cartesian form.

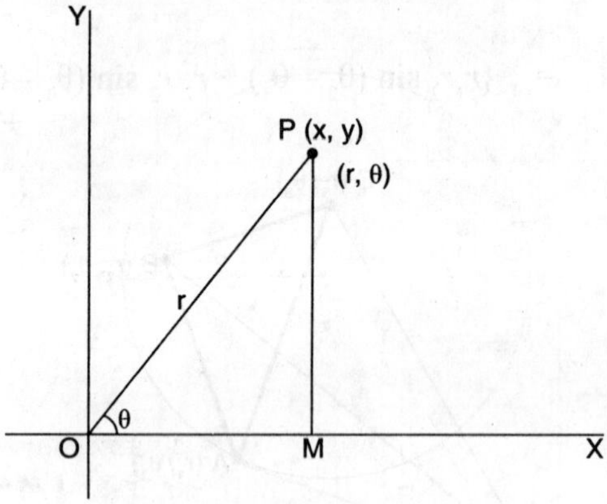

Fig. 1.9

Exercises

1. Find the lengths of the straight lines joining the pair of points whose polar coordinates are:

   (*i*) (3, 60°) and (5, 150°). (*ii*) (–6, 30°) and (4, 90°).

   (*iii*) (2, 40°) and (4, 100°). (*iv*) (2*a*, 30°) and (4*a*, 120°).

2. Prove that the points (0, 0°), (3, 90°) and (3, 30°) form an equilateral triangle.

3. Find the areas of the triangles with vertices

(*i*) (2, 60°), (3, 90°), (4, 120°). (*ii*) (–6, 30°), (4, 90°), (5, 150°).

(*iii*) (2*a*, 30°), (4*a*, 60°), (6*a*, 90°).

4. Transform the following equations to the corresponding polar coordinate equations:

(*i*) $3x + y = 0$. (*ii*) $x^2 + y^2 = 16$.

(*iii*) $y^2 = 4\,ax$. (*iv*) $(x^2 + y^2)^2 = 2\,a^2\,xy$.

5. Transform the following equations to the corresponding cartesian coordinate equations:

(*i*) $r^2 = a^2 \cos 2\theta$. (*ii*) $r = 8 \cos \theta$.

(*iii*) $r = \dfrac{4}{1+\cos\theta}$.

5. Locus

The path traced by a moving point under certain geometrical conditions is called the **locus** *of that point.* For instance, the locus of the point *P* which moves such that it remains at a constant distance 2 from the fixed point *A* is a circle, Fig. 1.10.

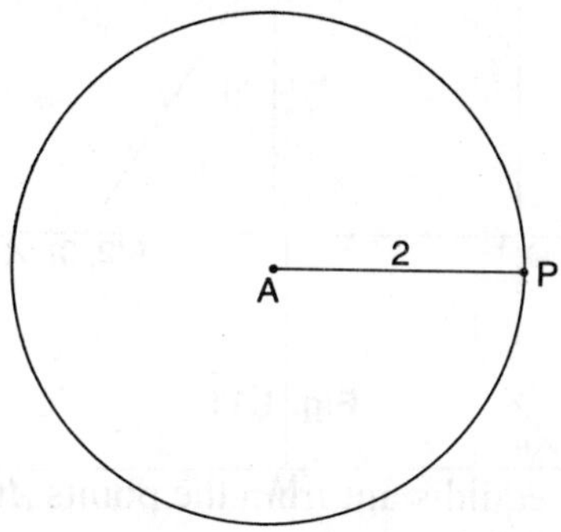

Fig. 1.10

The sciences of algebra and geometry correlated by means of the coordinate systems lead to a study of coordinate geometry. In plane coordinate geometry, the underlying feature is the correspondence between an equation in *x* and *y*, the coordinates of a variable point, and a geometric figure of the locus (path) of the variable point. In regard to this there are two problems to be studied. One is that we are given some geometrical facts which limits the positions of the moving point *P* and asked to find what relation is then satisfied by *x* and *y*, the coordinates of *P*. Conversely, an equation connecting *x* and *y* is given, and the problem is to find out on what geometrical locus *P* lies.

5.1 Equation of the locus. *An equation of a locus is a relation between x and y which is satisfied by the coordinates of all points of the locus and by no others.*

Illustration 1. $P(x, y)$ is equidistant from the point $A(2, 0)$ and y-axis. Find the equation of the locus of P.

Here $$PA = \sqrt{(x-2)^2 + y^2}.$$

It is given that

$$PA = \text{distance of } (x, y) \text{ from } y\text{-axis}$$

i.e. $$\sqrt{(x-2)^2 + y^2} = x$$

$\Rightarrow$ $$(x-2)^2 + y^2 = x^2$$

$\Rightarrow$ $$y^2 = 4(x-1)$$

This is the required equation of the locus of P.

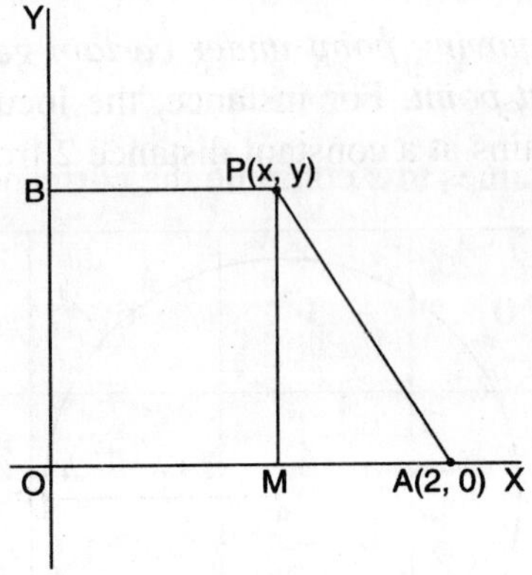

Fig. 1.11

Illustration 2. $P(x, y)$ is equidistant from the points $A(2, 3)$ and $B(3, -1)$. Find the equation of the locus of P.

Here $$PA = \sqrt{(x-2)^2 + (y-3)^2}$$

$$PB = \sqrt{(x-3)^2 + (y+1)^2}$$

Given that $PA = PB$.

$\therefore$ $$(x-2)^2 + (y-3)^2 = (x-3)^2 + (y+1)^2$$

$\Rightarrow$ $$2x - 8y + 3 = 0.$$

This is the required equation of the locus of P.

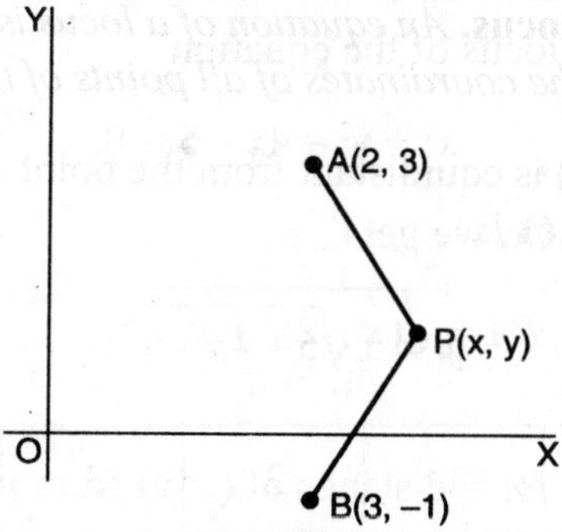

Fig. 1.12

5.2 Locus of the equation. *The locus (or graph) of an equation in x and y consists of all points whose coordinates satisfy the given equation.*

Illustration 1. Plot the locus of the equation $4x + 6y - 3 = 0$.

Solving for y in terms of x, we get

$$y = \frac{3 - 4x}{6}.$$

By giving different values to x compute the corresponding values of y.

x	0	3	6	9	12
y	$\frac{1}{2}$	$-\frac{3}{2}$	$-\frac{7}{2}$	$-\frac{11}{2}$	$-\frac{15}{2}$

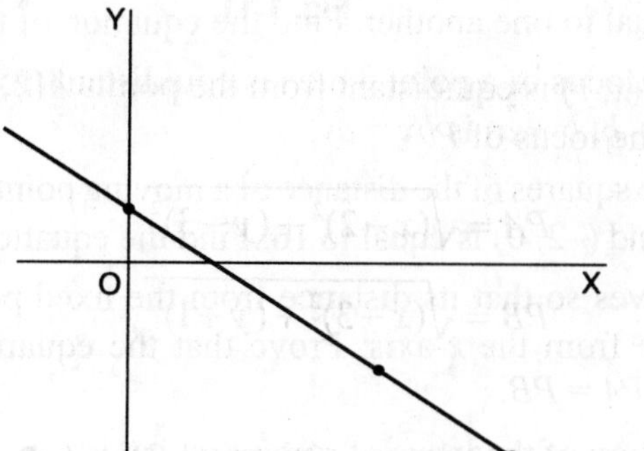

Fig. 1.13

Illustration 2. Plot the locus of the equation

$$x^2 + y^2 - 4x - 5 = 0.$$

Solving for y in terms of x, we get

$$y = \pm\sqrt{5 + 4x - x^2}$$

x	0	1	2	3	4	5
y	$\pm\sqrt{5}$	$\pm 2\sqrt{2}$	± 3	$\pm 2\sqrt{2}$	$\pm\sqrt{5}$	0

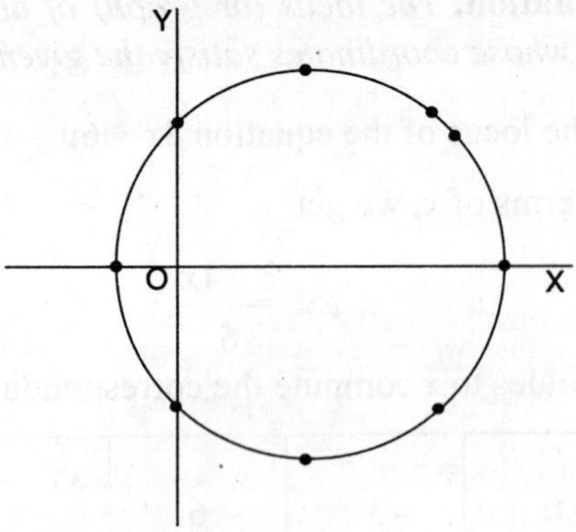

Fig. 1.14

Exercises

1. A point P moves so that its distances from the two points (3, 4) and (5, –2) are equal to one another. Find the equation of the locus of P.
2. Prove that the locus of a point which is equidistant from the point $(a + b, b - a)$ and $(a - b, b + a)$ is $bx = ay$.
3. The sum of the squares of the distance of a moving point from the two fixed points (2, 0) and (–2, 0) is equal to 16. Find the equation of its locus.
4. A point P moves so that its distance from the fixed point (0, 2) is equal to its distance from the x-axis. Prove that the equation of the locus is $x^2 = 4(y - 1)$.
5. Find the equation of the locus of a point which is at a distance 3 from the point (3, –1).
6. A point moves so that its distance from the x-axis is half of its distance from the origin. Find the equation of its locus.

7. Find the equation of the locus of a point whose distance from $(-1, 1)$ is equal to twice its distance from the x-axis.

8. Given $A(a, b)$ and $B(3a, 3b)$, show that if $P(x, y)$ is a point such that $PA = PB$, then $ax + by = 2(a^2 + b^2)$.

9. Find the locus of a point P which moves such that its distance from $(0, 3)$ is equal to the ordinate of P.

10. Find the locus of a point P which moves such that the difference of the distance from P to $(-5, 0)$ and $(5, 0)$ is numerically equal to 8.

The Straight Line

2

CHAPTER

1. Introduction

The simplest locus of a point in a plane is a straight line. In this chapter, we obtain the equations of a straight line, in various forms, and obtain a number of basic properties of the straight lines. Throughout the book we shall be calling a straight line simply a line.

2. Particular Lines

Before we come to a general discussion of the equations of a line, we examine one or two particular cases.

To find the equation of a line parallel to one of the coordinate axes.

Let AB be a line parallel to y-axis. Clearly AB is the locus of a point which moves such that its distance from y-axis remains fixed. Let $P(h, k)$ be the moving point. Then, by definition, P is at a distance a (say) from y-axis. Therefore,

$$h = a.$$

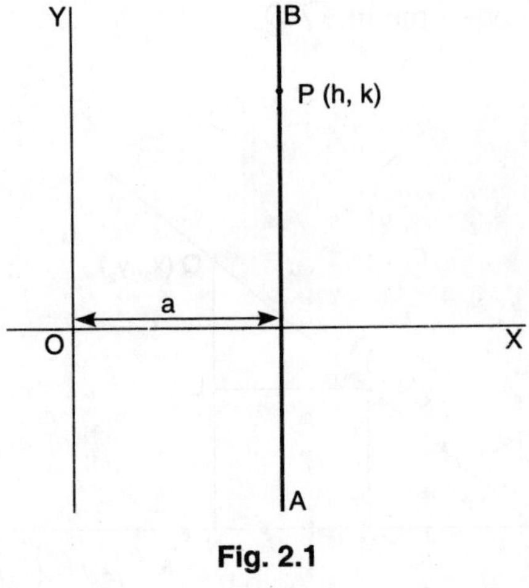

Fig. 2.1

Thus, the locus of (h, k) is $x = a$, which is the required equation of the line AB.

Similarly, the equation of the line parallel to x-axis which is at a distance b is

$$y = b.$$

Corollary. The equation of x-axis is $y = 0$ and that of y-axis is $x = 0$.

3. Slope of a Line

If any two points on a line are taken, their join makes a constant angle with a fixed direction, and the angle so formed is independent of the choice of the two points on the line. This is a precise way of saying that any line has a constant slope. It is customary to measure the angle α which a line makes with the positive direction of the x-axis. Then, the quantity $\tan \alpha$ is defined to be the **slope** of the line and is denoted by 'm'.

Note. The slope of a line is also sometimes referred to **gradient** of the line.

3.1 *To find the slope of the line joining the points P (x_1, y_1) and Q (x_2, y_2).*

Draw PM, QN parallel to y-axis and PL parallel to x-axis. Let the line joining P and Q make an angle α with x-axis. Then

$$\tan \alpha = \frac{LQ}{PL}.$$

But $$LQ = NQ - NL = NQ - MP = y_2 - y_1,$$

and $$PL = MN = ON - OM = x_2 - x_1,$$

$\therefore$ $$\tan \alpha = \frac{y_2 - y_1}{x_2 - x_1},$$

which is the required slope of the line PQ.

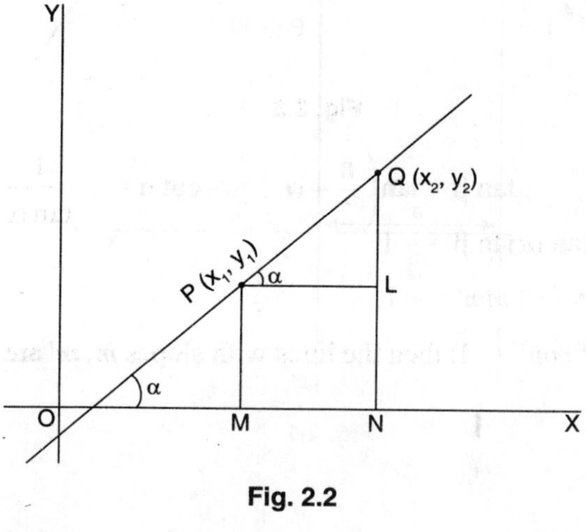

Fig. 2.2

Ex. Find the slope of the line joining the points (1, – 4) and (5, – 3).

3.2 Parallel and perpendicular lines. Two lines are said to be **parallel** if they make the same angle with x-axis and hence they have the same slope. Conversely, if two lines have the same slope, they make the same angle with x-axis and hence they are parallel. Thus, if m and m' be the slopes of any two parallel lines, then

$$m = m'.$$

Note. The lines parallel to x-axis have **zero slope** whereas the lines parallel to y-axis have the **slope** ∞.

Further consider any two perpendicular lines with slopes m, m' and making angles α, β with x-axis, respectively. Then

$$\tan\alpha = m, \tan\beta = m'.$$

From Fig. 2.3, it is clear that

$$\beta = \frac{\pi}{2} + \alpha.$$

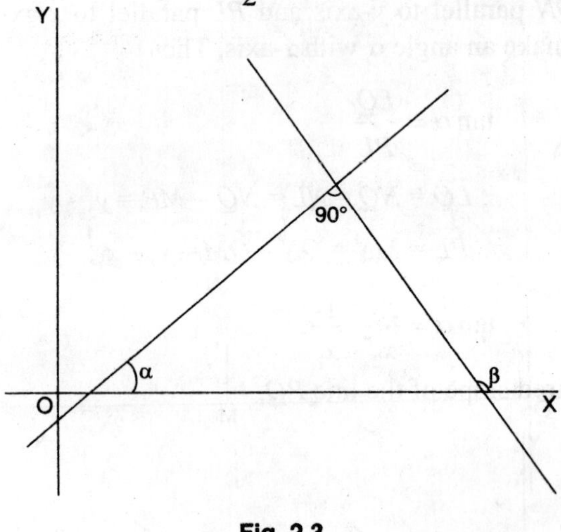

Fig. 2.3

$$\therefore \qquad \tan\beta = \tan\left(\frac{\pi}{2} + \alpha\right) = -\cot\alpha = -\frac{1}{\tan\alpha}$$

$$\Rightarrow \qquad \tan\alpha\tan\beta = -1$$

$$\Rightarrow \qquad mm' = -1.$$

Conversely, if $mm' = -1$, then the lines with slopes m, m' are perpendicular to each other.

4. Various Forms of the Equation of a Line

4.1 Slope intercept form. *To find the equation of a line which cuts off a given intercept on y-axis and is inclined to a given angle to the x-axis.*

Let AB be the line having slope m and making intercept c on y-axis. Let α be the angle which the line AB makes with the positive direction of x-axis so that $m = \tan \alpha$.

Let $P(x, y)$ be any point on the line AB. Draw PM parallel to y-axis and CN parallel to x-axis. Then

$$CN = OM - x$$

and
$$NP = MP - MN = MP - OC = y - c$$

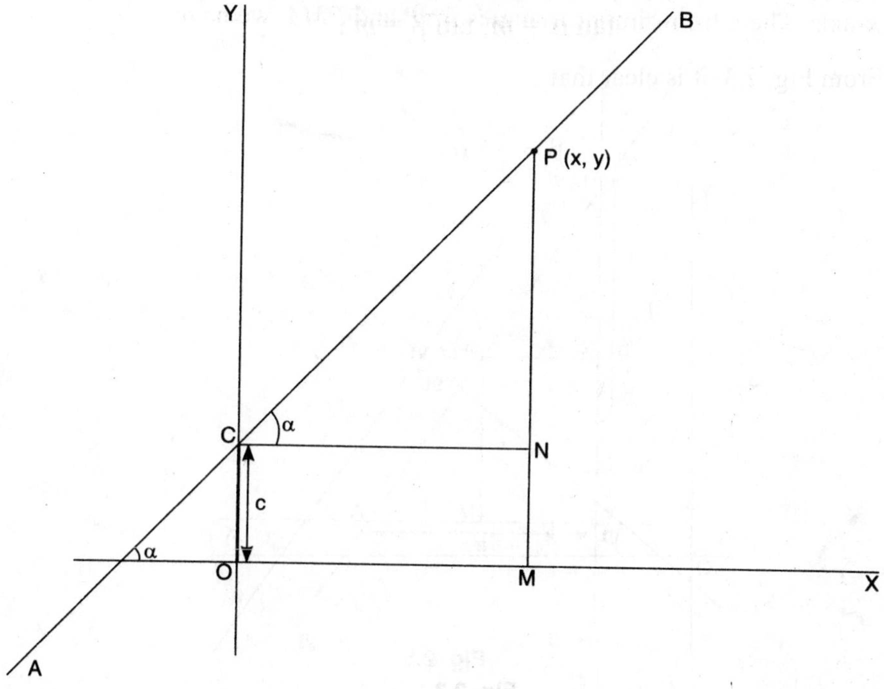

Fig. 2.4

In ΔPCN, $\angle PNC = 90°$. Therefore

$$\frac{NP}{CN} = \tan \alpha$$

$\Rightarrow$
$$\frac{y-c}{x} = \tan \alpha$$

$\Rightarrow$
$$y = x \tan \alpha + c.$$

Hence *the equation of the line having slope m and making an intercept c on y-axis is*

$$y = mx + c$$

Corollary. The equation of the line passing through the origin is $y = mx$.

4.2 Intercept form. *To find the equation of a line in terms of the intercept which it makes on the axes.*

Let AB be the line which cuts off intercepts a and b, respectively, on x-axis and y-axis so that

$$OA = a \text{ and } OB = b$$

Let $P(x, y)$ be any point on AB. Draw PM parallel to y-axis and PN parallel to x-axis. Then, from similar triangles BNP and PMA, we have

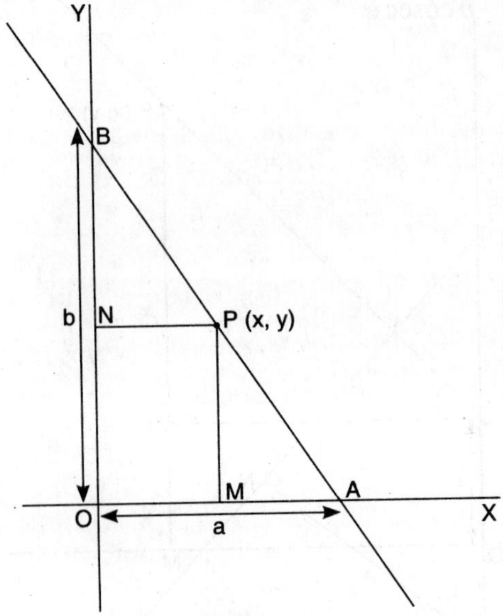

Fig. 2.5

$$\frac{NB}{MP} = \frac{NP}{MA}$$

$$\Rightarrow \quad \frac{OB - ON}{ON} = \frac{OM}{OA - OM}$$

$$\Rightarrow \quad \frac{b - y}{y} = \frac{x}{a - x}$$

$$\Rightarrow \quad bx + ay = ab.$$

Hence, *the equation of the line making intercepts a and b on the axes is*

$$\frac{x}{a}+\frac{y}{b}=1.$$

4.3 Normal form. *To find the equation of a line in terms of the length of the perpendicular on it from the origin and the angle which that perpendicular makes with x-axis.*

Let AB be the line whose perpendicular distance from the origin is p (say). Draw ON perpendicular to AB. Then $ON = p$. Let ON make an angle α with the positive direction of x-axis.

Now $OA = p \sec \alpha$ and $OB = p \operatorname{cosec} \alpha$. Clearly OA and OB are the intercepts of AB on the axes. Hence, by §4.2, we have

$$\frac{x}{p\sec\alpha}+\frac{y}{p\operatorname{cosec}\alpha}=1$$

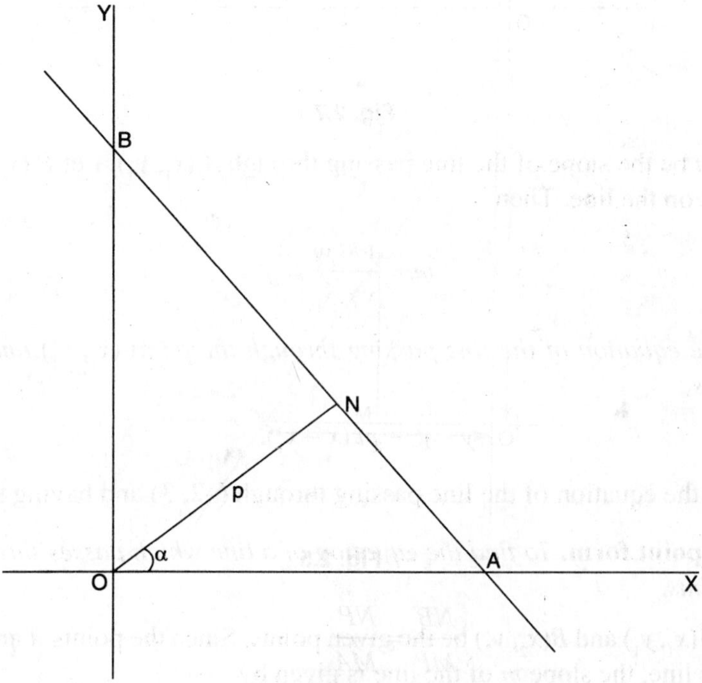

Fig. 2.6

$$\Rightarrow \qquad x\cos\alpha+y\sin\alpha=p.$$

This is the required equation of the line AB.

Note. Normal form is also referred to **'perpendicular form'**.

4.4 Point-slope form. *To find the equation of a line which passes through a given point and makes an angle with x-axis.*

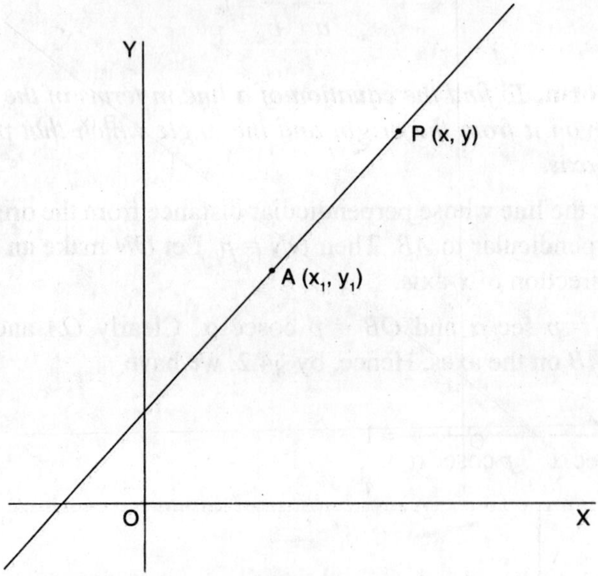

Fig. 2.7

Let m be the slope of the line passing through $A\ (x_1, y_1)$. Let $P\ (x, y)$ be any point on the line. Then

$$m = \frac{y - y_1}{x - x_1}.$$

Hence *the equation of the line passing through the point* (x_1, y_1) *and having slope m is*

$$y - y_1 = m\ (x - x_1).$$

Ex. Find the equation of the line passing through (–2, 3) and having slope 2.

4.5 Two-point form. *To find the equation of a line which passes through two given points.*

Let $A(x_1, y_1)$ and $B(x_2, y_2)$ be the given points. Since the points A and B both lie on the line, the slope m of the line is given by

$$m = \frac{y_2 - y_1}{x_2 - x_1}.$$

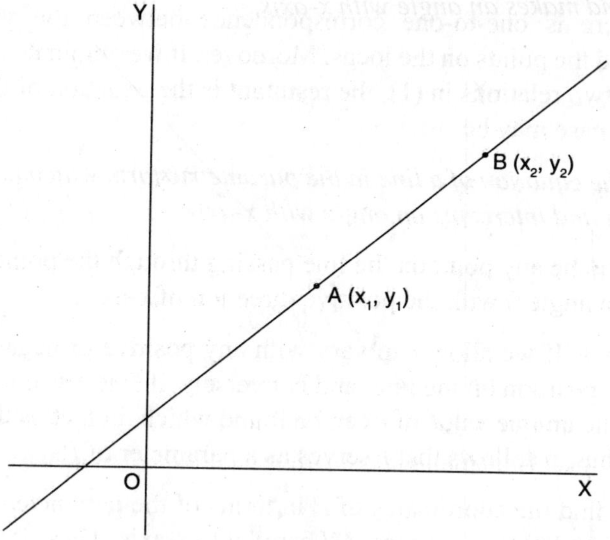

Fig. 2.8

The equation of the line through the point $A(x_1, y_1)$ having slope m is

$$y - y_1 = m\,(x - x_1).$$

Hence, *the line through the points* $A(x_1, y_1)$ *and* $B(x_2, y_2)$ *is*

$$y - y_1 = \frac{y_2 - y_1}{x_2 - x_1}(x - x_1).$$

Corollary. The equation of the line joining the origin to the point (x_1, y_1) is $y = \frac{y_1}{x_1} x$ *i.e.*, $xy_1 - x_1y = 0$.

Ex. Find the equation of the line passing through the points (1, 2) and (3, –4).

4.6 Parametric form. In coordinate geometry, very oftenly, we are interested to express the coordinates of a variable point on a line or on a curve in terms of a single variable called the **parameter**, viz., in the form

$$\left.\begin{aligned} x &= \phi(t) \\ y &= \psi(t) \end{aligned}\right\}. \qquad (1)$$

Here t is the parameter. In this regard we must have the following properties:

(*i*) Given a point on the locus there corresponds a unique value of the parameter.

(*ii*) Given a value of the parameter, there corresponds a unique point on the locus.

Thus there is one-to-one correspondence between the values of the parameter and the points on the locus. Moreover, if we eliminate the parameter between the two relations in (1), the resultant is the equation of the line or the curve, as the case may be.

To find the equation of a line in the parametric form which passes through a given point and intercepts an angle with x-axis.

Let $P(x, y)$ be any point on the line passing through the point $A(x_1, y_1)$ and inclined at an angle θ with the positive direction of x-axis.

Let $AP = r$. If we allow r to vary with any positive or negative values, P will take any position on the line; and conversely, if P is given to be any point on the line, the unique value of r can be found which, in fact, is the distance of P from A. Thus, it follows that r serves as a parameter of P.

Now, to find the coordinates of P in terms of the parameter r, let us draw AL and PM parallel to y-axis and AN parallel to x-axis. Then, it follows that

$$\begin{cases} OM = OL + LM = OL + AN \\ MP = MN + NP = LA + NP \end{cases}$$

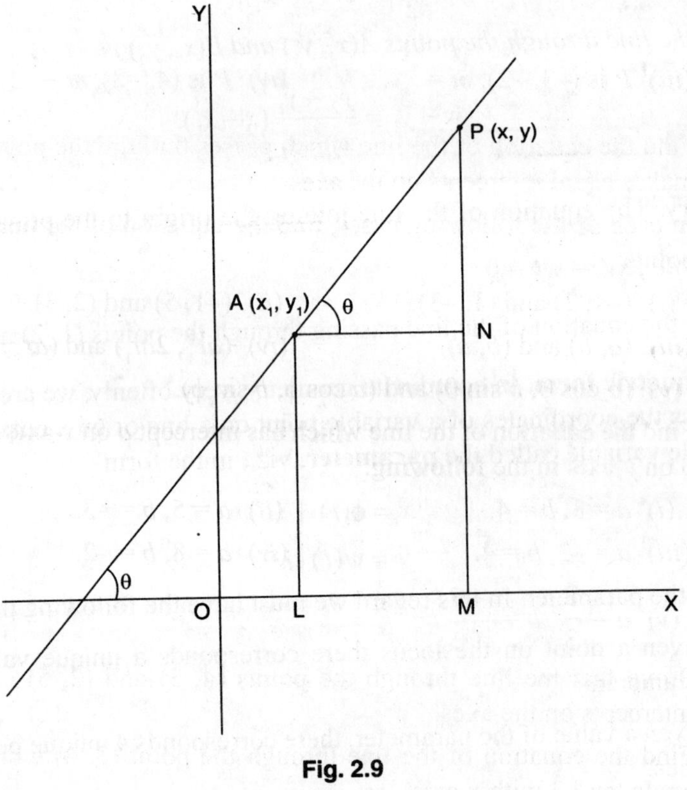

Fig. 2.9

$$\Rightarrow \begin{cases} x = x_1 + r\cos\theta \\ y = y_1 + r\sin\theta \end{cases}$$

Hence, *the equation of the line, in the parametric form, through A* (x_1, y_1) *and inclined at any angle* θ *is*

$$\frac{x - x_1}{\cos\theta} = \frac{y - y_1}{\sin\theta} = r. \qquad (2)$$

Remark. The parametric form (2) of the equation of a line is very useful. It is noted that the coordinates of any point P on the line are immediately expressed if we know to algebraic distance of P from the given point $A(x_1, y_1)$ measured along the line which, in fact, are

$$x = x_1 + r\cos\theta,\ y = y_1 + r\sin\theta.$$

Exercises

1. In each of the following cases, find the equation of the line through the point P with slope m:

   (*i*) P is (4, 1), $m = -\dfrac{1}{2}$. (*ii*) P is (1, 5), $m = 3$.

   (*iii*) P is (–3, –2), $m = \dfrac{1}{5}$. (*iv*) P is (4, –2), $m = -1$.

2. Find the equation of the line which passes through the point (3, –1) and makes equal intercepts on the axes.

3. In each of the following cases, find the equation of the line through the points:

   (*i*) (–4, 2) and (1, –3). (*ii*) (–1, 5) and (2, 3).

   (*iii*) (a, b) and (b, a). (*iv*) $(at_1^2, 2at_1)$ and $(at_2^2, 2at_2)$.

   (*v*) $(a\cos\theta, b\sin\theta)$ and $(a\cos\phi, b\sin\phi)$.

4. Find the equation of the line which has intercept a on x-axis and intercept b on y-axis in the following:

   (*i*) $a = 3, b = 4$. (*ii*) $a = 5, b = -3$.

   (*iii*) $a = -3, b = 4$. (*iv*) $a = 8, b = -2$.

   (*v*) $a = \dfrac{2}{3}, b = -\dfrac{5}{8}$.

5. Prove that the line through the points (4, 3) and (2, 5) cuts off equal intercepts on the axes.

6. Find the equation of the line through the point (3, 6) which makes an angle $\tan^{-1} 3$ with x-axis.

7. Find the equation of the line joining the points (3, 2) and (–1, –5). Prove that the point $(1, \frac{7}{2})$ lies on it, and find where it meets the axes.
8. What is the slope of the line $3x + 4y = 1$? Find the equations of the lines through the point (3, –1) which are parallel and perpendicular to it.
9. Find the equations of the medians of the triangle whose vertices are (4, 3), (–2, 5) and (2, –1).
10. Find the equation of the line through the point (2, –1) perpendicular to $4x - 3y = 6$.
11. Find the equation of the line with slope 3 which meets the x-axis in the same point as the line $x + 4y = 8$.

5. General Equation of First Degree

From the various forms of the equations of a line obtained so far in §4, we observe that the equation of a line is a first degree (or linear) equation in x and y, the coordinates of the moving point on the line. It is a natural temptation to examine the converse problem, viz., whether a first degree equation in x and y always represents a line.

5.1 *To prove that every equation of first degree in x and y represents a line.*

The general equation of first degree in x and y is

$$ax + by + c = 0, \qquad (1)$$

where a, b, c are constants.

Let (x_1, y_1), (x_2, y_2) and (x_3, y_3) be any three points lying on the locus of (1). Therefore

$$ax_1 + by_1 + c = 0 \qquad (2)$$

$$ax_2 + by_2 + c = 0 \qquad (3)$$

and $$ax_3 + by_3 + c = 0. \qquad (4)$$

On eliminating a, b, c from equations (2), (3) and (4), we get

$$\begin{vmatrix} x_1 & y_1 & 1 \\ x_2 & y_2 & 1 \\ x_3 & y_3 & 1 \end{vmatrix} = 0. \qquad (5)$$

Equation (5) shows that the area of the triangle with vertices (x_1, y_1), (x_2, y_2) and (x_3, y_3) is zero [§2.3, Chapter 1]. Hence the three points lying on the locus of (1) are collinear which shows that equation (1) represents a line.

5.2 Reduction to standard forms. *To reduce the general equation of first degree in standard forms of the line.*

Let
$$ax + by + c = 0 \tag{1}$$
be the general equation of first degree in x and y. From §5.1, equation (1) represents a line whose various forms have been obtained as follows:

(I) Slope-intercept form. Equation (1) can be written as

$$y = -\frac{a}{b}x - \frac{c}{b}, b \neq 0,$$

which is of the form $y = mx + k$, where

$$m = -\frac{a}{b} \text{ and } k = -\frac{c}{b}.$$

This is the **Slope-intercept form** of the line.

Note. The slope of the line $ax + by + c = 0$ is given by

$$m = -\frac{a}{b} = -\frac{\text{coefficient of } x}{\text{coefficient of } y}.$$

(II) Intercept form. On dividing equation (1) by $-c(c \neq 0)$, we get

$$\frac{a}{-c}x + \frac{b}{-c}y - 1 = 0$$

$$\Rightarrow \qquad \frac{x}{-\frac{c}{a}} + \frac{y}{-\frac{c}{b}} = 1,$$

which is the **intercept form** of the line. Clearly $-\frac{c}{a}$ and $-\frac{c}{b}$ are the intercepts on the axes.

Note. If $a \to 0$, or $b \to 0$ and $c \neq 0$, then the line (1) is parallel to one of the coordinate axes as the case may be. If both $a \to 0$ and $b \to 0$, the line (1) is parallel neither to x-axis nor to y-axis and in this case $c = 0$ which represents a line lying at infinity.

(III) Normal form. Equation (1) can also be written as

$$\frac{a}{\sqrt{a^2 + b^2}}x + \frac{b}{\sqrt{a^2 + b^2}}y = -\frac{c}{\sqrt{a^2 + b^2}}$$

This equation is of the form

$$x \cos \alpha + y \sin \alpha = p,$$

which is the **normal form**, where

$$\cos \alpha = \frac{a}{\sqrt{a^2+b^2}}, \sin \alpha = \frac{b}{\sqrt{a^2+b^2}} \text{ and } p = -\frac{c}{\sqrt{a^2+b^2}}.$$

Ex. Reduce the equation $2x + 3y + 4 = 0$ to various standard forms.

5.3 Intersection of two lines. *To find the coordinates of the point of intersection of two given lines.*

Let $$a_1 x + b_1 y + c_1 = 0 \tag{1}$$

and $$a_2 x + b_2 y + c_2 = 0 \tag{2}$$

be any two given lines. Let (x_1, y_1) be the point of intersection of (1) and (2). Then the point (x_1, y_1) lies on both (1) and (2). Therefore, we get

$$a_1 x_1 + b_1 y_1 + c_1 = 0 \tag{3}$$

and $$a_2 x_1 + b_2 y_1 + c_2 = 0. \tag{4}$$

On solving (3) and (4), we get

$$x_1 = \frac{b_1 c_2 - b_2 c_1}{a_1 b_2 - a_2 b_1} \text{ and } y_1 = \frac{c_1 a_2 - c_2 a_1}{a_1 b_2 - a_2 b_1}.$$

Thus the point of intersection is

$$\left(\frac{b_1 c_2 - b_2 c_1}{a_1 b_2 - a_2 b_1}, \frac{c_1 a_2 - c_2 a_1}{a_1 b_2 - a_2 b_1} \right).$$

Note. When $a_1 b_2 - a_2 b_1 = 0$, the coordinates of the point of intersection become infinite *i.e.* the point of intersection lies at infinity. This means that the two lines are parallel and do not intersect provided the lines are distinct.

Working rule. In order to obtain the point of intersection of two lines we simply solve the equations and obtain the values of x and y which will give the coordinates of the point of intersection of the lines.

Ex. Find the coordinates of the point of intersection of the lines

$$2x + 3y - 7 = 0$$

and $$3x + 4y + 6 = 0.$$

Exercises

1. Reduce the following equations into standard forms:
   (*i*) $x + y + 4 = 0$. (*ii*) $2x + 5y + 13 = 0$.
   (*iii*) $4x - 3y = 4$. (*iv*) $7x - 2y = 8$.
   (*v*) $3x + 4y = 1$.
2. Find the coordinates of the points of intersection of the following pair of lines:
   (*i*) $4x + y - 6 = 0$, $5x - 2y + 12 = 0$.
   (*ii*) $2x - y + 1 = 0$, $x + 3y - 10 = 0$.
   (*iii*) $2x + 3y + 4 = 0$, $4x + 3y + 2 = 0$.
3. Find the equation of the line passing through the point (2, 4) and the point of intersection of the lines $2x - y = 4$ and $3x + 2y = 13$.

6. Angle between Two Lines

To find the angle between two lines whose equations are given.

Let the equations of the lines be $y = mx + c$ and $y = m'x + c'$ making angles θ and θ', respectively with x-axis. Therefore

$$m = \tan\theta \text{ and } m' = \tan\theta'.$$

Let ϕ be the angle between the lines. Then, from the Fig. 2.10, we have

$$\phi = \theta - \theta'.$$

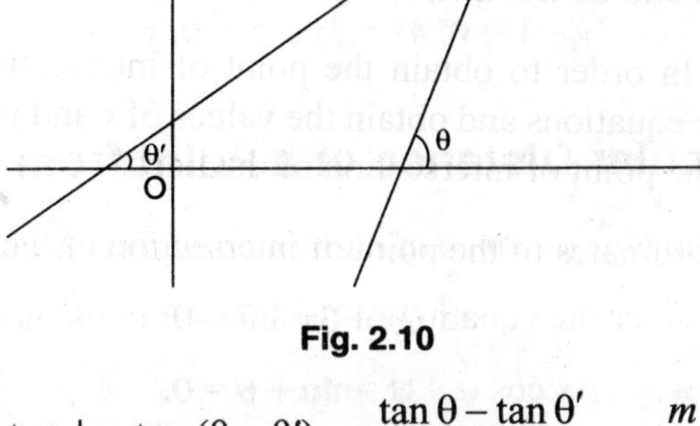

Fig. 2.10

$$\therefore \qquad \tan\phi = \tan(\theta - \theta') = \frac{\tan\theta - \tan\theta'}{1 + \tan\theta\tan\theta'} = \frac{m - m'}{1 + mm'}.$$

Thus the required angle is

$$\phi = \tan^{-1}\left(\frac{m - m'}{1 + mm'}\right).$$

Remark. Sometimes the value of tan ϕ may be negative which is, in fact, due to the factor $m - m'$ and it corresponds to the obtuse angle between the two lines. However, in numerical problems we oftenly consider the value of tan ϕ to be positive. Thus, consider the acute angle only.

Corollary 1. The lines $y = mx + c$ and $y = m'x + c'$ are parallel if $m = m'$.

Corollary 2. The lines $y = mx + c$ and $y = m'x + c'$ are perpendicular if $mm' = -1$.

Ex. 1. Prove that the angle between the lines $a_1x + b_1y + c_1 = 0$ and $a_2x + b_2y + c_2 = 0$ is $\tan^{-1}\left(\frac{a_2b_1 - a_1b_2}{a_1a_2 - b_1b_2}\right)$.

Ex. 2. The angle between the lines in which the origin lies, is the supplement of the angle between their perpendiculars.

(Hint: Consider the equations of the lines in normal form).

Exercises

1. Find the angle between the following pair of lines:
   (*i*) $4x - 3y + 1 = 0, x - y + 1 = 0$.
   (*ii*) $4x - 3y = 1, x - 5y = 4$.
   (*iii*) $5x - y = 9, x + 6y = 8$.
   (*iv*) $3x + y = 7, x + 2y + 9 = 0$.
2. Find the equation of the line through the point (–2, 3) which makes an angle of 45° with the line $3x + y + 5 = 0$.
3. Find all the angles of the triangle whose sides are
   $$2x - 3y - 1 = 0,\ 4x + 3y - 5 = 0,\ x + y + 2 = 0.$$

7. Perpendicular Distance of a Point from a Line

To find the perpendicular distance of a given point from a given line.

Let us first consider the equation of the line *AB* in the normal form

$$x \cos \alpha + y \sin \alpha - p = 0. \tag{1}$$

Let $P(x_1, y_1)$ be the given point from which perpendicular distance of the line AB is to be calculated.

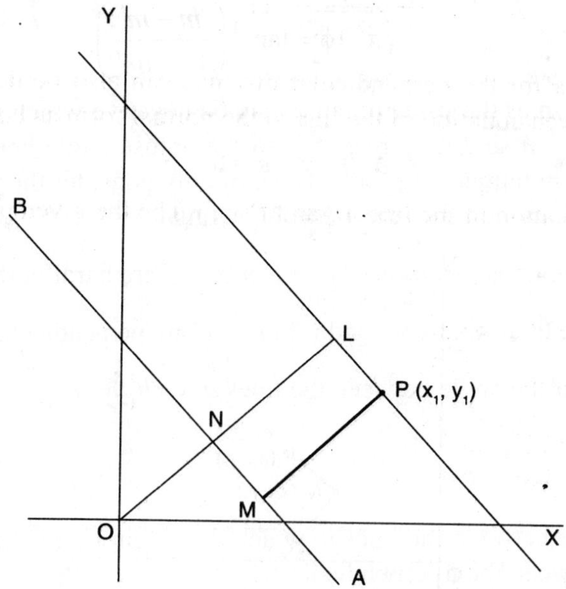

Fig. 2.11

Draw PM perpendicular from P to the line AB. Also draw a line parallel to AB passing through (x_1, y_1). Its equation is given by

$$x \cos \alpha + y \sin \alpha - p_1 = 0. \tag{2}$$

But it passes through (x_1, y_1).

$$\therefore \quad x_1 \cos \alpha + y_1 \sin \alpha - p_1 = 0. \tag{3}$$

Draw OL perpendicular from O to the line (2) which meets AB in the point N. From Fig. 2.11, we have

$$OL = p_1, \; ON = p.$$

Now $$MP = NL = OL - ON = p_1 - p.$$

Thus, *the perpendicular distance of the point $P(x_1, y_1)$ from AB is*

$$x_1 \cos \alpha + y_1 \sin \alpha - p.$$

Consider now the general equation of the line

$$ax + by + c = 0. \tag{4}$$

This equation should be written so that c is negative. This can be written in the normal form as

$$\frac{a}{\sqrt{a^2+b^2}}x + \frac{b}{\sqrt{a^2+b^2}}y + \frac{c}{\sqrt{a^2+b^2}} = 0.$$

Hence *the perpendicular distance of the point $P(x_1, y_1)$ from the line* (4) is

$$\frac{ax_1 + by_1 + c}{\sqrt{a^2 + b^2}} \square$$

The formula for the perpendicular distance can also be derived without reducing the given equation of the line to the normal form as below:

Let $$ax + by + c = 0 \qquad (1)$$

be the given equation of the line AB and $P(x_1, y_1)$ be the given point.

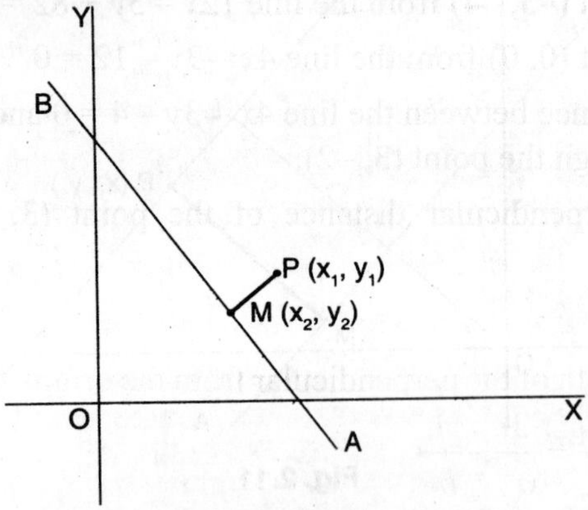

Fig. 2.12

Draw PM perpendicular from P to AB. Let the coordinates of the foot of the perpendicular M be (x_2, y_2) and let $MP = p$. The equation of the line passing through $P(x_1, y_1)$ and perpendicular to the line (1) *i.e.* equation of MP is

$$b(x - x_1) - a(y - y_1) = 0. \qquad (2)$$

But this passes through (x_2, y_2).

Therefore $$-b(x_1 - x_2) + a(y_1 - y_2) = 0. \qquad (3)$$

Since line (1) also passes through (x_2, y_2), we have

$$ax_2 + by_2 + c = 0. \qquad (4)$$

Equation (4) can also be written as

$$a(x_1 - x_2) + b(y_1 - y_2) = ax_1 + by_1 + c. \qquad (5)$$

On squaring and adding (3) and (5), we get

$$(a^2 + b^2)\{(x_1 - x_2)^2 + (y_1 - y_2)^2\} = (ax_1 + by_1 + c)^2.$$

Hence $$p = \sqrt{(x_1 - x_2)^2 + (y_1 - y_2)^2} = \pm\frac{(ax_1 + by_1 + c)}{\sqrt{a^2 + b^2}}.$$

The sign should be chosen so that the value of p is positive.

Working rule. In order to obtain the perpendicular distance of a point from a line, substitute the coordinates of the point in the expression of the equation and divide it by the quantity which is the square root of the sum of the squares of the coefficients of x and y in the equation of the line.

Exercises

1. Calculate the perpendicular distances of

(*i*) the point $(1, -2)$ from the line $3x - 4y = 2$;

(*ii*) the point $(-3, -4)$ from the line $12x - 5y + 82 = 0$;

(*iii*) the point $(0, 0)$ from the line $4x - 3y - 12 = 0$.

2. Find the distance between the line $4x - 3y - 4 = 0$ and a line parallel to it passing through the point $(3, -2)$.

3. Calculate perpendicular distance of the point $(3, -1)$ from the line $\frac{x}{2} - \frac{y}{3} = -1$.

4. If p is the length of the perpendicular from the origin to the line $\frac{x}{a} + \frac{y}{b} = 1$, prove that $\frac{1}{p^2} = \frac{1}{a^2} + \frac{1}{b^2}$.

5. If p and p' are the perpendiculars from the origin to the lines whose equations are $x \sec \theta + y \operatorname{cosec} \theta = a$ and $x \cos \theta - y \sin \theta = a \cos 2\theta$, prove that $4p^2 + p'^2 = a^2$.

6. Prove that the product of the perpendiculars from the two points $(\pm\sqrt{a^2 - b^2}, 0)$ to the line $\frac{x}{a} \cos \phi + \frac{y}{b} \sin \phi = 1$ is b^2.

8. Positive and Negative Sides of a Line

We know that the line $X'OX$ $(y = 0)$ divides the plane into two parts, one above the line $X'OX$ where $y > 0$ and one below the line $X'OX$ where $y < 0$. The boundary between the two regions is the line $y = 0$. In general, any line $ax + by + c = 0$ also divides the plane into two parts. It is natural curiosity to know if the expression $L \equiv ax + by + c$ has the same sign, positive or negative, for all the points on one side of the line.

To prove that the expression $L \equiv ax + by + c$ has the same sign for all the points lying on one side of the line $L = 0$.

Let $P(x_1, y_1)$ and $Q(x_2, y_2)$ be any two points in the plane and let their join meet the line $L = 0$ in R, where

$$\frac{QR}{PR} = \frac{\lambda}{\mu}.$$

Therefore the coordinates of the point R are

$$\left(\frac{\lambda x_1 + \mu x_2}{\lambda + \mu}, \frac{\lambda y_1 + \mu y_2}{\lambda + \mu}\right).$$

Since R lies on the line $L = 0$, we have

$$a\left(\frac{\lambda x_1 + \mu x_2}{\lambda + \mu}\right) + b\left(\frac{\lambda y_1 + \mu y_2}{\lambda + \mu}\right) + c = 0$$

$$\Rightarrow \qquad \frac{\lambda}{\mu} = -\frac{ax_1 + by_1 + c}{ax_2 + by_2 + c}.$$

If $ax_1 + by_1 + c$ and $ax_2 + by_2 + c$ have the same sign, the ratio $\frac{\lambda}{\mu}$ is negative and so R divides PQ externally [see Fig. 2.13(*a*)]. Hence, P and Q lie on the same side of the line $L = 0$. If $ax_1 + by_1 + c$ and $ax_2 + by_2 + c$ have opposite signs, the ratio $\frac{\lambda}{\mu}$ is positive and so R divides PQ internally [see Fig. 2.13(*b*)]. Hence, P and Q lie on the opposite sides of the line $L = 0$.

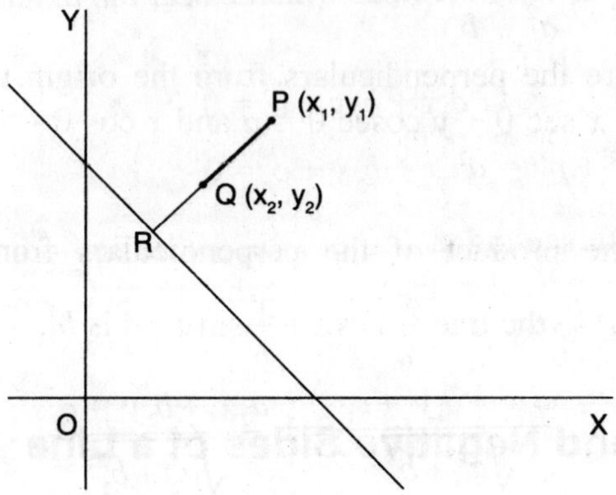

Fig. 2.13(a)

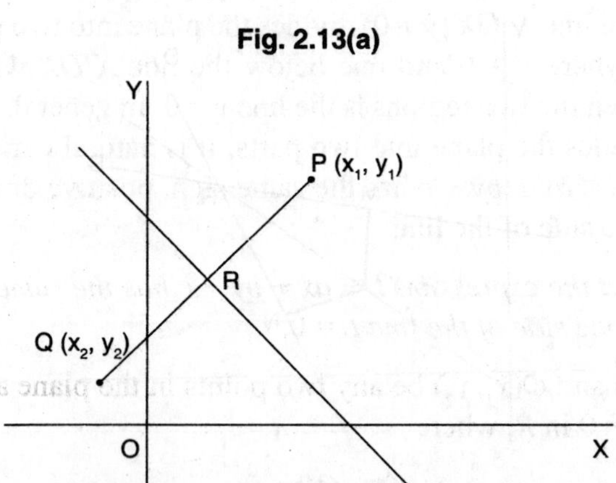

Fig. 2.13(b)

Thus, *the expression* $L \equiv ax + by + c$ *has the same sign for all the points lying on one side of the line* $L = 0$.

Corollary. The point (x_1, y_1) and the origin are on the same side of the line $ax + by + c = 0$ if and only if $ax_1 + by_1 + c$ and c have the same sign.

Remark. It may be noted that the sign of $ax_1 + by_1 + c$ has in itself no particular significance for deciding the *positive* or *negative side** of the line $ax + by + c = 0$ since $ax + by + c = 0$ and $-ax - by - c = 0$ represent the same line and as such the positive and negative regions of the plane would be interchanged. Therefore, we can not attach the term positive or negative infallibly to the two sides of the line.

9. Bisectors of Angles

The **bisector** *of the acute angle between the two lines is the locus of a point which moves so that the perpendicular distances of the point from the lines are equal in magnitude.*

To find the equations of the lines which bisect the angles between the two given lines.

Let
$$a_1x + b_1y + c_1 = 0 \qquad (1)$$
and
$$a_2x + b_2y + c_2 = 0 \qquad (2)$$

be any two lines intersecting at R. Let $P(x', y')$ be any point on either of the bisectors. Then perpendicular distances of the lines (1) and (2) from $P(x', y')$ must be equal in magnitude *i.e.*

$$\frac{a_1x' + b_1y' + c_1}{\sqrt{a_1^2 + b_1^2}} = \pm\frac{a_2x' + b_2y' + c_2}{\sqrt{a_2^2 + b_2^2}}.$$

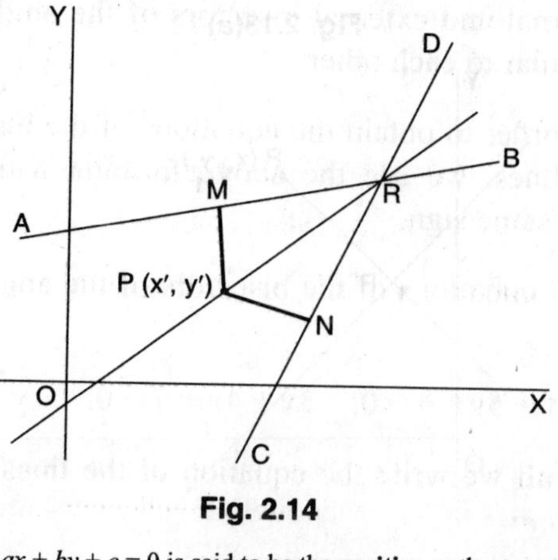

Fig. 2.14

* A side of the line $ax + by + c = 0$ is said to be the **positive** or the **negative side** according as the expression $ax + by + c$ has the positive or negative sign for the points on that side.

Hence, the point (x', y') lies on one or the other of the lines

$$\frac{a_1x+b_1y+c_1}{\sqrt{a_1^2+b_1^2}}=\pm\frac{a_2x+b_2y+c_2}{\sqrt{a_2^2+b_2^2}}. \qquad (3)$$

The two lines given by (3) are therefore the required bisectors.

Note. We can distinguish between the two bisectors given by (3) with the help of origin. Let us assume, without any loss of generality, that c_1 and c_2 both be positive for if it is not so it could be made on multiplying the equation by –1. Let (x', y') be any point on a bisector. Now, we have the following different cases:

Case I. When both the expressions $a_1x' + b_1y' + c_1$ and $a_2x' + b_2y' + c_2$ are positive. Then the point (x', y') and the origin lie on the same side of the lines

$$a_1x+b_1y+c_1=0 \qquad \text{and} \qquad a_2x+b_2y+c_2=0.$$

Thus the point (x', y') lies on the bisector of the angle which contains the origin. We call it **internal bisector** and its equation is

$$\frac{a_1x+b_1y+c_1}{\sqrt{a_1^2+b_1^2}}=\frac{a_2x+b_2y+c_2}{\sqrt{a_2^2+b_2^2}}.$$

Case II. When the expressions $a_1x' + b_1y' + c_1$ and $a_2x' + b_2y' + c_2$ have opposite signs. Then the point (x', y') and the origin lie on the same side of one line and on opposite side of the other. Thus, the point (x', y') lies on the bisector of the angle which does not contain the origin. We call it **external bisector** and its equation is

$$\frac{a_1x+b_1y+c_1}{\sqrt{a_1^2+b_1^2}}=-\frac{a_2x+b_2y+c_2}{\sqrt{a_2^2+b_2^2}}.$$

Remark. The internal and external bisectors of the angles between the two lines are perpendicular to each other.

Working rule. In order to obtain the equations of the bisectors of the angles between the two lines, we use the above formula with a remark that the equations have the same sign.

Example. Find the euqations of the bisectors of the angles between the pair of lines

$$12x + 5y - 4 = 0, \quad 3x + 4y + 7 = 0.$$

Solution. First of all we write the equation of the lines so that the constant terms are positive *i.e.*

$$-12x - 5y + 4 = 0, \quad 3x + 4y + 7 = 0.$$

The equation of the bisector of the angle between the lines which contain the origin is

$$\frac{-12x-5y+4}{\sqrt{(-12)^2+(-5)^2}} = \frac{3x+4y+7}{\sqrt{3^2+4^2}}$$

$$\Rightarrow \qquad 99x + 77y + 71 = 0.$$

The equation of the other bisector is

$$\frac{-12x-5y+4}{\sqrt{(-12)^2+(-5)^2}} = -\frac{3x+4y+7}{\sqrt{3^2+4^2}}$$

$$\Rightarrow \qquad 7x - 9y - 37 = 0.$$

Exercises

1. Show that the point (1, 2) is on the negative side of $3x - 4y - 1 = 0$ and on the positive side of $4x + 2y - 2 = 0$.
2. Prove that the points (3, –2) and (1, 2) are on opposite sides of the line $2x + 3y - 5 = 0$.
3. Find the equations of the bisectors of the angles between the pair of lines:

   (*i*) $2x + 3y = 1, \quad 2x - y + 3 = 0;$

   (*ii*) $3x - 4y = 5, \quad 4x - 3y = 3.$
4. Find the equation of the bisector of that angle between the lines $4x - 3y + 1 = 0$ and $12x + 5y + 13 = 0$ in which the origin lies.

10. Systems of Lines

We have expressed equations of the lines in various forms. Among these are the equations

$$y = mx + c, \quad \frac{x}{a} + \frac{y}{b} = 1 \text{ and } x \cos \alpha + y \sin \alpha = p.$$

Each of these equations has two constants which have geometrical significance. The constants of the first equation are m and c. When definite values are assigned to these letters, a line is completely determined. Other values of these, of course, determine other lines. Thus the quantities m and c are fixed for any particular line but change from line to line. These letters are called **parameters**. In the second equation a and b are the parameters whereas in the third equation α and p are the parameters.

A linear equation with only one parameter is obtained if the other parameter is replaced by a fixed value. The resulting equation represents all lines with a particular property if the remaining parameter is allowed to vary. Each value assumed by the parameter yields an equation which represents a definite line.

The collection of lines defined by a linear equation with one parameter is called a **family**, *or* **system of lines**. For example, if $m = 2$, the slope-intercept equation becomes

$$y = 2x + c.$$

This equation represents the family of parallel lines each of which has slope 2, where c is a parameter. There are, of course, infinitely many lines in the family.

10.1 System of lines parallel to a given line

Let the given line be

$$ax + by + c = 0 \quad (1)$$

Then, the system of lines parallel to the line (1) is

$$ax + by + \lambda = 0,$$

where λ is a parameter, since the slopes of the parallel lines are equal.

10.2 System of lines perpendicular to a given line

Let the given line be

$$ax + by + c = 0. \quad (1)$$

Then, the equation

$$bx - ay + \lambda = 0, \quad (2)$$

for any value of λ, represents a line perpendicular to the line (1) since the product of the slopes of the perpendicular lines is -1. If λ is a parameter, then the equation (2) represents a family of lines each of which is perpendicular to the line (1).

10.3 System of lines through a given point. Let the given point be $A(x_1, y_1)$. Then a system of lines through A is $y - y_1 = m\,(x - x_1)$, where m, the slope of the line, is a parameter. For different values of m, we have a different line of the system through the point $A(x_1, y_1)$.

A system of lines passing through a given point is called a **pencil** *of lines and the given point is called the* **vertex** *of the pencil.*

10.4 System of lines through the intersection of two lines. *To find the system of lines passing through the point of intersection of two given lines.*

Consider the two distinct non-parallel lines

$$L_1 \equiv a_1x + b_1y + c_1 = 0, \tag{1}$$

and
$$L_2 \equiv a_2x + b_2y + c_2 = 0. \tag{2}$$

Then, the equation

$$a_1x + b_1y + c_1 + \lambda\,(a_2x + b_2y + c_2) = 0, \tag{3}$$

where λ is parameter, represents a system of lines, through the point of intersection of the lines (1) and (2). To verify this statement we observe that:

(*i*) Equation (3) being linear in x and y, for any value of the parameters λ, represents a line.

(*ii*) If (x_1, y_1) is the point of intersection of the given lines (1) and (2), then $a_1x_1 + b_1y_1 + c_1 = 0$ and $a_2x_1 + b_2y_1 + c_2 = 0$.

Thus equation (3) is also satisfied by the coordinates (x_1, y_1). Therefore the line (3) passes through the point (x_1, y_1). Hence *the equation*

$$a_1x + b_1y + c_1 + \lambda\,(a_2x + b_2y + c_2) = 0$$

represents a system of lines through the point of intersection of the given lines.

Note 1. The system of lines (3) represents a pencil of lines with vertex as the point of intersection of the given lines (1) and (2). We refer to the given lines (1) and (2) as the **base lines** of the pencil. Clearly any two lines of the pencil can be chosen as base lines.

Note 2. If the lines (1) and (2) are parallel, equation (3) represents all lines parallel to (1) or (2) since all they have the same slope $-\dfrac{a_1}{b_1}$. In numerical examples, it is enough to choose $a_1x + b_1y + c_1 + \lambda = 0$ to represent all such lines which, in fact, is a system of lines parallel to $a_1x + b_1y + c_1 = 0$.

10.5 Concurrent lines. *To find the condition for the given three lines to be concurrent.*

Let the three lines be

$$a_1x + b_1y + c_1 = 0 \tag{1}$$

$$a_2x + b_2y + c_2 = 0 \tag{2}$$

and
$$a_3x + b_3y + c_3 = 0. \tag{3}$$

These lines, no two of which are assumed to be parallel, will be concurrent if any one of them (say) line (3) is a member of the pencil determined by the other lines (1) and (2) here; in other words, if line (3) can be expressed in the form of

$$a_1x + b_1y + c_1 + \lambda\,(a_2x + b_2y + c_2) = 0$$

for some suitable value of λ. This is the same thing as to find the values of two constants λ and μ such that

$$(a_1x + b_1y + c_1) + \lambda\,(a_2x + b_2y + c_2) + \mu\,(a_3x + b_3y + c_3) = 0$$

is satisfied.

Alternative test for concurrency

Let the lines represented by above equations are concurrent and (x_1, y_1) be the common point of intersection. Then

$$a_1x_1 + b_1y_1 + c_1 = 0 \tag{4}$$

$$a_2x_1 + b_2y_1 + c_1 = 0 \tag{5}$$

and $$a_3x_1 + b_3y_1 + c_3 = 0. \tag{6}$$

On eliminating x_1 and y_1 from (4), (5) and (6), we get

$$\begin{vmatrix} a_1 & b_1 & c_1 \\ a_2 & b_2 & c_2 \\ a_3 & b_3 & c_3 \end{vmatrix} = 0,$$

which is required condition for concurrency of the lines.

Exercises

1. Find the equation of the line which is perpendicular to $3x + 8y - 5 = 0$ and which passes through the point $(2, -1)$.
2. Prove that the three lines $5x + 3y - 7 = 0$, $3x - 4y - 10 = 0$ and $x + 2y = 0$ are concurrent.
3. Verify that the three lines $4x + 7y - 9 = 0$, $5x - 8y + 15 = 0$ and $9x - y + 6 = 0$ are concurrent.
4. Prove that the line through the point $(-4, 6)$ concurrent with the lines $3x - 2y + 3 = 0$ and $5x + 6y - 2 = 0$ passes through the origin.

11. Equation of a Line in Polar Coordinates

11.1 *To find the general equation of a line in the polar coordinates.*

Let O be the pole and OX the initial line. Let $P(r, \theta)$ be any point on the line. Draw perpendicular ON to the line.

Let $ON = p$ and $\angle XON = \alpha$.

Now in right angled triangle ONP, $ON = OP \cos NOP$, i.e.,

$$p = r\cos(\theta - \alpha).$$

This is the required equation of the line in polar coordinates.

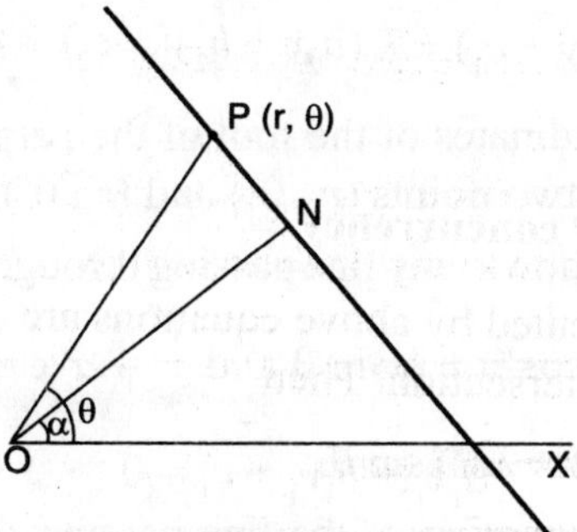

Fig. 2.15

Note. Transforming the equation to cartesian coordinates, this equation becomes the normal form of the line.

11.2 Polar equation of the line joining two points. *To find the polar equation of the line joining two given points.*

Let O be the pole and OX the initial line. Let $P(r_1, \theta_1)$ and $Q(r_2, \theta_2)$ be the given points. Let $R(r, \theta)$ be any point on the line joining the points P and Q.

Then area of ΔPOQ = area of ΔPOR + area of ΔROQ

i.e. $\frac{1}{2}OP.OQ \sin POQ = \frac{1}{2}OP.OR \sin POR + \frac{1}{2}OR.OQ \sin ROQ$

i.e. $\frac{1}{2}r_1 r_2 \sin(\theta_2 - \theta_1) = \frac{1}{2}rr_1 \sin(\theta - \theta_1) + \frac{1}{2}rr_2 \sin(\theta_2 - \theta)$

$$\Rightarrow \quad \frac{\sin(\theta_2 - \theta_1)}{r} = \frac{\sin(\theta - \theta_1)}{r_2} + \frac{\sin(\theta_2 - \theta)}{r_1}.$$

This is the required equation of the line PQ.

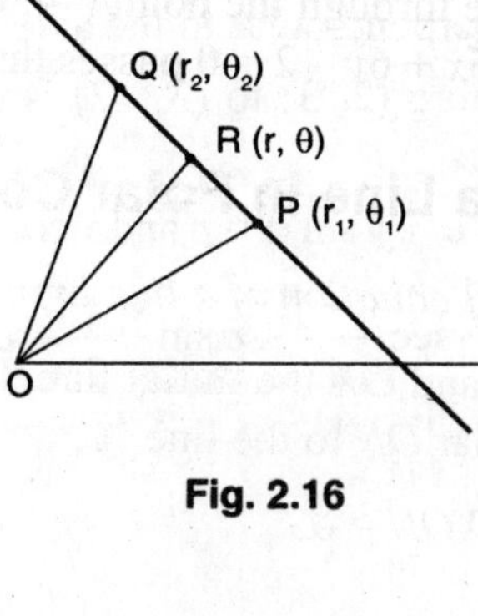

Fig. 2.16

Corollary. The equation of the line joining the point (r_1, θ_1) to the pole is $\theta = \theta_1$.

This equation can be obtained easily by taking $r_2 = 0$ and $\theta_2 = 0$ in the above equation.

Exercises

1. Find the polar coordinates of the foot of the perpendicular from the pole on the line joining two points (r_1, θ_1) and (r_2, θ_2).
2. Prove that the equation to any line passing through the point of intersection of the lines $\frac{l}{r} = a \cos \theta + b \sin \theta$ and $\frac{l'}{r} = a' \cos \theta + b' \sin \theta$ is $\frac{l + \lambda l'}{r} = (a + \lambda a') \cos \theta + (b + \lambda b') \sin \theta$.

   Hence obtain the equation of the line passing through the pole and the point of intersection of the two given lines.
3. A variable line cuts off from n given concurrent lines intercepts, the sum of the reciprocals of which is constant. Show that it always passes through a fixed point.

Miscellaneous Exercises

1. If the line $\lambda x + 6y + 5 = 0$ is parallel to the line $4x - 3y + 2 = 0$, what is the value of λ?
2. Find the perpendicular distance from the origin of the line $\frac{x}{3} + \frac{y}{4} = 1$.
3. Find the equation of the line through (2, 3) parallel to $5x - 12y = 3$.
4. Write down the equation of the lines through (–2, 3), respectively, parallel and perpendicular to $3x - 5y = 1$.
5. Find the lines through (4, 5) which cut the axes so that the intercepts are equal in magnitude.
6. The vertices of a triangle are (0, –2), (2, 0) and (3, –3). Prove that it is isosceles and find the equation of the median which bisects shortest side. Verify that this line is perpendicular to the side it bisects.
7. Show that the line joining (2, 3) to (3, –2) is perpendicular to the line joining (–5, 0) to (0, 1).
8. Show that the lines $x \cos \alpha + y \sin \alpha = p$ and $x \cos \beta + y \sin \beta = p$ intersect at

$$\left(p \cos \frac{\alpha + \beta}{2} \sec \frac{\alpha - \beta}{2}, p \sin \frac{\alpha + \beta}{2} \sec \frac{\alpha - \beta}{2} \right).$$

9. Find the coordinates of the circumcentre of the triangle formed by the lines

$$3x - y - 5 = 0,\ x + 2y - 4 = 0 \text{ and } 5x + 3y + 1 = 0.$$

10. Prove that the area of the triangle formed by the three lines $y = m_1x + c_1$, $y = m_2x + c_2$ and $y = m_3x + c_3$ is

$$\frac{1}{2}\left[\frac{(c_2 - c_3)^2}{m_2 - m_3} + \frac{(c_3 - c_1)^2}{m_3 - m_1} + \frac{(c_1 - c_2)^2}{m_1 - m_2}\right].$$

11. Prove that the medians of a triangle are concurrent.

12. Find λ such that the lines $x + y + 5 = 0$, $5x - y + 7 = 0$ and $\lambda x + 4y + 6 = 0$ are concurrent.

13. Three vertices of a parallelogram $ABCD$ are $A(1, 0)$, $B(-1, 0)$ and $C(2, 3)$. Find the coordinates of D.

14. Find the equation of the line joining (1, 1) to the point of intersection of the lines $2x + 4y - 5 = 0$ and $4x + 2y - 5 = 0$.

15. Prove that the locus of a point which moves so that the sum of the perpendiculars let fall from it upon two given lines is constant, is a line.

Change of Axes

3 CHAPTER

1. Introduction

In coordinate geometry, the coordinates of a point, the equation of a locus etc. are always with reference to some system of coordinate axes.

Sometimes while solving the problems in coordinate geometry, it is desirable as well as convenient to consider a figure under discussion and to change its equation, either temporarily or permanently, to a different system of axes such that an important set of points assume the coordinates of simple form. In general, we come across with the following three types of change of axes.

(*i*) **Translation of axes** (Change of origin only, the new axes being parallel to the old ones).

(*ii*) **Rotation of axes** (Change of directions of the axes without changing the origin).

(*iii*) **General transformation** (Translation and rotation of the axes together).

2. Translation of Axes

To change the origin of a system of coordinate axes without changing the directions of the axes.

Let OX and OY be the original axes and (x, y) be the coordinates of a point P referred to these axes.

Let $O'(\alpha, \beta)$ be the new origin and $O'X'$ and $O'Y'$ the coordinate axes parallel to the original axes. Let (x', y') be the coordinates of the point P with respect to new axes.

Draw PM and $O'N$ perpendiculars to OX, where PM meets $O'X'$ in M'.

Then $\qquad ON = \alpha,\ NO' = \beta,\ OM = x,\ MP = y.$

Also $\qquad O'M' = x'$ and $M'P = y'$.

Now $\qquad x = OM = ON + NM = ON + O'M' = \alpha + x'$

and $$y = MP = MM' + M'P = NO' + M'P = \beta + y'.$$

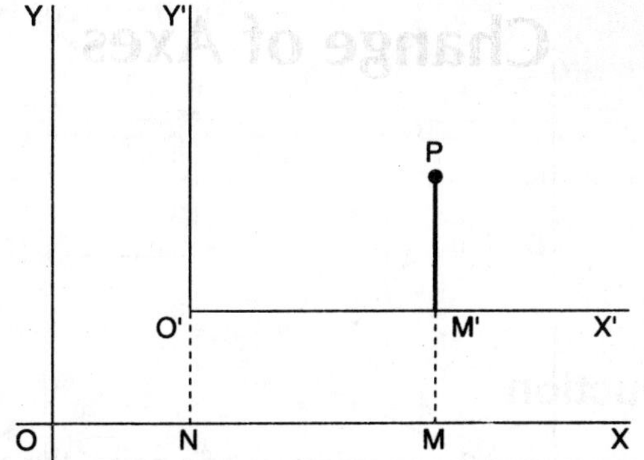

Fig. 3.1

Hence, *the new coordinates* (x', y') *of the point P are connected with the old one* (x, y) *by means of the relations*

$$x = x' + \alpha, y = y' + \beta.$$

Note. The pair of equations $x = x' + \alpha$, $y = y' + \beta$ giving the relationship between the two sets of coordinate axes is called the **translation of axes.**

Ex. Transform to parallel axes through the point (3, 5) the equation

$$x^2 + y^2 - 6x - 10y - 2 = 0.$$

3. Rotation of Axes

To change the directions of the axes without changing the origin.

Let OX and OY be the original axes and OX', OY' the new axes which are obtained by rotating the original axes and through an angle θ. Rotation is taken in anti-clockwise direction.

Let (x, y) be the coordinates of a point P with respect to original axes and (x', y') be coordinates of P with respect to new axes.

Draw PM and PN perpendiculars to OX and OX' and also NN' and NM' perpendiculars to OX and PM, respectively. Then

$$OM = x, MP = y, ON = x', NP = y' \text{ and } \angle M'PN = \theta.$$

Now, we have

$$x = OM = ON' - MN' = ON' - M'N = ON \cos\theta - NP \sin\theta$$
$$= x' \cos\theta - y' \sin\theta$$

and $\quad y = MP = MM' + M'P = N'N + M'P = ON \sin\theta + NP \cos\theta$

$$= x' \sin\theta + y' \cos\theta.$$

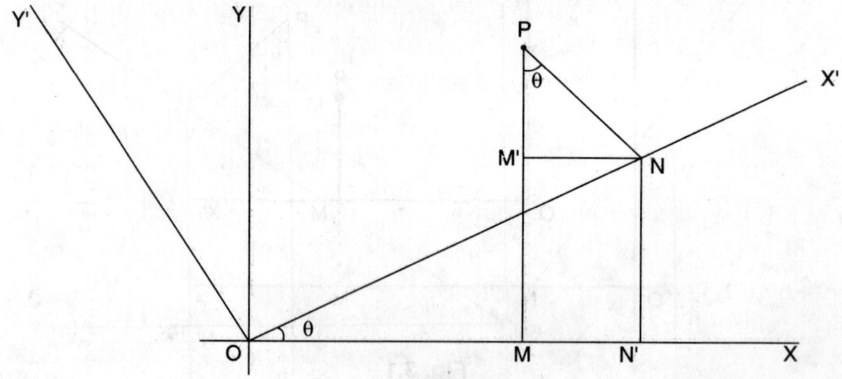

Fig. 3.2

Hence, the relations connecting the two systems of axes are given by

$$\left.\begin{aligned} x &= x'\cos\theta - y'\sin\theta \\ y &= x'\sin\theta + y'\cos\theta \end{aligned}\right\}. \qquad (1)$$

Note. An equivalent form of the relations (1) is

$$\left.\begin{aligned} x' &= x\cos\theta + y\sin\theta \\ y' &= -x\sin\theta + y\cos\theta \end{aligned}\right\}.$$

Ex. What does the equation

$$4x^2 + 2\sqrt{3}xy + 2y^2 = 1$$

become when the axes are turned through an angle of 30°?

4. General Transformation

To change the origin and directions of the axes together.

Let OX and OY be the original axes. Shifting the origin to O' (α, β) so that $O'X'$ and $O'Y'$ are parallel to the original axes. Further rotate $O'X'$ and $O'Y'$ through an angle θ so that we get the new axes as $O'X''$and $O'Y''$. Let (x, y) be the coordinates of any point P with respect to the axes OX and OY; (x', y') the coordinates of P with respect to $O'X'$ and $O'Y'$ and (x'', y'') the coordinates of P with respect to $O'X''$ and $O'Y''$.

Then $\quad OM = x,\ MP = y,\ O'M' = x',\ M'P = y',\ O'N'' = x'',\ N''P = y''.$

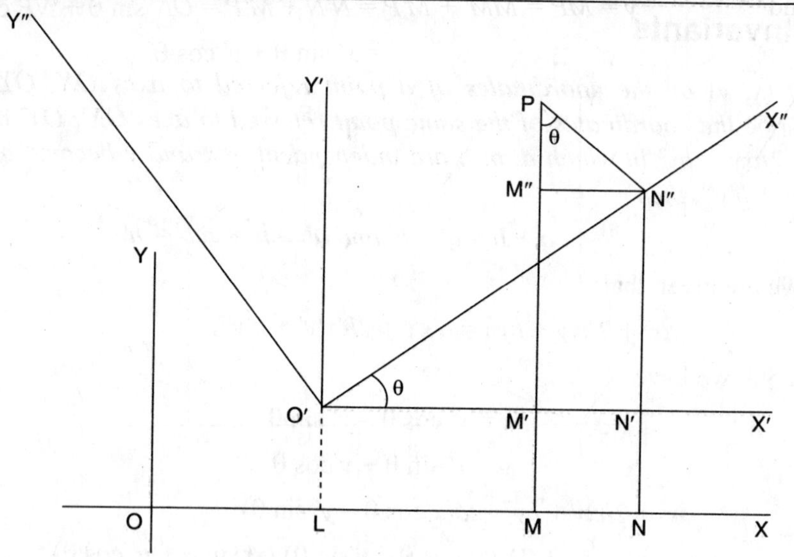

Fig. 3.3

Now, we have

$$x = OM = ON - MN = OL + LN - MN$$
$$= OL + O'N' - M''N''$$
$$= OL + O'N'' \cos\theta - N''P \sin\theta$$
$$= \alpha + x'' \cos\theta - y'' \sin\theta$$

and

$$y = MP = MM' + M'M'' + M''P$$
$$= O'L + N'N'' + M''P$$
$$= \beta + O'N'' \sin\theta + N''P \cos\theta$$
$$= \beta + x'' \sin\theta + y'' \cos\theta.$$

Hence, the new coordinates (x'', y'') of the point P are connected by the old one (x, y) by means of the relations

$$\left.\begin{aligned} x &= \alpha + x''\cos\theta - y''\sin\theta \\ y &= \beta + x''\sin\theta + y''\cos\theta \end{aligned}\right\}.$$

Ex. Transform the equation

$$x^2 - 4xy + y^2 + 8x + 2y - 5 = 0$$

to new axes through (2, 3) rotated through an angle of 45°.

5. Invariants

5.1 *If (x, y) be the coordinates of a point referred to axes OX, OY and (x', y') be the coordinates of the same point referred to axes OX', OY' and if $ax^2 + 2hxy + by^2$ in which a, b, h are independent of x and y become $a'x'^2 + 2h'x'y' + b'y'^2$, then*

$$a + b = a' + b' \text{ and } ab - h^2 = a'b' - h'^2.$$

We are given that

$$ax^2 + 2hxy + by^2 = a'x'^2 + 2h'x'y' + b'y'^2. \qquad (1)$$

From § 3, we have

$$x = x' \cos\theta - y' \sin\theta$$

and $$y = x' \sin\theta + y'\cos\theta.$$

Now $$ax^2 + 2hxy + by^2 = a(x'\cos\theta - y'\sin\theta)^2$$
$$+ 2h(x'\cos\theta - y'\sin\theta)(x'\sin\theta + y'\cos\theta)$$
$$+ b(x'\sin\theta + y'\cos\theta)^2$$
$$= (a\cos^2\theta + 2h\sin\theta\cos\theta + b\sin^2\theta)\, x'^2$$
$$+ \{-2a\sin\theta\cos\theta + 2h(\cos^2\theta - \sin^2\theta)$$
$$+ 2b\sin\theta\cos\theta\}\, x'y'$$
$$+ (a\sin^2\theta - 2h\sin\theta\cos\theta + b\cos^2\theta)\, y'^2 \qquad (2)$$

On comparing (1) and (2), we get

$$a' = a\cos^2\theta + 2h\sin\theta\cos\theta + b\sin^2\theta \qquad (3)$$

$$b' = a\sin^2\theta - 2h\sin\theta\cos\theta + b\cos^2\theta \qquad (4)$$

and $$h' = -a\sin\theta\cos\theta + h(\cos^2\theta - \sin^2\theta) + b\sin\theta\cos\theta. \qquad (5)$$

On adding (3) and (4), we get

$$a' + b' = a + b.$$

Equations (3), (4) and (5) respectively, can also be written as

$$2a' = (a + b) + (a - b)\cos 2\theta + 2h\sin 2\theta \qquad (6)$$

$$2b' = (a + b) - \{(a - b)\cos 2\theta + 2h\sin 2\theta\} \qquad (7)$$

and $$2h' = -(a - b)\sin 2\theta + 2h\cos 2\theta. \qquad (8)$$

Subtracting square of (8) from the product of (6) and (7), we get

$$4a'b' - 4h'^2 = (a + b)^2 - \{(a - b)\cos 2\theta + 2h\sin 2\theta\}^2$$
$$- \{-(a - b)\sin 2\theta + 2h\cos 2\theta\}^2$$
$$= (a + b)^2 - \{(a - b)^2 + 4h^2\}$$
$$= 4ab - 4h^2.$$

$$\Rightarrow \quad a'b' - h'^2 = ab - h^2 \; \square$$

Note. The quantities $a + b$ and $ab - h^2$ are called **invariants**. These invariants are of great importance in the development of the theory of the general equation of second degree.

Ex. Prove that if $ax^2 + 2hxy + by^2 = 1$ and $a'x^2 + 2h'xy + b'y^2 = 1$ represent the same conic and the axes are rectangular, then

$$(a - b)^2 + 4h^2 = (a' - b')^2 + 4h'^2.$$

5.2 Removal of the xy term. *To find the angle through which the axes be rotated so that the expression $ax^2 + 2hxy + by^2$ may become of the form $a'x'^2 + b'y'^2$.*

Let the axes be rotated through an angle θ. Then by §3, we have

$$x = x' \cos\theta - y' \sin\theta$$

and $$y = x' \sin\theta + y' \cos\theta.$$

$$\therefore \quad ax^2 + 2hxy + by^2 = (a \cos^2\theta + 2h \sin\theta \cos\theta + b \sin^2\theta)\, x'^2$$
$$+ \{-2(a - b) \sin\theta \cos\theta + 2h (\cos^2\theta - \sin^2\theta)\}\, x'y'$$
$$+ (a \sin^2\theta - 2h \sin\theta \cos\theta + b \cos^2\theta)\, y'^2. \qquad (1)$$

Thus, the expression $ax^2 + 2hxy + by^2$ will be of the form $a'x'^2 + b'y'^2$ if coefficient of $x'y'$ in right side of (1) is zero,

i.e., if $-2(a - b) \sin\theta \cos\theta + 2h (\cos^2\theta - \sin^2\theta) = 0$

i.e., if $-(a - b) \sin 2\theta + 2h \cos 2\theta = 0$

$$\Rightarrow \quad \tan 2\theta = \frac{2h}{a-b}.$$

Hence, the required angle is given by

$$\theta = \frac{1}{2} \tan^{-1} \frac{2h}{a-b} \; \square$$

Remark. In the special case where $ab - h^2 = 0$, the removal of the xy term in $ax^2 + 2hxy + by^2$ by means of invariants, we note that

$$a'b' = ab - h^2 = 0$$

$$\Rightarrow \quad a' = 0 \text{ or } b' = 0.$$

Thus, in this special case the removal of the xy term in

$$ax^2 + 2hxy + by^2$$

by turning the axes through an angle θ given by $\tan 2\theta = \dfrac{2h}{a-b}$ makes either $a' = 0$ or $b' = 0$.

Exercises

1. Simplify the following equations by changing to a new origin in each case:

   (*i*) $x^2 + y^2 + 2x - 4y + 1 = 0$; new origin $(-1, 2)$.

   (*ii*) $x^2 + 2y^2 - 6x + 16y + 39 = 0$; new origin $(3, -4)$.

2. Write down the equations for a rotation of axes through an angle $\dfrac{\pi}{4}$. Hence, prove that the curve $2xy = a^2$ can be transformed to $x^2 - y^2 = a^2$.
3. Transform the equation $x^2 - 6xy + 9y^2 + 4x + 8y + 15 = 0$ to new axes through $(-2, -1)$ rotated through an angle $\tan^{-1}\frac{1}{2}$.
4. Transform the equation $3x^2 - 24xy + 10y^2 + 6x + 52y = 0$ to new axes through $(3, 1)$ rotated through an angle $\tan^{-1}\frac{3}{4}$.
5. If (x, y) and (x', y') be the coordinates of the same point referred to two sets of rectangular axes with the same origin and if $ux + vy$, where u and v are independent of x and y, becomes $u'x' + v'y'$, then $u^2 + v^2 = u'^2 + v'^2$.
6. Show that the equation $x \cos \alpha + y \sin \alpha = p$, when the axes are turned through an angle α, becomes $x = p$. Interpret this fact.
7. What does the equation $(x-a)^2+(y-b)^2=c^2$ become when it is transferred to parallel axes through (*i*) the point $(a - c, b)$ (*ii*) the point $(a, b - c)$?
8. What does the equation $(a - b)(x^2 + y^2) - 2abx = 0$ become if the origin be moved to the point $\left(\dfrac{ab}{a \quad b}, 0\right)$?
9. Prove that the transformation of rectangular axes which converts

$$\frac{X^2}{p} + \frac{Y^2}{q} \text{ into } ax^2 + 2hxy + by^2 \text{ will convert } \frac{X^2}{p-\lambda} + \frac{Y^2}{p-\lambda}$$

$$\text{into } \frac{ax^2 + 2hxy + by^2 - \lambda(ab - h^2)(x^2 + y^2)}{1 - (a+b)\lambda + (ab - h^2)\lambda^2}.$$

Pair of Lines

4
CHAPTER

1. Introduction

Consider the equation

$$(lx + my + n)(l'x + m'y + n') = 0. \qquad (1)$$

This can be satisfied by the points whose coordinates satisfy

$$lx + my + n = 0 \text{ or } l'x + m'y + n' = 0. \qquad (2)$$

Thus, all the points which satisfy equation (1) lie on one or the other of the equations in (2); and conversely, any point whose coordinates satisfy any of the two equations in (2) satisfies the equation (1). The equations in (2) being linear, the equation (1), therefore, represents a pair of lines.

If we multiply the two factors in (1) together, then it takes the form

$$ll'x^2 + (lm' + l'm)\,xy + mm'y^2 + (ln' + l'n)\,x + (mn' + m'n)\,y + nn' = 0.$$

This is a second degree equation in x and y. Thus, we note that a pair of lines is given by a second degree equation in x and y.

Conversely, one is interested to know if the most general second degree equation in x and y *i.e.*

$$ax^2 + 2hxy + by^2 + 2gx + 2fy + c = 0 \qquad (3)$$

represents a pair of lines. It is not possible, in general. However, it would only be possible when the left hand expression in equation (3) can be decomposed into two real linear factors for which some relation among the coefficients a, b, c, f, g, h is required to be satisfied. Before, we deal with this problem in a later section, we come to a particular case when both the lines pass through the origin. In this case the equation (1) takes the form

$$(lx + my)\,(l'x + m'y) = 0$$

i.e.

$$ll'x^2 + (lm' + l'm)\,xy + mm'y^2 = 0.$$

This is a homogeneous* equation of degree two in x and y. The general

* An equation of the form $a_0x^n + a_1x^{n-1}y + a_2x^{n-2}y^2 + \ldots + a_ny^n = 0$, in which the sum of the powers of x and y in every term is the same (say n), is called a homogeneous equation (of degree n).

expression of this form (homogeneous form) is

$$ax^2 + 2hxy + by^2 = 0.$$

We shall now prove that this always represents a pair of lines through the origin, unlike the case for the most general second degree equation in x and y.

2. Homogeneous Equation of Second Degree

2.1 *To prove that the general homogeneous equation of second degree represents a pair of lines passing through the origin.*

The general homogeneous equation of second degree is

$$ax^2 + 2hxy + by^2 = 0. \tag{1}$$

This can be written as

$$(ax + hy)^2 - (h^2 - ab)y^2 = 0, \qquad a \neq 0$$

$$\Rightarrow \qquad \{ax + (h + \sqrt{h^2 - ab})y\}\{ax + (h - \sqrt{h^2 - ab})y\} = 0.$$

So equation (1) represents the lines

$$ax + (h + \sqrt{h^2 - ab})y = 0$$

and

$$ax + (h - \sqrt{h^2 - ab})y = 0,$$

through the origin. The lines are real and distinct, coincident or imaginary according as $h^2 >=$ or $< ab$, respectively.

If $a = 0$, the equation (1) becomes

$$2hxy + by^2 = 0,$$

which represents the lines $y = 0$ and $2hx + by = 0$.

Remark. Most of the times, the problems regarding the pair of lines through the origin require the technique of introducing separate equations for the lines. This we shall be tackling by considering that $y = mx$ and $y = m'x$ are the lines represented by the given equation

$$ax^2 + 2hxy + by^2 = 0$$

so that

$$ax^2 + 2hxy + by^2 \equiv b\,(y - mx)(y - m'x)$$

$$\Rightarrow \qquad \begin{cases} m + m' = -\dfrac{2h}{b} \\[2ex] mm' = \dfrac{a}{b}. \end{cases}$$

In the problems, we shall be using the equations $y = mx$ and $y = m'x$, and finally eliminate m and m' by the use of the relations involving $m + m'$ and mm'.

Illustration. Find the equation of the pair of lines through the origin which represents the lines perpendicular to the pair of lines $ax^2 + 2hxy + by^2 = 0$.

Solution. Let $y = mx$ and $y = m'x$ be the lines represented by $ax^2 + 2hxy + by^2 = 0$. Then $m + m' = -\dfrac{2h}{b}$, $mm' = \dfrac{a}{b}$.

Now, the lines through the origin and perpendicular to the given lines are $y = -\dfrac{1}{m}x$ and $y = -\dfrac{1}{m'}x$. Thus the required equation of the pair of lines is

$$(x + my)(x + m'y) = 0$$

i.e.

$$x^2 + (m + m')xy + mm'y^2 = 0.$$

Substituting the values of $m + m'$ and mm', we get

$$bx^2 - 2hxy + ay^2 = 0.$$

2.2 Angle between the lines $ax^2 + 2hxy + by^2 = 0$. *To find the angle between the lines given by the equation*

$$ax^2 + 2hxy + by^2 = 0.$$

Let the lines represented by the given equation of the pair of lines be

$$y = mx \text{ and } y = m'x.$$

Then

$$ax^2 + 2hxy + by^2 \equiv b(y - mx)(y - m'x)$$

$$\Rightarrow \quad m + m' = -\frac{2h}{b}, \quad mm' = \frac{a}{b}.$$

Let ϕ be the angle between the lines. Then

$$\tan\phi = \frac{m - m'}{1 + mm'} = \frac{\pm\sqrt{(m + m')^2 - 4mm'}}{1 + mm'}$$

$$= \frac{\pm\sqrt{\dfrac{4h^2}{b^2} - \dfrac{4a}{b}}}{1 + \dfrac{a}{b}}$$

$$= \frac{\pm 2\sqrt{h^2 - ab}}{a + b}.$$

The sign positive or negative is to be taken according as the angle ϕ between the lines is an acute or obtuse angle.

Remarks. (*i*) It is important, in the formula for the angle, that $h^2 - ab \geqslant 0$ for otherwise the expression for $\tan \phi$ would have no meaning.

(*ii*) If $h^2 - ab = 0$, $\phi = 0$ and the two lines given by $ax^2 + 2hxy + by^2 = 0$ would coincide.

(*iii*) The two lines given by $ax^2 + 2hxy + by^2 = 0$ are perpendicular if and only if $a + b = 0$, *i.e.* the sum of the coefficients of x^2 and y^2 is zero.

Ex. Find the angle between the lines $x^2 + 3xy + 2y^2 = 0$.

2.3 Bisectors of the angles between the lines $ax^2 + 2hxy + by^2 = 0$. *To find the equation of the pair of lines bisecting the angles between the lines* $ax^2 + 2hxy + by^2 = 0$.

Let $y = mx$ and $y = m'x$ be the two lines represented by

$$ax^2 + 2hxy + by^2 = 0. \tag{1}$$

Then, as before, we have

$$m + m' = -\frac{2h}{b}, \qquad mm' = \frac{a}{b}.$$

First method. Let θ and θ' be the angles which the two lines make with the positive direction of x-axis so that

$$m = \tan\theta \text{ and } m' = \tan\theta'.$$

Let $y = x\tan\phi$ be the equation of a bisector. Then

$$2\phi = \theta + \theta' \text{ or } \theta + \theta' + \pi$$

so that

$$\tan 2\phi = \tan(\theta + \theta')$$

$$\Rightarrow \quad \frac{2\tan\phi}{1 - \tan^2\phi} = \frac{\tan\theta + \tan\theta'}{1 - \tan\theta\tan\theta'}$$

$$\Rightarrow \quad \frac{2\dfrac{y}{x}}{1 - \dfrac{y^2}{x^2}} = \frac{m + m'}{1 - mm'}$$

$$= \frac{2h}{a - b}.$$

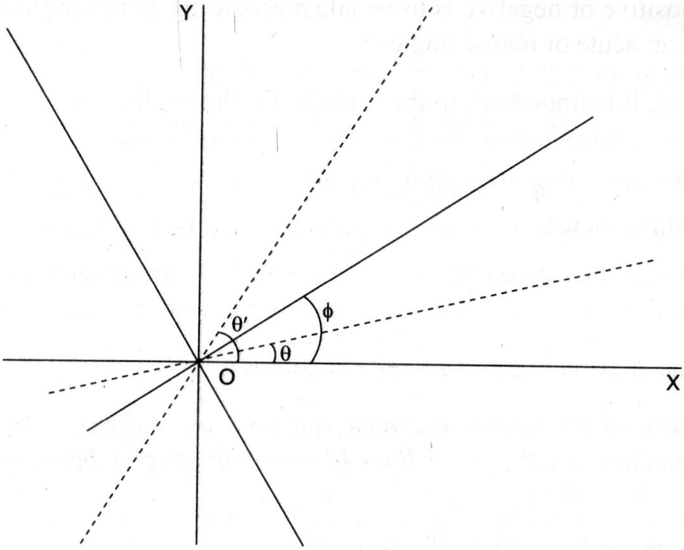

Fig. 4.1

Hence, *the equation of the bisectors of the angles between the lines* $ax^2 + 2hxy + by^2 = 0$ *is given by*

$$\frac{x^2 - y^2}{a - b} = \frac{xy}{h}.$$

Second method. The pair of bisectors of the angles between the lines represented by (1) is given by [Art. 9 Chapter II]

$$\frac{mx - y}{\sqrt{m^2 + 1}} = \pm \frac{m'x - y}{\sqrt{m'^2 + 1}}$$

$$\Rightarrow \quad (m'^2 + 1)(mx - y)^2 = (m^2 + 1)(m'x - y)^2$$

$$\Rightarrow \quad (m^2 - m'^2)(x^2 - y^2) - 2(m - m')(1 - mm')\, xy = 0$$

$$\Rightarrow \quad (m + m')(x^2 - y^2) - 2(1 - mm')\, xy = 0$$

$$\Rightarrow \quad -\frac{2h}{b}(x^2 - y^2) - 2\left(1 - \frac{a}{b}\right) xy = 0$$

$$\Rightarrow \quad \frac{x^2 - y^2}{a - b} = \frac{xy}{h}.$$

Ex. Find the equation of the bisectors of the angles between the lines represented by

$$6x^2 - 13xy + 5y^2 = 3.$$

2.4 Examples

Example 1. Show that the two lines represented by

$$x^2(\tan^2\theta + \cos^2\theta) - 2xy\tan\theta + y^2\sin^2\theta = 0 \quad (1)$$

make angles α, β with x-axis such that $\tan\alpha - \tan\beta = 2$.

Solution. Let $y = mx$ and $y = m'x$ be the lines given by (1). Then

$$m + m' = \frac{2\tan\theta}{\sin^2\theta} = \frac{2}{\sin\theta\cos\theta}$$

and
$$mm' = \frac{\tan^2\theta + \cos^2\theta}{\sin^2\theta} = \sec^2\theta + \cot^2\theta.$$

$$\therefore \quad \tan\alpha - \tan\beta = m - m' = \sqrt{(m+m')^2 - 4mm'}$$

$$= \sqrt{\frac{4}{\sin^2\theta\cos^2\theta} - 4(\sec^2\theta + \cot^2\theta)} = 2.$$

Example 2. Find the condition that one of the two lines $ax^2 + 2hxy + by^2 = 0$ may be perpendicular to one of the lines given by $a'x^2 + 2h'xy + b'y^2 = 0$.

Solution. Let $y = mx$ be one of the lines of $ax^2 + 2hxy + by^2 = 0$. Then

$$a + 2hm + bm^2 = 0. \quad (1)$$

Any line perpendicular to $y = mx$ passing through the origin is $x = -my$. Clearly $x = -my$ will satisfy $a'x^2 + 2h'xy + b'y^2 = 0$ if

$$a'm^2 - 2h'm + b' = 0. \quad (2)$$

Now, from (1) and (2), we get

$$\frac{m^2}{2hb' + 2h'a} = \frac{m}{aa' - bb'} = \frac{1}{-2h'b - 2ha'}$$

$$\Rightarrow \quad \begin{cases} m = \dfrac{aa' - bb'}{-2(h'b + ha')} \\[2ex] m^2 = \dfrac{hb' + h'a}{-(h'b + ha')} \end{cases}$$

$$\Rightarrow \qquad \frac{hb' + h'a}{-(h'b + ha')} = \frac{(aa' - bb')^2}{4(h'b + ha')^2}$$

$$\Rightarrow \qquad 4(ah' + hb')(ha' + bh') + (aa' - bb')^2 = 0.$$

Example 3. Prove that the lines represented by

$$ax^2 + 2hxy + by^2 + \lambda(x^2 + y^2) = 0$$

have the same pair of bisectors for all values of λ. Interpret the case for $\lambda = -(a + b)$.

Solution. The equation representing the bisectors of the given pair of lines is

$$\frac{x^2 - y^2}{(a + \lambda) - (b + \lambda)} = \frac{xy}{h}$$

i.e.

$$\frac{x^2 - y^2}{a - b} = \frac{xy}{h}.$$

Since the equation of the bisectors is free from λ, the given equation has the same pair of bisectors for all values of λ.

Consider now the case when $\lambda = -(a + b)$. The given equation reduces to

$$ax^2 + 2hxy + by^2 - (a + b)(x^2 + y^2) = 0$$

i.e.

$$bx^2 - 2hxy + ay^2 = 0.$$

This represents a pair of lines which are perpendicular to the lines given by

$ax^2 + 2hxy + by^2 = 0$ (See Illustration in § 2.1).

Example 4. Let the pairs of the lines

$$x^2 - 2pxy - y^2 = 0 \text{ and } x^2 - 2qxy - y^2 = 0$$

be such that each pair bisects the angles between the other **pair**. Prove that $pq = -1$.

Solution. The bisectors of the angles between the lines $x^2 - 2pxy - y^2 = 0$ are

$$\frac{x^2 - y^2}{1 - (-1)} = \frac{xy}{-p}, \; i.e. \quad px^2 + 2xy - py^2 = 0.$$

Since these bisectors are given to be represented by $x^2 - 2qxy - y^2 = 0$, we note that

$$\frac{p}{1} = \frac{2}{-2q} = \frac{-p}{-1}, \text{ i.e., } pq = -1.$$

Exercises

1. Form the equations which represent the following pair of lines:
   (*i*) $3x - y = 0, \quad x + 3y = 0$; (*ii*) $y = 0, \quad 4y = x$.
2. Show that each of the following equations represents a pair of lines, and find the angle between each pair:
   (*i*) $x^2 - 4y^2 = 0$. (*ii*) $x^2 - 2xy \cot 2\alpha - y^2 = 0$.
   (*iii*) $x^2 - 2xy \sec \alpha + y^2 = 0$. (*iv*) $2x^2 - 7xy + 3y^2 = 0$.
3. Find the value of λ for which the two lines $3x^2 - 8xy + \lambda y^2 = 0$ are perpendicular to one another.
4. Prove that, if the lines $ax^2 + 2hxy + by^2 = 0$ are perpendicular, then so are the lines $bx^2 + 2kxy + ay^2 = 0$.
5. Find the condition that one of the lines $ax^2 + 2hxy + by^2 = 0$ may coincide with one of the lines $a'x^2 + 2h'xy + b'y^2 = 0$.
6. Prove that the two lines
   $$(x^2 + y^2)(\cos^2\theta \sin^2\alpha + \sin^2\theta) = (x \tan\alpha - y \sin\theta)^2$$
   include an angle 2α.
7. Find the equations of the bisectors of the angles between the pair of lines:
   (*i*) $3x^2 + 8xy + 4y^2 = 0$; (*ii*) $15x^2 - xy - 6y^2 = 0$;
   (*iii*) $12x^2 - 7xy - y^2 = 0$.
8. Show that the angle between one of the lines given by $ax^2 + 2hxy + by^2 = 0$ and one of the lines $ax^2 + 2hxy + by^2 + \lambda(x^2 + y^2) = 0$ is equal to the angle between the other two lines of the systems.
9. Find the condition on m such that the bisectors of the angles between the lines $m^2x^2 + 2(m + 1)xy - y^2 = 0$ are the axes of coordinates.
10. If the bisectors of the angles between the lines $ax^2 + 2hxy + by^2 = 0$, coincide with the bisectors of the angles between the lines $\alpha x^2 + 2\lambda xy + \beta y^2 = 0$, prove that $h(\alpha - \beta) = \lambda(a - b)$.
11. Find the condition that one of the bisectors of the angles between the lines $ax^2 + 2hxy + by^2 = 0$ is the line $lx + my + n = 0$.

3. General Second Degree Equation

3.1 Condition that the most general second degree equations in x and y represent a pair of lines. *To find the necessary condition that the general equation of the second degree*

$$ax^2 + 2hxy + by^2 + 2gx + 2fy + c = 0 \qquad (1)$$

may represent a pair of lines.

First Method. Equation (1) may represent a pair of lines if the expression in the left hand side of it can be decomposed into two real linear factors in x and y. Suppose $a \neq 0$. Multiply equation (1) by a and arranging it in descending powers of x, we get

$$a^2x^2 + 2ax(hy + g) = -aby^2 - 2afy - ac.$$

Completing the perfect square on the left hand side, we have

$$(ax + hy + g)^2 = (h^2 - ab)y^2 + 2(gh - af)\,y + g^2 - ac$$

$$\Rightarrow \qquad ax + hy + g = \pm\sqrt{(h^2 - ab)y^2 + 2(gh - af)y + g^2 - ac} \cdot \qquad (2)$$

Clearly equation (2) represents two real linear equations if the quantity inside the square root forms a perfect square for which

$$4(gh - af)^2 = 4(h^2 - ab)(g^2 - ac).$$

This, after simplification, gives the required condition as

$$abc + 2fgh - af^2 - bg^2 - ch^2 = 0. \qquad (3)$$

If $a = 0$ and $b \neq 0$, multiply equation (1) by b and arrange the terms in descending powers of y. Proceeding similarly as in the previous case, we get the required condition.

In case $a = 0$ and $b = 0$, we must have $h \neq 0$ for otherwise the equation (1) will reduce to a linear equation. The equation (1) reduces to

$$hxy + gx + fy + \tfrac{1}{2}c = 0$$

$$\Rightarrow \qquad (hx + f)(hy + g) = fg - \tfrac{1}{2}ch.$$

Therefore, the condition for this equation to represent a pair of lines is that

$$fg - \tfrac{1}{2}ch = 0,$$

which is the same as (3) for the case $a = 0$ and $b = 0$.

Second Method. Let equation (1) represent a pair of lines intersecting at the point (α, β). Shifting the origin to the point (α, β) by the transformations

$$x = X + \alpha, y = Y + \beta,$$

equation (1) reduces to

$$a(X+\alpha)^2 + 2h(X+\alpha)(Y+\beta) + b(Y+\beta)^2 + 2g(X+\alpha) + 2f(Y+\beta) + c = 0$$

$$\Rightarrow aX^2 + 2hXY + bY^2 + 2(a\alpha + h\beta + g)X +$$

$$2(h\alpha + b\beta + f)Y + a\alpha^2 + 2h\alpha\beta + b\beta^2 + 2g\alpha + 2f\beta + c = 0. \qquad (4)$$

Equation (4) represents a pair of lines passing through the new origin. Therefore, it must be a second degree homogeneous equation in X and Y, and consequently the terms of the first degree in X, Y and the constant terms must disappear from equation (4) so that we are left only with

$$aX^2 + 2hXY + bY^2 = 0.$$

Thus, it is required that the following conditions are satisfied

$$a\alpha + h\beta + g = 0 \tag{5}$$

$$h\alpha + b\beta + f = 0 \tag{6}$$

and $$a\alpha^2 + 2h\alpha\beta + b\beta^2 + 2g\alpha + 2f\beta + c = 0. \tag{7}$$

Multiplying equation (5) by α, equation (6) by β and adding, and then subtracting this sum from equation (7), we get

$$g\alpha + f\beta + c = 0. \tag{8}$$

Now, eliminating α, β from the equations (5), (6) and (8), we obtain

$$\begin{vmatrix} a & h & g \\ h & b & f \\ g & f & c \end{vmatrix} = 0.$$

On solving the determinant, we get the condition

$$abc + 2fgh - af^2 - bg^2 - ch^2 = 0.$$

Third Method. Suppose equation (1) represents a pair of lines. Let the lines be

$$lx + my + n = 0 \text{ and } l'x + m'y + n' = 0.$$

Then

$$ax^2 + 2hxy + by^2 + 2gx + 2fy + c \equiv (lx + my + n)(l'x + m'y + n').$$

Comparing the coefficients of various terms on both the sides, we get

$$ll' = a,\ mm' = b,\ nn' = c,$$

$$lm' + l'm = 2h,\ mn' + m'n = 2f,\ ln' + l'n = 2g.$$

By multiplication, we get

$$\begin{aligned} 8fgh &= (lm' + l'm)(mn' + m'n)(ln' + l'n) \\ &= 2ll'mm'nn' + ll'(m^2n'^2 + m'^2n^2) \\ &\quad + mm'(l^2n'^2 + l'^2n^2) + nn'(l^2m'^2 + l'^2m^2) \\ &= 2abc + a(4f^2 - 2bc) + b(4g^2 - 2ac) + c(4h^2 - 2ab) \end{aligned}$$

$$\Rightarrow \qquad abc + 2fgh - af^2 - bg^2 - ch^2 = 0.$$

Note. The above condition is also sufficient.

Corollary. If the general equation of second degree

$$ax^2 + 2hxy + by^2 + 2gx + 2fy + c = 0$$

represents a pair of lines, the coordinates (α, β) of the point where they intersect are given by

$$\alpha = \frac{hf - bg}{ab - h^2}, \beta = \frac{gh - af}{ab - h^2}.$$

[**Hint.** The values of α and β are determined by solving the equations $lx + my + n = 0$ and $l'x + m'y + n' = 0$].

Note. If $h^2 = ab$, then the point of intersection is at infinity and as such the two lines would be parallel.

Ex. Show that the equation $x^2 - 4xy + y^2 + 6x - 3 = 0$ represents a pair of lines.

3.2 *If the general equation*

$$ax^2 + 2hxy + by^2 + 2gx + 2fy + c = 0 \tag{1}$$

represents a pair of lines, then the equation

$$ax^2 + 2hxy + by^2 = 0 \tag{2}$$

represents a pair of lines parallel to them through the origin.

Let $y = mx + c$ and $y = m'x + c'$ be the lines represented by (1) so that

$$ax^2 + 2hxy + by^2 + 2gx + 2fy + c \equiv b(y - mx - c)(y - m'x - c').$$

Then $\qquad ax^2 + 2hxy + by^2 \equiv b(y - mx)(y - m'x).$

Thus, equation (2) represents the lines $y = mx$ and $y = m'x$ through the origin, which are parallel to $y = mx + c$ and $y = m'x + c'$, respectively.

Corollary. If equation (1) represents a pair of lines, then

(*i*) the angle between these lines is given by

$$\tan \phi = \frac{2\sqrt{h^2 - ab}}{a + b},$$

(*ii*) the two lines are parallel if $h^2 = ab$,

(*iii*) the two lines are perpendicular if and only if

$$a + b = 0.$$

Ex. Calculate the angle between the lines

$$x^2 + 8xy + y^2 + 16x + 4y + 4 = 0.$$

3.3 Example. Find the value of λ so that the equation

$$2x^2 + xy - y^2 - 11x - 5y + \lambda = 0,$$

may represent a pair of lines

Solution. Here $a = 2, h = \dfrac{1}{2}, b = -1,$

$$g = -\frac{11}{2}, f = -\frac{5}{2}, c = \lambda.$$

The given equation represents a pair of lines if

$$abc + 2fgh - af^2 - bg^2 - ch^2 = 0$$

i.e. $$-2\lambda + \frac{55}{4} - \frac{50}{4} + \frac{121}{4} - \frac{\lambda}{4} = 0.$$

Hence $$\lambda = 14.$$

Exercises

1. Show that each of the following equations represents a pair of lines
   (*i*) $x^2 + 8xy + y^2 + 16x + 4y + 4 = 0;$
   (*ii*) $x^2 + 6xy + 9y^2 + 4x + 12y - 5 = 0;$
   (*iii*) $x^2 - 2xy + y^2 - 4x + 4y + 4 = 0;$
   (*iv*) $2x^2 + 3xy - 2y^2 + 5x - 10y - 12 = 0;$
   (*v*) $x^2 + xy - 6y^2 + 7x + 31y - 18 = 0.$
2. Find the value of λ so that the following equations may represent a pair of lines
   (*i*) $12x^2 - 10xy + 2y^2 + 11x - 5y + \lambda = 0;$
   (*ii*) $x^2 + 2\lambda xy + y^2 + 6x + 2y + 9 = 0;$
   (*iii*) $12x^2 + 36xy + \lambda y^2 + 6x + 6y + 3 = 0;$
   (*iv*) $\lambda x^2 + 2xy + \lambda y^2 + 4x + 4y + 3 = 0.$
3. Find the angle between the pair of lines
   (*i*) $x^2 - xy - 6y^2 - 7x + 31y - 18 = 0;$
   (*ii*) $4x^2 + 5xy + y^2 + 3y - 4 = 0;$
   (*iii*) $x^2 - xy - 6y^2 - 3x + 14y - 4 = 0.$

4. For what value of λ does the equation

$12x^2 + 7xy + \lambda y^2 + 13x - y + 3 = 0,$

represent a pair of lines, and what is then the angle between them?

4. Pair of Lines Joining the Origin to the Points of Intersection of a Curve and a Line

4.1 *To find the equation of the pair of lines joining the origin to the points of intersection of the curve*

$$ax^2 + 2hxy + by^2 + 2gx + 2fy + c = 0, \qquad (1)$$

and the line

$$lx + my + n = 0. \qquad (2)$$

Make the equations (1) and (2) homogeneous by introducing a third variable z in the following way:

$$ax^2 + 2hxy + by^2 + 2gxz + 2fyz + cz^2 = 0 \qquad (3)$$

$$lx + my + nz = 0. \qquad (4)$$

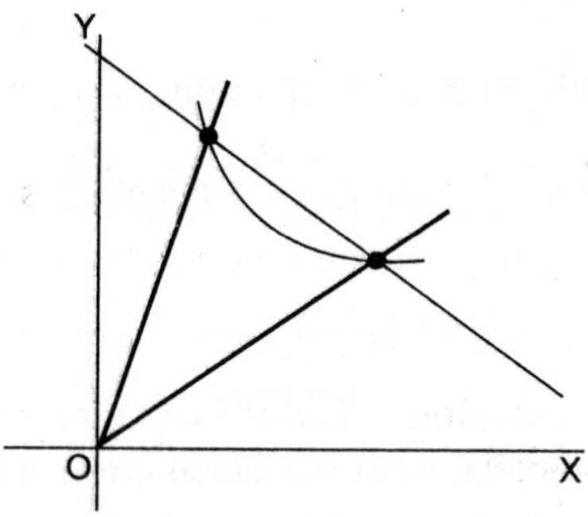

Now, eliminating z from (3) and (4), we get

$$ax^2 + 2hxy + by^2 - 2gx\left(\frac{lx+my}{n}\right)$$
$$-2fy\left(\frac{lx+my}{n}\right) + c\left(\frac{lx+my}{n}\right)^2 = 0$$

$$\Rightarrow n^2(ax^2 + 2hxy + by^2) - 2n(gx + fy)(lx + my)$$
$$+ c\,(lx + my)^2 = 0. \qquad (5)$$

This is a homogeneous equation of degree two in x and y and so it represents a pair of lines through the origin. Moreover, we note that if (x_1, y_1) is a point of intersection of (1) and (2), then

$$\begin{cases} ax_1^2 + 2hx_1y_1 + by_1^2 + 2gx_1 + 2fy_1 + c = 0 \\ lx_1 + my_1 + n = 0 \end{cases}$$

and so the point (x_1, y_1) also satisfies equation (5). Thus, equation (5) represents a pair of lines through the origin and the points of intersection of (1) and (2).

Working Rule. The equation of the pair of lines joining the origin to the points of intersection of a curve and a line is obtained by making the equation of the curve homogeneous with the help of the equation of the line.

4.2 Examples

Example 1. Find the equation to the pair of lines obtained by joining the origin to the points of intersection of the line $y = mx + c$ and the circle $x^2 + y^2 = a^2$, and prove that they are at right angles if $2c^2 = a^2(1 + m^2)$.

Solution. Make the equation of the circle and the equation of the line homogeneous by introducing a third variable z as follows:

$$\left.\begin{aligned} x^2 + y^2 &= a^2z^2 \\ y &= mx + cz \end{aligned}\right\}.$$

Eliminating z from the above pair of equations, we get

$$x^2 + y^2 = a^2\left(\frac{y - mx}{c}\right)^2$$

$$\Rightarrow \quad (c^2 - a^2m^2)x^2 + 2a^2m\,xy + (c^2 - a^2)y^2 = 0. \quad (1)$$

This is the equation representing a pair of lines through the origin which pass through the points of intersection of the given circle and the line.

Further, the lines given by (1) will be at right angles to each other if

$$\text{coeff. of } x^2 + \text{coeff. of } y^2 = 0$$

i.e. if $\quad c^2 - a^2m^2 + c^2 - a^2 = 0$

$$\Rightarrow \quad 2c^2 = a^2(1 + m^2).$$

Example 2. Prove that the lines joining the origin to the points of intersection of the two curves $ax^2 + 2hxy + by^2 + 2gx = 0$ and $a'x^2 + 2h'xy + b'y^2 + 2g'x = 0$ will be at right angles to one another, if

$$g'(a + b) = g(a' + b').$$

Solution. Any curve through the points of intersection of the given curves is

$$ax^2 + 2hxy + by^2 + 2gx + \lambda(a'x^2 + 2h'xy + b'y^2 + 2g'x) = 0$$

i.e. $$(a + \lambda a')x^2 + 2(h + \lambda h')xy + (b + \lambda b')y^2 + 2(g + \lambda g')x = 0. \quad (1)$$

In order that this equation should represent a pair of lines passing through the origin, it must be homogeneous equation of second degree, for which the coefficient of x must vanish *i.e.*

$$2(g + \lambda g') = 0$$

$$\Rightarrow \qquad \lambda = -\frac{g}{g'}. \quad (2)$$

The two lines represented by (1) will be at right angles if

Coeff. of x^2 + Coeff. of $y^2 = 0$

i.e. $$a + \lambda a' + b + \lambda b' = 0$$

$$\Rightarrow \qquad a + b + \lambda(a' + b') = 0$$

$$\Rightarrow \qquad (a + b) - \frac{g}{g'}(a' + b') = 0 \text{ (On using (2))}$$

$$\Rightarrow \qquad g'(a + b) = g\,(a' + b').$$

Example 3. If the equation $ax^2 + 2hxy + by^2 + 2gx + 2fy + c = 0$ represents a pair of lines, prove that the equation to the pair of the lines passing through the points where these meet the axes is

$$ax^2 - 2hxy + by^2 + 2gx + 2fy + c + \frac{4fg}{c}xy = 0.$$

Solution. If the equation

$$ax^2 + 2hxy + by^2 + 2gx + 2fy + c = 0 \quad (1)$$

represents a pair of lines, then

$$abc + 2fgh - af^2 - bg^2 - ch^2 = 0. \quad (2)$$

The joint equation of the axes is

$$xy = 0. \quad (3)$$

Any curve through the points of intersection of (1) and (3) is

$$ax^2 + 2hxy + by^2 + 2gx + 2fy + c + \lambda xy = 0$$

i.e. $$ax^2 + (2h + \lambda)xy + by^2 + 2gx + 2fy + c = 0. \quad (4)$$

This equation will represent a pair of lines if

$$abc + 2fg\left(\frac{2h + \lambda}{2}\right) - af^2 - bg^2 - c\left(\frac{2h + \lambda}{2}\right)^2 = 0$$

$$\Rightarrow\ abc + 2fgh - af^2 - bg^2 - ch^2 + fg\lambda - ch\lambda - \frac{c\lambda^2}{4} = 0$$

$$\Rightarrow\ fg\lambda - ch\lambda - \frac{c\lambda^2}{4} = 0 \qquad \text{(On using (2))}$$

$$\Rightarrow\ \lambda = \frac{4(fg - ch)}{c}.$$

Substituting this value of λ in (4), we get

$$ax^2 + \left\{2h + \frac{4(fg - ch)}{c}\right\}xy + by^2 + 2gx + 2fy + c = 0$$

$$\Rightarrow\ ax^2 - 2hxy + by^2 + 2gx + 2fy + c + \frac{4fg}{c}xy = 0.$$

Exercises

1. Prove that the lines joining the origin to the points common to $3x^2 + 5xy - 3y^2 + 2x + 3y = 0$ and $3x - 2y = 1$ are at right angles.
2. Show that the angle between the lines joining the origin to the intersections of the line $y = 3x + 2$ with the curve
$$x^2 + 2xy + 3y^2 + 4x + 8y - 11 = 0,$$
is $\tan^{-1} \dfrac{2\sqrt{2}}{3}$.
3. Show that all chords of the curve $3x^2 - y^2 - 2x + 4y = 0$ which subtend a right angle at the origin pass through a fixed point.
4. Find the angle between the lines joining the origin to the points of intersection of the line $x - 3y + 2 = 0$ and the curve
$$y^2 - 17xy + 16y - 12 = 0.$$
5. Show that the lines joining the origin to the points common to $x^2 + hxy - y^2 + gx + fy = 0$ and $fx - gy = \lambda$ are at right angles for all values of λ.
6. Prove that the lines joining the origin to the points of intersection of
$$ax^2 + 2hxy + by^2 + 2gx + 2fy + c = 0$$
and $lx + my + n = 0$ are perpendicular to each other if
$$n^2(a + b) - 2n(gl + fm) + c(l^2 + m^2) = 0.$$
7. If the circle $x^2 + y^2 + 2gx + 2fy + c = 0$ intercepts on the line $lx + my = 1$ a length which subtends a right angle at the origin, prove that
$$c(l^2 + m^2) + 2(gl + fm + 1) = 0.$$

Miscellaneous Exercises

1. The distance of a point (x_1, y_1) from each of the two lines which pass through the origin of coordinates is δ; prove that the two lines are given by
$$(x_1 y - x y_1)^2 = \delta^2(x^2 + y^2).$$
2. Show that the product of the perpendiculars from the point (x', y') to the lines $ax^2 + 2hxy + by^2 = 0$ is
$$\frac{ax'^2 + 2hx'y' + by'^2}{\sqrt{(a-b)^2 + 4h^2}}.$$
3. If p_1, p_2 be the perpendiculars from (x', y') to the lines $ax^2 + 2hxy + by^2 = 0$, prove that
$$(p_1^2 + p_2^2)\{(a-b)^2 + 4h^2\} = 2\,(a-b)(ax'^2 - by'^2) + 4h(a+b)xy + 4h^2(x'^2 + y'^2).$$
4. Prove that the pair of lines $ax^2 + 2hxy + by^2 = 0$ is equally inclined to the pair of lines
$$a^2x^2 + 2h(a+b)\,xy + b^2y^2 = 0.$$
5. Show that the lines given by $ax + by + c = 0$ and $(ax + by)^2 - 3\,(ay - bx)^2 = 0$ form an equilateral triangle.
6. For what values of λ is the conic
$$\lambda\,(x^2 + y^2 + 2gx + 2fy + c) + 2xy = 0$$
a pair of lines? If two of these pairs consist of parallel lines, what condition must be satisfied by g, f and c?
7. Prove that the lines joining the origin to the points of intersection of the line $kx + hy = 2hk$ with the curve $(x-h)^2 + (y-k)^2 = c^2$ are at right angles if $h^2 + k^2 = c^2$.
8. Find the equation of the pair of lines joining the origin to the points of intersection of the line $ax + by = a + 2b$ and the locus $x^2 - xy + y - 1 = 0$.
9. Find the condition of a chord $ax + by + c = 0$ of the curve $x^2 - 3y^2 - 4x + 2y = 0$ if it subtends a right angle at the origin.
10. Show that the area of the triangle formed by the lines $ax^2 + 2hxy + by^2 = 0$ and $lx + my + n = 0$ is
$$\frac{n^2\sqrt{h^2 - ab}}{am^2 - 2hlm + bl^2}.$$
11. Show that the lines $(a^2 - 3b^2)x^2 + 8ab\,xy + (b^2 - 3a^2)y^2 = 0$ and $ax + by + c = 0$ form an equilateral triangle whose area is $\dfrac{c^2}{(a^2 + b^2)\sqrt{3}}$.

12. Prove that the general equation $ax^2 + 2hxy + by^2 + 2gx + 2fy + c = 0$ represents two parallel lines if $h^2 = ab$ and $bg^2 = af^2$.

Prove also that the distance between them is $2\sqrt{\dfrac{g^2 - ac}{a(a+b)}}$.

13. If $ax^2 + 2hxy + by^2 + 2gx + 2fy + c = 0$ represents a pair of lines, prove that

(*i*) the lines are equidistant from the origin, if $f^4 - g^4 = c(bf^2 - ag^2)$;

(*ii*) the product of the perpendiculars from the origin on these lines is

$$\frac{c}{\sqrt{(a-b)^2 + 4h^2}};$$

(*iii*) the square of the distance of their point of intersection from the origin is

$$\frac{c(a+b) - f^2 - g^2}{ab - h^2};$$

(*iv*) the area of the triangle formed by these lines and the x-axis is

$$\frac{g^2 - ac}{a\sqrt{h^2 - ab}}.$$

14. If $ax^2 + 2hxy + by^2 + 2gx + 2fy + c = 0$ represents a pair of lines intersecting in (α, β), show that

(*i*) the equation of the pair of lines bisecting the angles between them is

$$\frac{(x-\alpha)^2 - (y-\beta)^2}{a-b} = \frac{(x-\alpha)(y-\beta)}{h};$$

(*ii*) the area of the triangle formed by their bisectors and the x-axis is

$$\frac{\sqrt{(a-b)^2 + 4h^2}}{2h} \cdot \frac{ca - g^2}{ab - h^2}.$$

15. The equation $ax^2 + 2hxy + by^2 + 2gx + 2fy + c = 0$ represents two lines at right angles to each other. Prove that the square of the distance of their point of intersection from the origin is

$$\frac{f^2 + g^2}{b^2 + h^2}.$$

16. Prove that the locus of a point which moves such that the distance between the feet of the perpendiculars from it on the lines $ax^2 + 2hxy + by^2 = 0$ is a constant $2k$ is $(x^2 + y^2)(h^2 - ab) = k^2 \{(a-b)^2 + 4h^2\}$.

17. Show that the orthocentre of the triangle formed by the lines

$$ax^2 + 2hxy + by^2 = 0 \text{ and } lx + my = 1$$

is a point (x', y') such that

$$\frac{x'}{l} = \frac{y'}{m} = \frac{a+b}{am^2 - 2hlm + bl^2}.$$

18. If $\qquad ax^2 + 2hxy + by^2 + 2gx + 2fy + c = 0$

and $\qquad ax^2 + 2hxy + by^2 - 2gx - 2fy + c = 0,$

each represents a pair of lines, prove that the area of the parallelogram enclosed by them is $\dfrac{2c}{\sqrt{h^2 - ab}}$.

19. If the lines $ax^2 + 2hxy + by^2 = 0$ be two sides of a parallelogram and the line $lx + my = 1$ be one of its diagonals, prove that the equation of the other diagonal is $y(bl - hm) = x(am - hl)$.

20. If two of the lines given by $ax^3 + 3bx^2y + 3cxy^2 + dy^3 = 0$ are at right angles, prove that $a^2 + 3ac + 3bd + d^2 = 0$.

21. Prove that the equation $a(x^4 + y^4) - 4bxy(x^2 - y^2) + 6cx^2y^2 = 0$ represents two pairs of lines which are at right angles, and if $2b^2 = a^2 + 3ac$, the two pairs will coincide.

The Circle

5 CHAPTER

1. Definition of the Circle

A **circle** *is the locus of a point which moves such that it remains at a fixed distance from a fixed point.*

The fixed point is called the **centre** of the circle and the fixed distance the **radius** of the circle.

1.1 Equation of a circle. *To find the equation of a circle with given centre and radius.*

Let $C(h, k)$ be the centre and r the radius of the circle.

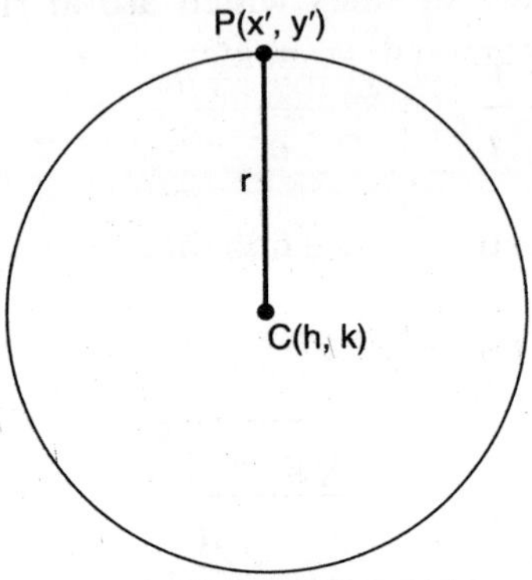

Fig. 5.1

Let $P(x', y')$ be any point on the locus (circle). Then, by definition

$$CP = r.$$

But $$CP = \sqrt{(x'-h)^2 + (y'-k)^2}.$$

$$\therefore \quad \sqrt{(x'-h)^2 + (y'-k)^2} = r$$

$$\Rightarrow \quad (x'-h)^2 + (y'-k)^2 = r^2$$

Hence, *the equation of the circle with centre* (h, k) *and radius* r *is*

$$(x-h)^2+(y-k)^2=r^2.$$

1.2 The general form of the equation. The equation of the circle is

$$(x-h)^2+(y-k)^2=r^2 \tag{1}$$

which on expanding takes the form of

$$x^2+y^2-2hx-2ky+h^2+k^2-r^2=0.$$

We note the following three things about this equation:

(*i*) This is a second degree equation in x and y;

(*ii*) The coefficients of x^2 and y^2 are equal;

(*iii*) There is no term containing the product xy.

Conversely, we consider the most general equation satisfying these three conditions:

$$ax^2+ay^2+2gx+2fy+c=0, \qquad (a \neq 0). \tag{2}$$

To prove that equation (2) *represents a circle.*

Equation (2) may be written as

$$\left(x+\frac{g}{a}\right)^2+\left(y+\frac{f}{a}\right)^2=\frac{g^2+f^2-ac}{a^2}.$$

By comparing it with equation (1) we find that it represents a circle with centre at the point $\left(-\frac{g}{a},-\frac{f}{a}\right)$ and radius given by

$$r=\frac{\sqrt{g^2+f^2-ac}}{a}.$$

Hence, equation (2) always represents a circle.

There will be no loss of generality if we choose $a = 1$ so that the general equation of the circle takes the form

$$x^2+y^2+2gx+2fy+c=0. \tag{3}$$

This circle has the centre $(-g, -f)$ and radius equal to

$$\sqrt{g^2+f^2-c}.$$

Remarks. (*i*) If $g^2+f^2>c$, equation (3) represents a circle with centre $(-g, -f)$ and real radius.

(*ii*) If $g^2 + f^2 = c$, then equation (3) represents a circle whose centre is $(-g, -f)$ and radius is zero *i.e.* the circle coincides with the centre. This circle is called the **point circle**.

(*iii*) If $g^2 + f^2 < c$, then equation (3) represents a circle with centre $(-g, -f)$ and radius imaginary. In this case there are no real points on the circle and we call it a **virtual circle**.

Corollary. The equation of the circle centred at the origin and of radius a is

$$x^2 + y^2 = a^2.$$

1.3 Circle with given diameter. *To find the equation of the circle described on the line joining the points $A(x_1, y_1)$ and $B(x_2, y_2)$ as diameter.*

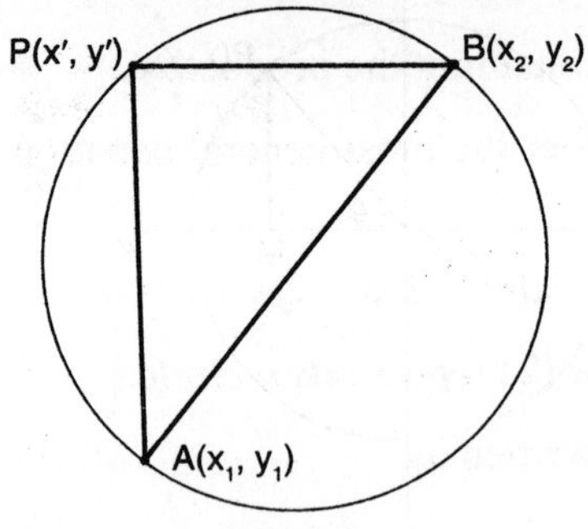

Fig. 5.2

Let $P(x', y')$ be any point on the circle. Then

$$m = \text{slope of the line } AP$$

$$= \frac{y' - y_1}{x' - x_1},$$

$$m' = \text{slope of the line } PB = \frac{y' - y_2}{x' - x_2}.$$

Since $\angle APB = 90°$, $mm' = -1$, we have

$$\left(\frac{y' - y_1}{x' - x_1}\right)\left(\frac{y' - y_2}{x' - x_2}\right) = -1$$

$$\Rightarrow \quad (x' - x_1)(x' - x_2) + (y' - y_1)(y' - y_2) = 0.$$

Hence, the required equation of the circle is

$$(x - x_1)(x - x_2) + (y - y_1)(y - y_2) = 0.$$

Ex. Find the equation to the circle of which the points (8, –2) and (2, –6) are the end points of a diameter.

1.4 Parametric form of the circle. *To find the parametric equations of a circle.*

Consider the equation of the circle

$$x^2 + y^2 = a^2. \tag{1}$$

Let $P(x, y)$ be any point on it. Let $\angle XOP = \theta$. Then, the coordinates (x, y) of P can be expressed as

$$\left.\begin{aligned} x &= a\cos\theta \\ y &= a\sin\theta \end{aligned}\right\} \tag{2}$$

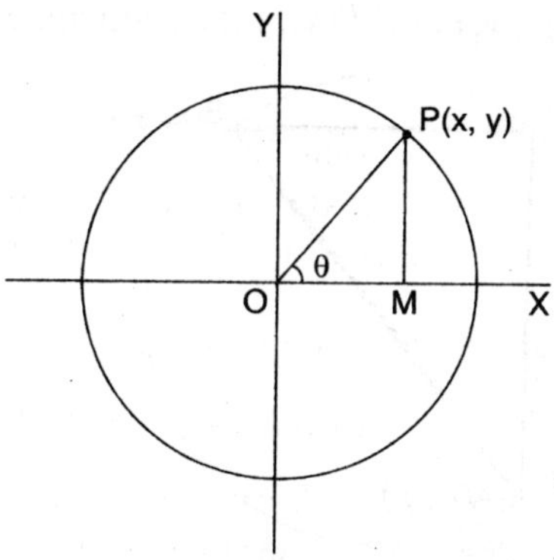

Fig. 5.3

Conversely, the point $(a \cos \theta, a \sin \theta)$ lies on the circle for all values of θ, $0 \leqslant \theta \leqslant 2\pi$. Hence equation (2) can be regarded as a parametric form of the circle given by equation (1), where θ is a parameter.

More generally, if we consider the circle with centre (h, k) and radius r, then its parametric form is

$$x = h + r\cos\theta$$
$$y = k + r\sin\theta.$$

1.5 Circle and a point. A circle divides the plane into two regions. A point may be either outside or inside the circle, if it is not on the circle itself which is the boundary between the two regions. We now proceed to find an algebraic condition for a point to be in a particular region determined by the circle.

To prove that the point (x_1, y_1) *lies outside, on or inside the circle* $x^2 + y^2 = a^2$ *according as* $x_1^2 + y_1^2 - a^2 >= \text{or} < 0$.

The distance between the centre of the circle and the point (x_1, y_1) is $> =$ or $<$ the radius of the circle *i.e.* $\sqrt{x_1^2 + y_1^2} > =$ or $< a$. Hence, the point (x_1, y_1) lies outside, on or inside the circle $x^2 + y^2 = a^2$ according as, respectively, $x_1^2 + y_1^2 - a^2 > =$ or < 0.

Ex. Prove that the point (x_1, y_1) lies outside, on or inside the circle $x^2 + y^2 + 2gx + 2fy + c = 0$ according as, respectively, $x_1^2 + y_1^2 + 2gx_1 + 2fy_1 + c >= \text{or} < 0$.

1.6 Examples

Example 1. Whatever be the value of α, prove that the locus of the intersection of the lines $x \cos \alpha + y \sin \alpha = a$ and $x \sin \alpha - y \cos \alpha = b$ is a circle.

Solution. The locus of the point of intersection of the given lines is obtained by eliminating α between them. For this squaring and adding the given equations, we get

$$(x \cos \alpha + y \sin \alpha)^2 + (x \sin \alpha - y \cos \alpha)^2 = a^2 + b^2$$

$$\Rightarrow \quad x^2 + y^2 = a^2 + b^2,$$

which is a circle.

Example 2. Prove that the circle on the chord $lx + my + n = 0$ of the circle $x^2 + y^2 = a^2$ as a diameter is $(l^2 + m^2)(x^2 + y^2 - a^2) + 2n(lx + my + n) = 0$.

Solution. Any circle through the intersection of the given circle and the chord is

$$x^2 + y^2 - a^2 + \lambda(lx + my + n) = 0. \quad (1)$$

The centre of the circle (1) is $(-\frac{1}{2}\lambda l, -\frac{1}{2}\lambda m)$. The chord $lx + my + n = 0$ will be a diameter of circle (1) if its centre lies on the chord *i.e.*

$$l(-\tfrac{1}{2}\lambda l) + m(-\tfrac{1}{2}\lambda m) + n = 0$$

$$\Rightarrow \quad \lambda = \frac{2n}{l^2 + m^2}.$$

Substituting this value of λ in (1), we get

$$(l^2 + m^2)(x^2 + y^2 - a^2) + 2n(lx + my + n) = 0.$$

Exercises

1. Find the equation to the circle:
   (*i*) whose radius is 4 and centre is (1, 2).
   (*ii*) whose radius is 3 and centre is (–3, –4).
   (*iii*) whose radius is $\sqrt{3}$ and centre is $(\frac{2}{3}, \frac{1}{3})$.
2. Find the coordinates of the centre and the radius of the circle whose equation is:
   (*i*) $x^2 + y^2 + 4x + 8y - 41 = 0$. (*ii*) $x^2 + y^2 - 4x - 16 = 0$.
   (*iii*) $3x^2 + 3y^2 + 6x - 5y = 0$.

3. Find the equation to the circle whose diameter is the line joining the points:
   (*i*) (5, 1) and (7, 5). (*ii*) (–4, –2) and (1, 1).
   (*iii*) (–2, 1) and (4, –3).
4. Find the equation to the circle which passes through the points:
   (*i*) (0, 0), (5, –7), (1, –3). (*ii*) (0, 0), $(a, 0)$, $(0, b)$.
   (*iii*) (3, 0), (4, 2), (0, 1). (*iv*) (–1, 1), (–2, 1), (4, 3).
5. Find the equation to the circle which passes through the points (–1, 2), (–4, 3) and has its centre on the line $4x - 3y = 5$.
6. Find the equation to the circle which passes through the origin and cuts off intercepts equal to 4 and 5 from the axes.
7. Find the equation to the circle
   (*i*) which touches the axes at a distance 4 from the origin.
   (*ii*) which passes through (5, 3) and has its centre at (–3, 1).
8. Find the equation to the circle circumscribing the triangle formed by the lines $x + y = 6$, $2x + y = 4$ and $x + 2y = 5$.
9. Find the value of f for which the circle $x^2 + y^2 - 4x + 2fy + 13 = 0$ has radius 4.
10. Prove that the locus of the point of intersection of the lines drawn through the points $(a, 0)$ and $(-a, 0)$ which include a constant angle θ is the circle $x^2 + y^2 - a^2 \pm 2ay \cot \theta = 0$.

    Hence, prove that the angles in the same segment of a circle are equal.

2. Tangent and Normal

Definition of tangent

Let P and Q be any two points on a curve such that P and Q are near to one another. Join the secant PQ. If $Q \to P$, then *the secant PQ is called the tangent at P to the curve.* In other words *the tangent at P is the limiting position of the secant PQ through P as Q tends to coincide with P.* Consequently, the tangent PT cuts the curve in two coincident points at P.

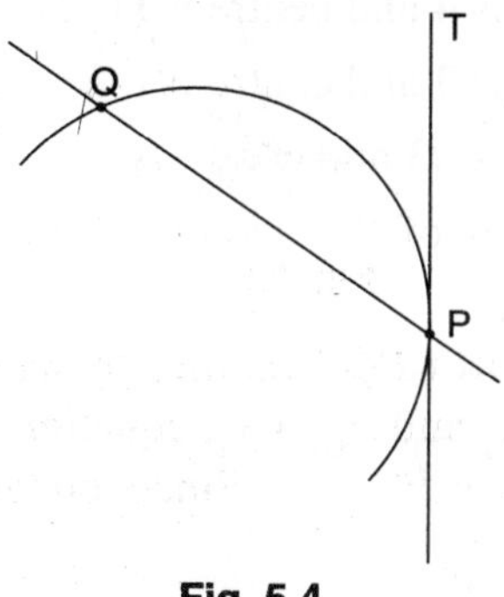

Fig. 5.4

Definition of normal

A line through the point P perpendicular to the tangent at P is called the normal to the curve at P.

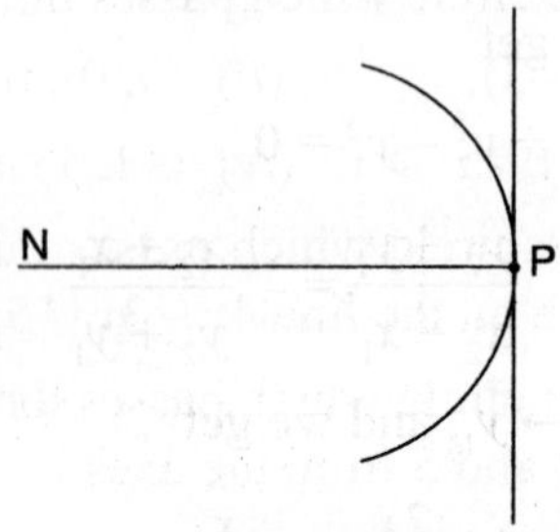

Fig. 5.5

2.1 Equation of the tangent at a point. *To find the equation of the tangent to a circle at a point on it.*

Let $$x^2 + y^2 = a^2 \qquad (1)$$

be the equation of the circle and $P(x_1, y_1)$ be the given point on it. Since (x_1, y_1) lies on (1), we have

$$x_1^2 + y_1^2 = a^2. \qquad (2)$$

Take $Q\ (x_2, y_2)$ any other point on (1). Then

$$x_2^2 + y_2^2 = a^2. \qquad (3)$$

The equation of the secant PQ is

$$y - y_1 = \frac{y_2 - y_1}{x_2 - x_1}(x - x_1). \qquad (4)$$

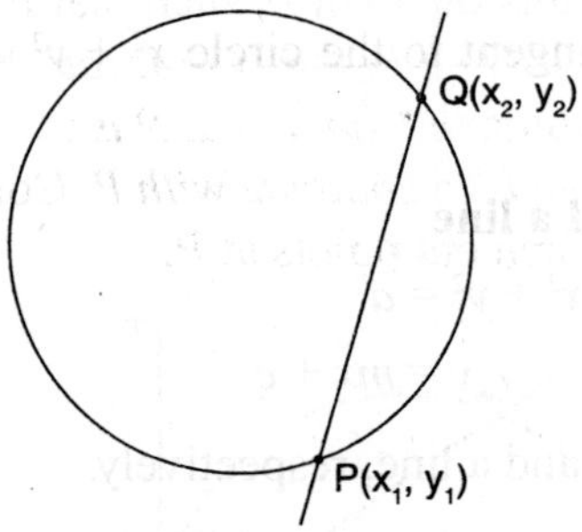

Fig. 5.6

As $Q \to P$, by definition, secant PQ becomes the tangent at P and therefore on taking the limit as $x_2 \to x_1$ and $y_2 \to y_1$, equation (4) would represent the equation of the tangent to (1) at $P(x_1, y_1)$. Since, on taking the limit, the value of $\dfrac{y_2 - y_1}{x_2 - x_1}$ is of the form $\dfrac{0}{0}$ which is an indeterminate form, we proceed to obtain the same as follows.

Subtracting (2) from (3), we get

$$x_2^2 - x_1^2 + y_2^2 - y_1^2 = 0$$

$$\Rightarrow \qquad \frac{y_2 - y_1}{x_2 - x_1} = -\frac{x_2 + x_1}{y_2 + y_1}.$$

Thus, as $Q \to P$, $x_2 \to x_1$ and $y_2 \to y_1$, and we get

$$\frac{y_2 - y_1}{x_2 - x_1} \to -\frac{2x_1}{2y_1} = -\frac{x_1}{y_1},$$

and hence as $Q \to P$, equation (4) becomes

$$y - y_1 = -\frac{x_1}{y_1}(x - x_1)$$

$$\Rightarrow \qquad xx_1 + yy_1 = x_1^2 + y_1^2.$$

Therefore, on using (2), we get

$$xx_1 + yy_1 = a^2.$$

Hence, *the equation of the tangent to the circle* $x^2 + y^2 = a^2$ *at the point* (x_1, y_1) *is* $xx_1 + yy_1 = a^2$. □

More generally, the equation of the tangent to the circle $x^2 + y^2 + 2gx + 2fy + c = 0$ at the point (x_1, y_1) on it is $xx_1 + yy_1 + g(x + x_1) + f(y + y_1) + c = 0$.

Working rule. The equation of the tangent at the point (x_1, y_1) to any circle can be obtained by writing xx_1 for x^2, yy_1 for y^2, $x + x_1$ for $2x$ and $y + y_1$ for $2y$.

Ex. Find the equation of the tangent to the circle $x^2 + y^2 - 4y - 1 = 0$ at the point (2, 1).

2.2 Intersection of a circle and a line

Let $\qquad x^2 + y^2 = a^2 \qquad (1)$

and $\qquad y = mx + c \qquad (2)$

be the equations of a circle and a line, respectively.

For the points of intersection we solve (1) and (2) simultaneously and we get

$$x^2 + (mx + c)^2 = a^2$$

i.e. $\qquad (1 + m^2)x^2 + 2mcx + c^2 - a^2 = 0. \qquad (3)$

Since equation (3) is quadratic in x it gives two values of x which may be real and distinct, coincident or imaginary. Accordingly there are two points of intersection of the circle (1) and the line (2). Hence a line always cuts a circle in two points which may be real and distinct, coincident or imaginary.

2.3 Condition of tangency. *To find the condition when the line $y = mx + c$ becomes a tangent to the circle $x^2 + y^2 = a^2$.*

First Method. It is clear from equation (3) in § 2.2 that the line $y = mx + c$ is a tangent to the circle $x^2 + y^2 = a^2$ if the roots of the equation (3) in § 2.2 are equal for which $m^2c^2 = (1 + m^2)(c^2 - a^2)$, *i.e.*, $c = a\sqrt{1+m^2}$, which is the required condition of tangency.

Second Method. Equation of the tangent to the given circle $x^2 + y^2 = a^2$ at any point (x_1, y_1) is $xx_1 + yy_1 = a^2$. The line $y = mx + c$ is a tangent to the given circle if

$$xx_1 + yy_1 - a^2 \equiv mx - y + c$$

$$\Rightarrow \qquad \frac{x_1}{m} = \frac{y_1}{-1} = \frac{-a^2}{c}$$

$$\Rightarrow \qquad x_1 = \frac{-a^2 m}{c}, y_1 = \frac{a^2}{c}.$$

But (x_1, y_1) lies on the given circle. Therefore

$$\left(-\frac{a^2 m}{c}\right)^2 + \left(\frac{a^2}{c}\right)^2 = a^2, \text{ i.e., } c = a\sqrt{1+m^2},$$

which is the required condition.

Third Method. The line $y = mx + c$ is a tangent to the circle $x^2 + y^2 = a^2$ if the perpendicular distance of the line from the centre of the circle is equal to the radius of the circle, *i.e.*, if $\frac{c}{\sqrt{1+m^2}} = a$ which gives $c = a\sqrt{1+m^2}$.

Remarks. (*i*) The line $y = mx + a\sqrt{1+m^2}$ is always a tangent to the circle $x^2 + y^2 = a^2$ whatever the value of m may be. In fact, it is a family of tangents to the circle $x^2 + y^2 = a^2$.

(*ii*) The line $y = mx + a\sqrt{1+m^2}$ is called the equation of the tangent in **'m' form** or in **slope form** to the circle $x^2 + y^2 = a^2$.

(*iii*) The point of contact of the tangent $y = mx + a\sqrt{1+m^2}$ to the circle

$x^2 + y^2 = a^2$ is

$$\left(\frac{-am}{\sqrt{1+m^2}}, \frac{a}{\sqrt{1+m^2}} \right).$$

Ex. Find the condition that the line $x \cos \alpha + y \sin \alpha = p$ may touch the circle $x^2 + y^2 = a^2$.

2.4 Equation of normal. *To find the equation of the normal to the circle $x^2 + y^2 = a^2$ at any point (x_1, y_1) on it.*

The equation of the tangent at the point (x_1, y_1) is

$$xx_1 + yy_1 = a^2.$$

Slope of the tangent $= -\dfrac{x_1}{y_1}$.

$\therefore$ Slope of the normal $= \dfrac{y_1}{x_1}$.

Hence, the equation of the normal at (x_1, y_1) to the circle $x^2 + y^2 = a^2$ is

$$y - y_1 = \frac{y_1}{x_1}(x - x_1), \text{ i.e., } xy_1 - x_1y = 0 . \square$$

Following similar lines, one can easily show that the equation of the normal to the circle $x^2 + y^2 + 2gx + 2fy + c = 0$ at the point (x_1, y_1) is $x(y_1 + f) - y(x_1 + g) - fx_1 + gy_1 = 0$.

Ex. Prove that the normal at any point to a circle passes through the centre of the circle.

2.5 Examples

Example 1. Find the equation to the circle whose centre is at the point (α, β) and which passes through the origin, and prove that the equation of the tangent at the origin is $\alpha x + \beta y = 0$.

Solution. Let r be the radius of the circle. Then equation of the circle with centre (α, β) and radius r is

$$(x - \alpha)^2 + (y - \beta)^2 = r^2. \quad (1)$$

It will pass through the origin if

$$\alpha^2 + \beta^2 = r^2. \quad (2)$$

Now, eliminating r from (1) and (2), we get

$$(x - \alpha)^2 + (y - \beta)^2 = \alpha^2 + \beta^2$$

$$\Rightarrow \qquad x^2 + y^2 - 2\alpha x - 2\beta y = 0. \qquad (3)$$

This is the required equation of the circle.

Further, equation of the tangent to the circle (3) at the origin is

$$x.0 + y.0 - \alpha(x + 0) - \beta(y + 0) = 0, \ \ i.e., \alpha x + \beta y = 0.$$

Example 2. Find the condition that the line $lx + my + n = 0$ may be normal to the circle $x^2 + y^2 + 2gx + 2fy + c = 0$.

Solution. The line will be a normal to the circle if it passes through the centre $(-g, -f)$ of the circle, *i.e.* $l(-g) + m(-f) + n = 0$, which gives $lg + mf = n$. This is the required condition.

Exercises

1. Find the equation of the tangent to the circle $x^2 + y^2 = 13$ at the point (2, 3).
2. Prove that the circle $x^2 + y^2 - 2ax - 2ay + a^2 = 0$ touches the axes.
3. Prove that the line $lx + my + n = 0$ touches the circle $x^2 + y^2 + 2gx + 2fy + c = 0$ if $(c - f^2)l^2 + 2fglm + (c - g^2)m^2 - 2n(gl + fm) + n^2 = 0$.
4. Prove that the tangent to a circle at any point on it is perpendicular to the radius at that point.
5. Find the equations of the two tangents to the circle $x^2 + y^2 = 4$ which make an angle of 60° with the x-axis.
6. Prove that the line $2x + y = 4$ is a tangent to the circle $x^2 + y^2 + 6x - 10y + 29 = 0$.
7. Find the condition that the line $lx + my + n = 0$ may touch the circle $(x - h)^2 + (y - k)^2 = r^2$.
8. A line moves so that the sum of the perpendiculars drawn to it from the points $(a, 0)$, $(-a, 0)$ is constant; show that it always touches a circle.
9. Prove that the line $x + y = 2$ touches the circles $x^2 + y^2 = 2$ and $x^2 + y^2 + 3x + 3y - 8 = 0$ at the same point.
10. Find the vertices of the triangle formed by the tangents at A, B and C to the circumcircle of the triangle whose vertices are $A(2, 3)$, $B(-2, 1)$ and $C(-3, -2)$.
11. For what value of λ does the line $4x + \lambda y + 7 = 0$ touch the circle $x^2 + y^2 - 6x + 4y - 12 = 0$.
12. Find the equation of the normal to the circle $x^2 + y^2 = 5$ at the point $(-1, 2)$.
13. Find the condition that the line $lx + my + n = 0$ may be normal to the circle $x^2 + y^2 = a^2$.

14. Find the equation of the normal to the circle $x^2 + y^2 - 2ax = 0$ at the point $(a(1 + \cos \alpha), a \sin \alpha)$.

3. Tangents from a Point

3.1 *To prove that from any point there can be drawn two tangents, real or imaginary, to a circle.*

Let the equation of the circle be

$$x^2 + y^2 = a^2. \tag{1}$$

Let $P(x_1, y_1)$ be the given point. Any tangent to the circle (1) is $y = mx + a\sqrt{1+m^2}$, where m is the slope and can take any value. Since the tangent passes through the given point $P(x_1, y_1)$, we have

$$y_1 = mx_1 + a\sqrt{1+m^2}$$

$$\Rightarrow \quad (y_1 - mx_1)^2 = a^2(1 + m^2)$$

$$\Rightarrow \quad (x_1^2 - a^2)m^2 - 2x_1y_1m + y_1^2 - a^2 = 0. \tag{2}$$

This is a quadratic equation in m giving two values of m (real and distinct, coincident or imaginary) corresponding to which there will be two tangents passing through the point $P(x_1, y_1)$. The two tangents will be real and distinct, coincident or imaginary according as the two roots of equation (2) are so. But the roots of equation (2) are real and distinct, coincident or imaginary according as $4x_1^2 y_1^2 - 4(x_1^2 - a^2)(y_1^2 - a^2) > = \text{ or } < 0$, *i.e.*, according as

$$x_1^2 + y_1^2 - a^2 > = \text{ or } < 0.$$

But this is the condition for the point (x_1, y_1) to be outside, on or inside the circle. Hence, from a point there can be drawn two tangents to a circle and these tangents will be real and distinct, coincident or imaginary according as the point lies outside, on or inside the circle.

Note. This result can also be obtained by considering the general equation of the circle.

3.2 Pair of tangents from a given point. *To find the equation of the pair of tangents drawn from an external point $A(x_1, y_1)$ to the circle $x^2 + y^2 = a^2$.*

First Method. Let $P(x', y')$ be any point on the locus (*i.e.* on either of the tangents). Then the equation of the line AP is

$$y - y_1 = \frac{y' - y_1}{x' - x_1}(x - x_1)$$

i.e. $x(y' - y_1) - y(x' - x_1) + x'y_1 - y'x_1 = 0.$ (1)

Since AP is a tangent which touches the circle at Q, the perpendicular distance of the line (1) from C is a, the radius of the circle *i.e.*

$$\frac{x'y_1 - y'x_1}{\sqrt{(x' - x_1)^2 + (y' - y_1)^2}} = a$$

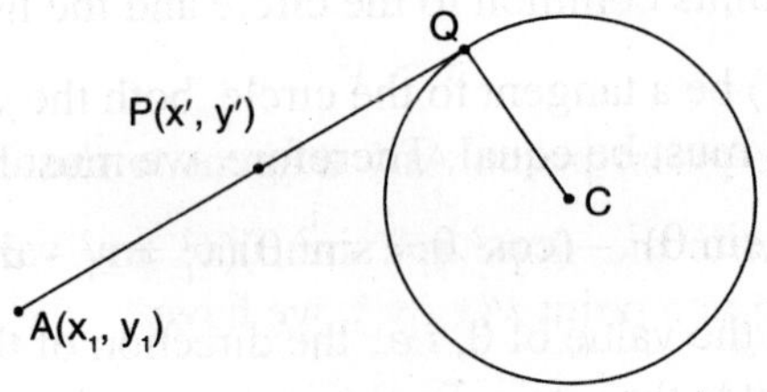

Fig. 5.7

$\Rightarrow$ $$(x'y_1 - y'x_1)^2 = a^2\{(x' - x_1)^2 + (y' - y_1)^2\}.$$

Hence, the locus of (x', y') is

$$(xy_1 - yx_1)^2 = a^2\{(x - x_1)^2 + (y - y_1)^2\}.$$

After simplification and rearrangement of the terms this can be written as

$$(x^2 + y^2 - a^2)(x_1^2 + y_1^2 - a^2) = (xx_1 + yy_1 - a^2)^2.$$

This is the required equation of the pair of tangents.

Hence, using the notations:

$$S \equiv x^2 + y^2 - a^2$$

$$S_1 \equiv x_1^2 + y_1^2 - a^2$$

$$T \equiv xx_1 + yy_1 - a^2,$$

we find that the equation of the pair of tangents drawn from the point (x_1, y_1) to the circle $x^2 + y^2 = a^2$ is $T^2 = SS_1$.

Second Method. The equations of a line through $A(x_1, y_1)$ making an angle θ with x-axis are

$$\frac{x - x_1}{\cos\theta} = \frac{y - y_1}{\sin\theta} = r, \quad (1)$$

r being the algebraic distance of the point $P(x, y)$ from $A(x_1, y_1)$ measured along the line. Therefore, the coordinates of a point on the line which is at a distance r from $A(x_1, y_1)$ are given by

$$x = x_1 + r\cos\theta$$

$$y = y_1 + r\sin\theta.$$

If this point be on the circle, we must have

$$(x_1 + r\cos\theta)^2 + (y_1 + r\sin\theta)^2 = a^2$$

i.e. $(\cos^2\theta + \sin^2\theta)r^2 + 2(x_1\cos\theta + y_1\sin\theta)\,r + x_1^2 + y_1^2 - a^2 = 0.$

But this is a quadratic equation in r giving two values of r corresponding to which there are two points common to the circle and the line (1).

Now, if the line (1) be a tangent to the circle, both the values of r given by the quadratic equation must be equal. Therefore, we must have

$$(x_1\cos\theta + y_1\sin\theta)^2 - (\cos^2\theta + \sin^2\theta)(x_1^2 + y_1^2 - a^2) = 0. \qquad (1)$$

This equation gives us the value of θ, *i.e.*, the direction of the line (1) so that it may become the tangent to the circle. To obtain the actual equation of the tangent (s), eliminate θ between (1) and (2). Thus, the equation of the tangent is

$$\left\{x_1\left(\frac{x-x_1}{r}\right) + y_1\left(\frac{y-y_1}{r}\right)\right\}^2$$

$$= \left\{\left(\frac{x-x_1}{r}\right)^2 + \left(\frac{y-y_1}{r}\right)^2\right\}\{x_1^2 + y_1^2 - a^2\}$$

$$\Rightarrow \quad (xx_1 + yy_1 - x_1^2 - y_1^2)^2$$

$$= \{x^2 + y^2 + x_1^2 + y_1^2 - 2(xx_1 + yy_1)\}\{x_1^2 + y_1^2 - a^2\}$$

$$\Rightarrow \quad (xx_1 + yy_1 - a^2)^2 = (x^2 + y^2 - a^2)(x_1^2 + y_1^2 - a^2).$$

Using the symbols S, S_1 and T as earlier, we find that the equation of the pair of tangents is $T^2 = SS_1$.

Note. The equation of the pair of tangents drawn from the point (x_1, y_1) to the circle given by the general equation is $x^2 + y^2 + 2gx + 2fy + c = 0$, then $T^2 = SS_1$, where

$$S \equiv x^2 + y^2 + 2gx + 2fy + c$$

$$S_1 \equiv x_1^2 + y_1^2 + 2gx_1 + 2fy_1 + c$$

$$T \equiv xx_1 + yy_1 + g(x + x_1) + f(y + y_1) + c.$$

3.3 Length of the tangent. *To find the length of the tangent drawn from an external point $A(x_1, y_1)$ to the circle $x^2 + y^2 + 2gx + 2fy + c = 0$.*

Let the tangent drawn from the point $A(x_1, y_1)$ to the given circle touch it at the point T.

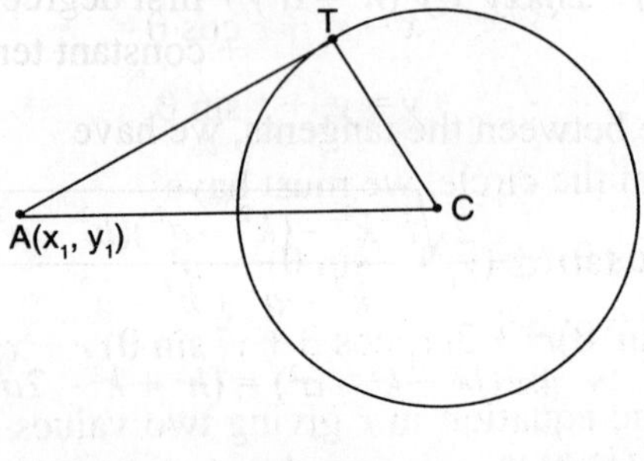

Fig. 5.8

Then CT = radius of the circle

$$= \sqrt{g^2 + f^2 - c}$$

and $AC = \sqrt{(x_1 + g)^2 + (y_1 + f)^2}$.

Now, in right angled triangle ACT, we have

$$AT^2 = AC^2 - CT^2$$
$$= (x_1 + g)^2 + (y_1 + f)^2 - (g^2 + f^2 - c)$$
$$= x_1^2 + y_1^2 + 2gx_1 + 2fy_1 + c.$$

Hence, *the length of the tangent from the point* (x_1, y_1) *to the circle* $x^2 + y^2 + 2gx + 2fy + c = 0$ *is* $\sqrt{x_1^2 + y_1^2 + 2gx_1 + 2fy_1 + c}$.

Working rule. After the equation of the circle being written in the form so that each of the coefficients of x^2 and y^2 is unity and the right side is zero, write x_1 for x and y_1 for y in the left side of the equation of the circle and then take the square root. This gives the length of the tangent from the point (x_1, y_1) to the circle.

Ex. Find the length of the tangent from the point (–3, 4) to the circle

$$x^2 + y^2 - 4x - 6y - 3 = 0.$$

3.4 Example. The angle between two tangents to the circle $x^2 + y^2 = a^2$ is constant and equal to α. Prove that the locus of their points of intersection is given by $4a^2(x^2 + y^2 - a^2) = (x^2 + y^2 - 2a^2)^2 \tan^2 \alpha$.

What happens to this locus if $\alpha = \pi/2$?

Solution. Let (h, k) be the point of intersection of the two tangents to the given circle. Then, the equation of the pair of tangents is

$$(x^2 + y^2 - a^2)(h^2 + k^2 - a^2) = (xh + yk - a^2)^2$$

$\Rightarrow \quad x^2(k^2 - a^2) - 2hkxy + y^2(h^2 - a^2)$ + first degree terms + constant terms = 0.

Since α is the angle between the tangents, we have

$$\tan \alpha = \frac{2\sqrt{h^2k^2 - (k^2 - a^2)(h^2 - a^2)}}{k^2 - a^2 + h^2 - a^2}$$

$$\Rightarrow \quad 4a^2(h^2 + k^2 - a^2) = (h^2 + k^2 - 2a^2) \tan^2 \alpha.$$

Hence, the locus of (h, k) is

$$4a^2(x^2 + y^2 - a^2) = (x^2 + y^2 - 2a^2) \tan^2 \alpha.$$

When $\alpha = \pi/2$, the locus becomes $x^2 + y^2 = 2a^2$.

Exercises

1. Prove that the pair of tangents from the point (1, 2) to the circle $x^2 + y^2 - 4x + 2y = 0$ are mutually perpendicular.
2. Find the equation of the pair of tangents from the origin to the circle $x^2 + y^2 + 2gx + 2fy + c = 0$.
3. From any point on the circle $x^2 + y^2 + 2gx + 2fy + c = 0$ tangents are drawn to the circle
$$x^2 + y^2 + 2gx + 2fy + c \sin^2 \alpha + (g^2 + f^2) \cos^2 \alpha = 0.$$
Prove that the angle between them is 2α.
4. Find the length of the tangent drawn from the point $(a + b, a - b)$ to the circle $x^2 + y^2 + 2bx - 3b^2 = 0$.
5. Prove that the locus of a point, which moves so that the tangents from it to the two circles
$$x^2 + y^2 - 5x - 3 = 0,\ 3x^2 + 3y^2 + 2x + 4y - 6 = 0$$
are equal, is a line.

4. Chord of Contact

The line through the points of contact of the tangents drawn from a point (x_1, y_1) to a circle (or a conic in general) is called the **chord of contact** *of tangents from the point (x_1, y_1) to the circle (or conic).*

4.1 Equation of the chord of contact. *To find the equation of the chord of contact of tangents drawn from the point (x_1, y_1) to the circle $x^2 + y^2 = a^2$.*

Let $A(x', y')$ and $B(x'', y'')$ be the points of contacts of the tangents drawn from the point $P(x_1, y_1)$ to the circle

$$x^2 + y^2 = a^2. \tag{1}$$

Then, the equations of the tangents at the points A and B to the circle (1) are

$$x'x + y'y = a^2$$

and
$$x''x + y''y = a^2.$$

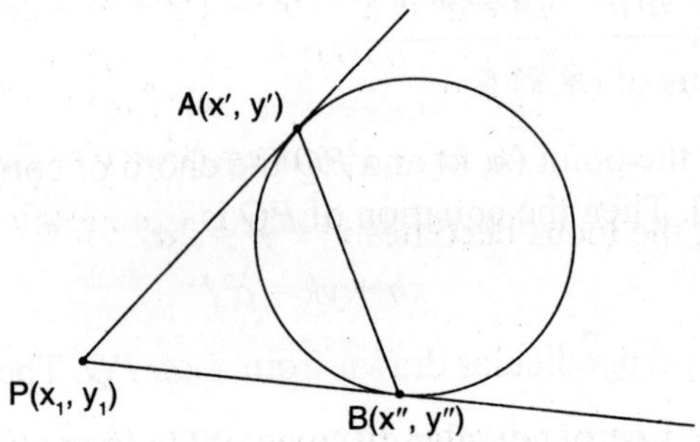

Fig. 5.9

But each of these tangents passes through the point $P(x_1, y_1)$. Therefore

$$x'x_1 + y'y_1 = a^2$$

and
$$x''x_1 + y''y_1 = a^2.$$

It is clear from these relations that (x', y') and (x'', y'') both satisfy the equation $xx_1 + yy_1 = a^2$. This is the required equation of the chord of contact of the tangents drawn from the point $P(x_1, y_1)$.

Remarks. (*i*) We note that the equation of the chord of contact of tangents drawn from a point (x_1, y_1) to a circle is of the same form as that the equation of the tangent at the point (x_1, y_1) to the circle. The reason is simple since in case the point $P(x_1, y_1)$ lies on the circle, the points A and B both coincide with P, and in this limiting position the chord of contact becomes the tangent to the circle at the point $P(x_1, y_1)$.

(*ii*) If the point $P(x_1, y_1)$ lies outside the circle, the situation is quite simple and clear geometrically also.

(*iii*) If the point $P(x_1, y_1)$ is inside the circle, the tangents drawn from the point $P(x_1, y_1)$ to the circle will be imaginary. But the chord of contact still exists and is real since x_1 and y_1 are real. Thus, there is a real line joining the imaginary points of contact of the two imaginary tangents drawn from the point inside the circle. This is consistent with the result that a line always meets a circle in two points, real or imaginary.

Thus the chord of contact of the tangents drawn from a point (x_1, y_1) to a circle always exists and is real whatever the position of the point P may be, outside, on or inside the circle.

4.2 Example. Tangents are drawn from the point (h, k) to the circle $x^2 + y^2 = a^2$. Prove that the area of the triangle formed by them and the line joining their points of contact is $\dfrac{a(h^2+k^2-a^2)^{3/2}}{h^2+k^2}$.

Solution. Let A be the point (h, k) and PQ the chord of contact of the tangents drawn from $A(h, k)$. Then the equation of PQ is

$$xh + yk = a^2. \tag{1}$$

Let AM be the perpendicular drawn from A on PQ. Then

$$AM = \text{perpendicular distance of (1) from } A(h, k)$$

$$= \frac{h^2+k^2-a^2}{\sqrt{h^2+k^2}}.$$

Also $\qquad AP = \text{length of the tangent} = \sqrt{h^2+k^2-a^2}.$

Now, in right angled triangle APM, we have

$$PM^2 = AP^2 - AM^2$$

$$= (h^2+k^2-a^2) - \frac{(h^2+k^2-a^2)^2}{h^2+k^2} = \frac{a^2(h^2+k^2-a^2)}{h^2+k^2}$$

Hence, the area of the triangle APQ is

$$2(\tfrac{1}{2}PM.AM)$$

$$= \frac{a\sqrt{h^2+k^2-a^2}}{\sqrt{h^2+k^2}} \cdot \frac{(h^2+k^2-a^2)}{\sqrt{h^2+k^2}}$$

$$= \frac{a(h^2+k^2-a^2)^{3/2}}{h^2+k^2}.$$

Exercises

1. Find the equation of the chord of contact of the point $(4, 3)$ with respect to the circle $x^2 + y^2 - 3x + 12y + 6 = 0$.
2. Find the equation of the chord of contact of the point $(a, -b)$ with respect to the circle $x^2 + y^2 + 2ax - 2by + a^2 - b^2 = 0$.

3. Prove that the chord of contact of the point (1, –2) with respect to the circles $x^2 + y^2 + 6y + 5 = 0$ and $x^2 + y^2 + 2x + 8y + 5 = 0$ coincide.
4. Find the condition that the chord of contact of tangents from the point (x_1, y_1) to the circle $x^2 + y^2 = a^2$ should subtend a right angle at the centre.
5. Find the locus of a point the chord of contact of the tangents from which subtends a right angle at the centre of the circle $x^2 + y^2 = a^2$.

5. Pole and Polar

The **Polar** *of a point with respect to a circle (or a conic in general) is the locus of the points of intersection of the tangents drawn at the end points of chords through that point and the point is said to be the* **pole** *of the polar.*

Let P be a point outside or inside the circle. Take three chords PA_1B_1, PA_2B_2 and PA_3B_3. Let the tangents at the ends of these chords meet at points T_1, T_2 and T_3, respectively. Then, the path of the points T's is, in fact, the polar of the point P. One may refer to Fig. 5.10(a) when the point P is outside the circle and to Fig. 5.10(b) when the point P is inside the circle.

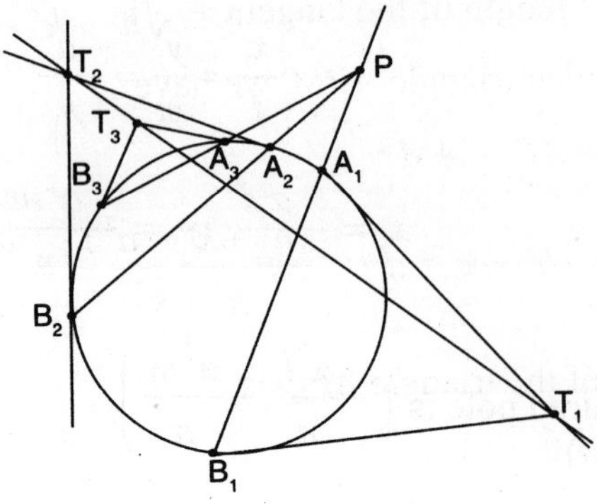

Fig. 5.10(*a*)

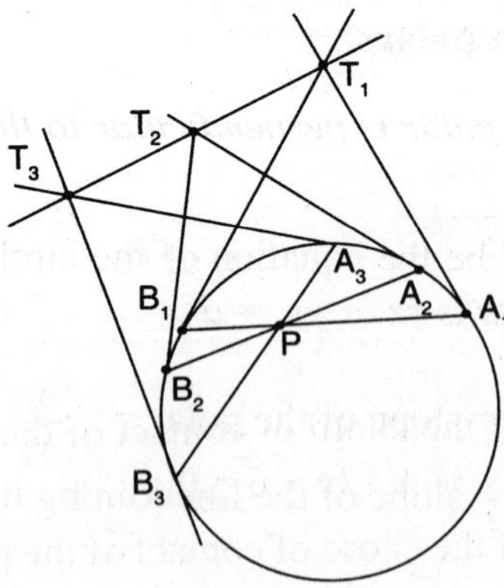

Fig. 5.10(*b*)

5.1 Equation of the polar. *To find the equation of the polar of the point* (x_1, y_1) *with respect to the circle* $x^2 + y^2 = a^2$.

Let $P(x', y')$ be any point on the locus, *i.e.*, the point of intersection of the tangents drawn at the extremities of a chord through the point (x_1, y_1). The equation of the chord of contact of tangents from (x', y') is

$$xx' + yy' = a^2.$$

But it passes through the point (x_1, y_1). Therefore

$$x_1x' + y_1y' = a^2.$$

Hence, the locus of (x', y') is $xx_1 + yy_1 = a^2$.

Note. The terms pole and polar are correlating and these terms here are nothing to do with those occurring in polar coordinates.

5.2 Pole of a line. *To find the pole of a line* $lx + my + n = 0$ *with respect to the circle* $x^2 + y^2 + a^2$.

Let (x_1, y_1) be the pole of the line $lx + my + n = 0$ with respect to the circle $x^2 + y^2 = a^2$. Then this line must be identical with the polar of (x_1, y_1) with respect to the circle $x^2 + y^2 = a^2$ *i.e.* with $xx_1 + yy_1 = a^2$. Therefore

$$\frac{x_1}{l} = \frac{y_1}{m} = \frac{-a^2}{n}$$

$$\Rightarrow \qquad x_1 = -\frac{a^2 l}{n},\ y_1 = -\frac{a^2 m}{n}.$$

Hence, the required pole is $\left(-\frac{a^2 l}{n}, -\frac{a^2 m}{n}\right)$.

Ex. Find the pole of the line $x + 2y - 1 = 0$ with respect to the circle $x^2 + y^2 = 3$.

5.3 Propositions on polars.

Proposition I. *The polar is perpendicular to the line joining the centre to the pole.*

Let $x^2 + y^2 = a^2$ be the equation of the circle and $P(x_1, y_1)$ the pole. Then, equation of the polar is $xx_1 + yy_1 = a^2$.

Now $\qquad m_1$ = slope of the polar $= -\frac{x_1}{y_1}$

m_2 = slope of the line joining the centre to the pole (x_1, y_1)

$$= \frac{y_1}{x_1}$$

We note that $m_1 m_2 = -1$. Hence the proposition follows.

Proposition II. *The pole and the point of intersection of the polar with the join of the centre and the pole are inverse points* with respect to the circle* $x^2 + y^2 = a^2$.

Let C be the centre of the circle and Q the point of intersection of the polar with CP, where $P(x_1, y_1)$ is the pole. Then, by Proposition *I*, we note that

$$CQ = \text{perpendicular distance of } C \text{ from the polar of } P$$

$$= \frac{a^2}{\sqrt{x_1^2 + y_1^2}}$$

and $$CP = \sqrt{x_1^2 + y_1^2}.$$

$\therefore \quad CP \,.\, CQ = a^2 = (\text{radius})^2.$

Hence, the points P and Q are the inverse points with respect to the circle $x^2 + y^2 = a^2$.

Proposition III. *If the polar of a point P passes through the point Q, then polar of Q passes through P.*

Let (x_1, y_1) and (x_2, y_2), respectively, be the coordinates of P and Q. Since the polar of the point P passes through the point Q, we have

$$x_1 x_2 + y_1 y_2 = a^2.$$

But this is the condition also that the point (x_1, y_1) should lie on the line

$$xx_2 + yy_2 = a^2,$$

which, in fact, is the polar of the point Q. This proves the proposition.

Two points such that each lies on the polar of the other are called **conjugate points.**

Proposition IV. *If the pole of a line L_1, lies on another line L_2, then the pole of L_2, lies on L_1.*

Let $P(x_1, y_1)$ and $Q(x_2, y_2)$ be the poles of the lines L_1 and L_2, respectively. In other words, L_1 and L_2 are the polars of P and Q, respectively. Since the polar of Q passes through P, by Proposition III, the polar of P (*i.e.* the line L_1) must pass through Q. Hence the proposition follows.

* Two points P and Q are said to be inverse points with respect to a circle if $CP.CQ = (\text{radius})^2$, C being the centre of the circle.

Two lines such that each contains the pole of the other are called **conjugate lines.**

Proposition V. *If the polars of any two points P and Q meet in R, then the polar of R is the line PQ.*

Let (x_1, y_1) and (x_2, y_2) be the coordinates of the points P and Q, respectively. Then, the polars of P and Q with respect to the circle $x^2 + y^2 = a^2$ are

$$xx_1 + yy_1 = a^2 \text{ and } xx_2 + yy_2 = a^2$$

By solving the simultaneous equations, the coordinate of R are given by

$$\left(\frac{a^2(y_2 - y_1)}{x_1y_2 - x_2y_1}, \frac{-a^2(x_2 - x_1)}{x_1y_2 - x_1y_1}\right).$$

Now, the polar of R is

$$\frac{xa^2(y_2 - y_1)}{x_1y_2 - x_2y_1} - \frac{ya^2(x_2 - x_1)}{x_1y_2 - x_2y_1} = a^2$$

$$\Rightarrow \qquad y - y_1 = \frac{y_2 - y_1}{x_2 - x_1}(x - x_1),$$

which, in fact, is the equation of the line PQ.

5.4 Examples

Example 1. Prove that the locus of a point whose polar with respect to the circle $x^2 + y^2 = a^2$ touches the circle $(x - c)^2 + (y - d)^2 = b^2$ is $b^2(x^2 + y^2) = (a^2 - cx - dy)^2$.

Solution. Let (h, k) be a point on the locus. Then its polar with respect to the circle $x^2 + y^2 = a^2$ is $xh + yk = a^2$. This line will touch the circle $(x - c)^2 + (y - d)^2 = b^2$ if the perpendicular distance of it from the centre of the circle is equal to the radius, *i.e.,* if

$$\frac{a^2 - ch - dk}{\sqrt{h^2 + k^2}} = b$$

$$\Rightarrow \qquad b^2(h^2 + k^2) = (a^2 - ch - dk)^2.$$

Hence, the locus of (h, k) is $b^2(x^2 + y^2) = (a^2 - cx - dy)^2$.

Example 2. Prove that the lines $l_1x + m_1y + n_1 = 0$ and $l_2x + m_2y + n_2 = 0$ are conjugate with respect to the circle $x^2 + y^2 = a^2$ if $(l_1l_2 + m_1m_2)a^2 = n_1n_2$.

Solution. Let (x_1, y_1) be the pole of the line $l_1x + m_1y + n_1 = 0$. Then its polar with respect to the given circle is $xx_1 + yy_1 = a^2$. But $l_1x + m_1y + n_1 = 0$ and

$xx_1 + yy_1 - a^2 = 0$ are identical. Therefore, we have

$$\frac{x_1}{l_1} = \frac{y_1}{m_1} = \frac{-a^2}{n_1},$$

which gives $x_1 = -\frac{a^2 l_1}{n_1}$ and $y_1 = -\frac{a^2 m_1}{n_1}$. The given lines are conjugate if the point (x_1, y_1), pole of $l_1x + m_1y + n_1 = 0$, lies on $l_2x + m_2y + n_2 = 0$, *i.e.*, if

$$l_2\left(-\frac{a^2 l_1}{n_1}\right) + m_2\left(-\frac{a^2 m_1}{n_1}\right) + n_2 = 0, \text{ i.e., if } (l_1 l_2 + m_1 m_2)a^2 = n_1 n_2.$$

Exercises

1. Find the polar of the point (3, –1) with respect to the circle $x^2 + y^2 = 4$.
2. Find the polar of the point (1, –1) with respect to the circle $x^2 + y^2 + 4x + 8y - 5 = 0$.
3. Find the pole of the line $2x + 3y - 6 = 0$ with respect to the circle $x^2 + y^2 = 5$.
4. Find the pole of the line $3x - 2y - 5 = 0$ with respect to the circle $x^2 + y^2 + 2x - 4y - 4 = 0$.
5. Find the locus of a point whose polar with respect to the circle $x^2 + y^2 = a^2$ touches the circle $(x - a)^2 + y^2 = a^2$.
6. Prove that the polar of a given point with respect to anyone of the circles $x^2 + y^2 - 2kx + c^2 = 0$, where k is variable, always passes through a fixed point, whatever the value of k be.
7. Prove that the polar of the point (p, q) with respect to the circle $x^2 + y^2 = a^2$ touches $(x - c)^2 + (y - d)^2 = b^2$, if $b^2(p^2 + q^2) = (a^2 - cp - dq)^2$.
8. Find the locus of a point P which is such that its polar with respect to one circle touches a second circle.

6. Chord with Given Middle Point

6.1 *To find the equation of the chord of the circle $x^2 + y^2 = a^2$ in terms of its middle point (x_1, y_1).*

First Method. The equations of a line through the point (x_1, y_1) are

$$\frac{x - x_1}{\cos\theta} = \frac{y - y_1}{\sin\theta} = r, \tag{1}$$

θ being its inclination with the x-axis and r the algebraic distance of (x, y) from (x_1, y_1) measured along the line. Then, any point on the line has the coordinates $(x_1 + r\cos\theta, y_1 + r\sin\theta)$ and this point will be common to the circle if

$$(x_1 + r\cos\theta)^2 + (y_1 + r\sin\theta)^2 = a^2$$

i.e. $$(\cos^2\theta + \sin^2\theta)r^2 + 2(x_1\cos\theta + y_1\sin\theta)r + x_1^2 + y_1^2 - a^2 = 0. \quad (2)$$

This equation being quadratic in r, gives two values of r. Since (x_1, y_1) is the middle point of the chord, the two values of r given by equation (2) must be equal in magnitude and opposite in sign, *i.e.*, the sum of the roots of equation (2) must be zero. Therefore

$$x_1\cos\theta + y_1\sin\theta = 0. \quad (3)$$

This equation gives us the suitable value of θ so that the line (1) may be the chord of the circle having (x_1, y_1) as the middle point. Therefore, the equation of the chord is

$$x_1\left(\frac{x - x_1}{r}\right) + y_1\left(\frac{y - y_1}{r}\right) = 0$$

i.e. $$xx_1 + yy_1 = x_1^2 + y_1^2.$$

Hence, *the equation of the chord of the circle $x^2 + y^2 = a^2$ in terms of its middle point is $xx_1 + yy_1 = x_1^2 + y_1^2$.*

Using the notation of § 3.2, the equation of the chord can be written as

$$T = S_1.$$

Second Method. The equation of any line through the point $M(x_1, y_1)$ is $y - y_1 = m(x - x_1)$, where m is the slope of the line. The point (x_1, y_1) being the middle point of the chord PQ, CM is perpendicular to PQ, where C is the centre of the circle. Therefore

$$m = -\frac{1}{\text{Slope of } CM} = -\frac{x_1}{y_1}.$$

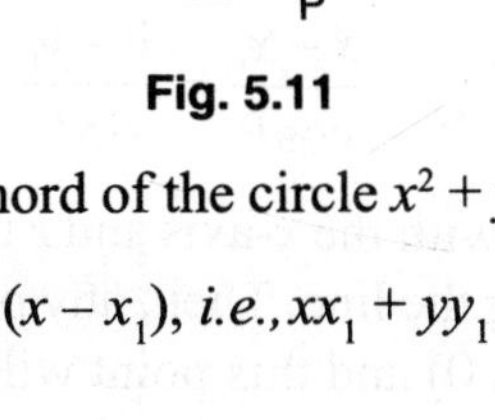

Fig. 5.11

Hence, the equation of the chord of the circle $x^2 + y^2 = a^2$ in terms of its middle point (x_1, y_1) is $y - y_1 = -\dfrac{x_1}{y_1}(x - x_1)$, *i.e.*, $xx_1 + yy_1 = x_1^2 + y_1^2$.

Note. The second method is applicable only for the circle whereas the first method will be applicable alike for other conics like parabola, ellipse and hyperbola also.

Ex. Find the equation of the chord of the circle $x^2 + y^2 + 2gx + 2fy + c = 0$ in terms of its middle point.

6.2 Example. Find the locus of the middle points of chords of the circle $x^2 + y^2 = a^2$ which pass through the fixed point (h, k).

Solution. Let (x', y') be the middle point of the chord of the given circle. Then its equation is $xx' + yy' = x'^2 + y'^2$. But this passes through (h, k). Therefore

$$hx' + ky' = x'^2 + y'^2.$$

Hence, the locus of (x', y') is $x^2 + y^2 - hx - ky = 0$.

Exercises

1. Find the equation of the chord of the circle $x^2 + y^2 - 4x + 3y - 1 = 0$ whose middle point is $(-2, 1)$.
2. Find the middle point of the chord of the circle $x^2 + y^2 = a^2$, lying along the line $lx + my + n = 0$.
3. Find the locus of the middle points of chords of the circle $x^2 + y^2 = a^2$, which subtend a right angle at the point $(c, 0)$.
4. Tangents are drawn to a circle from a point which always lies on a given line. Prove that the locus of the middle point of the chord of contact is another circle.
5. Find the locus of the middle points of the chords of a circle which arc of constant length.

7. Polar Equation of a Circle

To find the polar equation of a circle in terms of the given radius and the coordinates of the centre.

Let $C(r_1, \theta_1)$ be the centre and a the radius of the circle. Let $P(r, \theta)$ be any point on the circle.

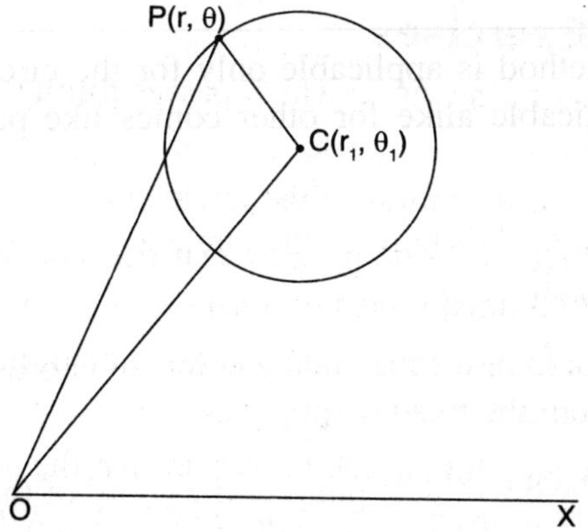

Fig. 5.12

Then, from triangle *OCP*, we have

$$CP^2 = OC^2 + OP^2 - 2OC.OP \cos COP$$

But $CP = a$, $OC = r_1$, $OP = r$ and $\angle COP = \theta - \theta_1$.

Therefore $a^2 = r_1^2 + r^2 - 2rr_1 \cos(\theta - \theta_1)$,

which is the polar equation of the circle.

Corollary 1. The equation of the circle passing through the pole is $r = 2a \cos(\theta - \theta_1)$.

Corollary 2. The equation of the circle passing through the pole and having the centre on the initial line is $r = 2a \cos \theta$.

Exercises

1. Find the equation to the circle described on the line joining the points (r_1, θ_1) and (r_2, θ_2) as diameter.
2. Find the coordinates of the centre of the circle $r = A \cos \theta + B \sin \theta$.
3. Find the polar equation of a circle, the initial line being a tangent.
4. Find the equation to the chord joining the points on the circle $r = 2a \cos \theta$ whose vectorial angles are θ_1 and θ_2, and deduce the equation to the tangent at the point θ_1.
5. Find the condition that the line $\frac{1}{r} = A \cos \theta + B \sin \theta$ may touch the circle $r = 2a \cos \theta$.
6. Prove that for all values of α, the line $r \cos(\theta - \theta_1) = a + r_1 \cos \theta_1$ is a tangent to the circle $a^2 = r_1^2 + r^2 - 2rr_1 \cos \theta$.

Miscellaneous Exercises

1. Find the equation of the circle which passes through three points (0, 1), (1, 0) and (2, 1).
2. Find the centre and the radius of the circle $2(x^2 + y^2) + 6x - 7y = 0$.
3. Show that the locus of a point such that the sum of the squares of its distances from two fixed points is constant, is a circle.
4. Show that the locus of a point such that the ratio of its distances from two given points is constant, is a circle.
5. Find the locus of a point which moves so that the square of the tangent drawn from it to the circle $x^2 + y^2 = a^2$ is equal to c times its distance from the line $lx + my + n = 0$.
6. Find the locus of a point whose distance from a fixed point is in a constant ratio to the tangent drawn from it to a given circle.
7. Prove that the locus of a point which moves so that the chords of contact of the tangents from the point to two fixed circles are perpendicular, is a circle.
8. Prove that the locus of a point which moves so that the sum of the squares of its distances from the three vertices of a triangle is constant, is a circle whose centre is at the centroid of the triangle.
9. Find the equation of the circle inscribed in the triangle the equations of whose sides are $x = 1$, $2y = 5$ and $3x - 4y = 5$.
10. Prove that the circle whose centre is (3, 5) and which touches the y-axis is $x^2 + y^2 - 6x - 10y + 25 = 0$. Find the equation of the other tangent from the origin, and the coordinates of the point of contact.
11. Find the coordinates of the poles of the lines $3x - 11y - 13 = 0$, $8x + y - 2 = 0$ and $3x + 2y + 1 = 0$ with respect to the circle $x^2 + y^2 - 4x + 3 = 0$, and show that they are collinear.
12. Find the length of the common chord of the circles whose equations are $(x - a)^2 + y^2 = a^2$ and $x^2 + (y - b)^2 = b^2$, and prove that the equation to the circle whose diameter is this common chord is $(a^2 + b^2)(x^2 + y^2) = 2ab(bx + ay)$.
13. Find the locus of a point which moves so that its polars with respect to two fixed circles are mutually orthogonal.
14. Find the locus of a point of intersection of the tangent to a given circle and the perpendicular let fall on this tangent from a fixed point on the circle.
15. Prove that if the length of the tangent from (h, k) to the circle $x^2 + y^2 = 6$ be twice the length of the tangent from (h, k) to the circle $x^2 + y^2 + 3x + 3y = 0$, then $h^2 + k^2 + 4h + 4k + 2 = 0$.
16. A point moves so that the square of its distance from a fixed point varies as its perpendicualr distance from a fixed line. Prove that it describes a circle.
17. A point moves so that the sum of the square of its distances from the four sides of a square is constant. Prove that the locus of the point is a circle.

18. Prove that the locus of the mid-points of the chords of the circle $x^2 + y^2 + 2gx + c = 0$ which pass through the origin is the circle $x^2 + y^2 + gx = 0$.

19. The line $y = mx + c$ cuts off a chord of length $2d$ from the circle $x^2 + y^2 = a^2$. Prove that $c^2 = (a^2 - d^2)(1 + m^2)$.

20. Prove that the circle on the chord $x \cos \alpha + y \sin \alpha = p$ of the circle $x^2 + y^2 = a^2$ as diameter is $x^2 + y^2 - a^2 - 2p\,(x \cos \alpha + y \sin \alpha - p) = 0$.

21. Find the locus of a point the polars of which with respect to two given circles make a given angle with one another.

22. Find the locus of the foot of the perpendicular let fall from the origin upon any chord of the circle $x^2 + y^2 + 2gx + 2fy + c = 0$, which subtends a right angle at the origin.

Find also the locus of the middle points of these chords.

23. A tangent is drawn to the circle $(x - a)^2 + y^2 = b^2$ and a perpendicular tangent to the circle $(x + a)^2 + y^2 = c^2$. Find the locus of their point of intersection, and prove that the bisector of the angle between them always touches one or other of two fixed circles.

24. Prove that the distances of two points from the centre of a circle are proportional to the perpendicular drawn from one point on the polar of the other.

25. Find the locus of the poles of the line $\frac{x}{a} + \frac{y}{b} = 1$ with respect to the circles which touch the coordinate axes.

26. Prove that the locus of the middle points of a system of parallel chords of a circle is a line passing through the centre.

27. O is a fixed point and P any point on a given circle; OP is joined and on it a point Q is taken so that $OP \cdot OQ = k^2$, a constant quantity. Prove that the locus of Q is a circle which becomes a line when O lies on the original circle.

28. The distances from the origin of the centres of three circles $x^2 + y^2 - 2\lambda x = c^2$ (where c is a constant and λ a variable) are in geometrical progression. Prove that the lengths of the tangents drawn to them from any point on the circle $x^2 + y^2 = c^2$ are also in geometrical progression.

29. From any point on one given circle tangents are drawn to another given circle. Prove that the locus of the middle point of the chord of contact is a third circle.

30. Prove that the equation of a line meeting the circle $x^2 + y^2 = a^2$ in two point at equal distances d from a point (x_1, y_1) on the circumference is $xx_1 + yy_1 - a^2 + \frac{1}{2}d^2 = 0$. Apply this to find the equation of the tangent at (x_1, y_1).

Systems of Circles

CHAPTER 6

1. Intersection of Two Circles

Let the equations of the two circles be

$$S_1 \equiv x^2 + y^2 + 2g_1x + 2f_1y + c_1 = 0 \quad (1)$$

$$S_2 \equiv x^2 + y^2 + 2g_2x + 2f_2y + c_2 = 0. \quad (2)$$

Now, consider the equation $S_1 + \lambda S_2 = 0$, where λ is an arbitrary constant. This equation takes the form of

$$(1 + \lambda)x^2 + (1 + \lambda)y^2 + 2(g_1 + \lambda g_2)x + 2(f_1 + \lambda f_2)y + c_1 + \lambda c_2 = 0. \quad (3)$$

The coordinates (x, y) which satisfy (1) and (2), also satisfy (3). Therefore, equation (3) represents, in general, a curve (circle or line) through the points of intersection of (1) and (2).

Case I. $\lambda = -1$. Equation (3) becomes a linear equation in x and y, and as such represents a line through the common points of the circles (1) and (2).

Note. There is only one line through the points of intersection of the two circles. As such there are two points of intersection, real or imaginary, of two circles. When the two points are real and distinct, the line $S_1 - S_2 = 0$ is the common chord of the two circles whereas in case the two points are coincident, the line $S_1 - S_2 = 0$ is the common tangent of the circles $S_1 = 0$ and $S_2 = 0$. Finally, when these circles do not meet in real points, still the equation $S_1 - S_2 = 0$ represents a real line which of course passes through the imaginary common points of the circles $S_1 = 0$ and $S_2 = 0$. In this case this line may represent some other locus.

Case II. $\lambda \neq -1$. Equation (3) represents circles through the common points of the circles $S_1 = 0$ and $S_2 = 0$. In this case it gives a family of such circles since for each value of $\lambda (\neq -1)$ there is a member of the family.

1.1 Angle of intersection of two circles.

Let there be any two circles intersecting at points P and Q. It is well known from the elementary geometry that the angles between the tangents to the two circles at point P is the same as at Q.

The **angle of intersection** *of any two circles is the common value of the angle between the two tangents at their point of intersection.*

Two circles are said to cut **orthogonally** *if their angle of intersection is a right angle.*

1.2 Condition for orthogonal intersection of two circles.

To find a necessary and sufficient condition for the circles

$$x^2 + y^2 + 2g_1x + 2f_1y + c_1 = 0 \qquad (1)$$

and

$$x^2 + y^2 + 2g_2x + 2f_2y + c_2 = 0 \qquad (2)$$

to cut orthogonally.

Let A and B be the centres of the two circles, and P the point of intersection. Suppose the two circles cut orthogonally. Then the tangents at P to the two circles are at right angles. But the tangent to a circle is always at right angles to its radius.

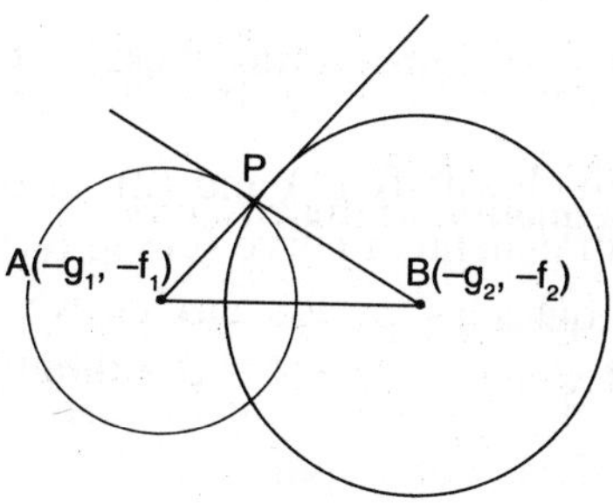

Fig. 6.1

Therefore, $\angle BPA$ is a right angle and so

$$AB^2 = AP^2 + PB^2.$$

But $$AB^2 = (g_1 - g_2)^2 + (f_1 - f_2)^2$$

$$AP^2 = g_1^2 + f_1^2 - c_1$$

and $$PB^2 = g_2^2 + f_2^2 - c_2.$$

$\therefore$ $$(g_1 - g_2)^2 + (f_1 - f_2)^2 = g_1^2 + f_1^2 - c_1 + g_2^2 + f_2^2 - c_2$$

$\Rightarrow$ $$2g_1g_2 + 2f_1f_2 = c_1 + c_2.$$

This is the required necessary condition for the circles (1) and (2) to cut orthogonally. This condition can also be proved to be sufficient by working the same steps backwards.

Ex. Prove that the circles $2x^2 + 2y^2 - 7x + 5 = 0$ and $x^2 + y^2 - 6x + 4y + 8 = 0$ cut orthogonally.

1.3 Examples

Example 1. Find the equation of the circle which cuts orthogonally each of the circles $x^2 + y^2 + 2x + 17y + 4 = 0$, $x^2 + y^2 + 7x + 6y + 11 = 0$, and $x^2 + y^2 - x + 22y + 3 = 0$.

Solution. Let the equation of the circle be

$$x^2 + y^2 + 2gx + 2fy + c = 0. \qquad (1)$$

It will cut orthogonally the given circle if

$$2g + 17f = c + 4, \qquad (2)$$

$$7g + 6f = c + 11, \qquad (3)$$

$$-g + 22f = c + 3. \qquad (4)$$

Subtracting (2) from (3) and (4) from (2), we get

$$5g - 11f = 7 \text{ and } 3g - 5f = 1,$$

which give $g = -3$ and $f = -2$. Substituting these values of g and f in (2), we get $c = -44$.

Hence, the required equation of the circle is $x^2 + y^2 - 6x - 4y - 44 = 0$.

Example 2. Find the equations of the circles which intersect the circles $x^2 + y^2 - 6y + 1 = 0$ and $x^2 + y^2 - 4y + 1 = 0$ orthogonally and touch the line $3x + 4y + 5 = 0$.

Solution. Let the equation of the circle be

$$x^2 + y^2 + 2gx + 2fy + c = 0. \qquad (1)$$

It will cut the given circles orthogonally if $-6f = c + 1$ and $-4f = c + 1$, which give $f = 0$ and $c = -1$. Therefore, the circle (1) becomes

$$x^2 + y^2 + 2gx - 1 = 0. \qquad (2)$$

This circle will touch the line $3x + 4y + 5 = 0$ if the distance of the line from the centre of the circle = the radius of the circle, *i.e.*

$$\frac{-3g + 5}{\sqrt{3^2 + 4^2}} = \sqrt{g^2 - 1}$$

$$\Rightarrow \qquad g = 0 \text{ or } -\frac{15}{8}.$$

Substituting these values of g in (2), the required equations of the circles become $x^2 + y^2 = 1$; $4(x^2 + y^2) - 15x - 4 = 0$.

Example 3. Prove that the two circles which pass through the two points $(0, a)$ and $(0, -a)$ and touch the line $y = mx + c$ will cut orthogonally if $c^2 = a^2(2 + m^2)$.

Solution. Let the equation of the circle be

$$x^2 + y^2 + 2gx + 2fy + k = 0. \quad (1)$$

It will pass through $(0, a)$ and $(0, -a)$ if

$$a^2 + 2fa + k = 0 \text{ and } a^2 - 2fa + k = 0.$$

On solving these equations, we get

$$f = 0 \text{ and } k = -a^2.$$

Therefore, the equation (1) becomes

$$x^2 + y^2 + 2gx - a^2 = 0. \quad (2)$$

This circle will touch the line $y = mx + c$ if the distance of the line from the centre of the circle equal to the radius of the circle, *i.e.*

$$\frac{-gm + c}{\sqrt{m^2 + 1}} = \sqrt{g^2 + a^2}$$

$$\Rightarrow \quad g^2 + 2gcm + a^2(1 + m^2) - c^2 = 0.$$

Let g_1, g_2 be the roots of this equation. Then

$$g_1 + g_2 = -2cm \text{ and } g_1g_2 = a^2(1 + m^2) - c^2;$$

and the equations of the two circles are

$$x^2 + y^2 + 2g_1x - a^2 = 0, \text{ and } x^2 + y^2 + 2g_2x - a^2 = 0.$$

These circles will cut orthogonally if

$$2g_1g_2 = -2a^2$$

$$\Rightarrow \quad a^2(1 + m^2) - c^2 = -a^2$$

$$\Rightarrow \quad c^2 = a^2(2 + m^2).$$

Exercises

1. Prove that the circles $x^2 + y^2 - 8x - 2y + 16 = 0$ and $3x^2 + 3y^2 - 14x + 23y - 15 = 0$ cut orthogonally.
2. Find the equation of the circle which passes through the origin and cuts orthogonally each of the circles $x^2 + y^2 - 6x + 8 = 0$ and $x^2 + y^2 - 2x - 2y - 7 = 0$.
3. Find the equation of the circle which passes through $(1, 1)$ and cuts orthogonally each of the circles $x^2 + y^2 - 8x - 2y + 16 = 0$ and $x^2 + y^2 - 4x - 4y - 1 = 0$.

4. Find the equation of the circle which passes through (3, 0), cuts orthogonally the circle $x^2 + y^2 - 6x + 4y - 3 = 0$ and touches y-axis.

5. Find the equation to the circle cutting orthogonally the three circles

$$x^2 + y^2 - 2x + 3y - 7 = 0,$$
$$x^2 + y^2 + 5x - 5y + 9 = 0,$$
and
$$x^2 + y^2 + 7x - 9y + 29 = 0.$$

6. Find the equation of the circle which cuts orthogonally each of the three circles

$$x^2 + y^2 = a^2,$$
$$(x - c)^2 + y^2 = a^2,$$
$$x^2 + (y - b)^2 = a^2.$$

7. Find the equation to the circle passing through the points (1, 0), (4, 0) and (0, 2). Prove that this circle cuts orthogonally the circle $x^2 + y^2 = 4$.

8. Prove that the two circles which pass through two points $(a, 0)$ and $(-a, 0)$ and touch the line $lx + my + n = 0$ will cut orthogonally if $a^2(2l^2 + m^2) = n^2$.

9. Find the locus of the centre of the circle which cuts two given circles orthogonally.

10. Prove that any circle which passes through the point $(b, 0)$ and cuts the circle $x^2 + y^2 = a^2$ orthogonally also passes through the point $\left(\frac{-a^2}{b}, 0\right)$.

11. Prove that any circle which passes through the point (α, β) and cuts the circle $x^2 + y^2 = c^2$ orthogonally also passes through the point $\left(\frac{c^2\alpha}{\alpha^2 + \beta^2}, \frac{c^2\beta}{\alpha^2 + \beta^2}\right)$.

12. A circle has its centre at the point (1, 2) and passes through the point (0, 3). Find its equation and also that of the circle whose centre is at the point (4, 3) and which cuts the first circle at right angles.

13. A and B are the centres of the circles $x^2 + y^2 + 4x - 2y + 4 = 0$, $x^2 + y^2 - 2x + 6y + 1 = 0$. Find the equation of the circle whose centre lies on AB and which cuts both the circles orthogonally.

14. If the equations of two circles with radii a, a' are $S = 0$, $S' = 0$; prove that the circles $\frac{S}{a} \pm \frac{S'}{a} = 0$ will intersect orthogonally.

15. AB is a diameter of a circle. Prove that the polar of A with respect to any circle which cuts the first circle orthogonally passes through B.

2. Radical Axis

Radical axis *of two circles is the locus of a point which moves so that the lengths of the tangents drawn from it to the two circles are equal.*

2.1 Equation of the radical axis. *To find the equation of the radical axis of two circles.*

Let the equations of the two circles be

$$x^2 + y^2 + 2g_1x + 2f_1y + c_1 = 0 \quad (1)$$

and

$$x^2 + y^2 + 2g_2x + 2f_2y + c_2 = 0. \quad (2)$$

Let $P(x', y')$ be the moving point on the locus from which the lengths of the tangents drawn to the circles (1) and (2) are equal, *i.e.*, $PQ = PR$.

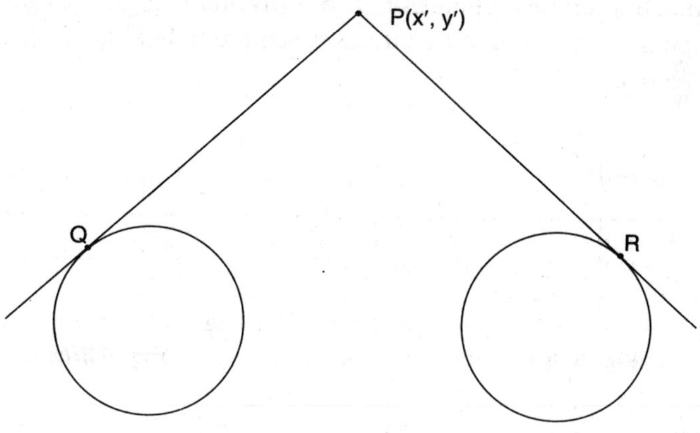

Fig. 6.2

But PQ = length of the tangent to the circle (1) from $P(x', y')$

$$= \sqrt{x'^2 + y'^2 + 2g_1x' + 2f_1y' + c_1},$$

and PR = length of the tangent to the circle (2) from $P(x', y')$

$$= \sqrt{x'^2 + y'^2 + 2g_2x' + 2f_2y' + c_2}.$$

$$\therefore \quad \sqrt{x'^2 + y'^2 + 2g_1x' + 2f_1y' + c_1} = \sqrt{x'^2 + y'^2 + 2g_2x' + 2f_2y' + c_2}$$

$$\Rightarrow 2(g_1 - g_2)x' + 2(f_1 - f_2)y' + c_1 - c_2 = 0.$$

Hence, the locus of (x', y') is $2(g_1 - g_2)x + 2(f_1 - f_2)y + c_1 - c_2 = 0$.

This is the required equation of the radical axis of the circles (1) and (2).

Remark. The equation of the radical axis being linear in x and y, represents a line.

Working rule. In order to obtain the radical axis of two circles, reduce the equations of the circles to the standard forms in such a way that the coefficients of x^2 and y^2 are unity in the equations of the circles and then subtract one equation from the other. Thus, the radical axis of the two circles

$$S_1 \equiv x^2 + y^2 + 2g_1x + 2f_1y + c_1 = 0, \text{ and } S_2 \equiv x^2 + y^2 + 2g_2x + 2f_2y + c_2 = 0$$

is $S_1 - S_2 = 0$.

Note 1. Radical axis is a line through the common points, real or imaginary, of the two circles (compare the equation of the radical axis with that of the line through the points of intersection of the two circles as obtained in § 1). As such, the concept of the radical axis seems to be wider than that of the common chord or the common tangent of the two circles. Fig. 6.3 (*a*) shows the case of two intersecting circles and the radical axis coincides with the common chord of the circles. Fig. 6.3 (*b*) and Fig. 6.3 (*c*) explain the cases of non-intersecting circles whereas Fig. 6.3 (*d*) and Fig. 6.3 (*e*) point out the intermediate case of two touching circles when the radical axis coincides with the common tangent.

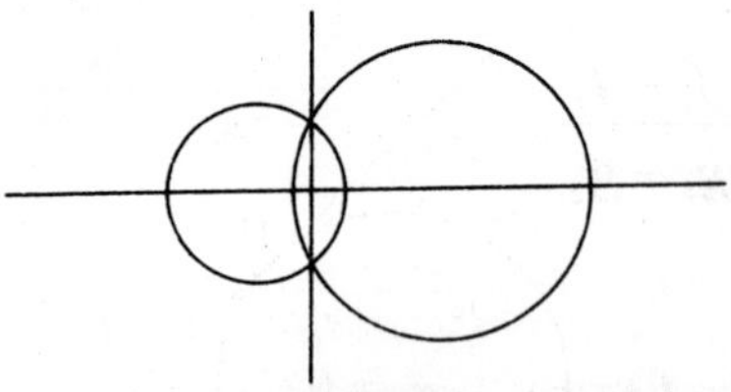

Fig. 6.3(*a*)

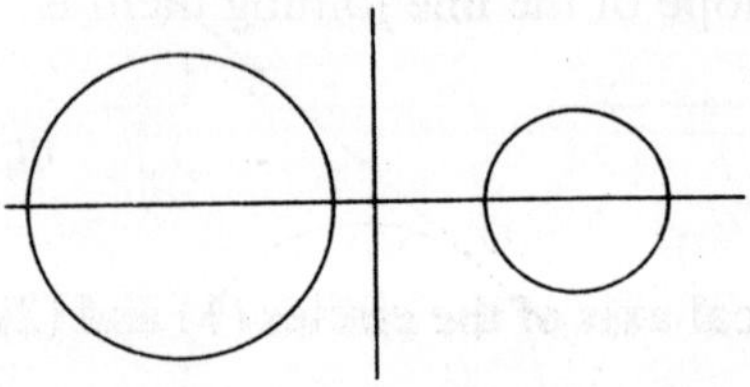

Fig. 6.3(*b*)

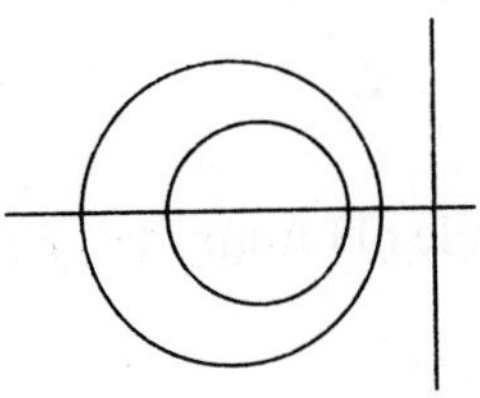

Fig. 6.3(*c*)

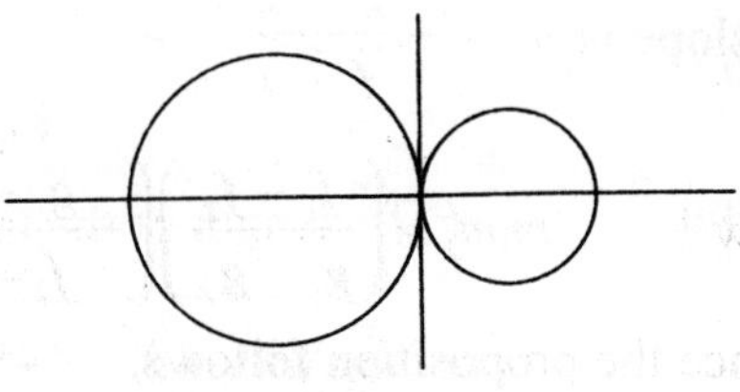

Fig. 6.3(*d*)

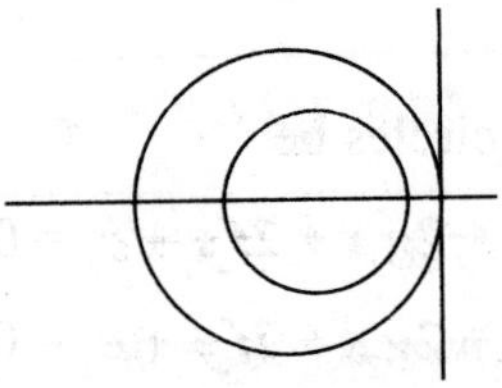

Fig. 6.3(*e*)

Note 2. In case of the circles having a common chord, Fig. 6.3(*a*), the points which lie inside both the circles *i.e.* the points on the common chord do not strictly speaking belong to the locus (*i.e.* radical axis) since real tangents cannot be drawn from them to the circles. As such the lengths of the tangents from these points to the circles have no meaning. This type of anomaly may be avoided by introducing the concept of the **power** of a point (x_1, y_1) with respect to the circle $x^2 + y^2 + 2gx + 2fy + c = 0$ to be the expression $x_1^2 + y_1^2 + 2gx_1 + 2fy_1 + c$ irrespective of the fact that this expression may be positive or negative, and then defining the radical axis as the locus of a point whose powers with respect to the two circles are equal.

2.2 Propositions on radical axis

Proposition I. *The radical axis of two circles is perpendicular to the line joining their centres.*

Let the equations of the two circles be

$$x^2 + y^2 + 2g_1x + 2f_1y + c_1 = 0 \qquad (1)$$

and

$$x^2 + y^2 + 2g_2x + 2f_2y + c_2 = 0. \qquad (2)$$

The centres of these circles are $(-g_1, -f_1)$ and $(-g_2, -f_2)$, respectively. Therefore, the slope of the line joining them is

$$m_1 = \frac{f_1 - f_2}{g_1 - g_2}.$$

Radical axis of the circles (1) and (2) is

$$2(g_1 - g_2)x + 2(f_1 - f_2)y + c_1 - c_2 = 0.$$

Its slope is $m_2 = -\dfrac{g_1 - g}{f_1 - f}$

Now $$m_1m_2 = \left(\frac{f_1 - f_2}{g_1 - g_2}\right)\left(-\frac{g_1 - g_2}{f_1 - f_2}\right) = -1.$$

Hence the proposition follows.

Proposition II. *The radical axes of three circles taken in pairs meet in a point.*

Let the equation of the circles be

$$x^2 + y^2 + 2g_1x + 2f_1y + c_1 = 0$$

$$x^2 + y^2 + 2g_2x + 2f_2y + c_2 = 0$$

and

$$x^2 + y^2 + 2g_3x + 2f_3y + c_3 = 0.$$

The equations of the radical axes of these circles taken in pairs are

$$2(g_1 - g_2)x + 2(f_1 - f_2)y + c_1 - c_2 = 0 \quad (1)$$

$$2(g_2 - g_3)x + 2(f_2 - f_3)y + c_2 - c_3 = 0 \quad (2)$$

and $$2(g_3 - g_1)x + 2(f_3 - f_1)y + c_3 - c_1 = 0. \quad (3)$$

On adding (1) and (2), we get

$$2(g_1 - g_3)\,x + 2(f_1 - f_3)\,y + c_1 - c_3 = 0.$$

This is the same as equation (3) which shows that the coordinates of the point of intersection of the lines (1) and (2) also satisfy equation (3). As such, all the three lines meet in a point.

The point of concurrence of the three radical axes of three circles taken in pairs is called the **radical centre** *of the three circles.*

Note. To find the radical centre of any given three circles, it is enough to find the point of intersection of any two of the three radical axes.

Proposition III. *The difference of the squares of the lengths of the tangents to two circles from any point in their plane varies as the distance of the point from their radical axis.*

Let equations of the circle be

$$x^2 + y^2 + 2g_1x + 2f_1y + c_1 = 0 \quad (1)$$

and $$x^2 + y^2 + 2g_2x + 2f_2y + c_2 = 0. \quad (2)$$

Let $P(x_1, y_1)$ be any point in the plane of the circles (1) and (2). The equation of the radical axis of these circles is

$$2(g_1 - g_2)\,x + 2\,(f_1 - f_2)\,y + c_1 - c_2 = 0.$$

The perpendicular distance from the point $P(x_1, y_1)$ to the radical axis is given by

$$d = \frac{2(g_1 - g_2)x_1 + 2(f_1 - f_2)y_1 + c_1 - c_2}{\sqrt{4(g_1 - g_2)^2 + 4(f_1 - f_2)^2}}.$$

Now, the difference of the squares of the lengths of the tangents drawn from the point $P(x_1, y_1)$ to the circles (1) and (2) is

$$(x_1^2 + y_1^2 + 2g_1x_1 + 2f_1y_1 + c_1) - (x_1^2 + y_1^2 + 2g_2x_1 + 2f_2y_1 + c_2)$$

$$= 2(g_1 - g_2)\,x_1 + 2\,(f_1 - f_2)\,y_1 + c_1 - c_2$$

$$= d\sqrt{4(g_1 - g_2)^2 + 4(f_1 - f_2)^2},$$

which is proportional to d.

Proposition IV. *If two circles cut a third circle orthogonally, the radical axis of the two circles passes through the centre of the third circle.*

Let the equations of the two circles be

$$x^2 + y^2 + 2g_1x + 2f_1y + c_1 = 0 \quad (1)$$

and $$x^2 + y^2 + 2g_2x + 2f_2y + c_2 = 0. \quad (2)$$

The radical axis of the circles (1) and (2) is

$$2(g_1 - g_2)\,x + 2(f_1 - f_2)\,y + c_1 - c_2 = 0. \quad (3)$$

Let the equation of the third circle be

$$x^2 + y^2 + 2gx + 2fy + c = 0. \quad (4)$$

Since the circles (1) and (2) cut the circle (4) orthogonaly, we have

$$2gg_1 + 2ff_1 = c + c_1 \quad (5)$$

and $$2gg_2 + 2ff_2 = c + c_2. \quad (6)$$

Subtracting (6) from (5), we get

$$2g\,(g_1 - g_2) + 2f\,(f_1 - f_2) = c_1 - c_2. \quad (7)$$

The radical axis given by (3) will pass through the centre $(-g, -f)$ of circle (4) if $-2(g_1 - g_2)\,g - 2\,(f_1 - f_2)\,f + c_1 - c_2 = 0$, which is true by virtue of (7).

Hence the proposition follows.

2.3 Examples

Example 1. Find the radical centre of the circles

$$x^2 + y^2 + x + 2y + 3 = 0, \quad (1)$$

$$x^2 + y^2 + 2x + 4y + 5 = 0, \quad (2)$$

$$x^2 + y^2 - 7x - 8y - 9 = 0. \quad (3)$$

Solution. The radical axis of the circles (1) and (2) is

$$x + 2y + 2 = 0, \quad (4)$$

and the radical axis of the circles (1) and (3) is

$$4x + 5y + 6 = 0. \quad (5)$$

On solving (4) and (5), we get $x = y = -\frac{2}{3}$. Hence the required radical centre is $(-\frac{2}{3}, -\frac{2}{3})$.

Example 2. Find the equation of the circle whose diameter is the common chord of the circles

$$x^2 + y^2 + 2x + 3y + 1 = 0 \text{ and } x^2 + y^2 + 4x + 3y + 2 = 0.$$

Solution. The equation of the common chord (radical axis) of the given circles is $2x + 1 = 0$. Now, any circle through the extremities of this chord is

$$x^2 + y^2 + 2x + 3y + 1 + \lambda(2x + 1) = 0$$

i.e. $$x^2 + y^2 + 2(1 + \lambda)x + 3y + 1 + \lambda = 0. \qquad (1)$$

The common chord $2x + 1 = 0$ will be the diameter of circle (1) if its centre $(-1-\lambda, -\frac{3}{2})$ lie on the chord, *i.e.*, $2(-1-\lambda) + 1 = 0$ which gives $\lambda = -\frac{1}{2}$.

Substituting the value of λ in (1), the required equation of the circle becomes $x^2 + y^2 + x + 3y + \frac{1}{2} = 0$.

Example 3. If the circles

$$x^2 + y^2 + 2gx + 2fy = 0 \text{ and } x^2 + y^2 + 2g'x + 2f'y = 0$$

touch each other, prove that $f'g = fg'$.

Solution. The radical axis of the given circles is $(g - g')x + (f - f')y = 0$. The two circles will touch if the radical axis become the tangent to any of the circles, *i.e.*

$$\frac{(g-g')(-g)+(f-f')(-f)}{\sqrt{(g-g')^2+(f-f')^2}} = \sqrt{g^2+f^2}$$

$$\Rightarrow \qquad f'g = fg',$$

Aliter. Since the given circles pass through the origin, in case the circles touch each other, the origin will be the point of contact and the tangents at the origin will coincide. But the tangents at the origin are given by

$$gx + fy = 0 \text{ and } g'x + f'y = 0.$$

Hence, we get

$$\frac{g}{g'} = \frac{f}{f'}, \text{ i.e., } f'g = fg'.$$

Example 4. Prove that if the four points of intersection of the circles $x^2 + y^2 + ax + by + c = 0$ and $x^2 + y^2 + a'x + b'y + c' = 0$ by the lines $Ax + By + C = 0$ and $A'x + B'y + C' = 0$, respectively, are concyclic, then

$$\begin{vmatrix} a-a' & b-b' & c-c' \\ A & B & C \\ A' & B' & C' \end{vmatrix} = 0.$$

Solution. Let the four points of intersection lie on the circle

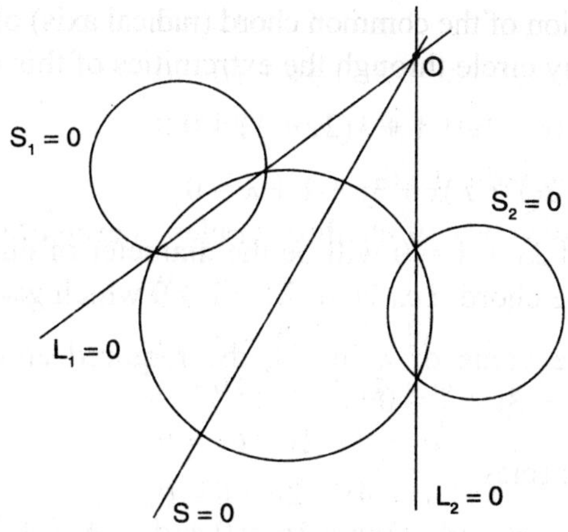

Fig. 6.4

$$S = x^2 + y^2 + 2gx + 2fy + d = 0,$$

and let the given circles and the lines be named

$$S_1 = 0, S_2 = 0, L_1 = 0, L_2 = 0, \text{ respectively.}$$

The radical axis of $S_1 = 0$ and $S = 0$ is

$$L_1 \equiv Ax + By + C = 0. \tag{1}$$

The radical axis of $S_2 = 0$ and $S = 0$ is

$$L_2 \equiv A'x + B'y + C' = 0. \tag{2}$$

The radical axis of $S_1 = 0$ and $S_2 = 0$ is

$$S_1 - S_2 \equiv (a - a')x + (b - b')\,y + c - c' = 0. \tag{3}$$

But the lines (1), (2) and (3) are concurrent. Therefore

$$\begin{vmatrix} a - a' & b - b' & c - c' \\ A & B & C \\ A' & B' & C' \end{vmatrix} = 0.$$

Hence the result.

Exercises

1. Find the radical axis of each of the following pairs of circles:

(*i*) $x^2 + y^2 + 2x + 3y + 1 = 0$ and $x^2 + y^2 + 4x + 3y + 2 = 0$.

(*ii*) $x^2 + y^2 - 3x - 4y + 5 = 0$ and $3x^2 + 3y^2 - 7x + 8y + 11 = 0$.

2. Prove that the three circles

$$x^2 + y^2 + 3x + 6y + 12 = 0,$$
$$x^2 + y^2 + 2x + 8y + 16 = 0,$$
and
$$x^2 + y^2 + 12y + 24 = 0,$$

have a common radical axis.

3. Find the radical centre of the three circles in each of the following:

(i)
$$x^2 + y^2 = 9,$$
$$x^2 + y^2 - 2x - 2y - 5 = 0,$$
and
$$x^2 + y^2 + 4x + 6y - 19 = 0.$$

(ii)
$$x^2 + y^2 - 2x + 6y = 0,$$
$$x^2 + y^2 - 4x - 2y + 6 = 0,$$
and
$$x^2 + y^2 - 12x + 2y + 30 = 0.$$

(iii)
$$x^2 + y^2 + 4x + 7 = 0,$$
$$x^2 + y^2 + y = 0,$$
and
$$2x^2 + 2y^2 + 3x + 5y + 9 = 0.$$

4. Prove that the locus of the points such that the difference of the squares of the lengths of the tangents from them to two given circles is constant, is a line parallel to their radical axis.

5. Find the radical axis of the circles

$$x^2 + y^2 + 2g_1x + 2f_1y + c_1 = 0 \text{ and } x^2 + y^2 + 2g_2x + 2f_2y + c_2 = 0.$$

Prove that the first of these circles will bisect the circumference of the other if $2g_2(g_1 - g_2) + 2f_2\,(f_1 - f_2) = c_1 - c_2$.

6. Prove that the circles

$$x^2 + y^2 + 2ax + c = 0 \text{ and } x^2 + y^2 + 2by + c = 0$$

touch if $\dfrac{1}{a^2} + \dfrac{1}{b^2} = \dfrac{1}{c}$.

3. Coaxal Circles

Let
$$S_1 \equiv x^2 + y^2 + 2g_1x + 2f_1y + c_1 = 0$$
and
$$S_2 \equiv x^2 + y^2 + 2g_2x + 2f_2y + c_2 = 0$$

be the equations of any two circles.

Now, consider the equation

$$S_1 + \lambda S_2 = 0, \tag{1}$$

where λ is a constant.

The value $\lambda = -1$ corresponds to the radical axis $S_1 - S_2 = 0$ of the two circles $S_1 = 0$ and $S_2 = 0$ whereas any other value of λ yields a circle. Thus equation (1) represents a system of circles, except when $\lambda = -1$. Consider now two distinct values λ_1 and λ_2 of λ such that none of them is -1. Then

$$S_1 + \lambda_1 S_2 = 0 \text{ and } S_1 + \lambda_2 S_2 = 0$$

are the equations of the two circles of the family (1), and their radical axis is given by

$$\frac{S_1 + \lambda_1 S_2}{1+\lambda_1} - \frac{S_1 + \lambda_2 S_2}{1+\lambda_2} = 0$$

$$\Rightarrow \quad (\lambda_2 - \lambda_1)(S_1 - S_2) = 0$$

$$\Rightarrow \quad S_1 - S_2 = 0 \qquad (\because \lambda_1 \neq \lambda_2).$$

Thus, we note that any two distinct circles of the family (1) have the radical axis $S_1 - S_2 = 0$ which is independent of λ.

A system of circles is said to be **coaxal** *if every pair of circles of the system has the same radical axis.*

Thus $S_1 + \lambda S_2 = 0$ represents a coaxal system of circles.

To prove that the centres of all circles of a coaxal system are collinear and lie on a line perpendicular to the common radical axis.

We note that the centre of the coaxal system $S_1 + \lambda S_2 = 0$ is

$$\left(-\frac{g_1 + \lambda g_2}{1+\lambda}, -\frac{f_1 + \lambda f_2}{1+\lambda} \right).$$

This is a point which lies on the line joining the centres of the circles $S_1 = 0$ and $S_2 = 0$. Thus, the centres of all the circles in the system are collinear. It is trivial to see that the line joining the centres $(-g_1, -f_1)$ and $(-g_2, -f_2)$ is perpendicular to the common radical axis $S_1 - S_2 = 0$.

3.1 Standard form of the equation to a coaxal system.

To find the equation of a coaxal system of circles in its standard form.

Let us choose the line of centres as the x-axis and the common radical axis (perpendicular to the line of centres) as the y-axis.

Since the centre of each circle lies on the x-axis, the y-coordinate of it is zero. As such any two circles from the system may be represented by the equations

$$\left.\begin{aligned} x^2 + y^2 + 2gx + c = 0 \\ x^2 + y^2 + 2g'x + c' = 0 \end{aligned}\right\}. \qquad (1)$$

The radical axis of these circles is

$$2(g - g')\,x + c - c' = 0.$$

But the radical axis is $x = 0$. Therefore, $c' = c$.

Hence, the equation of any circle in the system takes the form

$$x^2 + y^2 + 2gx + c = 0,$$

where the constant term c is fixed for the whole system, and the coefficient g is a parameter which varies from one circle to another in the family.

3.2 Intersection of circles of a coaxal system

Let
$$x^2 + y^2 + 2gx + c = 0$$

be the equation of a coaxal system of circles with centres on x-axis and common radical axis as y-axis. The points of intersection of the system with its radical axis are given by (taking $x = 0$) $y^2 + c = 0$, *i.e.*, $y = \pm\sqrt{-c}$.

Thus, $(0, \sqrt{-c})$ and $(0, -\sqrt{-c})$ are the coordinates of the points where the radical axis meets a circle in the system. Since these coordinates are independent of g (the parameter of the system), every circle of the system meets the radical axis in the same two points.

The points $(0, \sqrt{-c})$ *and* $(0, -\sqrt{-c})$ *are called the* **common points** *of the co-axal system.*

Now, there are three cases:

Case I. When c is negative, the common points are real and different [see Fig. 6.5(a)]. In this case the system is said to be of **intersecting species.**

Case II. When $c = 0$, the common points are coincident [see Fig. 6.5(b)] and hence every circle of the system touches y-axis (the radical axis) at the origin. In other words any two circles of the system touch at the origin.

Case III. When c is positive, the common points are imaginary [see Fig. 6.5(c)] and the system is said to be of **non-intersecting species.**

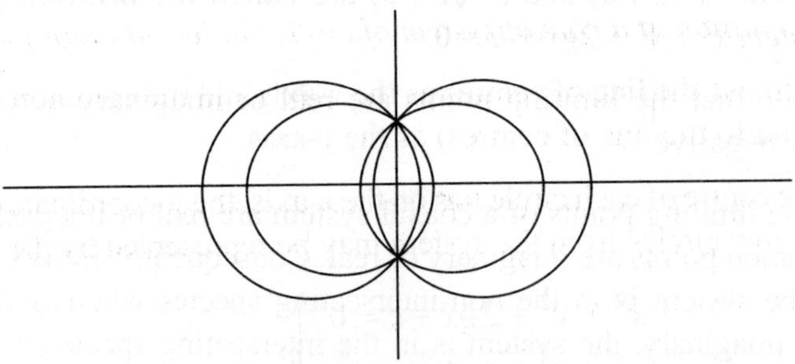

Fig. 6.5(a)

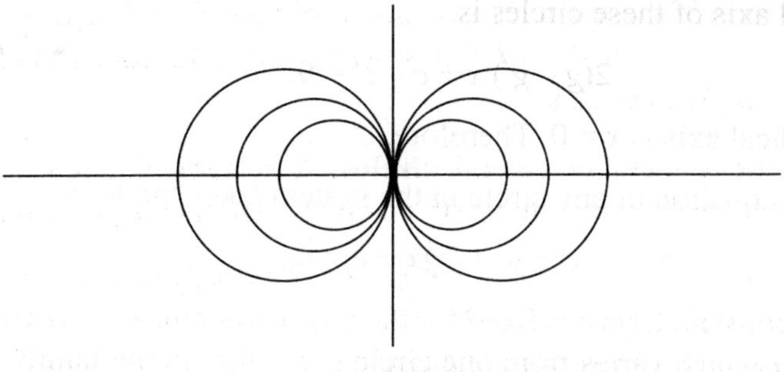

Fig. 6.5(*b*)

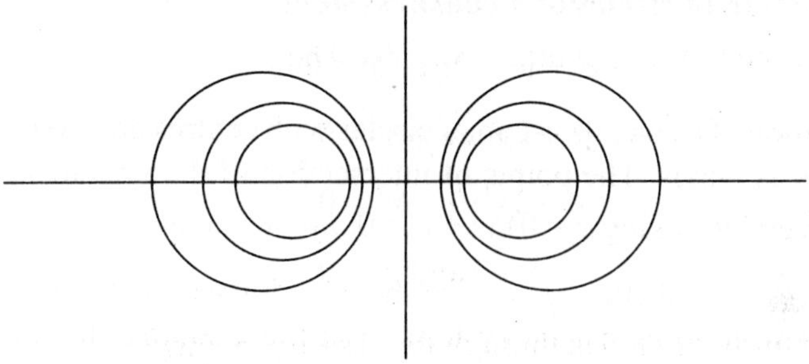

Fig. 6.5(*c*)

3.3 Limiting points of a coaxal system.

The equation $x^2 + y^2 + 2gx + c = 0$ gives any circle of a coaxal system which can be written as

$$(x + g)^2 + y^2 = (\sqrt{g^2 - c})^2.$$

This equation represents a circle whose centre is the point $(-g, 0)$ and radius = $\sqrt{g^2 - c}$. If the radius of the circle is zero *i.e.* $g^2 - c = 0$, the circle becomes a point circle. But in this case $g = \pm\sqrt{c}$.

Thus, at the particular points $(\sqrt{c}, 0)$ and $(-\sqrt{c}, 0)$, we have point circles which are in the system.

The points $(\sqrt{c}, 0)$ and $(-\sqrt{c}, 0)$ *are called the* **limiting points** *of the coaxal system* $x^2 + y^2 + 2gx + c = 0$.

We note that the limiting points are real or imaginary according as c is positive or negative.

Note 1. The limiting points of a coaxal system are real or imaginary according as the common points are imaginary or real. Consequently, if the limiting points are real, the system is in the non-intersecting species whereas if the limiting points are imaginary, the system is in the intersecting species. Hence, we see that, except in the second case of § 3.2 which is a compromise between the two, there are either limiting points or common points but not both simultaneously.

Note 2. The circle coaxal with two given circles $S_1 = 0$ and $S_2 = 0$ can be written as $S_1 + \lambda S_2 = 0$ or $S_1 + \lambda L = 0$, where $L = 0$ is the equation of the radical axis of the two given circles.

3.4 Conjugate system of coaxal circles. A system of circles through the limiting points, real or imaginary, of a coaxal system is of particular interest.

To find the system of circles through the limiting points $(\pm\sqrt{c}, 0)$ *of the coaxal system* $x^2 + y^2 + 2gx + c = 0$.

Let the circle given by the equation

$$x^2 + y^2 + 2g'x + 2f'y + c' = 0$$

pass through the limiting points $(\pm\sqrt{c}, 0)$. Then

$$\left.\begin{aligned} c + 2g'\sqrt{c} + c' = 0 \\ c - 2g'\sqrt{c} + c' = 0 \end{aligned}\right\}$$

$$\Rightarrow \qquad c' = -c \text{ and } g' = 0.$$

Hence, the family of circles through the limiting points $(\pm\sqrt{c}, 0)$ is given by the equation

$$x^2 + y^2 + 2fy - c = 0,$$

where f (written for f') is the parameter of the family.

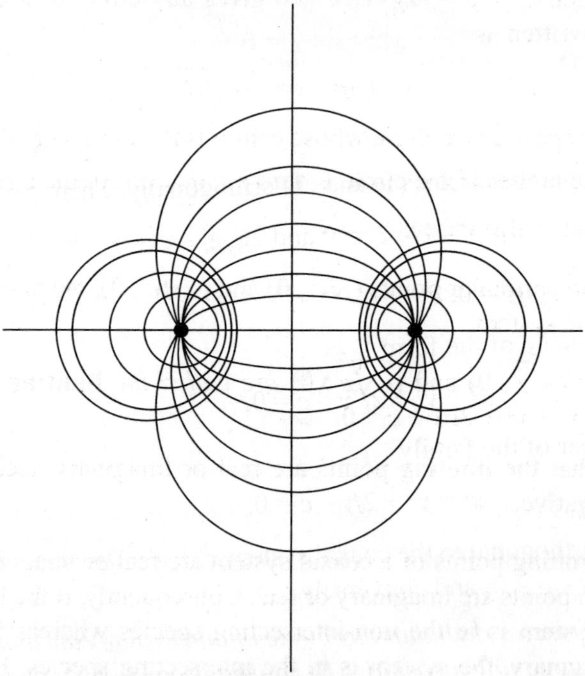

Fig. 6.6

Remark. The system of circles $x^2 + y^2 + 2fy - c = 0$ represents another coaxal system of circles having y-axis as the line of centres and x-axis as the common radical axis. The limiting points of the system are $(0, \pm\sqrt{-c})$ and the common points are $(\pm\sqrt{c}, 0)$.

We note several things relating the two coaxal systems

$$x^2 + y^2 + 2gx + c = 0 \text{ and } x^2 + y^2 + 2fy - c = 0,$$

which are as follows:

(*i*) The line of centres of either system is the common radical axis of the other system.

(*ii*) The limiting points of the either system are the common points of the other.

(*iii*) If either system is of the intersecting species, the other must be of the non-intersecting species.

(*iv*) Every member of one system cuts orthogonally every member of the other.

To prove that any circle which cuts any two given circles of the coaxal system $x^2 + y^2 + 2gx + c = 0$ orthogonally is a circle of the system $x^2 + y^2 + 2fy - c = 0$.

Let the given two circles of the coaxal system be represented by the equations

$$x^2 + y^2 + 2g_1x + c = 0 \quad (1)$$

and

$$x^2 + y^2 + 2g_2x + c = 0. \quad (2)$$

Let

$$x^2 + y^2 + 2g'x + 2f'y + c' = 0 \quad (3)$$

be a circle which cuts the circles (1) and (2) orthogonally. Then, we have

$$2g_1g' - c - c' = 0 \text{ and } 2g_2g' - c - c' = 0$$

$$\Rightarrow \qquad g' = 0,\ c' = -c.$$

Thus, circle (3) must be of the form

$$x^2 + y^2 + 2f'y - c = 0,$$

which is the member of the family

$$x^2 + y^2 + 2fy - c = 0.$$

Hence, any circle orthogonal to the coaxal system $x^2 + y^2 + 2gx + c = 0$ is in the family $x^2 + y^2 + 2fy - c = 0$; and conversely.

Two systems of circles such that any circle of one system cuts any circle of the other system orthogonally are said to be **conjugate systems** (*or* **orthogonal systems**) *of circles.*

Note. Through each point of the plane there is just one member of each system of the conjugate systems.

Remark. The orthogonal property of two systems of curves appears in everyday life, for example, in an electric of magnetic field, lines of force and equipotentials form orthogonal systems.

3.5 Examples

Example 1. Find the radical axis of the circles

$$x^2 + y^2 + 4x - 3 = 0 \text{ and } x^2 + y^2 + 6x - 8y + 7 = 0.$$

Also, find the equation to the circle coaxal with these two circles and passing through the point (1, 1).

Solution. The radical axis of the given circles is

$$(x^2 + y^2 + 4x - 3) - (x^2 + y^2 + 6x - 8y + 7) = 0$$

i.e. $$x - 4y + 5 = 0.$$

The equation to the circle coaxal with the given circles is

$$x^2 + y^2 + 4x - 3 + \lambda(x - 4y + 5) = 0.$$

But this circle passes through (1, 1). Therefore, $\lambda = -\dfrac{3}{2}$.

Hence, the required equation of the circle is

$$x^2 + y^2 + 4x - 3 - \tfrac{3}{2}(x - 4y + 5) = 0, \text{ i.e., } 2x^2 + 2y^2 + 5x + 12y - 21 = 0.$$

Example 2. Find the limiting points of the coaxal system of circles determined by

$$x^2 + y^2 - 6x - 4y + 3 = 0 \text{ and } x^2 + y^2 + 10x + 4y - 1 = 0.$$

Solution. The equation to the coaxal system containing the given circles is

$$x^2 + y^2 - 6x - 4y + 3 + \lambda(x^2 + y^2 + 10x + 4y - 1) = 0$$

$$\Rightarrow \quad x^2 + y^2 - \frac{6 - 10\lambda}{1+\lambda}x - \frac{4 - 4\lambda}{1+\lambda}y + \frac{3 - \lambda}{1+\lambda} = 0. \quad (1)$$

Centre of (1) is $\left(\dfrac{3 - 5\lambda}{1+\lambda}, \dfrac{2 - 2\lambda}{1+\lambda}\right)$ and the radius of (1) is

$$\sqrt{\left(\frac{3 - 5\lambda}{1+\lambda}\right)^2 + \left(\frac{2 - 2\lambda}{1+\lambda}\right)^2 - \frac{3 - \lambda}{1+\lambda}}.$$

For limiting points of the coaxal system (1), radius of (1) should be zero, *i.e.*,

$$\left(\frac{3-5\lambda}{1+\lambda}\right)^2+\left(\frac{2-2\lambda}{1+\lambda}\right)^2-\frac{3-\lambda}{1+\lambda}=0$$

$$\Rightarrow \qquad 3\lambda^2-4\lambda+1=0$$

$$\Rightarrow \qquad \lambda=\tfrac{1}{3},\,1.$$

Putting these values of λ in the coordinates of the centre of (1), we get (1, 1) and (–1, 0), which are the required limiting points.

Example 3. Find the equation to the circle which passes through the origin and belongs to the coaxal system of which the limiting points are (1, 2) and (4, 3).

Solution. The equations to the point circles are

$$(x-1)^2+(y-2)^2=0 \text{ and } (x-4)^2+(y-3)^2=0. \qquad (1)$$

The equation to the circle coaxal with these circles is

$$(x-1)^2+(y-2)^2+\lambda\{(x-4)^2+(y-3)^2\}=0. \qquad (2)$$

But this circle passes through the origin. Therefore, $\lambda=-\frac{1}{5}$.

Putting this value of λ in equation (2), we get $2x^2+2y^2-x-7y=0$.

This is the required equation of the circle.

Exercises

1. Find the coordinates of the limiting points of each of the pairs of circles:
   (*i*) $x^2+y^2+2x+5=0$ and $x^2+y^2+2y+5=0$.
   (*ii*) $x^2+y^2+2x+4y+7=0$ and $x^2+y^2+4x+2y+5=0$.
   (*iii*) $x^2+y^2+2x-6y=0$ and $2x^2+2y^2-10y+5=0$.
   (*iv*) $x^2+y^2-6x+12y+5=0$ and $3x^2+3y^2+10x-20y+15=0$.
2. The point (2, 1) is a limiting point of a coaxal system of the circles of which $x^2+y^2-6x-4y+3=0$ is a member. Find the equation of the radical axis and the coordinates of the other limiting point.
3. A system of coaxal circles is defined by one of the limiting points (–1, 2) and the circle $x^2+y^2+18x+4y-35=0$. Find the coordinates of the second limiting point and also the equation of the other circle of the system which has the same radius as the given circle.
4. Find the general equation of the system of circles any pair of which have the same radical axis as the circles

$$x^2 + y^2 + x - 5y - 3 = 0 \text{ and } x^2 + y^2 + 3x + 4y + 6 = 0.$$

Prove that the equation of the member of the system which passes through the origin is $3x^2 + 3y^2 + 5x - 6y = 0$.

5. Find the coordinates of the limiting points of the system of circles coaxal with the circles $x^2 + y^2 - 6x - 6y + 4 = 0$ and $x^2 + y^2 - 2x - 4y + 3 = 0$. Find also the equations of the circles of this coaxal system which touch the line $x + y - 5 = 0$.

6. Prove that the limiting points of the coaxal system of circles

$$x^2 + y^2 + 2gx + c + \lambda(x^2 + y^2 + 2fy + c') = 0$$

and real if $(c - c')^2 > 4(f^2g^2 - f^2c - g^2c')$.

7. Find the equation to the circle which is coaxal with the circles

$$x^2 + y^2 + 3x - 4y + 5 = 0 \text{ and } x^2 + y^2 - 5x + 2y - 1 = 0$$

and which passes through the point (3, 1). Find also the radical axis of the system.

8. Find the equation to the circle which belongs to the coaxal system of which the limiting points are (1, –2), (2, 3) and which passes through the origin.

9. Find the equation of the circle which belongs to the coaxal system of which the limiting points are (1, –1), (2, 0) and which passes through the origin.

10. Prove that the polar of one limiting point of a coaxal system with respect to any circle of the system passes through the other limiting point.

11. The circle $x^2 + y^2 + 4x - 6y + 3 = 0$ is one of the circles of a coaxal system having as radical axis the line $2x - 4y + 1 = 0$. Find the circle of the system which touches the line $x + 3y - 2 = 0$.

12. Prove that the locus of points, the lengths of the tangents drawn from them are in a given ratio is a circle coaxal with the given circles.

13. Find the limiting points of the coaxal system given by

$$x^2 + y^2 + 2\lambda\,(x + y - 4) - 6 = 0.$$

Find the equations of the two circles through these points which have radius 3.

14. Prove that the polar of a limiting point of a coaxal system with respect to any circle of the system is the same for all circles of the system.

15. The equation to a circle of a given coaxal system is

$$x^2 + y^2 + 2gx + 2fy + c = 0$$

and the origin is a limiting point of this system. Prove that the equation to the orthogonal system is

$$(x^2 + y^2)(g + \mu f) + c(x + \mu y) = 0.$$

Miscellaneous Exercises

1. Show that the condition for the two circles

$$x^2+y^2+2g_1x+2f_1y+c_1=0 \text{ and } x^2+y^2+2g_2x+2f_2y+c_2=0$$

should touch is

$$(2g_1g_2+2f_1f_2-c_1-c_2)^2=4(g_1^2+f_1^2-c_1)(g_2^2+f_2^2-c_2).$$

2. Find the equation to the circle through the points of intersection of the circles $x^2+y^2=1$ and $x^2+y^2+2x+4y+1=0$, which touches the line $x+2y+5=0$.

3. Find the radical axis of the circles $x^2+y^2+6x+2y+1=0$ and $x^2+y^2-6x-2y+1=0$. Also, find the equation to the circle coaxal with these two circles and passing through the point (1, 1).

4. The circle $x^2+y^2+2x+4y+1=0$ belongs to a coaxal system and a limiting point is (0, –2). Find the equation of the radical axis and coordinates of the other limiting point.

5. Prove that the circles $x^2+y^2-4y-5=0$ and $x^2+y^2-6x-12y+29=0$ intersect each other at right angles. Find the equation of their common chord and that of the tangent to the first circle at each point of intersection.

6. Prove that the circles $x^2+y^2+2x-8y+8=0$ and $x^2+y^2+10x-2y+22=0$ touch each other. Find the coordinates of the point of contact P and the equation of the common tangent at that point. Find the equation to the circle which touches each of these circles at P and passes through the point (3, 7).

7. If two circles cut orthogonally, prove that the polar of any point P on the first circle with respect to the second passes through the other end of the diameter of the first circle which goes through P.

Hence, prove that the equation of the circle which cuts the three circles

$$x^2+y^2+2g_1x+2f_1y+c_1=0$$
$$x^2+y^2+2g_2x+2f_2y+c_2=0$$
$$x^2+y^2+2g_3x+2f_3y+c_3=0$$

orthogonally is

$$\begin{vmatrix} x+g_1 & y+f_1 & g_1x+f_1y+c_1 \\ x+g_2 & y+f_2 & g_2x+f_2y+c_2 \\ x+g_3 & y+f_3 & g_3x+f_3y+c_3 \end{vmatrix}=0.$$

8. Prove that the common tangent to two circles of a coaxal system subtends a right angle at either limiting point of the system.

9. Prove that the equation to the circle cutting orthogonally the circles

$$(x-a)^2+(y-b)^2=b^2,$$
$$(x-b)^2+(y-a)^2=a^2,$$

and $$(x-a-b-c)^2+y^2=ab+c^2,$$

is $$x^2+y^2-2(a+b)x-(a+b)y+a^2+3ab+b^2=0.$$

10. Prove that the polar lines of a fixed point P with respect to the circles of a given coaxal system pass through a fixed point Q.

11. Find the limiting points of the system of circles

$$x^2+y^2+2gx+c+\lambda(x^2+y^2+2fy+c')=0$$

and show that the square of the distance between them is

$$\frac{(c-c')^2-4f^2g^2+4f^2c+4g^2c'}{f^2+g^2}.$$

12. Prove that if a circle cuts two of a coaxal system of circles at right angles, it will cut them all at right angles.

13. Prove that the equations of two given circles can always be put in the form

$$x^2+y^2+ax+b=0,\quad x^2+y^2+a'x+b=0$$

and that one of the circles will be within the other if aa' and b are both positive.

14. If A, B, C be the centres of three coaxal circles and t_1, t_2, t_3 be the tangents to them from any point, show that $BC{\cdot}t_1^2+CA{\cdot}t_2^2+AB{\cdot}t_3^2=0$.

15. Prove that the locus of the centre of a circle which cuts three given circles at the same angle is a line.

16. If polars of a point P with respect to two given circles meet in Q, prove that the radical axis of the circles bisects PQ.

17. Prove that as λ varies the circles $x^2+y^2+2ax+2by+c+\lambda(ax-by+1)=0$ form a coaxal system, and find the equations of the radical axis and the line of centres. Find the equations of the circles which are orthogonal to all the circles of the given system.

18. Find the equations of the two circles of radius $\frac{1}{2}$ each of which is orthogonal to both of the circles $x^2+y^2+2x=0$ and $x^2+y^2-2x-2y=0$.

19. A coaxal system is defined by the circles

$$x^2+y^2+2ax+2by+c=0\quad(a^2+b^2>c),$$

and $$(a^2+b^2)(x^2+y^2)+2acx+2bcy+c^2=0.$$

Prove that the origin is one of the limiting points and find the equation of the orthogonal coaxal system of circles.

20. The polars of a point P with respect to two fixed circles meet in the point Q. Prove that the circle on PQ as diameter passes through two fixed points and cuts both the given circles at right angles.

21. Prove that the tangents drawn from any point of a fixed circle of a coaxal system to two other fixed circles of the system are in a constant ratio.

22. Prove that the circles with respect to which the given line $lx + my + n = 0$ is polar of the origin form a coaxal system with radical axis $2lx + 2my + n = 0$.

Find the coordinates of the limiting points of this system and the equation of the system of circles which cuts the system orthogonally.

23. Prove that the limiting points of a system of coaxal circles are inverse points with respect to every circle of the system.

24. Prove that the limiting points of the system

$$x^2 + y^2 + 2gx + c + \lambda(x^2 + y^2 + 2fy + k) = 0$$

subtend a right angle at the origin if $\frac{c}{g^2} + \frac{k}{f^2} = 2$.

The Parabola

7 CHAPTER

1. Introduction

The parabola, ellipse and hyperbola are cases of curves called conic sections, or simply conics. The name is derived from the fact that they may be obtained as sections made by a plane with a double right circular cone, and they were first studied in this way. In this book, however, we shall discuss them as loci by means of equations.

In the present chapter we shall study the conic section—parabola. This curve may already have been met in other branches of mathematics. A projectile, for example, a ball or bullet, travels in a path which is approximately a parabola. The paths of some comets are nearly parabolic. Cables of some suspension bridges hang in the form of a parabola. The surface generated by revolving a parabola about its axis is called a paraboloid of revolution. A reflecting surface in this form has the property that light emanating at the focus is reflected in the direction of the axis. This kind of surface is used in headlights, in some telescopes and in devices to reflect sound waves. A comparatively recent application of parabolic metal surfaces is found in radar and other microwave equipments. The surfaces reflect radio waves in the same way that light is reflected and are used in directing outgoing beams and also in receiving incoming beams.

2. Parabola

The locus of a point which moves such that it is equidistant from a fixed point and a fixed line is called a **parabola**.

The fixed point is called the **focus** and the fixed line the **directrix**.

Ex. Find the equation of the parabola whose focus is $(-1, 1)$ and directrix is $x + 3y + 3 = 0$.

The line passing through the focus and perpendicular to the directrix is called the **axis** of the parabola.

2.1 Standard equation of a parabola. *To find the equation of a parabola in its standard form.*

Let S be the focus and ZK the directrix. Draw SZ perpendicular from the focus S to the directrix ZK.

Take A as the middle point of ZS. Then, by definition, ZS is the axis and the point A lies on the parabola. The point A is called the **vertex** of the parabola.

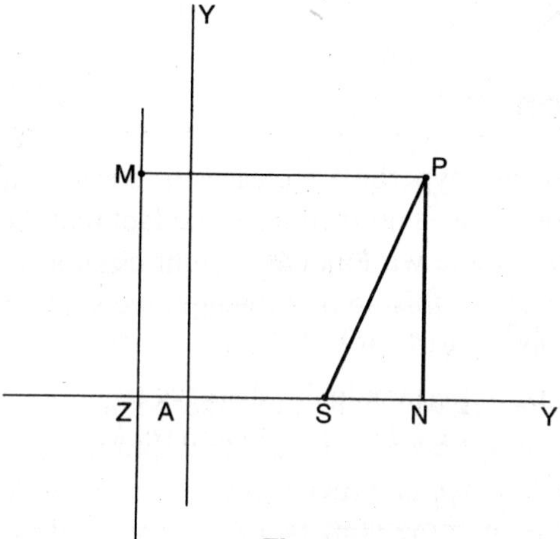

Fig. 7.1

Take the point A as origin of coordinates, the x-axis along the line AS and y-axis along the perpendicular to AS at A, as shown in Fig. 7.1.

Let $AS = a$. Then, the coordinates of the focus S are $(a, 0)$ and the equation of the directrix is $x = -a$. Let $P(h, k)$ be a moving point on the locus (parabola). Then, by definition of the parabola $SP = MP$.

But $\qquad SP = \sqrt{(h-a)^2 + k^2}$

and $\qquad MP = ZN = ZA + AN = a + h.$

Therefore $\qquad (h-a)^2 + k^2 = (a+h)^2$, *i.e.*, $k^2 = 4ah$.

Hence, the locus of (h, k) is $y^2 = 4ax$.

2.2 General equation of a parabola. *To find the equation of the parabola whose focus is* (α, β) *and equation of the directrix is* $ax + by + c = 0$.

Let $P(h, k)$ be a moving point on the locus (parabola). Then, by definition

$$SP = MP.$$

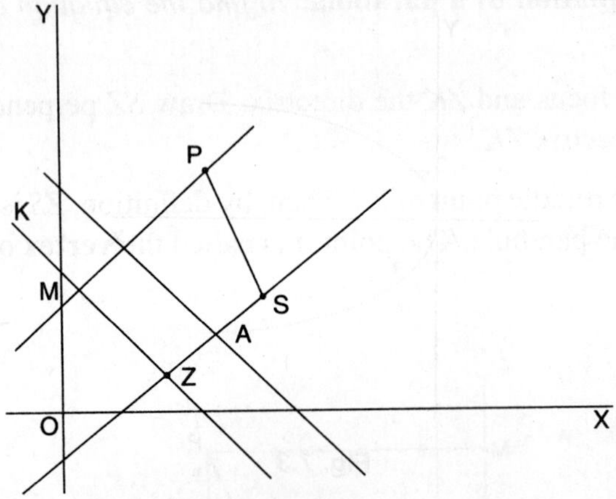

Fig. 7.2

But $$SP = \sqrt{(h-\alpha)^2 + (k-\beta)^2}$$

and MP = perpendicular distance of $ax + by + c = 0$ from (h, k)

$$= \frac{ah + bk + c}{\sqrt{a^2 + b^2}}.$$

Therefore $$(h - \alpha)^2 + (k - \beta)^2 = \frac{(ah + bk + c)^2}{a^2 + b^2}.$$

Hence, the locus of (h, k) is

$$(x - \alpha)^2 + (y - \beta)^2 = \frac{(ax + by + c)^2}{a^2 + b^2}.$$

This is the required equation of the parabola.

2.3 Tracing of a parabola. *To trace the parabola $y^2 = 4ax$, where a is positive.*

We have the following observations regarding this curve:

(*i*) No part of the curve lies on the left of y-axis since when x is negative, the corresponding values of y are imaginary.

(*ii*) The curve is symmetrical about x-axis since for each positive value of x, there are two equal and opposite values of y.

(*iii*) x-axis meets the curve only at the point A.

(*iv*) y increases as x increases and y tends to infinity as x tends to infinity. Hence, the curve extends upto infinity on the right of y-axis.

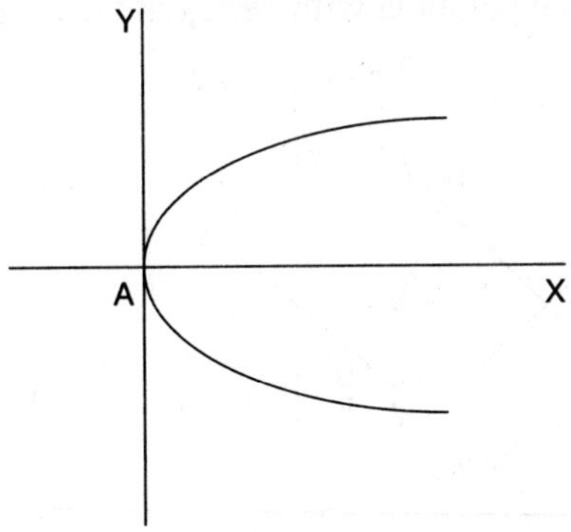

Fig. 7.3

2.4 Standard forms of parabola.

(*i*) The equation of the parabola with vertex at the origin and focus at $(a, 0)$ is

$$y^2 = 4ax.$$

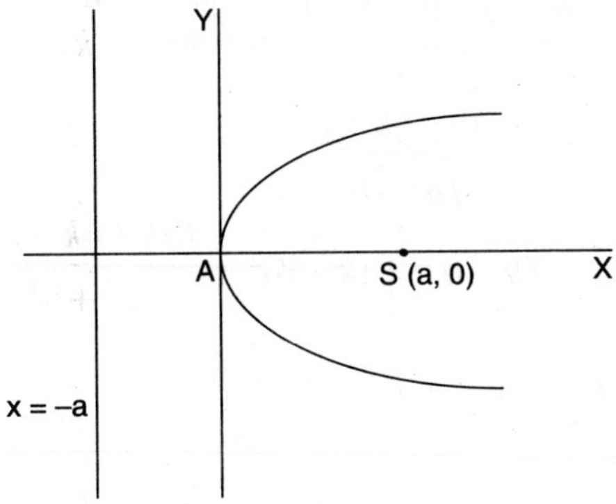

Fig. 7.4 (*a*)

(*ii*) The equation of the parabola with vertex at the origin and focus at $(-a, 0)$ is

$$y^2 = -4ax.$$

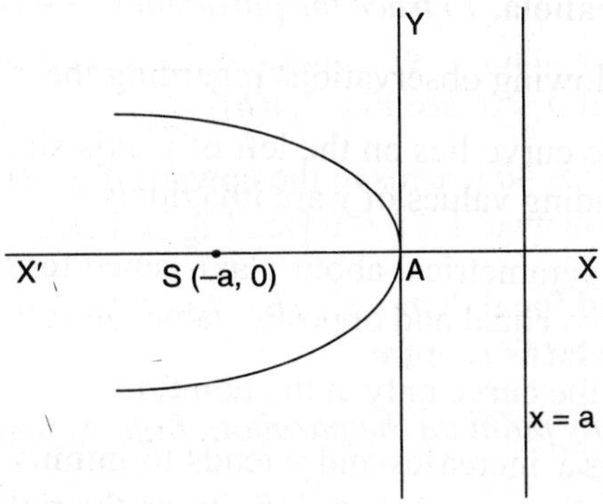

Fig. 7.4(*b*)

(*iii*) The equation of the parabola with vertex at the origin and focus at (0, *b*) is

$$x^2 = 4by.$$

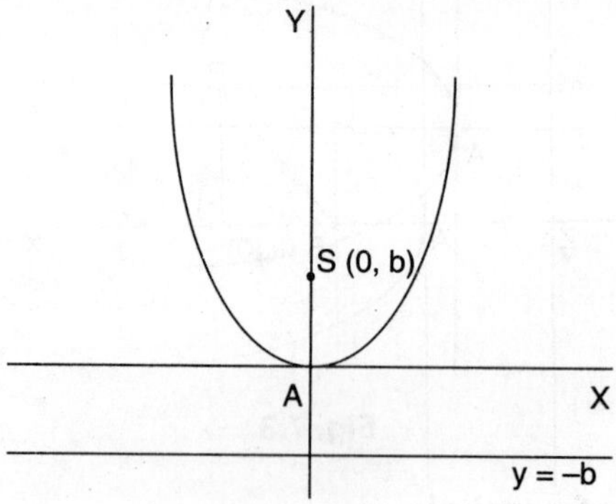

Fig. 7.4(*c*)

(*iv*) The equation of the parabola with vertex at the origin and focus at (0, –*b*) is

$$x^2 = -4by.$$

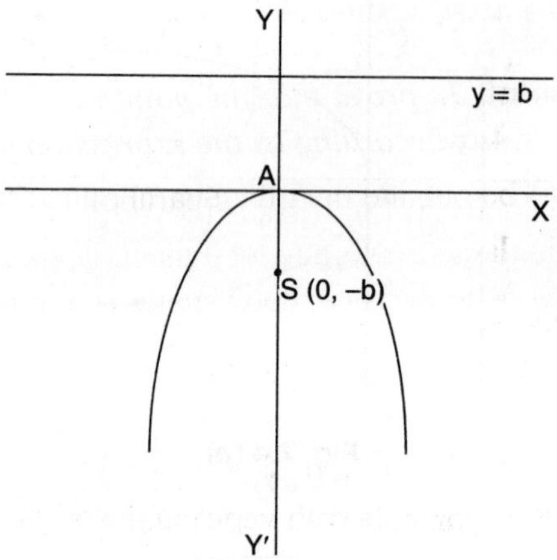

Fig. 7.4(*d*)

Remarks. (*i*) The parabola $y^2 = 4ax$ opens to the right if $a > 0$ [see Fig. 7.4(*a*)] and opens to the left if $a < 0$ [see Fig. 7.4(*b*)].

(*ii*) The parabola $x^2 = 4by$ opens to the upward if $b > 0$ [see Fig. 7.4(*c*)] and opens to the downward if $b < 0$ [see Fig. 7.4(*d*)].

2.5 Latus rectum and focal distance. *The chord through focus parallel to the directrix is called the* **latus rectum**.

The distance of any point on the parabola from its focus is called the **focal distance** *of the point.*

Let the equation of the parabola be

$$y^2 = 4ax.$$

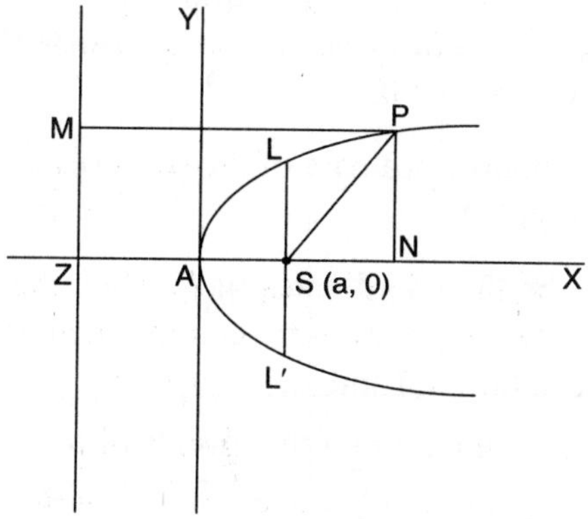

Fig. 7.5

Then, latus rectum = $LSL' = 2LS = 4a$, since $LS = 2a$.

Focal distance of any point $P(x, y)$ on the parabola is

$$SP = MP = ZN = ZA + AN = a + x.$$

2.6 Parabola and a point. *To prove that the point* (x_1, y_1) *lies outside, on or inside* the parabola* $y^2 = 4ax$ *according as the expression* $y_1^2 - 4ax_1 > = or < 0$.

Let the point (x_1, y_1) be outside the given parabola.

Draw PM perpendicular to x-axis and let it meet the parabola in the point N. Let ordinate of the point N be y_2. Then, coordinates of the point N are (x_1, y_2).

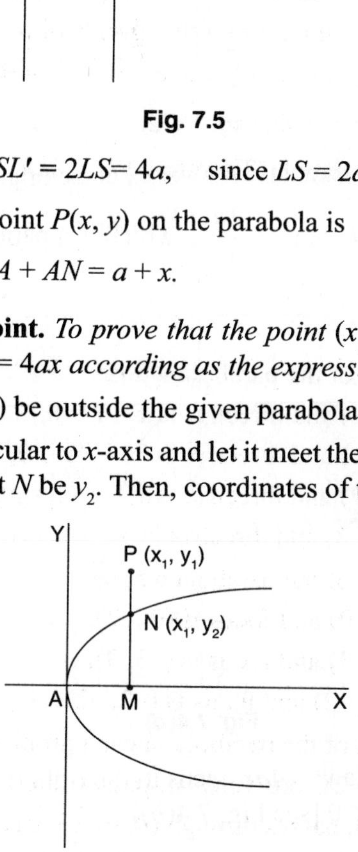

Fig. 7.6

The point $P(x_1, y_1)$ lies outside the parabola if

$$MP > MN, \text{ i.e., if } MP^2 > MN^2.$$

But $MP^2 = y_1^2$ and $MN^2 = y_2^2 = 4ax_1$.

* By the points inside a parabola, we mean the points lying in the region, determined by parabola which contains the focus.

Thus, the point (x_1, y_1) lies outside the parabola if $y_1^2 - 4ax_1 > 0$.

Similarly, we can prove that the point (x_1, y_1) lies inside the parabola if $y_1^2 - 4ax_1 < 0$ whereas the point (x_1, y_1) lies on the parabola if $y_1^2 - 4ax_1 = 0$.

Hence, *the point (x_1, y_1) lies outside, on or inside the parabola $y^2 = 4ax$ according as $y_1^2 - 4ax_1 > = $ or < 0.*

2.7 Example. *PQ* is a double ordinate of the parabola $y^2 = 4ax$. Find the locus of its points of trisection.

Solution. Let (x_1, y_1) be the coordinates of *P*. Then the coordinates of *Q* are $(x_1, -y_1)$. Let *M* and *N* be the points of trisection. Then the coordinates of these points are, respectively, $(x_1, \frac{1}{3}y_1)$ and $(x_1, -\frac{1}{3}y_1)$.

Let (h, k) be the coordinates of either points of trisection. Then

$$h = x_1 \text{ and } k = \pm\tfrac{1}{3}y_1$$

$$\Rightarrow \qquad x_1 = h \text{ and } y_1 = \pm 3k.$$

But (x_1, y_1) lies on the parabola. Therefore $9k^2 = 4ah$.

Hence, the locus of (h, k) is $y^2 = \dfrac{4a}{9}x$, which is a parabola.

Exercises

1. Find the equation of the parabola whose
   (*i*) focus is (5, 2) and directrix is $x - 1 = 0$;
   (*ii*) focus is (2, 1) and directrix is $3x + 4y = 0$;
   (*iii*) focus is (1, 2) and directrix is $x + y - 2 = 0$;
   (*iv*) focus is (4, –3) and directrix is $3x - 4y = 6$.
2. Find the equation of the parabola whose
   (*i*) vertex is (0, 0) and focus is (0, –2);
   (*ii*) vertex is (3, 3) and focus is (–3, 3);
   (*iii*) vertex is (5, –2) and focus is (–2, –2).
3. Find the equation of the parabola in each problem:
   (*i*) vertex (–1, –2), axis vertical; passes through (3, 6).
   (*ii*) axis vertical, passes through (0, 0), (3, 0) and (–1, 4).
   (*iii*) axis horizontal, passes through (0, 4), (0, –1) and (6, 1).
4. Find the vertex, focus, directrix and length of the latus rectum of the following parabolas:
   (*i*) $5x^2 + 24y = 0$. (*ii*) $y^2 + 2x - 4y + 3 = 0$.
   (*iii*) $x^2 - 4x - 5y - 1 = 0$. (*iv*) $y^2 + 4x + 2y - 8 = 0$.

5. A double ordinate of the parabola $y^2 = 4ax$ is of length $8a$. Prove that the lines from the vertex to its two ends are at right angles.
6. A variable circle is described to pass through $(a, 0)$ and touch the line $x + y = 0$. Prove that the locus of the centre of the circle is a parabola.
7. If a circle be drawn so as always to touch a given line and also a given circle, prove that the locus of its centre is a parabola.

3. Tangent and Normal

3.1 Equation of tangent at a point. *To find the equation of the tangent to a parabola at a point on it.*

Let
$$y^2 = 4ax \tag{1}$$
be the equation of the parabola and $P(x_1, y_1)$ be the given point on it. Since, (x_1, y_1) lies on (1), we have
$$y_1^2 = 4ax_1. \tag{2}$$
Take $Q(x_2, y_2)$ any other point on (1). Then
$$y_2^2 = 4ax_2. \tag{3}$$
The equation of the secant PQ is
$$y - y_1 = \frac{y_2 - y_1}{x_2 - x_1}(x - x_1). \tag{4}$$

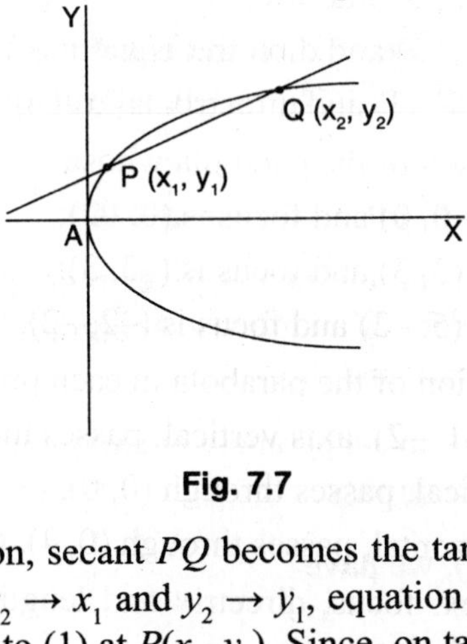

Fig. 7.7

As $Q \to P$, by definition, secant PQ becomes the tangent at P and therefore on taking the limit as $x_2 \to x_1$ and $y_2 \to y_1$, equation (4) would represent the equation of the tangent to (1) at $P(x_1, y_1)$. Since, on taking the limit, the value of $\frac{y_2 - y_1}{x_2 - x_1}$ is of the form $\frac{0}{0}$ which is an indeterminate form, we proceed to obtain the same as follows.

Subtracting (2) from (3), we get

$$y_2^2 - y_1^2 = 4a(x_2 - x_1),\ i.e.,\ \frac{y_2 - y_1}{x_2 - x_1} = \frac{4a}{y_2 + y_1}.$$

Thus, as $Q \to P$, $x_2 \to x_1$ and $y_2 \to y_1$, we get

$$\frac{y_2 - y_1}{x_2 - x_1} \to \frac{4a}{2y_1} = \frac{2a}{y_1}$$

and hence as $Q \to P$, equation (4) becomes

$$y - y_1 = \frac{2a}{y_1}(x - x_1),\ i.e.,\ yy_1 - y_1^2 = 2ax - 2ax_1.$$

On using (2), we get

$$yy_1 - 4ax_1 = 2ax - 2ax_1.$$

Hence, *the equation of the tangent to the parabola* $y^2 = 4ax$ *at the point* (x_1, y_1) *is* $yy_1 = 2a(x + x_1)$.

Working rule. The equation of the tangent to the parabola $y^2 = 4ax$ at the point (x_1, y_1) is obtained by writting yy_1 for y^2 and $x + x_1$ for $2x$ in the equation of the parabola.

Corollary. The tangent to the parabola $y^2 = 4ax$ at the vertex is the y-axis.

3.2 Intersection of a parabola and a line

Let $y^2 = 4ax$ and $y = mx + c$ be the equations of a parabola and a line, respectively. For the points of intersection, we solve them simultaneously and we get

$$m^2x^2 + 2(mc - 2a)\,x + c^2 = 0. \tag{1}$$

Since, equation (1) is quadratic in x it gives two values of x which may be real and distinct, coincident or imaginary. Accordingly there are two points of intersection of the parabola and the line.

Hence, a line always cuts a parabola in two points which may be real and distinct, coincident or imaginary.

Note. From equation (1), we have

$$x = \frac{2a - mc \pm \sqrt{4a^2 - 4amc}}{m^2}.$$

If $m = 0$, x becomes infinite *i.e.* if the line is parallel to x-axis then it meets the parabola at infinity. Further, if we take $m = 0$ in equation (1), then one root of equation (1) is real and finite. Thus, we conclude that *a line parallel to the axis of the parabola meets the curve in two points of which one is finite and the other infinite.*

3.3 Condition of tangency. *To find the condition when the line $y = mx + c$ becomes tangent to the parabola $y^2 = 4ax$.*

First Method. It is clear, from equation (1) in §3.2, that the line $y = mx + c$ is tangent to the parabola $y^2 = 4ax$ if the roots of the equation (1) in §3.2 are equal for which $4(mc - 2a)^2 = 4m^2c^2$, *i.e.,* $c = \frac{a}{m}$. This is the required condition.

Second Method. The equation of the tangent to the given parabola $y^2 = 4ax$ at any point (x_1, y_1) is $yy_1 = 2a(x + x_1)$. The line $y = mx + c$ is tangent to the given parabola if

$$2ax - yy_1 + 2ax_1 \equiv mx - y + c$$

$$\Rightarrow \qquad \frac{2a}{m} = \frac{-y_1}{-1} = \frac{2ax_1}{c}$$

$$\Rightarrow \qquad x_1 = \frac{c}{m}, y_1 = \frac{2a}{m}.$$

But (x_1, y_1) lies on the given parabola. Therefore

$$\frac{4a^2}{m^2} = 4a\frac{c}{m}, \text{ i.e., } c = \frac{a}{m}.$$

This is the required condition.

Remarks. (*i*) The line $y = mx + \frac{a}{m}$ is always a tangent to the parabola $y^2 = 4ax$ whatever the value of m may be. In fact, it is a family of tangents to the parabola $y^2 = 4ax$, where m stands for a parameter.

(*ii*) The point of contact of the tangent $y = mx + \frac{a}{m}$ to the parabola $y^2 = 4ax$ is $\left(\frac{a}{m^2}, \frac{2a}{m}\right)$.

(*iii*) The line $y = mx + \frac{a}{m}$ is called the equation of the tangent in **'m' form** or in **slope form** to the parabola $y^2 = 4ax$.

Ex. Show that the equation of the tangent to the parabola $y^2 = 4ax$ which makes an angle θ which its axis is $y = x \tan\theta + a \cot\theta$.

3.4 Equation of the normal. *To find the equation of the normal to a parabola at a given point on it.*

Let $y^2 = 4ax$ be the equation of the parabola and $P(x_1, y_1)$ be the given point on it. The equation of the tangent at the point $P(x_1, y_1)$ is $yy_1 = 2a(x + x_1)$.

$$\text{Slope of the tangent} = \frac{2a}{y_1}.$$

$$\text{Slope of the normal} = -\frac{y_1}{2a}.$$

Hence, the equation of the normal at the point $P(x_1, y_1)$ to the parabola $y^2 = 4ax$ is

$$y - y_1 = -\frac{y_1}{2a}(x - x_1) \text{ i.e., } xy_1 + 2ay = (2a + x_1)y_1 \quad \square$$

If we denote the slope of the normal by m, *i.e.*, $m = -\dfrac{y_1}{2a}$. Then

$$y_1 = -2am \text{ and } x_1 = \frac{y_1^2}{4a} = am^2.$$

Hence, the equation of the normal in terms of m is

$$y = mx - 2am - am^3.$$

Note. The line $y = mx - 2am - am^3$ is called the equation of the normal in **'m' form** or in **slope form**. In fact, this represents a system of normals to the parabola $y^2 = 4ax$, where m stands for the parameter.

Remark. The coordinates of the foot of the normal $y = mx - 2am - am^3$ to the parabola $y^2 = 4ax$ are $(am^2, -2am)$.

3.5 *To prove that the three normals can be drawn to a parabola from an external point and that the algebraic sum of the ordinates of the feet of these normals is zero.*

The equation of the normal at the point $(am^2, -2am)$ to the parabola $y^2 = 4ax$ is $y = mx - 2am - am^3$. Let it pass through a fixed point (h, k). Then

$$am^3 + (2a - h)\,m + k = 0. \tag{1}$$

Equation (1) is cubic in m, therefore it has three roots m_1, m_2 and m_3 (say). Hence, through any point, three normals can be drawn to a parabola of which one at least must be real.

Further, by theory of equations, we have

$$m_1 + m_2 + m_3 = 0$$

$$\Rightarrow \quad -2am_1 - 2am_2 - 2am_3 = 0$$

$$\Rightarrow \quad y_1 + y_2 + y_3 = 0,$$

where y_1, y_2, y_3 are the ordinates of the feet of the normals.

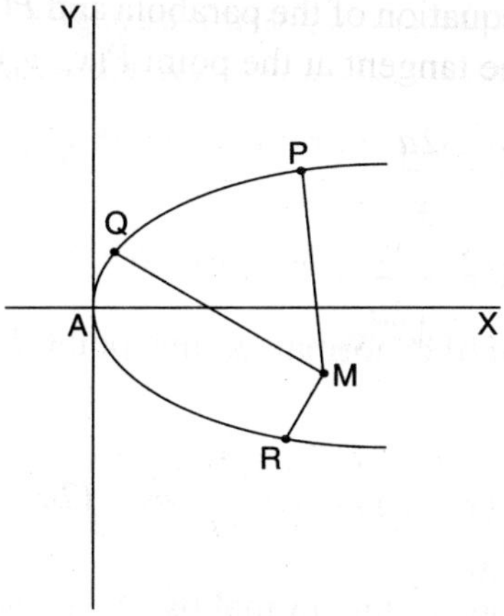

Fig. 7.8

The feet of the normals drawn from a point are called **co-normal points.**

In Fig. 7.8 three normals are drawn from the point M. The feet of the normals are P, Q and R. Thus, P, Q and R are co-normal points.

3.6 Circle through co-normal points. *To find the equation to the circle through the three points the normals at which meet in a point.*

Let the normals at the points P, Q, R to the parabola $y^2 = 4ax$ meet in the point (h, k). Then, we have

$$am^3 + (2a - h)\,m + k = 0. \tag{1}$$

Taking $-2am = y$, the ordinates of P, Q, R are the roots of the equation

$$y^3 + 4a\,(2a - h)\,y - 8a^2k = 0. \tag{2}$$

Let equation of the circle passing through P, Q, R be

$$x^2 + y^2 + 2gx + 2fy + c = 0. \tag{3}$$

Multiply equation (3) by $16a^2$ and put y^2 for $4ax$. Then the ordinates of the points of intersection of the circle and the parabola are the roots of the equation

$$y^4 + 8a\,(2a + g)\,y^2 + 32a^2fy + 16a^2c = 0. \tag{4}$$

If y_1, y_2, y_3, y_4 be the roots of the equation (4), then by theory of equations, we have

$$y_1 + y_2 + y_3 + y_4 = 0.$$

But from equation (2), we have

$$y_1 + y_2 + y_3 = 0,$$

where y_1, y_2 and y_3 are the ordinates of P, Q and R, respectively. Therefore $y_4 = 0$ which shows that the circle passing through P, Q and R passes through the vertex of the parabola, for all values of h and k. Thus $c = 0$.

Now, from equation (4) the ordinates of P, Q and R are the roots of the equation

$$y^3 + 8a(2a + g)y + 32a^2 f = 0. \qquad (5)$$

On comparing (2) and (5), we get

$$2(2a + g) = 2a - h \text{ and } -8a^2 k = 32a^2 f$$

$$\Rightarrow \quad 2g = -(2a + h) \text{ and } 2f = -\frac{k}{2}$$

Hence, *the equation of the circle passing through the points the normals at which meet in the point* (h, k) *is*

$$x^2 + y^2 - (2a + h)x - \tfrac{1}{2}ky = 0.$$

3.7 Examples

Example 1. Two lines are at right angles to one another, and one of them touches the parabola $y^2 = 4a(x + a)$ and the other to $y^2 = 4a'(x + a')$. Prove that the locus of the point of intersection of the lines is $x + a + a' = 0$.

Solution. Any tangents to the given parabolas are

$$y = m(x + a) + \frac{a}{m} \qquad (1)$$

and

$$y = m'(x + a') + \frac{a'}{m'}. \qquad (2)$$

Since, the tangents are at right angles, we have

$$mm' = -1. \qquad (3)$$

Substituting the value of m' from (3) in (2), we get

$$y = -\frac{1}{m}(x + a') - a'm. \qquad (4)$$

Subtracting (4) from (1), we obtain

$$0 = \left(m + \frac{1}{m}\right)x + \left(m + \frac{1}{m}\right)a + \left(m + \frac{1}{m}\right)a'$$

$$\Rightarrow \quad x + a + a' = 0.$$

Example 2. Find the locus of the foot of the perpendicular drawn from the vertex on a tangent to the parabola $y^2 = 4ax$.

Solution. Any tangent to the given parabola is

$$y = mx + \frac{a}{m}. \tag{1}$$

The equation of the perpendicular drawn from the vertex (0, 0) to the tangent (1) is

$$y = -\frac{1}{m}x. \tag{2}$$

The locus of the foot of the perpendicular is obtained by eliminating m between (1) and (2). Hence, the required locus is

$$y = -\frac{x}{y}\cdot x + \left(-\frac{y}{x}\right)a\text{, i.e., } (x + a)y^2 + x^3 = 0.$$

Example 3. If the tangent to the parabola $y^2 = 4ax$ meets the axis in T and the tangent at the vertex A in Y, and the rectangle $TAYQ$ is completed. Prove that the locus of Q is the parabola $y^2 + ax = 0$.

Solution. Any tangent to the parabola $y^2 = 4ax$ is

$$y = mx + \frac{a}{m}.$$

This tangent meets the axis in $T\left(-\frac{a}{m^2}, 0\right)$ and the tangent at the vertex in $Y\left(0, \frac{a}{m}\right)$. Let (h, k) be the coordinates of Q. Then

$$h = -\frac{a}{m^2} \text{ and } k = \frac{a}{m}.$$

Eliminating m from these equations, we get $k^2 + ah = 0$.

Hence, the locus of (h, k) is $y^2 + ax = 0$.

Example 4. Prove that the locus of the middle point of the portion of a normal to the parabola $y^2 = 4ax$ intercepted between the curve and the axis is another parabola. Find the vertex and the latus rectum of this parabola.

Solution. The equation of the normal at $(am^2, -2am)$ to the given parabola is

$$y = mx - 2am - am^3.$$

This normal meets the axis in the point $(2a + am^2, 0)$. Let (x', y') be the middle point of the given portion of the normal. Then

$$x' = \frac{2a + am^2 + am^2}{2} = a + am^2 \text{ and } y' = \frac{0 + (-2am)}{2} = -am.$$

Eliminating m between these equations, we get $y'^2 = a(x' - a)$.

Hence, the locus of (x', y') is $y^2 = a(x - a)$. This equation represents a parabola whose vertex and latus rectum are, respectively, $(a, 0)$ and a.

Example 5. The normals to the parabola $y^2 = 4ax$ from a point P meet the axis in A, B, C. If B is the middle point of AC, prove that the locus of P is

$$27ay^2 = 2(x - 2a)^3.$$

Solution. Any normal to the given parabola is $y = mx - 2am - am^3$.

Let it pass through the point $P(h, k)$. Then

$$am^3 + (2a - h)\, m + k = 0.$$

Let m_1, m_2, m_3 be the roots of the equation. Then, by the theory of equations, we have

$$m_1 + m_2 + m_3 = 0, \tag{1}$$

$$m_1m_2 + m_2m_3 + m_3m_1 = \frac{2a - h}{a} \tag{2}$$

and

$$m_1m_2m_3 = -\frac{k}{a}. \tag{3}$$

The normals through (h, k) are

$$y = m_1x - 2am_1 - am_1^3,$$

$$y = m_2x - 2am_2 - am_2^3,$$

and

$$y = m_3x - 2am_3 - am_3^3.$$

These normals meet the axis in $A\ (2a + am_1^2,\ 0)$, $B(2a + am_2^2,\ 0)$ and $C(2a + am_3^2, 0)$, respectively. Since, B is the middle point of AC, we have

$$2a + am_2^2 = \frac{2a + am_1^2 + 2a + am_3^2}{2}$$

$$\Rightarrow \quad m_1m_3 = -\frac{m_2^2}{2}. \tag{4}$$

Now from (1), (2) and (4), we have

$$m_2^2 = \frac{2}{3}\left(\frac{h - 2a}{a}\right). \tag{5}$$

Further, from (3) and (4), we get

$$m_2^3 = \frac{2k}{a}. \tag{6}$$

Eliminating m_2 between (5) and (6), we get

$$m_2^6 = \left(\frac{2k}{a}\right)^2 = \left\{\frac{2}{3}\left(\frac{h-2a}{a}\right)\right\}^3 = m_2^6, \text{ i.e., } 27ak^2 = 2(h-2a)^3.$$

Hence, the locus of (h, k) is $27ay^2 = 2(x - 2a)^3$.

Example 6. Prove that the locus of the points such that two of the three normals to the parabola $y^2 = 4ax$ from them coincide is $27ay^2 = 4(x - 2a)^3$.

Solution. Any normal to the given parabola is $y = mx - 2am - am^3$.

Let it pass through the point (h, k). Then

$$am^3 + (2a - h)\, m + k = 0.$$

Let m_1, m_2, m_3 be the roots of the equation. Then, by theory of equations, we have

$$m_1 + m_2 + m_3 = 0 \tag{1}$$

$$m_1m_2 + m_2m_3 + m_3m_1 = \frac{2a-h}{a} \tag{2}$$

and

$$m_1m_2m_3 = \frac{-k}{a}. \tag{3}$$

Since, any two normals coincide, we can take $m_1 = m_2$. Now, equations (1), (2) and (3) give

$$m_3 = -2m_1, m_1^2 = \frac{h-2a}{3a}, m_1^3 = \frac{k}{2a}.$$

Thus

$$m_1^6 = \frac{(h-2a)^3}{27a^3} = \frac{k^2}{4a^2}, \text{ i.e., } 27ak^2 = 4(h-2a)^3.$$

Hence, the locus of (h, k) is $27ay^2 = 4\,(x - 2a)^3$.

Exercises

1. Find the equations of the tangents and the normals to the parabola $y^2 = 4ax$ at the ends of its latus rectum.

2. Prove that the line $lx + my + n = 0$ touches the parabola $y^2 = 4ax$ if $ln = am^2$.

3. Prove that the line $y = mx + c$ touches the parabola $y^2 = 4a(x + a)$ if $c = ma + \dfrac{a}{m}$.

4. Prove that the tangent to the parabola $y^2 = 4ax$ at the point (x_1, y_1) is perpendicular to the tangent at the point $\left(\dfrac{a^2}{x_1}, -\dfrac{4a^2}{y_1}\right)$.

5. Prove that the normal chord at the point whose ordinate is equal to its abscissa subtends a right angle at the focus.

6. Two equal parabolas have the same vertex and their axes are at right angles. Prove that the common tangent touches each at the end of a latus rectum.

7. Prove that the chord of the parabola $y^2 = 4ax$ whose equation is $y - x\sqrt{2} + 4a\sqrt{2} = 0$ is a normal to the curve, and that its length is $6\sqrt{3}a$.

8. Prove that the chord of the parabola $y^2 = 4ax$ which is normal at the point whose abscissa is $2a$ subtends a right angle at the vertex.

9. PNP' is a double ordinate of the parabola $y^2 = 4ax$. Prove that the locus of the point of intersection of the normal at P and the line through P' parallel to the axis is equal parabola $y^2 = 4a(x - 4a)$.

10. The normal at any point P meets the axis in G and the tangent at the vertex in G'. If A is the vertex and the rectangle $AGQG'$ is completed, prove that the equation of the locus of Q is $x^3 = 2ax^2 + ay^2$.

11. Find the locus of the foot of the perpendicular drawn from a fixed point on any tangent to a parabola.

12. Prove that a normal chord of a parabola which subtends a right angle at the vertex makes an angle θ with the axis, where $\tan\theta = \sqrt{2}$.

13. From the point where any normal to the parabola $y^2 = 4ax$ meets the axis is drawn a line perpendicular to this normal. Prove that this line always touches an equal parabola.

14. If three normals from a point to the parabola $y^2 = 4ax$ cut the axis in points whose distances from the vertex are in arithmetical progression. Prove that the point lies on the curve $27ay^2 = 2(x - 2a)^3$.

15. Find the locus of a point which is such that two of the normals drawn from it to the parabola $y^2 = 4ax$ are at right angles.

16. Find the locus of a point P when three normals drawn from it are such that one bisects the angle between the other two.

4. Tangents from a Point

4.1 *To prove that from any point there can be drawn two tangents, real or imaginary, to a parabola.*

Let the equation of the parabola be

$$y^2 = 4ax. \quad (1)$$

Let $P(x_1, y_1)$ be the given point. Any tangent to the parabola (1) is

$$y = mx + \frac{a}{m},$$

where m is the slope and can take any value. Since, tangent passes through the given point $P(x_1, y_1)$, we have

$$m^2x_1 - my_1 + a = 0. \quad (2)$$

This is a quadratic equation in m, giving two values of m (real and distinct, coincident or imaginary) corresponding to which there will be two tangents passing through the point $P(x_1, y_1)$. The two tangents will be real and distinct, coincident or imaginary according as the roots of the equation (2) are so. In fact the roots of the equation (2) are real and distinct, coincident or imaginary according as $y_1^2 - 4ax_1 > = \text{ or } < 0$.

But this is the condition for the point (x_1, y_1) to be outside, on or inside the parabola.

Hence, from a point there can be drawn two tangents to a parabola and these tangents will be real and distinct, coincident or imaginary according as the point lies outside, on or inside the parabola.

4.2 Pair of tangents from a given point. *To find the equation of the pair of tangents drawn from an external point* $P(x_1, y_1)$ *to the parabola* $y^2 = 4ax$.

Let $Q(x, y)$ be any point on the locus (*i.e.* on either of the tangents). Then equations of the line PQ are

$$\frac{x - x_1}{\cos\theta} = \frac{y - y_1}{\sin\theta} = r, \quad (1)$$

θ being its inclination from the axis of the parabola and r the algebraic distance of the point $Q(x, y)$ from $P(x_1, y_1)$ measured along the line. Therefore, a point on the line and at a distance r from $P(x_1, y_1)$ is given by

$$x = x_1 + r\cos\theta,\ y = y_1 + r\sin\theta.$$

If this point be on the parabola, we must have

$$(y_1 + r\sin\theta)^2 = 4a(x_1 + r\cos\theta)$$

i.e. $r^2 \sin^2 \theta + 2r(y_1 \sin \theta - 2a \cos \theta) + y_1^2 - 4ax_1 = 0.$

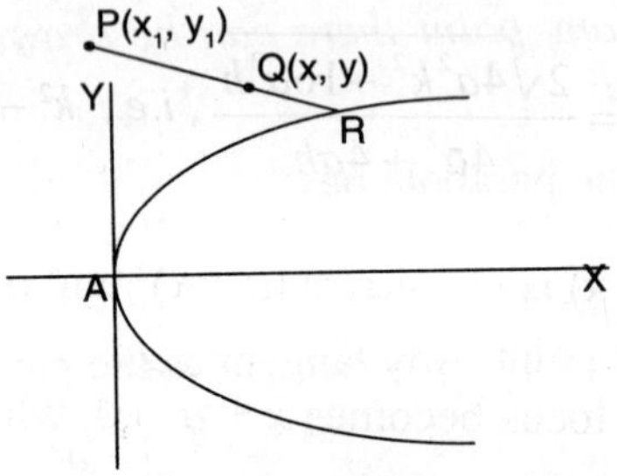

Fig. 7.9

This is a quadratic equation in r giving two values of r corresponding to which there are two points common to the parabola and the line (1).

Now, if the line (1) be a tangent to the parabola, both the values of r given by the quadratic equation must be equal. Therefore, we must have

$$(y_1 \sin \theta - 2a \cos \theta)^2 - (y_1^2 - 4ax_1) \sin^2 \theta = 0. \qquad (2)$$

This equation gives us the value of θ *i.e.*, the direction of the line (1) so that it may become the tangent to the parabola. To obtain the actual equation of the tangent (s), eliminate θ between (1) and (2). Thus, the equation of the tangent is

$$\left\{ y_1 \left(\frac{y - y_1}{r} \right) - 2a \left(\frac{x - x_1}{r} \right) \right\}^2 = \left(\frac{y - y_1}{r} \right)^2 (y_1^2 - 4ax_1)$$

$$\Rightarrow (yy_1 - 2ax - 2ax_1)^2 = (y^2 - 4ax)(y_1^2 - 4ax_1).$$

Hence, the equation of the pair of tangents is

$$T^2 = SS_1,$$

where $S \equiv y^2 - 4ax$, $S_1 \equiv y_1^2 - 4ax_1$, and $T \equiv yy_1 - 2ax - 2ax_1$.

Ex. Find the equation of the tangents drawn from the point $(-2, 0)$ to the parabola $y^2 = 8x$ and prove that they are perpendicular.

4.3 Example. The angle between two tangents to the parabola $y^2 = 4ax$ is constant and equal to α. Prove that the locus of their point of intersection is given by

$y^2 - 4ax = (a + x)^2 \tan^2 \alpha$. What happens to this locus if $\alpha = \dfrac{\pi}{2}$?

Solution. Let (h, k) be the point of intersection of the two tangents to the given parabola. Then, the equation of the pair of tangents is

$$(y^2 - 4ax)(k^2 - 4ah) = (ky - 2ax - 2ah)^2$$

$$\Rightarrow \quad 4a^2x^2 - 4akxy + 4ahy^2 + \text{first degree terms} + \text{constant terms} = 0.$$

Since, α is the angle between the tangents, we have

$$\tan\alpha = \frac{2\sqrt{4a^2k^2 - 16a^3h}}{4a^2 + 4ah}, \text{ i.e., } k^2 - 4ah = (a+h)^2 \tan^2\alpha.$$

Hence, the locus of (h, k) is $y^2 - 4ax = (a + x)^2 \tan^2 \alpha$.

When $\alpha = \frac{\pi}{2}$, the locus becomes $x + a = 0$, which is the equation of the directrix.

Exercises

1. Find the equation of the locus of the point of intersection of two tangents to a parabola which make an angle of 45° with one another.
2. Prove that the locus of the point of intersection of two tangents to a parabola at points on the curve whose ordinates are in a constant ratio is a parabola.
3. The two tangents from a point P to the parabola $y^2 = 4ax$ make angles α, β will x-axis. Find the locus of P when

   (*i*) $\tan\alpha + \tan\beta$ is constant. (*ii*) $\tan^2\alpha + \tan^2\beta$ is constant.
4. Prove that the two tangents which make equal angles, respectively with the axis and the directrix but are not at right angles, intersect on the latus rectum.
5. A pair of tangents are drawn which are equally inclined to a line whose inclination to the axis is α. Prove that the locus of their point of intersection is the line $y = (x - a) \tan 2\alpha$.

5. Chord of Contact

5.1 Equation of the chord of contact. *To find the equation of the chord of contact of the tangents drawn from the point* (x_1, y_1) *to the parabola* $y^2 = 4ax$.

Let $Q(x', y')$ and $R(x'', y'')$ be the points of contacts of the tangents drawn from the point $P(x_1, y_1)$ to the parabola

$$y^2 = 4ax. \quad (1)$$

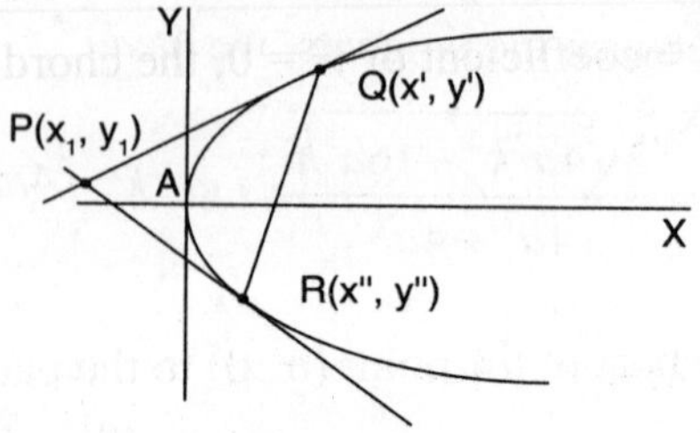

Fig. 7.10

Then, the equations of the tangents at the points Q and R to the parabola (1) are, respectively

$$y'y = 2a(x' + x) \text{ and } y''y = 2a(x'' + x).$$

But each of these tangents passes through the point $P(x_1, y_1)$. Therefore

$$y'y_1 = 2a(x' + x_1) \text{ and } y''y_1 = 2a(x'' + x_1).$$

It is clear from these relations that the points (x', y') and (x'', y'') both satisfy the equation

$$yy_1 = 2a(x + x_1).$$

This is the required equation of the line which passes through the points $Q(x', y')$ and $R(x'', y'')$, the points of contacts of the tangents drawn from the point $P(x_1, y_1)$ to the parabola (1).

Remarks. (*i*) If the point $P(x_1, y_1)$ lies inside the parabola, the two tangents will be imaginary, but the equation of the chord of contact is real since x_1 and y_1 are real.

(*ii*) If the point $P(x_1, y_1)$ lies on the parabola, then the chord of contact becomes the tangent at that point.

Working rule. The equation of the chord of contact of the tangents drawn from an external point (x_1, y_1) to the parabola $y^2 = 4ax$ is obtained by substituting yy_1 for y^2 and $x + x_1$ for $2x$ in the equation of the parabola.

5.2 Example. Prove that if tangents be drawn to the parabola $y^2 = 4ax$ from a point on the line $x + 4a = 0$, their chord of contact will subtend a right angle at the vertex.

Solution. Any point on the line $x + 4a = 0$ is $(-4a, k)$. The equation of the chord of contact of the tangents drawn from the point $(-4a, k)$ to the given parabola is $ky = 2a(x - 4a)$. The equation of the lines joining the vertex to the points of intersection of the parabola and the chord of contact is

$$y^2 = 4ax\left(\frac{2ax - ky}{8a^2}\right), \text{ i.e., } 2ax^2 - kxy - 2ay^2 = 0.$$

Since, coefficient of x^2 + coefficient of $y^2 = 0$, the chord of contact subtends a right angle at the vertex.

Exercises

1. Tangents are drawn from the point (α, β) to the parabola $y^2 = 4ax$. Prove that the length of their chord of contact is $\dfrac{(\beta^2 - 4a\alpha)^{1/2}(\beta^2 + 4a^2)^{1/2}}{a}$.
2. Tangents are drawn at the ends of a normal chord of the parabola $y^2 = 4ax$. Prove that the locus of their point of intersection is $(x + 2a)\, y^2 + 4a^3 = 0$.
3. If two tangents to a parabola make equal angles with a fixed line, prove that the chord of contact must pass through a fixed point.
4. A variable chord PQ of the parabola passes through the fixed point R. Prove that the tangents at P, Q meet on a fixed line, and show that this line is the chord of contact of the tangents drawn from the point R.

6. Pole and Polar

6.1 Equation of the polar. *To find the equation of the polar of the point* (x_1, y_1) *with respect to the parabola* $y^2 = 4ax$.

Let $P(x', y')$ be any point on the locus *i.e.* the point of intersection of the tangents drawn at the extremities of a chord through the point (x_1, y_1).

The equation of the chord of contact of the tangents from the point (x', y') is

$$yy' = 2a\,(x + x').$$

But it passes through the point (x_1, y_1). Therefore

$$y_1 y' = 2a(x_1 + x').$$

Hence, the locus of (x', y') is $yy_1 = 2a(x + x_1)$.

6.2 Pole of a line. *To find the pole of the line* $lx + my + n = 0$ *with respect to the parabola* $y^2 = 4ax$.

Let (x_1, y_1) be the pole of the line $lx + my + n = 0$ with respect to the parabola $y^2 = 4ax$. Then this line must be identical with the polar of (x_1, y_1) with respect to the parabola $y^2 = 4ax$, *i.e.*, with $yy_1 = 2a(x + x_1)$. Therefore

$$\frac{2a}{l} = \frac{-y_1}{m} = \frac{2ax_1}{n}, \text{ i.e., } x_1 = \frac{n}{l}, y_1 = -\frac{2am}{l}.$$

Hence, the required pole is $\left(\frac{n}{l}, \frac{-2am}{l}\right)$.

Ex. Prove that the polar of the focus of a parabola is the directrix.

6.3 Examples

Example 1. Prove that the locus of poles of normal chords of the parabola $y^2 = 4ax$ is $(x + 2a)y^2 + 4a^3 = 0$.

Solution. Any normal chord of the given parabola is

$$y = mx - 2am - am^3. \quad (1)$$

Let (h, k) be the pole of (1). Then, its polar with respect to the given parabola is

$$ky = 2a(x + h). \quad (2)$$

But equations (1) and (2) represent the same line. Therefore, on comparing (1) and (2), we get

$$\frac{m}{2a} = \frac{1}{k} = \frac{-2am - am^3}{2ah}, \text{ i.e., } m = \frac{2a}{k} \text{ and } -2m - m^3 = \frac{2h}{k}.$$

Therefore $$-\frac{4a}{k} - \frac{8a^3}{k^3} = \frac{2h}{k}, \text{ i.e., } -2ak^2 - 4a^3 = hk^2.$$

Hence, locus of (h, k) is $(x + 2a)y^2 + 4a^3 = 0$.

Example 2. Prove that the locus of the poles of tangents to the parabola $y^2 = 4ax$ with respect to the circle $x^2 + y^2 - 2ax = 0$ is the circle $x^2 + y^2 - ax = 0$.

Solution. Any tangent to the given parabola is

$$y = mx + \frac{a}{m}. \quad (1)$$

Let (h, k) be the pole of (1). Then, its polar with respect to the circle $x^2 + y^2 - 2ax = 0$ is $hx + ky - a(x + h) = 0$, *i.e.*,

$$(h - a)x + ky - ah = 0. \quad (2)$$

But equations (1) and (2) represent the same line. Therefore, on comparing (1) and (2), we get

$$\frac{h-a}{m} = \frac{k}{-1} = \frac{-ah}{\frac{a}{m}}$$

$$\Rightarrow \qquad m = -\left(\frac{h-a}{k}\right) \text{ and } m = \frac{k}{h}.$$

Therefore $$-\left(\frac{h-a}{k}\right) = \frac{k}{h}, \text{ i.e., } -h^2 + ah = k^2.$$

Hence, the locus of (h, k) is $x^2 + y^2 - ax = 0$.

Exercises

1. Prove that the locus of the poles of chords of the parabola $y^2 = 4ax$ which subtend a constant angle α at the vertex is the curve
$$(x + 4a)^2 = 4(y^2 - 4ax) \cot^2\alpha.$$
Hence or otherwise prove that if the constant angle is a right angle the locus is a line perpendicular to the axis of the parabola.
2. Prove that the polar of $(-a, 2a)$ with respect to the circle $x^2 + y^2 - 2ax - 3a^2 = 0$ touches the parabola $y^2 = 4ax$.
3. Prove that the polar of any point on the circle $x^2 + y^2 - 2ax - 3a^2 = 0$, with respect to the circle $x^2 + y^2 + 2ax - 3a^2 = 0$, will touch the parabola $y^2 = -4ax$.
4. Prove that the locus of the poles of tangents to the parabola $y^2 = 4ax$ with respect to the parabola $y^2 = 4bx$ is the parabola $y^2 = \frac{4b^2}{a}x$.
5. A point P is such that the line drawn through it perpendicular to its polar with respect to the parabola $y^2 = 4ax$ touches the parabola $x^2 = 4by$. Prove that the locus of P is the line $2ax + by + 4a^2 = 0$.
6. The middle point of a chord of a parabola is on a fixed line perpendicular to the axis of the parabola. Prove that the locus of the pole of the chord is another parabola.
7. Prove that the locus of the poles of the chords which subtend a right angle at the fixed point (h, k) is
$$ax^2 - hy^2 + (4a^2 + 2ah)x - 2aky + a(h^2 + k^2) = 0.$$

7. Chord with Given Middle Point

7.1 *To find the equation of a chord of the parabola* $y^2 = 4ax$ *in terms of its middle point* (x_1, y_1).

First Method

The equations of a line through the point (x_1, y_1) are

$$\frac{x - x_1}{\cos\theta} = \frac{y - y_1}{\sin\theta} = r, \tag{1}$$

θ being its inclination with the axis of the parabola and r the algebraic distance of (x, y) from (x_1, y_1) measured along the line. Then any point on the line has the coordinates $(x_1 + r\cos\theta, y_1 + r\sin\theta)$ and this point will be common to the parabola if

$$(y_1 + r\sin\theta)^2 = 4a(x_1 + r\cos\theta),$$

i.e.
$$r^2\sin^2\theta + 2r(y_1\sin\theta - 2a\cos\theta) + y_1^2 - 4ax_1 = 0. \tag{2}$$

This equation being quadratic in r, gives two values of r. Since, (x_1, y_1) is the middle point of the chord, the two values of r given by the equation (2) must be equal in magnitude and opposite in sign *i.e.* the sum of the roots of equation (2) must be zero, *i.e.*

$$y_1\sin\theta - 2a\cos\theta = 0. \tag{3}$$

This equation gives us the suitable value of θ so that the line (1) may be the chord of the parabola having (x_1, y_1) as the middle point, Therefore, the equation of the chord is

$$y_1\left(\frac{y - y_1}{r}\right) - 2a\left(\frac{x - x_1}{r}\right) = 0,\ i.e.,\ yy_1 - 2ax = y_1^2 - 2ax_1.$$

Second Method

Let $P(x', y')$ and $Q(x'', y'')$ be the end points of the chord whose middle point is (x_1, y_1). Then

$$x_1 = \frac{x' + x''}{2} \text{ and } y_1 = \frac{y' + y''}{2}.$$

Slope of PQ is $\dfrac{y'' - y'}{x'' - x'}$. Hence, its equation is

$$y - y_1 = \frac{y'' - y'}{x'' - x'}(x - x_1). \tag{1}$$

Since, $P(x', y')$ and $Q(x'', y'')$ lie on the given parabola, we have

$$y'^2 = 4ax' \text{ and } y''^2 = 4ax''.$$

which give

$$y''^2 - y'^2 = 4a(x'' - x')$$

$$\Rightarrow \quad \frac{y'' - y'}{x'' - x'} = \frac{4a}{y'' + y'} = \frac{4a}{2y_1}.$$

Therefore, equation (1) becomes

$$y - y_1 = \frac{2a}{y_1}(x - x_1),\ i.e.,\ yy_1 - 2ax = y_1^2 - 2ax_1.$$

Hence, *the equation of the chord of the parabola* $y^2 = 4ax$ *in terms of its middle point* (x_1, y_1) *is* $yy_1 - 2ax = y_1^2 - 2ax_1$. This equation of the chord can be written as $T = S_1$, where $T \equiv yy_1 - 2a(x + x_1)$ and $S_1 \equiv y_1^2 - 4ax_1$.

7.2 *To find the locus of the middle points of a system of parallel chords of a parabola.*

Let the equation of the parabola be $y^2 = 4ax$ and m the slope of the chords in the given system of parallel chords.

Let (x_1, y_1) be any point on the locus, *i.e.,* the middle point of one of the chords parallel to the line $y = mx$. Then, the equation of the chord is $yy_1 - 2ax = y_1^2 - 2ax_1$. Therefore

$$m = \frac{2a}{y_1},\ i.e.,\ y_1 = \frac{2a}{m}.$$

Hence, *the locus of the middle points of a system of parallel chords having slope m is* $y = \frac{2a}{m}$. This is the equation of a line parallel to x-axis, *i.e.,* the axis of the parabola $y^2 = 4ax$.

7.3 Examples

Example 1. Find the locus of the middle points of chords of the parabola $y^2 = 4ax$ which pass through the fixed point (h, k).

Solution. Let (x', y') be the middle point of the chord of the given parabola. Then its equation is

$$yy' - 2ax = y'^2 - 2ax'.$$

This will pass through the point (h, k) if

$$ky' - 2ah = y'^2 - 2ax'.$$

Hence, the locus of (x', y') is $y^2 - ky = 2a(x - h)$, which represents a parabola.

Example 2. Prove that the locus of the middle points of normal chords of the parabola $y^2 = 4ax$ is

$$\frac{y^2}{2a} + \frac{4a^3}{y^2} = x - 2a.$$

Solution. Any normal chord of the given parabola is

$$y = mx - 2am - am^3. \tag{1}$$

Let (x', y') be the middle point of (1). Then equation of the chord with middle point (x', y') is

$$yy' - 2ax = y'^2 - 2ax'. \tag{2}$$

But equations (1) and (2) represent the same line. Therefore, on comparing (1) and (2), we get

$$\frac{m}{2a} = \frac{1}{y'} = \frac{-2am - am^3}{y'^2 - 2ax'}$$

$$\Rightarrow \qquad m = \frac{2a}{y'} \text{ and } -2am - am^3 = \frac{y'^2 - 2ax'}{y'}.$$

Therefore

$$\frac{4a^2}{y'} - \frac{8a^4}{y'^3} = \frac{y'^2 - 2ax'}{y'}$$

$$\Rightarrow \qquad \frac{y'^2}{2a} + \frac{4a^3}{y'^2} = x' - 2a.$$

Hence, locus of (x', y') is $\dfrac{y^2}{2a} + \dfrac{4a^3}{y^2} = x - 2a.$

Exercises

1. Find the locus of the middle points of chords of the parabola $y^2 = 4ax$ which
   (*i*) pass through vertex (0, 0).
   (*ii*) pass through a fixed point $(a, 0)$.
   (*iii*) subtend a right angle at the vertex.
   (*iv*) subtend a constant angle α at the vertex.
2. Two parabolas have a common focus and their axes in opposite directions. Prove that the locus of the middle points of chords of either which touch the other is another parabola.
3. A tangent to the parabola $y^2 = 4bx$ meets the parabola $y^2 = 4ax$ in P and Q. Prove that the equation to the locus of the midpoint of PQ is $y^2 (2a - b) = 4a^2x$.
4. The line $y = m(x + a)$ meets the parabola in two points P and Q. Find the coordinates of the mid-point of PQ in terms of a and m, and prove that the locus of this point is the parabola $y^2 = 2a(x + a)$.

5. Prove that the locus of the middle point of a variable chord of the parabola $y^2 = 4ax$ such that the focal distances of its extremities are in the ratio 2:1 is

$$9(y^2 - 2ax)^2 = 4a^2(2x - a)(4x + a).$$

8. Parametric Coordinates

Sometimes it is convenient to express the coordinates of a point on the parabola in terms of a single variable. If we substitute $y = 2at$ in the equation of the parabola $y^2 = 4ax$, we obtain $x = at^2$. This is, as t varies the point $(at^2, 2at)$ always lies on the parabola $y^2 = 4ax$. Accordingly, the parabola can be represented by the equations

$$x = at^2, y = 2at.$$

These equations are called **parametric equations** of the parabola. The point $(at^2, 2at)$ is referred to the point 't' of the curve.

For the parabola $y^2 = 4ax$, one can easily have the following:

(*i*) The equation of the chord joining the points t_1 and t_2 is

$$(t_1 + t_2)y = 2(x + at_1t_2).$$

(*ii*) The equation of the tangent at the point t is

$$ty = x + at^2.$$

(*iii*) The equation of the normal at the point t is

$$tx + y = 2at + at^3.$$

8.1 Propositions on the parabola

Proposition I. *The tangent at any point of a parabola bisects the angle between the focal chord through the point and the perpendicular on the directrix from the point.*

Let $P(at^2, 2at)$ be any point on the parabola $y^2 = 4ax$.

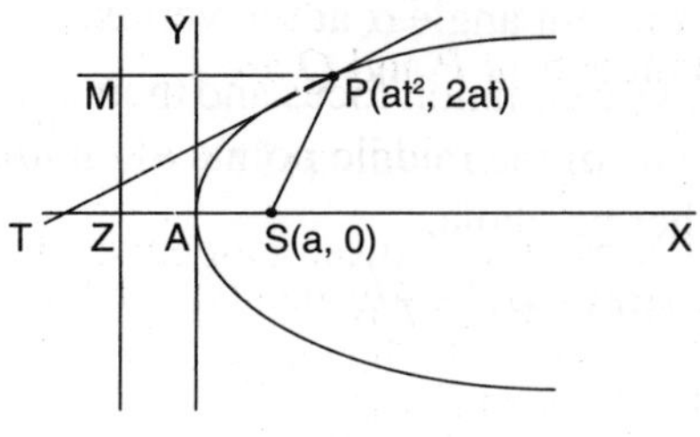

Fig. 7.11

Then, equation of the tangent PT at $P(at^2, 2at)$ is $ty = x + at^2$.

Slope of the tangent $PT = \frac{1}{t}$.

Slope of the focal chord $SP = \frac{2at - 0}{at^2 - a} = \frac{2t}{t^2 - 1}$.

Now $$\tan SPT = \frac{\frac{2t}{t^2-1} - \frac{1}{t}}{1 + \frac{2t}{(t^2-1)} \cdot \frac{1}{t}} = \frac{1}{t}.$$

Thus $$\tan PTS = \tan SPT = \frac{1}{t}$$

$\Rightarrow$ $$\angle PTS = \angle SPT.$$

But $$\angle PTS = \angle TPM.$$

Hence $$\angle TPM = \angle SPT.$$

This prove the proposition.

Proposition II. *The tangents at the extremities of a focal chord of a parabola intersect at right angles on the directrix.*

Let $P(at_1^2, 2at_1)$ and $Q(at_2^2, 2at_2)$ be the extremities of a focal chord of the parabola

$$y^2 = 4ax.$$

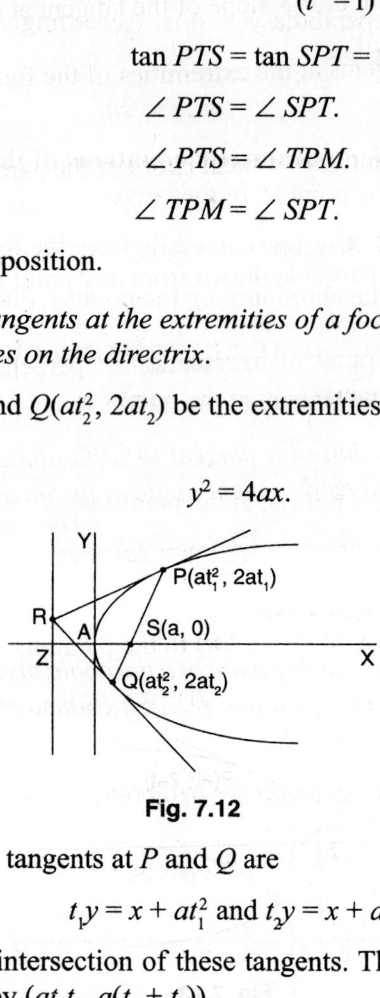

Fig. 7.12

Then, equations of the tangents at P and Q are

$$t_1 y = x + at_1^2 \text{ and } t_2 y = x + at_2^2.$$

Let R be the point of intersection of these tangents. Then, the coordinates of the point R are given by $(at_1t_2, a(t_1 + t_2))$.

Further, equation of the chord PQ is

$$(t_1 + t_2)y = 2(x + at_1t_2).$$

This chord will pass through focus $(a, 0)$ if $0 = 2(a + at_1t_2)$, *i.e.*, if $t_1t_2 = -1$.

Hence, the coordinates of the point R become $\left\{-a, \frac{a(t_1^2-1)}{t_1}\right\}$.

It is easy to see that the point R lies on the directrix of the parabola.

Further, slopes of the tangents at P and Q are $\frac{1}{t_1}$ and $\frac{1}{t_2}$, respectively.

Now, Slope of the tangent at P × slope of the tangent at Q *is* $\frac{1}{t_1}\cdot\frac{1}{t_2}=-1$.

This shows that the tangents at the extremities of the focal chord PQ intersect at right angles.

Remarks. (*i*) The extremities of a focal chord of the parabola may be taken as the point 't' and '$-\frac{1}{t}$'.

(*ii*) The tangents to a parabola drawn from any point on the directrix are at right angles.

(*iii*) The locus of the point of intersection of perpendicular tengents to a parabola is the directrix.

Proposition III. *The portion of a tangent to a parabola cut off between the directrix and the curve subtends a right angle at the focus.*

Let $P(at^2, 2at)$ be any point on the parabola

$$y^2 = 4ax. \qquad (1)$$

The equation of the tangent at $P(at^2, 2at)$ to the parabola (1) is

$$ty = x + at^2. \qquad (2)$$

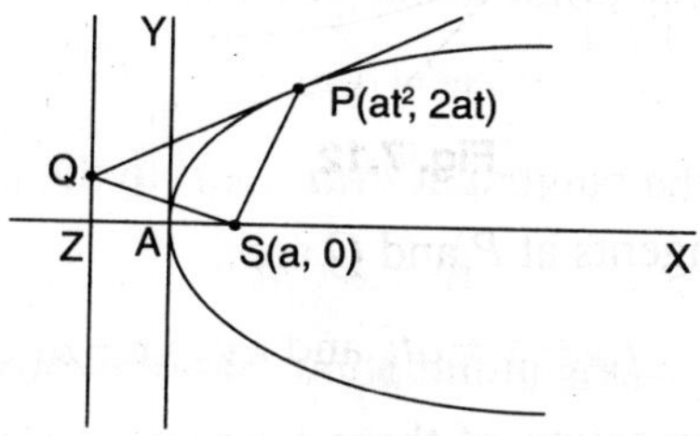

Fig. 7.13

Equation of the directrix of the parabola (1) is

$$x + a = 0. \qquad (3)$$

By solving (2) and (3) simultaneously, the coordinates of the point of intersection Q of these lines are given by $(-a, a(t^2-1)t^{-1})$.

Now $$\text{Slope of } QS = \frac{t^2-1}{2t}.$$

$$\text{Slope of } SP = \frac{2t}{t^2-1}.$$

$$\text{The slope of } QS \times \text{slope of } SP = -\left(\frac{t^2-1}{2t}\right)\times\left(\frac{2t}{t^2-1}\right) = -1.$$

This shows that QS and SP are at right angles. Hence, *the portion PQ of the tangent at P subtends a right angle at the focus.*

Ex. Prove that any tangent to parabola and the perpendicular to it from the focus meet on the tangent at the vertex.

Definition. Let the tangent and the normal at any point P on the parabola meet the axis in the points T and G, respectively. Draw PM perpendicular to the axis. Then TM is called the **subtangent** and MG the **subnormal** of the point P.

Proposition IV. *The subtangent of a point on a parabola is bisected at the vertex and the subnormal is of constant length.*

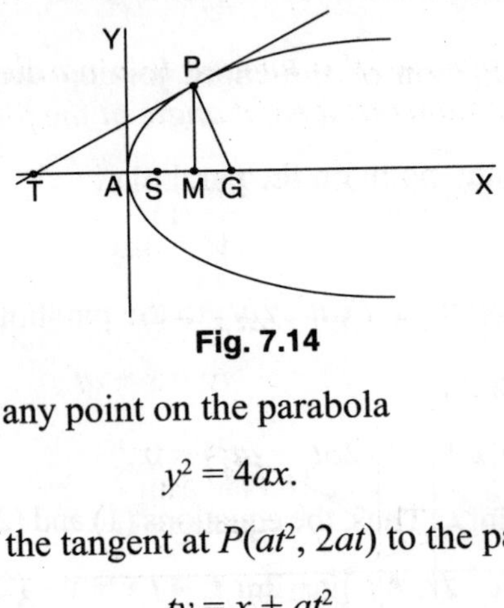

Fig. 7.14

Let $P(at^2, 2at)$ be any point on the parabola

$$y^2 = 4ax. \tag{1}$$

Then, the equation of the tangent at $P(at^2, 2at)$ to the parabola (1) is

$$ty = x + at^2. \tag{2}$$

The line (2) meets the x-axis in the point T (say) whose coordinates are given by $(-at^2, 0)$. Therefore

$$\text{Subtangent} = TM = TA + AM = at^2 + at^2 = 2at^2,$$

But $$TA = AM = at^2.$$

Hence, *the subtangent is bisected at the vertex A.*

Further, the equation of the normal at $P(at^2, 2at)$ is

$$tx + y = 2at + at^3.$$

This normal meets the x-axis in the point G (say) whose coordinates are given by $(2a + at^2, 0)$. Therefore

$$\text{Subnormal} = MG = AG - AM = 2a + at^2 - at^2 = 2a = \text{constant}.$$

8.2 Examples

Example 1. The normal at a point t_1 on the parabola $y^2 = 4ax$ meets it again at the point t_2. Prove that $t_2 = -t_1 - \dfrac{2}{t_1}$.

Solution. First Method. The equation of the normal at the point t_1 to the given parabola $y^2 = 4ax$ is

$$t_1x + y = 2at_1 + at_1^3.$$

But this normal meets the parabola again at the point t_2. Therefore

$$t_1 \cdot at_2^2 + 2at_2 = 2at_1 + at_1^3$$

$$\Rightarrow \qquad t_2 = -t_1 - \frac{2}{t_1}.$$

Second Method. Equation of the chord joining the points t_1 and t_2 on the parabola $y^2 = 4ax$ is

$$y - 2at_1 = \frac{2at_2 - 2at_1}{at_2^2 - at_1}(x - at_1^2)$$

$$\Rightarrow \qquad 2x - (t_1 + t_2)y + 2at_1t_2 = 0. \qquad (1)$$

The normal at the point t_1

$$t_1x + y - (2at_1 + at_1^3) = 0 \qquad (2)$$

passes through the point t_2. Thus, the equations (1) and (2) represent the same line.

Therefore $\qquad 2x - (t_1 + t_2)y + 2at_1t_2 \equiv t_1x + y - (2at_1 + at_1^3)$

$$\Rightarrow \qquad t_2 = -t_1 - \frac{2}{t_1}.$$

Example 2. If the point $(at^2, 2at)$ is one extremity of a focal chord of the parabola $y^2 = 4ax$, find the coordinates of the other extremity, and show that the length of the chord is $a\left(t + \dfrac{1}{t}\right)^2$.

Solution. Let $(at_1^2, 2at_1)$ be the coordinates of the other extremity of the focal chord. Then, $tt_1 = -1$. Hence, the coordinates of the other extremity are $\left(\frac{a}{t^2}, \frac{-2a}{t}\right)$. Thus

$$\text{Length of the chord} = \sqrt{\left(at^2 - \frac{a}{t^2}\right)^2 + \left(2at + \frac{2a}{t}\right)^2}$$

$$= a\left(t + \frac{1}{t}\right)^2.$$

Example 3. Prove that the locus of the points of intersection of the tangents to the parabola $y^2 = 4ax$ which intercept a fixed length l on the directrix is

$$(y^2 - 4ax)(x + a)^2 = l^2x^2.$$

Solution. Let the two tangents to the given parabola be

$$t_1y = x + at_1^2 \text{ and } t_2y = x + at_2^2.$$

Let (h, k) be the point of intersection of these tangents. Then

$$h = at_1t_2 \text{ and } k = a(t_1 + t_2).$$

These tangents cut the directrix $x + a = 0$ in the points

$$M\left(-a, \frac{at_1^2 - a}{t_1}\right) \text{ and } N\left(-a, \frac{at_2^2 - a}{t_2}\right),$$

respectively. But intercept $MN = l$ (given). Therefore

$$\sqrt{(-a+a)^2 + \left(\frac{at_2^2 - a}{-t_2} - \frac{at_1^2 - a}{t_1}\right)^2} = l$$

$$\Rightarrow (k^2 - 4ah)(h + a)^2 = l^2h^2.$$

Hence, the locus of (h, k) is $(y^2 - 4ax)(x + a)^2 = l^2x^2$.

Example 4. Prove that the locus of the point of intersection of two tangents to the parabola $y^2 = 4ax$ which with the tangent at the vertex form a triangle of constant area c^2, is the curve $(y^2 - 4ax)\,x^2 = 4c^4$.

Solution. The equations of the tangents at the points $(at_1^2, 2at_1)$ and $(at_2^2, 2at_2)$

to the given parabola are

$$t_1 y = x + at_1^2, \text{ and } t_2 y = x + at_2^2.$$

Let (h, k) be the point of intersection of these tangents. Then

$$h = at_1 t_2 \qquad \text{and} \qquad k = a(t_1 + t_2).$$

These tangents meet the tangent at the vertex (*i.e.* $x = 0$) at $(0, at_1)$ and $(0, at_2)$, respectively. Now

$$\frac{1}{2}\begin{vmatrix} h & k & 1 \\ 0 & at_1 & 1 \\ 0 & at_2 & 1 \end{vmatrix} = c^2$$

$$\Rightarrow \qquad \tfrac{1}{2}h(at_1 - at_2) = c^2$$

$$\Rightarrow \qquad ah\sqrt{(t_1 + t_2)^2 - 4t_1 t_2} = 2c^2$$

$$\Rightarrow \qquad a^2h^2\left(\frac{k^2}{a^2} - \frac{4h}{a}\right) = 4c^4$$

$$\Rightarrow \qquad h^2(k^2 - 4ah) = 4c^4.$$

Hence, the locus of (h, k) is $(y^2 - 4ax)\, x^2 = 4c^4$.

Example 5. Find the locus of the point of intersection of normals at the ends of a focal chord of the parabola $y^2 = 4ax$.

Solution. Let $(at_1^2, 2at_1)$ and $(at_2^2, 2at_2)$ be the ends of the focal chord. Then $t_1 t_2 = -1$. The equations of the normals at these points are

$$t_1 x + y = 2at_1 + at_1^3, \qquad (1)$$

$$t_2 x + y = 2at_2 + at_2^3. \qquad (2)$$

Let (h, k) be the point of intersection of (1) and (2). Then

$$h = 2a + a\,(t_1^2 + t_2^2 + t_1 t_2) \text{ and } k = -at_1 t_2(t_1 + t_2) = a(t_1 + t_2).$$

Now $\qquad h = 2a + a\,\{(t_1 + t_2)^2 - t_1 t_2\}$

$$\Rightarrow \qquad h - 2a = a\left(\frac{k^2}{a^2} + 1\right)$$

$$\Rightarrow \qquad k^2 = a\,(h - 3a).$$

Hence, the locus of (h, k) is $y^2 = a\,(x - 3a)$.

Example 6. Prove that the semi-latus rectum is a harmonic mean between the segments of any focal chord of a parabola.

Solution. Let $P(at_1^2, 2at_1)$ and $Q(at_2^2, 2at_2)$ be the ends of any focal chord of the parabola $y^2 = 4ax$. Then $t_1t_2 = -1$. Focus of the parabola is $S(a, 0)$. Now

$$SP = \sqrt{(at_1^2 - a)^2 + (2at_1 - 0)^2} = a(t_1^2 + 1),$$

and similarly $$SQ = a(t_2^2 + 1).$$

Therefore, the harmonic mean of SP and SQ

$$= \frac{2SP \cdot SQ}{SP + SQ} = \frac{2a^2(t_1^2 + 1)(t_2^2 + 1)}{a(t_1^2 + 1) + a(t_2^2 + 1)}$$

$$= \frac{2a(t_1^2 + t_2^2 + 2)}{t_1^2 + t_2^2 + 2} \qquad \because t_1t_2 = -1$$

$= 2a =$ semi-latus rectum.

Exercises

1. Find the equation of the chord joining two points $(at_1^2, 2at_1)$ and $(at_2^2, 2at_2)$ on the parabola $y^2 = 4ax$. Prove that this chord will pass through the focus of the parabola if $t_1t_2 = -1$.
2. Prove that the tangent drawn at one extremity of a focal chord of the parabola $y^2 = 4ax$ is parallel to the normal drawn at the other extremity.
3. Find the locus of the point of intersection of two normals to the parabola $y^2 = 4ax$ which are at right angles to one another.
4. Prove that the locus of the point of intersection of two tangents to a parabola which intercept a given distance $4c$ on the tangent at the vertex is an equal parabola.
5. Prove that the locus of the middle points of chords of the parabola $y^2 = 4ax$ which are of given length l is $(y^2 - 4ax)(y^2 + 4a^2) + a^2l^2 = 0$.
6. Prove that the tangents at the extremities of a focal chord of a parabola intersect at right angles on the directrix.
7. Find the equation of the normal to $y^2 = 4ax$ at the point $(at^2, 2at)$ and find the point where this normal meets the parabola again.
8. Find the locus of the mid-point of the chord QR of the parabola $y^2 = 4ax$ when the tangents at its extremities Q and R are perpendicular to each other. Also, find the locus of the point of intersection of the normals to the parabola at Q and R.

9. Prove that if the difference of the squares of the perpendiculars on a moving line from two fixed points is constant, the line touches a fixed parabola.

10. If the perpendiculars be drawn on any tangent to a parabola from two fixed points on the axis, which are equidistant from the focus, prove that the difference of their squares is constant.

11. The normals at two points P, Q on the parabola $y^2 = 4ax$ intersect on the curve. Prove that the ordinates of P, Q are the roots of the equation $y^2 + ky + 8a^2 = 0$.

12. Two lines AP, AQ are drawn through the vertex of a parabola at right angles to one another, meeting the curve in P, Q. Prove that the line PQ cuts the axis in a fixed point.

13. If the circle $x^2 + y^2 + 2gx + 2fy + c = 0$ cut the parabola in four points, prove that the algebraic sum of the ordinates of those points will be zero.

14. If PSQ be a focal chord of a parabola, and PA meets the directrix in M; prove that MQ will be parallel to the axis of the parabola.

15. Prove that if two tangents to a parabola intercept a constant length on any fixed tangent, the locus of their point of intersection is another equal parabola.

16. From any point on the latus rectum of a parabola perpendiculars are drawn to the tangents at its extremities. Prove that the line joining the feet of these perpendiculars touches the parabola.

17. Two parabolas have the same axis, tangents are drawn from points on the first to the second. Prove that the middle points of the chords of contact with the second lie on a fixed parabola.

18. If the normals at two points on a parabola intersect on a curve, the line joining the points will pass through a fixed point on the axis.

19. If through a fixed point any chord of a parabola be drawn, and normals be drawn at the ends of the chord. Prove that the locus of the point of intersection of the normals is another parabola.

20. Prove that the locus of the point of intersection of the tangents to the parabola $y^2 = 4ax$ at $(at^2, 2at)$, $\left(\frac{a}{t^2}, \frac{2a}{t}\right)$ is a line parallel to y-axis.

21. If a chord of the parabola $y^2 = 4ax$ subtends a right angle at the vertex, the tangents at its extremities meet on the line $x + 4a = 0$.

22. The normal of the parabola $y^2 = 4ax$ at a point $P(at^2, 2at)$ makes an angle θ with the x-axis and meets the parabola again at Q. Prove that

$$PQ = 4a \sec\theta \operatorname{cosec}^2\theta.$$

23. Prove that, if the difference of the ordinates of two points on a parabola

is constant, then the locus of the point of intersection of the tangents at these points is an equal parabola.

24. Find the locus of the point of intersection of the normals at two points on a parabola when the chord joining them subtends a right angle at the vertex.

25. Prove that the length of the intercept on the normal at the point $(at^2, 2at)$ made by the circle which is described on the focal distance of the given point as diameter is $a\sqrt{1+t^2}$.

26. Prove that the locus of the middle points of all tangents drawn from points on the directrix to the parabola is $y^2(2x + a) = a(3x + a)^2$.

27. Two equal parabolas have the same focus and their axes are at right angles. A normal to one is perpendicular to a normal to the other. Prove that the locus of the point of intersection of these normals is another parabola.

28. Prove that the area of the triangle formed by three points on a parabola is twice the area of the triangle formed by the tangents at these points.

29. Prove that the circle described on any focal chord of a parabola as diameter touches the directrix.

30. A circle on any focal chord of a parabola as diameter cuts the curve again in P and Q. Prove that PQ passes through a fixed point.

31. The normals at the extremities of a chord of the parabola $y^2 = 4ax$ meet on the parabola. Prove that the middle point of the chord lies on the parabola $y^2 = 2a(x + 2a)$.

9. Diameter

A line parallel to the axis of the parabola is called a **diameter** *of the parabola.*

As we have seen in § 7.2 that the locus of the middle points of a system of parallel chords to a parabola is a line parallel to the axis of the parabola, it is a diameter of the parabola. Thus

$$y = \frac{2a}{m}$$

is a diameter of parabola $y^2 = 4ax$, where m is the slope of the parallel chords of the parabola. In fact, $y = \frac{2a}{m}$ represents the family of diameters of the parabola $y^2 = 4ax$, m being the parameter.

On the other hand, it is noted that any line $y = k$ parallel to x-axis (the axis of the parabola) represents the locus of the middle points of a system of parallel chords having slope $m = \frac{2a}{k}$. Therefore, the diameter of a parabola can alternatively be defined as *the locus of the middle points of a system of parallel chords of the parabola.*

9.1 Propositions on diameters

Proposition I. *To prove that the tangent at the extremity of a diameter of a parabola is parallel to the chords which are bisected by that diameter.*

Let the equation of the parabola be $y^2 = 4ax$.

Let $y = mx + c$ be a member of a system of parallel chords of the parabola. Then equation of the diameter is $y = \dfrac{2a}{m}$.

The coordinates of the point of intersection of the parabola and the diameter are $\left(\dfrac{a}{m^2}, \dfrac{2a}{m}\right)$. The equation of the tangent at $\left(\dfrac{a}{m^2}, \dfrac{2a}{m}\right)$ to the parabola is

$$y \cdot \frac{2a}{m} = 2a\left(x + \frac{a}{m^2}\right), \text{ i.e., } y = mx + \frac{a}{m}.$$

But the line $y = mx + \dfrac{a}{m}$ is parallel to each member of the system of parallel chords. Hence, the proposition follows.

Proposition II. *To prove that the tangents at the ends of any chord of a parabola meet on the diameter which bisects the chord.*

Let the equation of the parabola be

$$y^2 = 4ax. \tag{1}$$

Let

$$y = mx + c \tag{2}$$

be the equation of any chord of the parabola (1).

Let (x_1, y_1) be the point of intersection of the tangents at the extremities of the chord (2). Then equation of the chord of contact of (x_1, y_1) with respect to the parabola (1) is

$$yy_1 = 2a(x + x_1). \tag{3}$$

But equations (2) and (3) represent the same line. Thus, we have

$$mx - y + c \equiv 2ax - yy_1 + 2ax_1$$

$$\Rightarrow \quad \frac{2a}{m} = \frac{-y_1}{-1} = \frac{2ax_1}{c}$$

$$\Rightarrow \quad x_1 = \frac{c}{m}, y_1 = \frac{2a}{m}.$$

Hence, locus of (x_1, y_1) is $y = \dfrac{2a}{m}$, which is the equation of the diameter.

Hence, the proposition follows.

Exercises

1. If the diameter through any point O of a parabola meet any chord in P, and the tangents at the ends of that chord meet the diameter in Q, Q'. Prove that $OP^2 = OQ \cdot OQ'$.
2. The normal at a point P of a parabola meets the curve again in Q, and T is the pole of PQ. Prove that T lies on the diameter passing through the other end of the focal chord passing through P, and that PT is bisected by the directrix.
3. Prove that a diameter intersects a parabola at the point of contact of the conjugate tangent line.
4. Prove that the normal at any point of a parabola bisects the angle between the focal chord and the diameter through that point.

Miscellaneous Exercises

1. Find the equations of the two parabolas whose latus rectum is 6 and the axis and the tangent at the vertex are the lines whose equations are

$$3x + 4y + 1 = 0,\ 4x - 3y = 0.$$

respectively.

2. A parabola is drawn to pass through A and B, the ends of a diameter of a given circle of radius a, and to have as directrix a tangent to a concentric circle of radius b; the axes being AB and a perpendicular diameter. Prove that the locus of the focus of the parabola is

$$\frac{x^2}{b^2} + \frac{y^2}{b^2 - a^2} = 1.$$

3. Prove that the locus of a point, which moves, so that its distance from a fixed line is equal to the length of the tangent drawn from it to a given circle, is a parabola. Find the position of the focus and the directrix.
4. Prove that the parabolas $y^2 = 4ax$ and $x^2 = 4by$ cut one another at an angle

$$\tan^{-1} \frac{3a^{1/3}b^{1/3}}{2(a^{2/3} + b^{2/3})}.$$

5. Two parabolas have a common focus S and axes in opposite directions. Prove that they intersect at right angles. Prove also that the distance

between their vertices is equal to the distance, of either point of intersection from S.

6. Find the condition that the line $lx + my + n = 0$ should be a normal to the parabola $y^2 = 4ax$.

7. Prove that the circle circumscribing the triangle formed by any three tangents to a parabola passes through the focus.

8. If the normals to the parabola $y^2 = 4ax$ at three points $P \cdot Q$ and R meet in a point M and S is the focus, prove that $SP \cdot SQ \cdot SR = a \cdot SM^2$.

9. Prove that the shortest normal chord of the parabola $y^2 = 4ax$ is $6a\sqrt{3}$ and that its inclination to the axis is $\tan^{-1}\sqrt{2}$.

10. If from the vertex of a parabola a pair of chords is drawn at right angles to one another, and with these chords as adjacent sides a rectangle is constituted, prove that the locus of the outer corner of the rectangle is a parabola.

11. Prove that the distance between a tangent to the parabola $y^2 = 4ax$ and the parallel normal is a cosec θ $\sec^2\theta$, where θ is the inclination of either to the axis of the parabola.

12. A line PQ meets the parabola $y^2 = 4ax$ in R such that PQ is bisected at R. If the coordinates of P are (x_1, y_1), prove that the locus of Q is the parabola

$$(y + y_1)^2 = 8a(x + x_1).$$

13. The normals at the points Q, R on the parabola $y^2 = 4ax$ meet it again at the point P. Prove that the locus of the

    (*i*) circumcentre of the triangle PQR is the parabola $2y^2 = a(x - a)$;

    (*ii*) orthocentre of the triangle PQR is the parabola $y^2 = a(x + 6a)$.

14. Prove that the locus of the point of intersection of the normals at the ends of a system of parallel chords of a parabola is a line, which is normal to the curve.

15. The normals at P, Q, R on the parabola $y^2 = 4ax$ meet in a point on the line $y = k$. Prove that the sides of the triangle PQR touch the parabola $x^2 = 2ky$.

16. Tangents are drawn to the parabola $y^2 = 4ax$ at points whose abscissae are in the ratio $m^2 : 1$. Prove that the locus of their points of intersection is the curve $y^2 = (m^{1/2} + m^{-1/2})^2 ax$.

17. Prove that the length of any focal chord of a parabola is four times the distance of the focus from the point where the diameter bisecting the chord meets the parabola.

18. If P, Q, R are the feet of the normals drawn from a point to the parabola $y^2 = 4ax$, prove that

    (*i*) the centroid of the triangle PQR lies on the axis of the parabola;

    (*ii*) the circumcircle of the triangle PQR passes through the vertex of the parabola.

19. Find the coordinates of the point of intersection of the tangents $y = mx + \frac{a}{m}$, $y = m'x + \frac{a}{m'}$. Prove that the locus of their intersection is a line whenever mm' is constant; and that, when $mm' = -1$, this line is the directrix.

20. The perpendicular TN from any point T on its polar with respect to a parabola meets the axis in M. Prove that if $TN.TM$ is constant the locus of T is a parabola.

21. If the normals at two points of a parabola be inclined to the axis at angles θ, ϕ such that $\tan\theta \tan\phi = 2$. Prove that the locus of their intersection is a parabola.

22. Prove that the locus of the feet of the perpendiculars from the point $(a, 0)$ to the normals of the parabola is $y = a(x - a)$.

23. From a fixed point P on the parabola $y^2 = 4ax$ chords PQ, PQ' are drawn making equal angles ϕ with the tangent at P. Prove that for all values of ϕ, QQ' will pass through the same point R. Prove further that if P moves along the parabola, the locus of R is $(x + 2a)y^2 + 4a^3 = 0$.

24. A parabola is drawn with its axis parallel to a fixed line and passes through a fixed point and touches a given line. Prove that the locus of its vertex is a parabola.

25. From any point P on a parabola $y^2 = 4ax$, lines PQ and PR are drawn normal to the parabola at Q and R, respectively. Prove that QR passes through the point $(-2a, 0)$.

26. A chord PQ of a parabola is normal at P, and subtends a right angle at the vertex. If S is the focus, prove that $SQ = 3SP$.

27. If the variable chord PQ of the parabola $y^2 = 4ax$ subtends a right angle at the vertex, prove that it cuts the axis at a fixed point. Prove also that the tangents at P and Q meet on a fixed line.

28. If two points of the axis of a parabola are equidistant from the focus, prove that the difference of the squares of their distances from any tangent is independent of its position.

29. Prove that the orthocentre of any triangle formed by three tangents to a parabola lies on the directrix.

30. If the normals at P and Q meet on the parabola, prove that the point of intersection of the tangents at P and Q lies either on a certain line, which is parallel to the tangent at the vertex, or on the curve whose equation is $(x + 2a)y^2 + 4a^3 = 0$.

31. Prove that the locus of the centre of the circle, which passes through the vertex of a parabola and through its intersections with a normal chord, is the parabola $2y^2 = a(x - a)$.

32. The sides of a triangle touch a parabola, and two of its angular points lie on another parabola with its axis in the same direction. Prove that the locus of the third angular point is another parabola.

33. PP' is any one of a system of a parallel chords of a parabola, O is a point on PP' such that $PO \cdot OP'$ is constant. Prove that the locus of O is a parabola.

34. If a tangent to a parabola cut two given parallel lines in P, Q, the locus of the point of intersection of the other tangents from P, Q to the curve will be a parabola.

35. TP, TQ are tangents to a parabola, and p_1, p_2, p_3 are the lengths of the perpendiculars from P, T, Q respectively on any other tangent to the curve. Prove that $p_1 p_3 = p_2^2$.

36. A circle cuts the parabola $y^2 = 4ax$ at right angles and passes through the focus. Prove that its centre lies on the curve $y^2(2x + a) = a(3x + a)^2$.

37. The product of the tangents drawn from a point P to the parabola $y^2 = 4ax$ is equal to the product of the focal distance of P and the latus rectum. Prove that the locus of P is the parabola $y^2 = 4a(x + a)$.

38. Prove that if a chord of a parabola meets the axis in a fixed point, the normals at its extremities intersect on another fixed parabola. When does this parabola coincide with the original curve?

The Ellipse

8 CHAPTER

1. Definition

*An **ellipse** is the locus of a point which moves such that its distance from a fixed point is in a constant ratio (<1) to the perpendicular distance of it from a fixed line.*

The fixed point is called the **focus**, constant ratio the **eccentricity** and the fixed line the **directrix**.

1.1 Standard equation of an ellipse. *To find the equation of an ellipse in standard form.*

Let S be the focus, ZK the directrix and SZ the perpendicular from S to the directrix. Let e be the eccentricity of the ellipse. Let A divide SZ such that

$$SA : AZ = e : 1.$$

Then, A lies on the curve. Since, $e < 1$, there is another point A' on ZS produced such that

$$SA' : ZA' = e : 1$$

Thus $SA = e.AZ$ and $SA' = e.\ ZA'$.

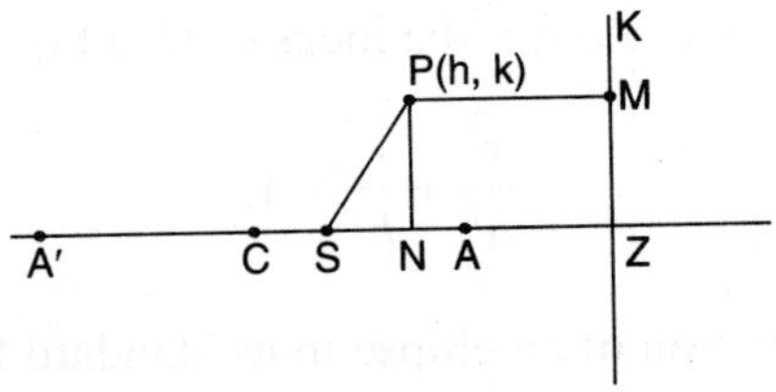

Fig. 8.1

Let C be the middle point of AA' and let $AA' = 2a$. Then

$$SA + SA' = e(AZ + ZA')$$

$$\Rightarrow \qquad AA' = e.2CZ$$

$$\Rightarrow \qquad 2a = e.2CZ$$

$$\Rightarrow \qquad CZ = \frac{a}{e}.$$

Further

$$SA' - SA = e(ZA' - AZ)$$

$$\Rightarrow \qquad SA' + SA - 2SA = e.AA'$$

$$\Rightarrow \qquad 2CA - 2SA = e.AA'$$

$$\Rightarrow \qquad CS = ae.$$

Let us take C to be the origin, CA the x-axis and a line through C perpendicular to CA the y-axis.

Let $P(h, k)$ be any point on the locus (ellipse). Draw PM perpendicular from P to the directrix ZK. Then, by definition of ellipse, we have

$$SP = e.PM.$$

But $$SP = \sqrt{(h - ae)^2 + k^2}$$

and $$PM = NZ = CZ - CN$$

$$= \frac{a}{e} - h.$$

Therefore $$(h - ae)^2 + k^2 = e^2 \left(\frac{a}{e} - h\right)^2$$

$$\Rightarrow \qquad \frac{h^2}{a^2} + \frac{k^2}{a^2(1 - e^2)} = 1.$$

Since, $e < 1$, $1 - e^2$ is positive, so that we may write

$$a^2(1 - e^2) = b^2$$

and choose b to be positive. Hence, the locus of (h, k) is

$$\frac{x^2}{a^2} + \frac{y^2}{b^2} = 1.$$

This is the required equation of an ellipse in its standard form.

1.2 General equation of an ellipse. *To find the general equation of an ellipse whose focus is* (α, β) *and the directrix is the line* $ax + by + c = 0$.

Let e be the eccentricity of the ellipse. Let $P(h, k)$ be any point on the locus (ellipse). Then, by definition

$$SP = e.PM.$$

$$\Rightarrow \qquad (h-\alpha)^2+(k-\beta)^2=e^2\left(\frac{ah+bk+c}{\sqrt{a^2+b^2}}\right)^2$$

Hence, locus of (h, k) is

$$(x-\alpha)^2+(y-\beta)^2=\frac{e^2(ax+by+c)^2}{a^2+b^2}.$$

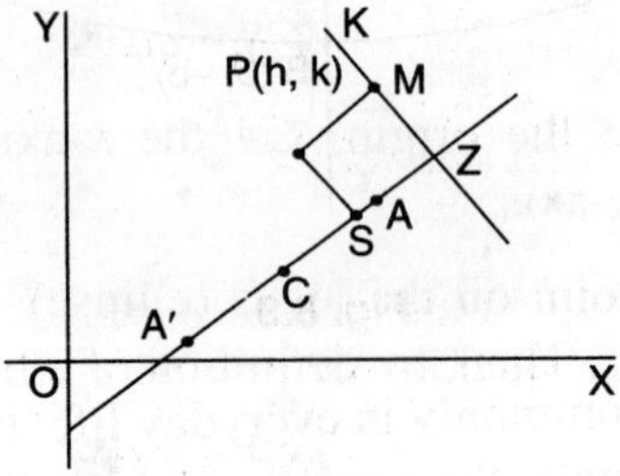

Fig. 8.2

Ex. Find the equation of an ellipse whose focus is $(-1, 1)$, directrix is $x-y+3=0$ and eccentricity is $\frac{1}{2}$.

1.3 Shape of an ellipse. *To trace the ellipse*

$$\frac{x^2}{a^2}+\frac{y^2}{b^2}=1.$$

In order to sketch the graph of the given ellipse we notice the following:

(*i*) The curve is symmetrical about both axes and the origin since if (x, y) lies on the curve, the points $(-x, y)$, $(x, -y)$ and $(-x, -y)$ also lie on the curve.

(*ii*) The curve meets the x-axis in the points $A(a, 0)$, $A'(-a, 0)$ and the y-axis in the points $B(0, b)$, $B'(0, -b)$.

(*iii*) Writting the equation of the ellipse in either of the forms

$$y=\pm\frac{b}{a}\sqrt{a^2-x^2} \text{ or } x=\pm\frac{a}{b}\sqrt{b^2-y^2},$$

we see that y is real if and only if $-a \leqslant x \leqslant a$ whereas x is real if and only if $-b \leqslant y \leqslant b$. Thus, the curve is limited and bounded. In fact, it lies entirely inside the rectangle $\{(x, y) : -a \leqslant x \leqslant a, -b \leqslant y \leqslant b\}$.

Therefore, in view of the points noticed above, we need only to examine its shape in the first quadrant. We find that as x increases from 0 to a, y decreases from b to 0. Hence, the ellipse has the shape as shown in Fig. 8.3.

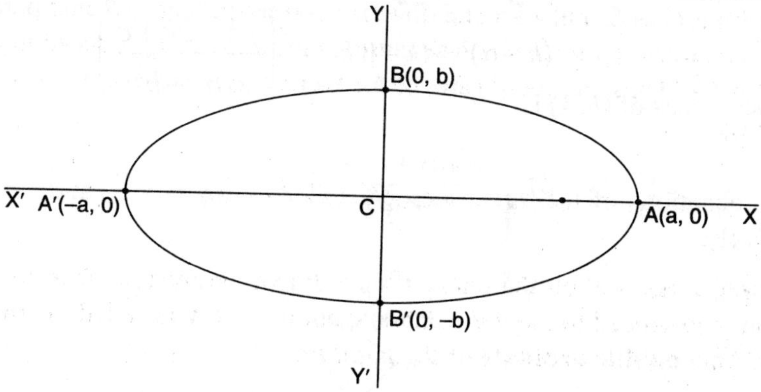

Fig. 8.3

Note. The ellipse occurs commonly in everyday life. It is the shape seen when any circle is viewed obliquely, the section of a circular cylinder by any plane not perpendicular to the axis of the cylinder. It is also the general path of any planet, including our own, in its orbit round the sun.

The **points** A and A' are called the **vertices** of the ellipse.

The two lines AA' and BB' with respect to which the curve is symmetrical are called the axes of the ellipse. The axis AA' is of length $2a$ and is called the **major axis** whereas the axis BB' is of length $2b$ and is called the **minor axis** since $a > b$. □

Since, the curve is symmetrical about the origin, the point C (the origin) bisects every chord through it. Therefore, the point C is referred to the **centre** of the ellipse.

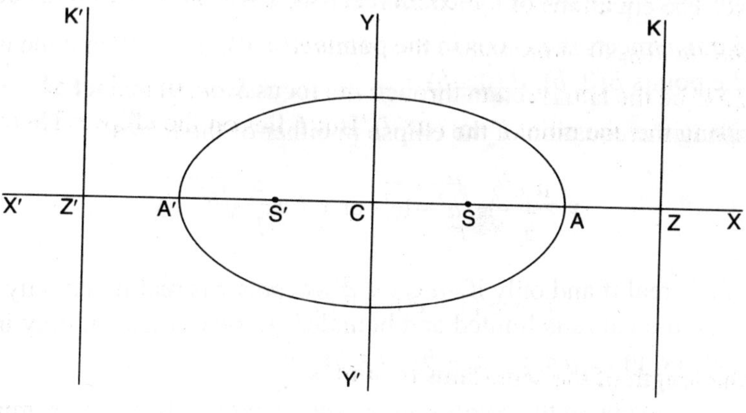

Fig. 8.4

Further, the symmetry of the curve about its minor axis (y-axis) verifies that there must be a second focus S' situated on the major axis at the same distance from C as S, and a second directrix corresponding to S' and parallel to the original directrix which would be cutting the major axis produced in Z' such that $CZ' = CZ$. Thus, the coordinates of the foci S and S' are $(ae, 0)$ and $(-ae, 0)$, respectively.

The equations of the directrices ZK and $Z'K'$ are $x = \dfrac{a}{e}$ and $x = -\dfrac{a}{e}$, respectively. □

If from a point P on the curve PN be drawn perpendicular to the major axis, and is produced to meet the curve again in P', PN is called the **ordinate** and PNP' the **double ordinate** of the point P.

A double ordinate through a focus is called a **latus rectum***.

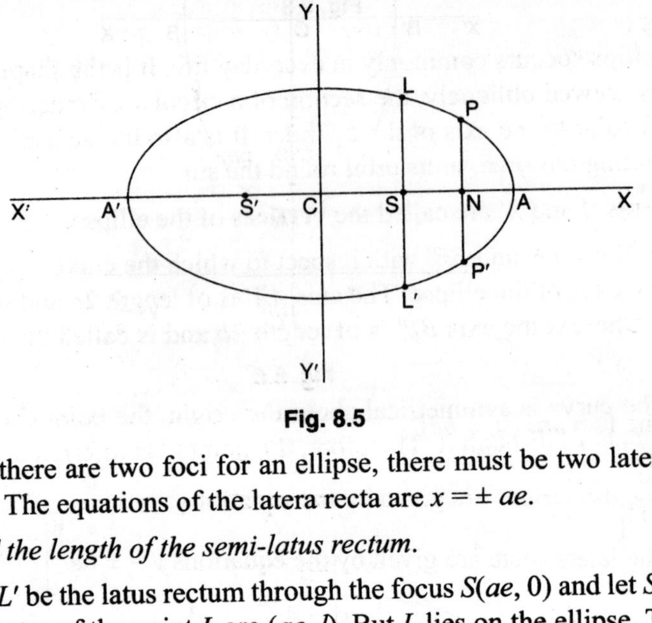

Fig. 8.5

Since, there are two foci for an ellipse, there must be two latera recta for the ellipse. The equations of the latera recta are $x = \pm\, ae$.

To find the length of the semi-latus rectum.

Let LSL' be the latus rectum through the focus $S(ae, 0)$ and let $SL = l$. Then the coordinates of the point L are (ae, l). But L lies on the ellipse. Therefore

$$\frac{a^2e^2}{a^2} + \frac{l^2}{b^2} = 1$$

$$\Rightarrow \qquad l^2 = b^2(1 - e^2) = \frac{b^4}{a^2}.$$

Hence, the length of the semi-latus rectum is $\dfrac{b^2}{a}$.

Note. If $b > a$, then for the ellipse $\dfrac{x^2}{a^2} + \dfrac{y^2}{b^2} = 1$, we note the following:

* In general latus rectum of an ellipse is the chord through a focus perpendicular to the major axis.

(*i*) The major axis is along y-axis and is of length $2b$.

(*ii*) The minor axis is along x-axis and is of length $2a$.

(*iii*) The eccentricity is given by $a^2 = b^2(1 - e^2)$

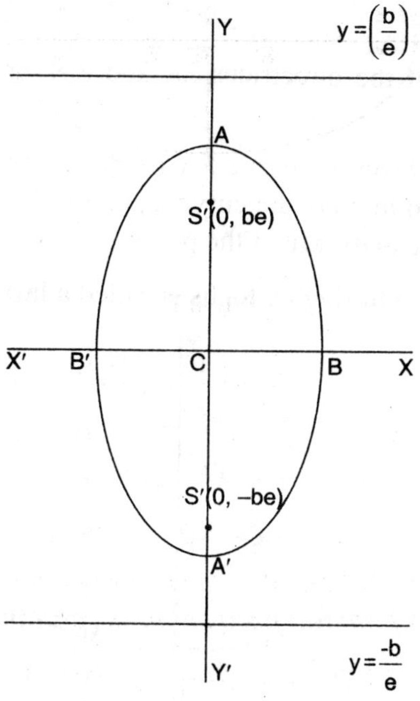

Fig. 8.6

(*iv*) The foci are $(0, \pm\, be)$.

(*v*) The directrices are given by the equations $y = \pm\, \dfrac{b}{e}$.

(*vi*) The latera recta are given by the equations $y = \pm\, be$.

(*vii*) The length of the semi-latus rectum is $\dfrac{a^2}{b}$.

1.4 Focal distances of a point. The distances of a point P on the ellipse from foci, *viz.*, SP and $S'P$ are called **focal distances** of P.

To prove that the sum of the focal distances of any point P on the ellipse $\dfrac{x^2}{a^2} + \dfrac{y^2}{b^2} = 1$, *from foci S and S′, is constant and is equal to 2a.*

Let (x, y) be the coordinates of the point P. We know that

$$SP = e.PM = e.NZ = e.(CZ - CN)$$

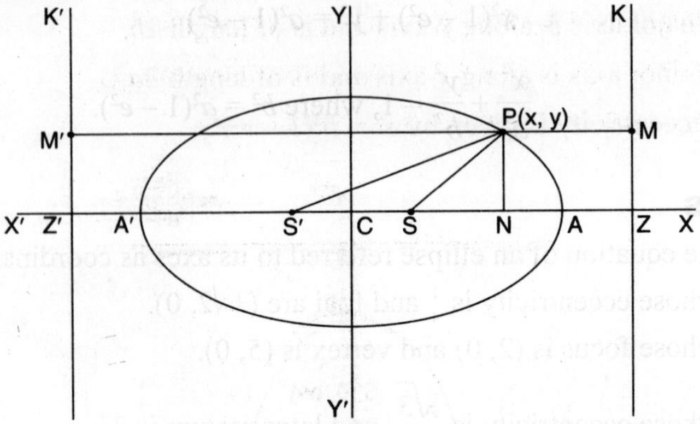

Fig. 8.7

$$= e \cdot \left(\frac{a}{e} - x\right) = a - ex.$$

and similarly $\qquad S'P = e.M'P = a + ex.$

Thus $\qquad SP + S'P = 2a.$

An ellipse may also be defined as *the locus of a point which moves so that the sum of its distances from two fixed points is constant.*

To find the equation of the curve from the above definition.

Let the distance between the two fixed points S and S' be $2ae$.

Let C be the middle point of SS'. Taking CS as x-axis and a line perpendicular to it through C as y-axis. Let $P(h, k)$ be the moving point so that

$$SP + S'P = \text{constant} = 2a \text{ (say)}.$$

But $\qquad SP = \sqrt{(h-ae)^2 + k^2}$ and $S'P = \sqrt{(h+ae)^2 + k^2}$.

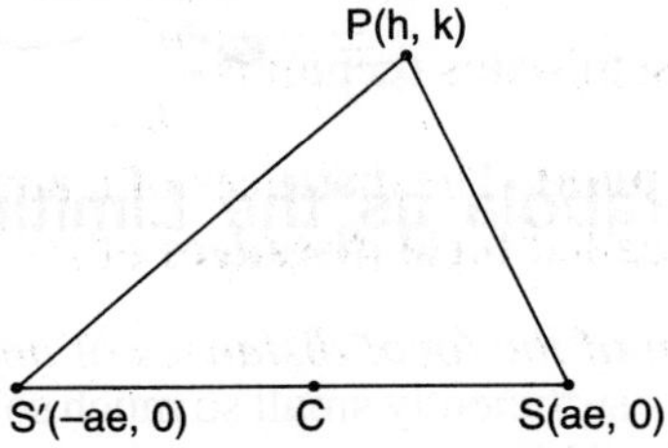

Fig. 8.8

Therefore $\qquad \sqrt{(h-ae)^2 + k^2} + \sqrt{(h+ae)^2 + k^2} = 2a$

$\Rightarrow \qquad h^2(1 - e^2) + k^2 = a^2(1 - e^2).$

Hence, the locus of (h, k) is

$$x^2(1-e^2)+y^2=a^2(1-e^2)$$

$$\Rightarrow \qquad \frac{x^2}{a^2}+\frac{y^2}{b^2}=1, \text{ where } b^2=a^2(1-e^2).$$

Exercises

1. Find the equation of an ellipse referred to its axes as coordinate axes

(*i*) whose eccentricity is $\frac{1}{4}$ and foci are $(\pm\sqrt{2}, 0)$.

(*ii*) whose focus is (2, 0) and vertex is (5, 0).

(*iii*) whose eccentricity is $\frac{\sqrt{3}}{2}$ and latus rectum is 2.

(*iv*) which passes through (3, 5) and $\left(7, \frac{5}{3}\right)$.

(*v*) whose end of the minor axis is (5, 0) and length of the latus rectum is $\frac{50}{13}$.

2. Find the eccentricities, foci, directrices and latera recta of the following ellipses:

(*i*) $4x^2+9y^2=144$. (*ii*) $x^2\tan^2\alpha+y^2\sec^2\alpha=1$.

(*iii*) $\frac{x^2}{169}+\frac{y^2}{25}=1$. (*iv*) $x^2+4y^2=9$.

(*v*) $25x^2+4y^2=100$.

3. Find the locus of a point which moves so that the sum of its distances from (–4, 3) and (4, 3) is 12.

4. The perimeter of a triangle is 20, and the points (–2, –3) and (–2, 3) are two of the vertices of it. Find the equation of the locus of the third vertex.

2. Circle and Parabola as the Limiting Cases of the Ellipse

We shall see that by taking e sufficiently small so much so that by taking $e \to 0$ and keeping a constant the ellipse tends to become a circle while taking e sufficiently near to 1 and moving the centre off to a great distance in such a way that the vertex A' and focus at S'' remains unchanged the ellipse becomes the parabola.

2.1 The circle as a limiting case of the ellipse. It is clear that for $b = a$, the equation of the ellipse $\frac{x^2}{a^2}+\frac{y^2}{b^2}=1$ takes the form $x^2 + y^2 = a^2$ which is a circle of radius a, and in this case the eccentricity $e = 0$.

Further, we note that if e is made sufficiently small keeping a unchanged, the *quantity* ae becomes very small and so the foci ($\pm ae$, 0) approach the centre of the ellipse. Also b^2 [$=a^2(1-e^2)$] approaches to a^2. Thus, as the foci approach the centre, the ellipse tends to become more circular in appearance, and when e is sufficiently small, rather, nearly equal to zero the ellipse becomes almost a circle. Hence, we say that a circle of radius a is the limiting case of an ellipse whose major axis is of length $2a$ and eccentricity tends to zero.

2.2 The parabola as the limiting case of the ellipse. Shifting the origin to the vertex $A'(-a, 0)$, the equation of the ellipse

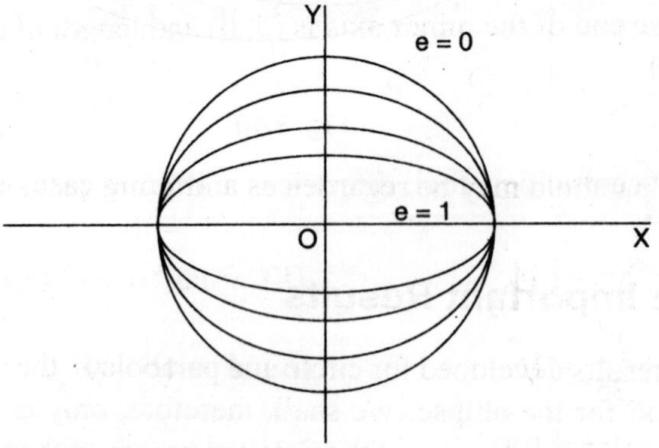

Fig. 8.9

$$\frac{x^2}{a^2}+\frac{y^2}{b^2}=1$$

takes the form

$$\frac{(X-a)^2}{a^2}+\frac{Y^2}{b^2}=1$$

i.e.
$$\frac{X^2}{a}-2X+\frac{Y^2}{a(1-e^2)}=0. \quad (1)$$

Now $A'S' = a(1 - e)$. Let us denote the distance $A'S'$ by d so that $a = \frac{d}{1-e}$. Thus, the equation (1) becomes

$$\frac{X^2(1-e)}{d}-2X+\frac{Y^2}{d(1+e)}=0.$$

Now, suppose A' and S' remain fixed so that d remains constant and finite. Make e sufficiently near to 1 so that a becomes large and large, *i.e.*, the centre moves off very far. In these circumstances the above equation approximates to $Y^2 = 4dX$. This represents a parabola. Thus, under the limiting circumstances described above, the ellipse approximates to a parabola.

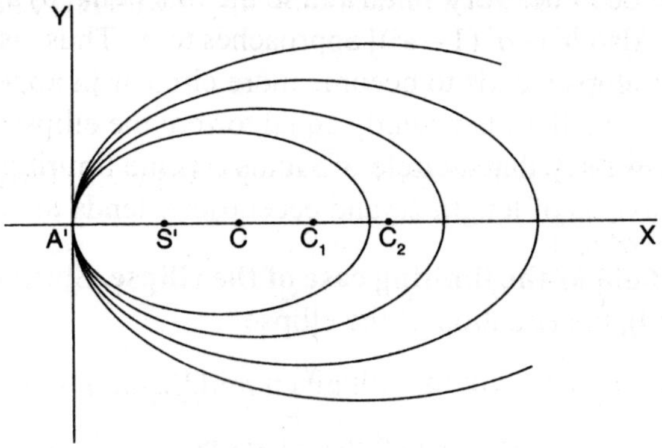

Fig. 8.10

Hence, a parabola may be regarded as a limiting case of the ellipse.

3. Some Important Results

Many of the results developed for circle and parabola in the preceding chapters also hold good for the ellipse. We shall, therefore, only enumerate them and the reader is advised to work out the detailed proofs of them.

Let the equation of the ellipse be

$$\frac{x^2}{a^2} + \frac{y^2}{b^2} = 1.$$

(*i*) The equation of the tangent at the point (x_1, y_1) is $\dfrac{xx_1}{a^2} + \dfrac{yy_1}{b^2} = 1.$

(*ii*) The line $y = mx + c$ is tangent to the ellipse if $c = \sqrt{a^2m^2 + b^2}$.

(*iii*) The equation of the normal at (x_1, y_1) to the ellipse is

$$\frac{x - x_1}{\dfrac{x_1}{a^2}} = \frac{y - y_1}{\dfrac{y_1}{b^2}}.$$

(*iv*) The equation of the pair of tangents from (x_1, y_1) to the ellipse is

$$T^2 = SS_1,$$

where $S \equiv \frac{x^2}{a^2}+\frac{y^2}{b^2}-1$, $S_1 \equiv \frac{x_1^2}{a^2}+\frac{y_1^2}{b^2}-1$ and $T \equiv \frac{xx_1}{a^2}+\frac{yy_1}{b^2}-1$.

(*v*) The equation of the chord of contact of the tangents drawn from (x_1, y_1) to the ellipse is $\frac{xx_1}{a^2}+\frac{yy_1}{b^2}=1$.

(*vi*) The equation of the polar of the point (x_1, y_1) with respect to the ellipse is

$$\frac{xx_1}{a^2}+\frac{yy_1}{b^2}=1.$$

(*vii*) The pole of the line $lx + my + n = 0$ with respect to the ellipse is

$$\left(-\frac{a^2l}{n}, -\frac{b^2m}{n}\right).$$

(*viii*) The equation of the chord with given middle point (x_1, y_1) is $T = S_1$. where S_1 and T are as defined in (*iv*) above.

(*ix*) The locus of the middle points of all chords parallel to $y = mx$ of the ellipse is $y = -\frac{b^2}{a^2m}x$, which is a line through the centre of the ellipse.

Ex.1. Prove that the tangent at the point (x_1, y_1) to the ellipse is parallel to the tangent at the point $(-x_1, -y_1)$.

Ex. 2. Prove that the polar of the focus $(ae, 0)$ is the corresponding directrix.

3.1 Examples

Example 1. If the normal at one end of a latus rectum of an ellipse passes through one extremity of the minor axis, prove that the eccentricity of the curve is given by the equation $e^4 + e^2 - 1 = 0$.

Solution. Let the equation of the ellipse be

$$\frac{x^2}{a^2}+\frac{y^2}{b^2}=1.$$

The coordinates of one end of a latus rectum of the ellipse are $\left(ae, \frac{b^2}{a}\right)$. Then, the equation of the normal at the point $\left(ae, \frac{b^2}{a}\right)$ to the ellipse is

$$x - ey = ae^3.$$

This normal will pass through the extremity $(0, -b)$ of the minor axis if

$$e^4 + e^2 - 1 = 0.$$

Example 2. A variable chord subtends a right angle at the centre of the ellipse. Find the locus of the point of intersection of the tangents at the ends.

Solution. Let the equation of the ellipse be

$$\frac{x^2}{a^2} + \frac{y^2}{b^2} = 1.$$

Let (h, k) be the point of intersection of the tangents drawn at the ends of the variable chord. Then it is the chord of contact and hence its equation is

$$\frac{xh}{a^2} + \frac{yk}{b^2} = 1.$$

The equation of the lines joining the centre (origin) to the points of intersections of the ellipse and the chord of contact is

$$\frac{x^2}{a^2} + \frac{y^2}{b^2} = \left(\frac{xh}{a^2} + \frac{yk}{b^2}\right)^2$$

i.e.
$$\left(\frac{1}{a^2} - \frac{h^2}{a^4}\right)x^2 - \frac{2hk}{a^2b^2}xy + \left(\frac{1}{b^2} - \frac{k^2}{b^4}\right) = 0.$$

These lines are at right angles if

coeff. of x^2 + coeff. of $y^2 = 0$, *i.e.*, if $\frac{1}{a^2} - \frac{h^2}{a^4} + \frac{1}{b^2} - \frac{k^2}{b^4} = 0.$

Hence, the locus of (h, k) is $\frac{x^2}{a^4} + \frac{y^2}{b^4} = \frac{1}{a^2} + \frac{1}{b^2}.$

Example 3. Find the locus of the poles of tangents to the ellipse $\frac{x^2}{a^2} + \frac{y^2}{b^2} = 1$ with respect to the parabola $y^2 = 4ax$.

Solution. Any tangent to the given ellipse is

$$y = mx + \sqrt{a^2m^2 + b^2}. \qquad (1)$$

Let (h, k) be the pole of (1). Then, its polar with respect to the parabola $y^2 = 4ax$ is

$$ky = 2a(x + h). \qquad (2)$$

But equations (1) and (2) are identical. Therefore, we have

$$\frac{m}{2a}=\frac{1}{k}=\frac{\sqrt{a^2m^2+b^2}}{2ah}$$

$$\Rightarrow \qquad m=\frac{2a}{k} \text{ and } \sqrt{a^2m^2+b^2}=\frac{2ah}{k}.$$

Thus $$a^2\left(\frac{4a^2}{k^2}\right)+b^2=\frac{4a^2h^2}{k^2}, \text{ i.e., } b^2k^2=4a^2(h^2-a^2)$$

Hence, the locus of (h, k) is $b^2y^2 = 4a^2(x^2 - a^2)$.

Example 4. Prove that the locus of the middle points of chords of the ellipse $\frac{x^2}{a^2}+\frac{y^2}{b^2}=1$, which subtend a right angle at the centre, is

$$\frac{x^2}{a^4}+\frac{y^2}{b^4}=\left(\frac{a^2+b^2}{a^2b^2}\right)\left(\frac{x^2}{a^2}+\frac{y^2}{b^2}\right)^2.$$

Solution. Let (h, k) be the middle point of the chord of the given ellipse. Then, its equation is

$$\frac{xh}{a^2}+\frac{yk}{b^2}=\frac{h^2}{a^2}+\frac{k^2}{b^2}.$$

The equation of the lines joining the centre of the ellipse to the points of intersections of the ellipse and the chord is

$$\frac{x^2}{a^2}+\frac{y^2}{b^2}=\left(\frac{\frac{xh}{a^2}+\frac{yk}{b^2}}{\frac{h^2}{a^2}+\frac{k^2}{b^2}}\right)^2$$

$$\Rightarrow \quad \left\{\frac{1}{a^2}\left(\frac{h^2}{a^2}+\frac{k^2}{b^2}\right)^2-\frac{h^2}{a^4}\right\}x^2-\frac{2hk}{a^2b^2}xy+$$

$$\left\{\frac{1}{b^2}\left(\frac{h^2}{a^2}+\frac{k^2}{b^2}\right)^2-\frac{k^2}{b^4}\right\}y^2=0. \qquad (1)$$

But the chord subtends a right angle at the centre. Therefore, the lines represented by (1) are at right angles so that coeff. of x^2 + coeff. of $y^2 = 0$, *i.e.*,

$$\frac{h^2}{a^4}+\frac{k^2}{b^4}=\left(\frac{1}{a^2}+\frac{1}{b^2}\right)\left(\frac{h^2}{a^2}+\frac{k^2}{b^2}\right)^2.$$

Hence, the locus of (h, k) is

$$\frac{x^2}{a^4}+\frac{y^2}{b^4}=\left(\frac{a^2+b^2}{a^2b^2}\right)\left(\frac{x^2}{a^2}+\frac{y^2}{b^2}\right)^2.$$

Example 5. The perpendicular from the centre of an ellipse $x^2/a^2 + y^2/b^2 = 1$ on the polar of a point with respect to the ellipse is a constant c. Prove that the locus of the point is the ellipse $\frac{x^2}{a^4}+\frac{y^2}{b^4}=\frac{1}{c^2}$.

Solution. Let (h, k) be the point on the locus. Then, its polar with respect to the given ellipse is

$$\frac{xh}{a^2}+\frac{yk}{b^2}=1.$$

Since, c is the perpendicular distance of the polar from the centre of the ellipse, we have

$$\frac{1}{\sqrt{\frac{h^2}{a^4}+\frac{k^2}{b^4}}}=c, \text{ i.e., } \frac{h^2}{a^4}+\frac{k^2}{b^4}=\frac{1}{c^2}.$$

Hence, the locus of (h, k) is $\frac{x^2}{a^4}+\frac{y^2}{b^4}=\frac{1}{c^2}$.

Example 6. From a point on the circle $x^2 + y^2 = a^2$, tangents are drawn to the ellipse $x^2/a^2 + y^2/b^2 = 1$. Prove that the locus of the middle points of the chords of contact is $\left(\frac{x^2}{a^2}+\frac{y^2}{b^2}\right)^2=\frac{x^2+y^2}{a^2}$.

Solution. Let (h, k) be the middle point of the chord of contact of the ellipse. Then, its equation is

$$\frac{xh}{a^2}+\frac{yk}{b^2}=\frac{h^2}{a^2}+\frac{k^2}{b^2}$$

$$\Rightarrow \quad \frac{x}{a^2}\left(\frac{h}{\frac{h^2}{a^2}+\frac{k^2}{b^2}}\right)+\frac{y}{b^2}\left(\frac{k}{\frac{h^2}{a^2}+\frac{k^2}{b^2}}\right)=1.$$

Clearly this is the chord of contact of the tangents drawn from the point

$$\left(\frac{h}{\frac{h^2}{a^2}+\frac{k^2}{b^2}}, \frac{k}{\frac{h^2}{a^2}+\frac{k^2}{b^2}}\right)$$

to the given ellipse. But this point lies on the cricle $x^2 + y^2 = a^2$. Therefore

$$\frac{h^2}{\left(\frac{h^2}{a^2}+\frac{k^2}{b^2}\right)^2}+\frac{k^2}{\left(\frac{h^2}{a^2}+\frac{k^2}{b^2}\right)^2} = a^2, \text{ i.e., } \left(\frac{h^2}{a^2}+\frac{k^2}{b^2}\right)^2 = \frac{h^2+k^2}{a^2}.$$

Hence, the locus of (h, k) is

$$\left(\frac{x^2}{a^2}+\frac{y^2}{b^2}\right)^2 = \frac{x^2+y^2}{a^2}.$$

Example 7. Prove that the locus of the middle points of chords of constant length $2c$ of the ellipse $\frac{x^2}{a^2}+\frac{y^2}{b^2} = 1$ is

$$\left(\frac{x^2}{a^2}+\frac{y^2}{b^2}-1\right)\left(\frac{x^2}{a^4}+\frac{y^2}{b^4}\right)+\frac{c^2}{a^2b^2}\left(\frac{x^2}{a^2}+\frac{y^2}{b^2}\right) = 0.$$

Solution. Let (h, k) be the middle point of the chord of length $2c$. Let α be the inclination of this chord with x-axis. Then $(h + c \cos \alpha, k + c \sin \alpha)$ and $(h - c \cos \alpha, k - c \sin \alpha)$ are the coordinates of the extremities of the chord. But these points lie on the given ellipse. Therefore

$$\frac{(h+c\cos\alpha)^2}{a^2}+\frac{(k+c\sin\alpha)^2}{b^2} = 1 \qquad (1)$$

and $$\frac{(h-c\cos\alpha)^2}{a^2}+\frac{(k-c\sin\alpha)^2}{b^2} = 1. \qquad (2)$$

Subtracting (2) from (1), we get

$$\frac{4ch\cos\alpha}{a^2}+\frac{4ch\sin\alpha}{b^2} = 0, \text{ i.e., } \tan\alpha = -\frac{\frac{h}{a^2}}{\frac{k}{b^2}}$$

$$\Rightarrow \qquad \sin\alpha = -\frac{\frac{h}{a^2}}{\sqrt{\frac{h^2}{a^4}+\frac{k^2}{b^4}}}, \cos\alpha = -\frac{\frac{k}{b^2}}{\sqrt{\frac{h^2}{a^4}+\frac{k^2}{b^4}}}.$$

On adding (1) and (2), we get

$$2\left[\frac{h^2}{a^2}+\frac{k^2}{b^2}+c^2\left(\frac{\cos^2\alpha}{a^2}+\frac{\sin^2\alpha}{b^2}\right)\right]=2$$

$$\Rightarrow \qquad \frac{h^2}{a^2}+\frac{k^2}{b^2}+c^2\left[\frac{1}{a^2}\cdot\frac{\frac{k^2}{b^4}}{\left(\frac{h^2}{a^4}+\frac{k^2}{b^4}\right)}+\frac{1}{b^2}\cdot\frac{\frac{h^2}{a^4}}{\left(\frac{h^2}{a^4}+\frac{k^2}{b^4}\right)}\right]=1$$

$$\Rightarrow \qquad \left(\frac{h^2}{a^2}+\frac{k^2}{b^2}-1\right)\left(\frac{h^2}{a^4}+\frac{k^2}{b^4}\right)+\frac{c^2}{a^2b^2}\left(\frac{h^2}{a^2}+\frac{k^2}{b^2}\right)=0.$$

Hence, the locus of (h, k) is

$$\left(\frac{x^2}{a^2}+\frac{y^2}{b^2}-1\right)\left(\frac{x^2}{a^4}+\frac{y^2}{b^4}\right)+\frac{c^2}{a^2b^2}\left(\frac{x^2}{a^2}+\frac{y^2}{b^2}\right)=0.$$

Exercises

1. Find the equation of the tangent to the ellipse $\frac{x^2}{a^2}+\frac{y^2}{b^2}=1$ which makes an angle of 60° with x-axis.

2. Prove that the line $x\cos\alpha + y\sin\alpha = p$ is a tangent to the ellipse $\frac{x^2}{a^2}+\frac{y^2}{b^2}=1$ if $a^2\cos^2\alpha + b^2\sin^2\alpha = p^2$.

3. Prove that the line $lx + my + n = 0$ is a tangent to the ellipse $\frac{x^2}{a^2}+\frac{y^2}{b^2}=1$ if $a^2l^2 + b^2m^2 = n^2$.

4. Find the locus of the foot of the perpendicular drawn from a focus to the tangents to the ellipse $\frac{x^2}{a^2}+\frac{y^2}{b^2}=1$.

5. Prove that the product of the perpendiculars drawn from the points $(c, 0)$, $(-c, 0)$ on the tangents to the ellipse $\frac{x^2}{a^2}+\frac{y^2}{b^2}=1$ is equal to b^2 if $c^2 = a^2 - b^2$.

6. Prove that the locus of the middle points of the chords of contact of tangents at right angles is the curve $\left(\frac{x^2}{a^2}+\frac{y^2}{b^2}\right)^2 = \frac{x^2+y^2}{a^2+b^2}$.

7. If a variable chord PQ of the ellipse $\frac{x^2}{a^2}+\frac{y^2}{b^2}=1$ subtends a right angle at the centre of the ellipse, prove that the locus of the pole of PQ with respect to the ellipse is $\frac{x^2}{a^4}+\frac{y^2}{b^4}=\frac{1}{a^2}+\frac{1}{b^2}$.

8. If the product of the perpendiculars from the foci upon the polar of P is constant and equals c^2, prove that the locus of P is the ellipse $b^4x^2(c^2 + a^2e^2) + c^2a^4y^2 = a^4b^4$.

9. Prove that the locus of the poles of tangents to the ellipse $\frac{x^2}{a^2}+\frac{y^2}{b^2}=1$ with respect to the circle $x^2 + y^2 = a^2$ is the ellipse $a^2x^2 + b^2y^2 = a^4$.

10. Prove that the locus of the middle points of chords of the ellipse $\frac{x^2}{a^2}+\frac{y^2}{b^2}=1$ passing through a fixed point (h, k) is $\frac{x(x-h)}{a^2}+\frac{y(y-k)}{b^2}=0$.

11. From a point P tangents are drawn to the ellipse $\frac{x^2}{a^2}+\frac{y^2}{b^2}=1$ and the chord of contact touches the circle $x^2 + y^2 = c^2$ $(c < b < a)$. Prove that the locus of P is the ellipse $\frac{x^2}{a^4}+\frac{y^2}{b^4}=\frac{1}{c^2}$.

12. Prove that the locus of the foot of the perpendicular on a varying tangent to an ellipse from either of its foci is a concentric circle.

13. If θ be the angle between the tangents drawn from the point (x_1, y_1) to the ellipse $\frac{x^2}{a^2}+\frac{y^2}{b^2}=1$, prove that

$$\tan\theta = \frac{2ab\left(\frac{x_1^2}{a^2}+\frac{y_1^2}{b^2}-1\right)^{\frac{1}{2}}}{x_1^2+y_1^2-a^2-b^2}.$$

14. The tangent to an ellipse at any point meets the tangents at the ends of the major axis in P and Q. Prove that the circle on PQ as diameter passes through the foci of the ellipse.

15. Tangents are drawn from a point $P(x_1, y_1)$ to the ellipse $\frac{x^2}{a^2}+\frac{y^2}{b^2}=1$ to touch it at Q and R. Prove that if S be either of the foci

$$\frac{SP^2}{SQ \cdot SR}=\frac{x_1^2}{a^2}+\frac{y_1^2}{b^2}.$$

16. The circle, having as diameter the line joining the foci of the ellipse whose major axis has length $2a$, meets the minor axis in P and Q. Prove that the sum of the squares of the perpendiculars from P and Q on any tangent to the ellipse is $2a^2$.

17. The perpendicular from any point P on the ellipse $\frac{x^2}{a^2}+\frac{y^2}{b^2}=1$ to the major axis meets the ellipse again the point Q. C is the centre of the ellipse and the normal at P meets CQ at the point R. Find the coordinates of R and prove that the locus of R is the ellipse

$$\frac{x^2}{a^2}+\frac{y^2}{b^2}=\frac{(a^2-b^2)^2}{(a^2+b^2)^2}.$$

18. If the normal at a point P on the ellipse $\frac{x^2}{a^2}+\frac{y^2}{b^2}=1$ meets the major axis at Q. If R is the point of intersection of the line through Q parallel to the y-axis and the line joining P to the centre of the ellipse, prove that the locus of R is

$$\frac{x^2}{a^2}+\frac{y^2}{b^2}=\frac{(a^2-b^2)^2}{a^4}.$$

19. Prove that the perpendicular from the focus of an ellipse whose centre is C on the polar of any point P will meet the line CP on the directrix.

20. If the normal at any point P cut the major axis in G, prove that, for different positions of P, the locus of the middle point of PG will be an ellipse.

21. Find the equations to the tangents at the ends of the latera recta of the ellipse $\frac{x^2}{a^2}+\frac{y^2}{b^2}=1$, and prove that they pass through the intersections of the axis and the directrices.

22. Tangents are drawn from any point on the ellipse $\frac{x^2}{a^2}+\frac{y^2}{b^2}=1$ to the circle $x^2+y^2=r^2$. Prove that the chords of contact are tangents to the ellipse $a^2x^2+b^2y^2=r^4$.

23. Find the locus of the poles of tangents to the ellipse $\frac{x^2}{a^2}+\frac{y^2}{b^2}=1$ with respect to the concentric ellipse $\frac{x^2}{\alpha^2}+\frac{y^2}{\beta^2}=1$.

4. Director Circle

To find the locus of the point of intersection of tangents to an ellipse $\frac{x^2}{a^2}+\frac{y^2}{b^2}=1$ *which meet at right angles.*

Let (h, k) be any point on the locus. The equation of a tangent to the ellipse is

$$y = mx + \sqrt{a^2m^2+b^2}\ .$$

This will pass through the point (h, k) if $k = mh + \sqrt{a^2m^2+b^2}$, *i.e.*,

$$(h^2 - a^2)m^2 - 2hkm + k^2 - b^2 = 0. \qquad (1)$$

Equation (1) is quadratic in m. This gives two roots say m_1 and m_2. Thus, there are two tangents passing through (h, k). These tangents are at right angles if $m_1m_2 = -1$, *i.e.* if $h^2 + k^2 = a^2 + b^2$. Hence, the locus of (h, k) is $x^2 + y^2 = a^2 + b^2$.

This is the required equation of the locus which represents a circle.

The circle $x^2 + y^2 = a^2 + b^2$ *is called the* **director circle** *or* **orthoptic circle** *of the ellipse* $\frac{x^2}{a^2}+\frac{y^2}{b^2}=1$.

5. Auxiliary Circle

The circle described on the major axis of an ellipse as diameter is called the **auxiliary circle**.

If the equation of the ellipse is

$$\frac{x^2}{a^2}+\frac{y^2}{b^2}=1, \qquad (1)$$

then, the equation of the auxiliary circle will be

$$x^2 + y^2 = a^2. \qquad (2)$$

Thus, if any ordinate NP of the ellipse be produced to meet the auxiliary circle in Q, we have from (1) and (2)

$$\frac{CN^2}{a^2}+\frac{NP^2}{b^2}=1 \text{ and } CN^2 + NQ^2 = a^2.$$

Therefore $\frac{a^2 - NQ^2}{a^2} + \frac{NP^2}{b^2} = 1$, *i.e.*, $\frac{NQ}{NP} = \frac{a}{b}$.

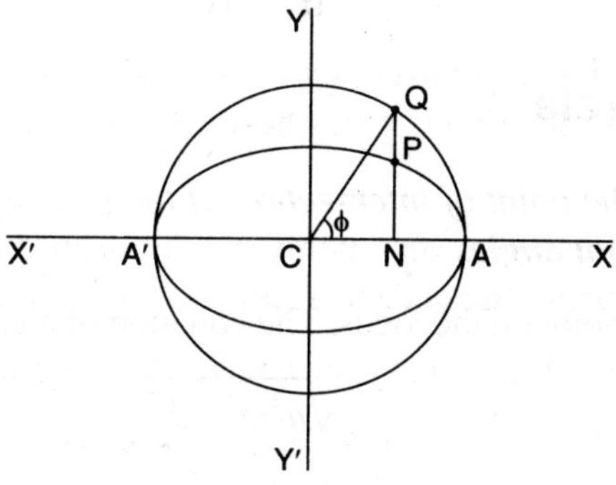

Fig. 8.11

Hence, the ordinates of the circle and of the ellipse are in a constant ratio to one another.

Ex. Perpendiculars are drawn from the points on a given circle upon a diameter. Prove that the locus of the points which divide these perpendiculars in a constant ratio is an ellipse of which the given circle is the auxiliary circle.

One may define the ellipse alternatively as follows:

Given a circle and from each point on it drawn perpendiculars upon a diameter. The locus of the points dividing these perpendiculars in a given ratio is an ellipse, of which the given circle is the auxiliary circle.

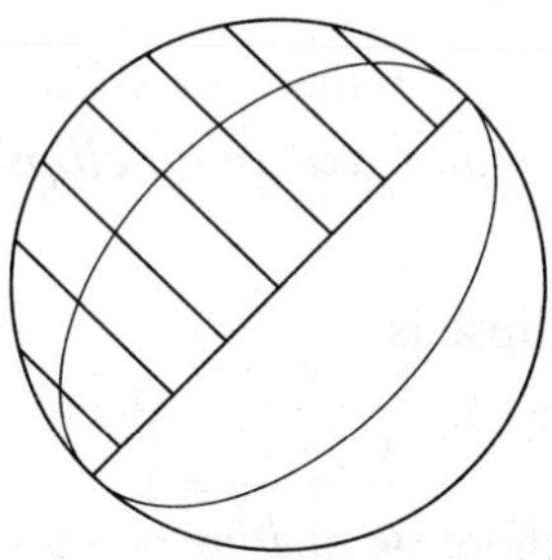

Fig. 8.12

6. Eccentric Angles

Let P be a point on an ellipse and Q the point on the auxiliary circle such that Q lies on the ordinate produced through the point P. Then, the angle ACQ is called the eccentric angle of P. The eccentric angle is usually denoted by 'ϕ'.

Now, we introduce a very important and useful method of representing any point P on ellipse in terms of the eccentric angle of that point.

We note that $CN = a \cos \phi$. To find the y-coordinate of the point P, we have

$$\frac{a^2 \cos^2 \phi}{a^2} + \frac{PN^2}{b^2} = 1, \text{ i.e., } PN = \pm\, b \sin \phi.$$

It is easy to see that the positive sign must be taken in all cases, for the y-coordinate of P it is positive if ϕ lies between 0 and π, and negative if ϕ lies between π and 2π.

Thus, the coordinates of any point on the ellipse can be expressed as $(a \cos \phi, b \sin \phi)$, ϕ being the eccentric angle of the point.

The point $(a \cos \phi, b \sin \phi)$ is also referred to the point ϕ. The equations

$$x = a \cos \phi, y = b \sin \phi$$

can be regarded as the parametric form of the ellipse $\frac{x^2}{a^2} + \frac{y^2}{b^2} = 1$, where ϕ is taken as a parameter.

6.1 *To find the equation of the chord joining two points whose eccentric angles are given.*

Let ϕ_1, ϕ_2 be the eccentric angles of the two points A and B on the ellipse

$$\frac{x^2}{a^2} + \frac{y^2}{b^2} = 1.$$

Then, the coordinates of the points are $(a \cos \phi_1, b \sin \phi_1)$ and $(a \cos \phi_2, b \sin \phi_2)$, respectively. The equation of the chord AB is

$$y - b \sin \phi_1 = \frac{b \sin \phi_2 - b \sin \phi_1}{a \cos \phi_2 - a \cos \phi_1}(x - a \cos \phi_1)$$

$$\Rightarrow bx \cos \frac{\phi_1 + \phi_2}{2} + ay \sin \frac{\phi_1 + \phi_2}{2} = ab \cos \frac{\phi_1 - \phi_2}{2}.$$

Hence, the equation of the chord joining the points ϕ_1 and ϕ_2 on the ellipse is

$$\frac{x}{a} \cos \frac{\phi_1 + \phi_2}{2} + \frac{y}{b} \sin \frac{\phi_1 + \phi_2}{2} = \cos \frac{\phi_1 - \phi_2}{2}.$$

Corollary 1. The equation of the tangent at the point $(a \cos \phi, b \sin \phi)$ to the ellipse $\frac{x^2}{a^2} + \frac{y^2}{b^2} = 1$ is $\frac{x}{a} \cos \phi + \frac{y}{b} \sin \phi = 1$.

Corollary 2. The equation of the normal at the point $(a \cos \phi, b \sin \phi)$ to the ellipse $\frac{x^2}{a^2}+\frac{y^2}{b^2}=1$ is $ax \sec \phi - by \operatorname{cosec} \phi = a^2 - b^2$.

6.2 *To prove that from any point four normals can be drawn to an ellipse.*

Let the equation of the ellipse be $\frac{x^2}{a^2}+\frac{y^2}{b^2}=1$. Let (x_1, y_1) be any point. The equation of the normal to the ellipse at the points ϕ is

$$\frac{ax}{\cos\phi}-\frac{by}{\sin\phi}=a^2-b^2.$$

Let it pass through (x_1, y_1). Then

$$\frac{ax_1}{\cos\phi}-\frac{by_1}{\sin\phi}=a^2-b^2$$

$$\Rightarrow \quad \frac{ax_1}{\cos^2\frac{\phi}{2}-\sin^2\frac{\phi}{2}}-\frac{by_1}{2\sin\frac{\phi}{2}\cos\frac{\phi}{2}}=\frac{a^2-b^2}{\cos^2\frac{\phi}{2}+\sin^2\frac{\phi}{2}}$$

$$\Rightarrow \quad \frac{ax_1}{1-\tan^2\frac{\phi}{2}}-\frac{by_1}{2\tan\frac{\phi}{2}}=\frac{a^2-b^2}{1+\tan^2\frac{\phi}{2}}$$

$$\Rightarrow \quad by_1 \tan^4\frac{\phi}{2}+2(ax_1+a^2-b^2)\tan^3\frac{\phi}{2}$$

$$+\,2(ax_1-a^2+b^2)\tan\frac{\phi}{2}-by_1=0. \qquad (1)$$

Equation (1) is of fourth degree in $\tan\frac{\phi}{2}$ and hence gives four values of $\tan\frac{\phi}{2}$ for a given point (x_1, y_1). If t_1 be one of the four values. Then, taking

$$\tan\frac{\phi}{2}=t_1, \text{ we get } \phi=2n\pi+2\tan^{-1}t_1.$$

All these values of ϕ give the same point on the ellipse, for by adding any multiple of 2π to the eccentric angle we reach the same point. Hence, each value of $\tan\frac{\phi}{2}$ gives just one point on the ellipse, real or imaginary.

Thus, there are four normals that can be drawn from a point (x_1, y_1) to the ellipse.

7. Propositions on Ellipse

Proposition I. *The portion of the tangent between the point of contact and the directrix subtends a right angle at the corresponding focus.*

Let $P(x_1, y_1)$ be any point on the ellipse

$$\frac{x^2}{a^2}+\frac{y^2}{b^2}=1.$$

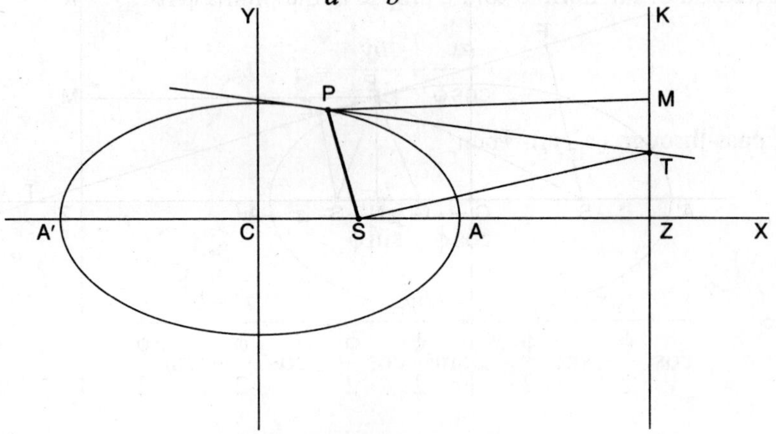

Fig. 8.13

The equation of the tangent at $P(x_1, y_1)$ to the ellipse is $\frac{xx_1}{a^2}+\frac{yy_1}{b^2}=1$.

Equation of the directrix ZK is $x=\frac{a}{e}$.

The coordinates of the point of intersection of the tangent and the directrix are

$$\left(\frac{a}{e}, \frac{a(1-e^2)(ae-x_1)}{ey_1}\right).$$

Now $\quad m_1 = \text{slope of } SP = \frac{y_1-0}{x_1-ae} = -\frac{y_1}{ae-x_1}.$

$$m_2 = \text{slope of } ST = \frac{\frac{a(1-e^2)(ae-x_1)}{ey_1}}{\frac{a}{e}-ae} = \frac{ae-x_1}{y_1}.$$

Therefore $\quad m_1 m_2 = -\left(\frac{y_1}{ae-x_1}\right)\left(\frac{ae-x_1}{y_1}\right) = -1.$

Hence $\angle TSP$ is a right angle.

Proposition II. *The tangent and the normal at any point of an ellipse bisect the angles between the focal radii to that point.*

Let the equation of the ellipse be $\dfrac{x^2}{a^2}+\dfrac{y^2}{b^2}=1.$

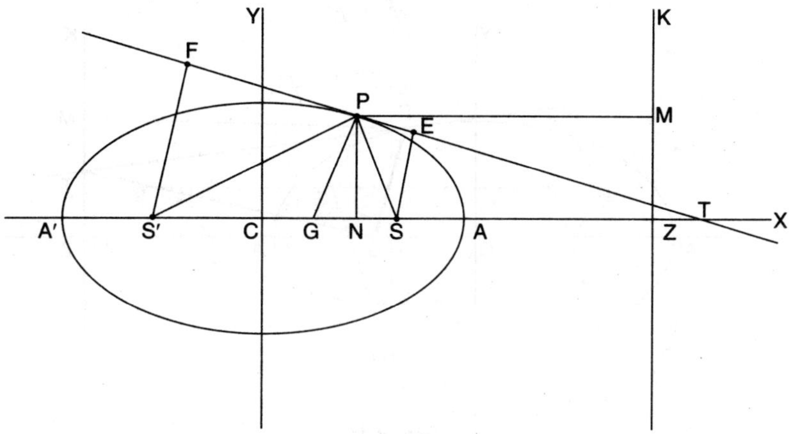

Fig. 8.14

Let $P(x_1, y_1)$ be any point on the ellipse. Let the tangent and normal at $P(x_1, y_1)$ meet the major axis in the points T and G respectively.

The equation to the normal PG is

$$\frac{x-x_1}{\dfrac{x_1}{a^2}}=\frac{y-y_1}{\dfrac{y_1}{b^2}}.$$

The coordinates of G are $(e^2x_1, 0)$. Therefore

$$GS = CS - CG = ae - e^2x_1 = e(a - ex_1).$$

Also $$SP = e\cdot PM = e\cdot NZ = e(CZ - CN) = e\left(\frac{a}{e}-x_1\right)$$

$$= a - ex_1.$$

So $$GS = e\cdot SP.$$

Similarly $$S'G = e.S'P.$$

Thus $$SP : S'P = GS : S'G.$$

Hence, by geometry, PG bisects the angle SPS'.

It follows that the tangent PT bisects the exterior angle between the focal radii.

Proposition III. *If the normal at any point P meet the major and minor axes in G and H, and if CM be the perpendicular upon this normal, then*

$$PM \cdot PG = b^2 \text{ and } PM \cdot PHJ = a^2.$$

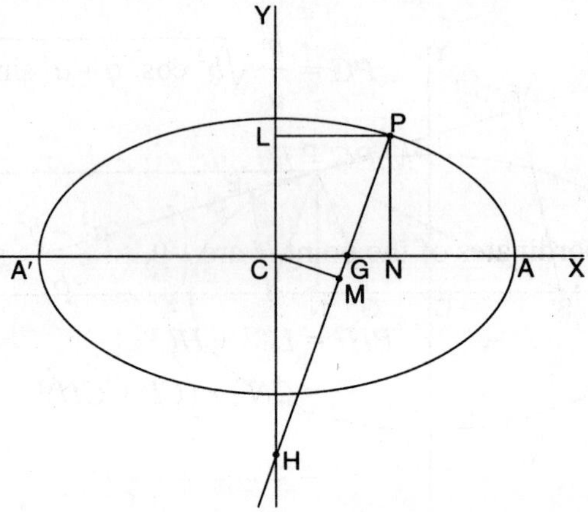

Fig. 8.15

Let the equation of the ellipse be $\frac{x^2}{a^2} + \frac{y^2}{b^2} = 1$.

The equation of the tangent at $P(a \cos \phi, b \sin \phi)$ to the ellipse is

$$\frac{x}{a} \cos \phi + \frac{y}{b} \sin \phi = 1$$

$\therefore$ PM = perpendicular distance of the tangent from C

$$= \frac{ab}{\sqrt{b^2 \cos^2 \phi + a^2 \sin^2 \phi}}.$$

The equation of the normal at $P(a \cos \phi, b \sin \phi)$ is

$$\frac{ax}{\cos \phi} - \frac{by}{\sin \phi} = a^2 - b^2$$

Hence, the coordinates of G are $\left(\frac{a^2 - b^2}{a} \cos \phi, 0\right)$.

Now $$PG^2 = GN^2 + NP^2$$

$$= (CN - CG)^2 + NP^2$$

$$= \left(a\cos\phi - \frac{a^2 - b^2}{a}\cos\phi \right)^2 + b^2\sin^2\phi$$

$$= \frac{b^2}{a^2}(b^2\cos^2\phi + a^2\sin^2\phi)$$

$$\Rightarrow \qquad PG = \frac{b}{a}\sqrt{b^2\cos^2\phi + a^2\sin^2\phi}.$$

$$\therefore \qquad PM.PG = b^2$$

Further, the coordinates of the point H are $\left(0, -\frac{a^2 - b^2}{b}\sin\phi\right)$

Now
$$PH^2 = LP^2 + HL^2$$
$$= CN^2 + (CL + CH)^2$$
$$= a^2\cos^2\phi + \left(b\sin\phi + \frac{a^2 - b^2}{b}\sin\phi \right)^2$$
$$= \frac{a^2}{b^2}(b^2\cos^2\phi + a^2\sin^2\phi)$$

$$\Rightarrow \qquad PH = \frac{a}{b}\sqrt{b^2\cos^2\phi + a^2\sin^2\phi}.$$

Hence
$$PM \cdot PH = a^2.$$

7.1 Examples

Example 1. Prove that the locus of the poles of normal chords of the ellipse $\frac{x^2}{a^2} + \frac{y^2}{b^2} = 1$ is the curve

$$\frac{a^6}{x^2} + \frac{b^6}{y^2} = (a^2 - b^2)^2.$$

Solution. The equation of the normal at the point (a cos θ, b sin θ) on the given ellipse is

$$ax\sec\theta - by\,\text{cosec}\,\theta = a^2 - b^2. \qquad (1)$$

Let (h, k) be the pole of the line (1). Then, its polar with respect to the given ellipse is

$$\frac{xh}{a^2} + \frac{yk}{b^2} = 1. \qquad (2)$$

Since, (1) and (2) represent the same line, we have

$$-ax \sec\theta + by \operatorname{cosec}\theta + a^2 - b^2 \equiv \frac{xh}{a^2} + \frac{yk}{b^2} - 1$$

$$\Rightarrow \qquad \frac{-a\sec\theta}{\frac{h}{a^2}} = \frac{b\operatorname{cosec}\theta}{\frac{k}{b^2}} = \frac{a^2 - b^2}{-1}$$

$$\Rightarrow \qquad \cos\theta = \frac{a^3}{h(a^2 - b^2)}, \sin\theta = \frac{-b^3}{k(a^2 - b^2)}.$$

$$\therefore \qquad \cos^2\theta + \sin^2\theta = \frac{a^6}{h^2(a^2 - b^2)^2} + \frac{b^6}{k^2(a^2 - b^2)^2}$$

$$\Rightarrow \qquad \frac{a^6}{h^2} + \frac{b^6}{k^2} = (a^2 - b^2)^2.$$

Hence, locus of (h, k) is

$$\frac{a^6}{x^2} + \frac{b^6}{y^2} = (a^2 - b^2)^2.$$

Example 2. If α, β, γ, δ be the eccentric angles of four points on the ellipse such that the normals at them are concurrent, prove that $\alpha + \beta + \gamma + \delta$ is an odd multiple of π.

Solution. Let the normals be concurrent at (x_1, y_1). Then $\tan\frac{\alpha}{2}$, $\tan\frac{\beta}{2}$, $\tan\frac{\gamma}{2}$ and $\tan\frac{\delta}{2}$ are the roots of the equation (1) in § 6.2. By theory of equations, we have

$$s_1 = \Sigma \tan\frac{\alpha}{2} = \frac{-2(ax_1 + a^2 - b^2)}{by_1},$$

$$s_2 = \Sigma \tan\frac{\alpha}{2} \tan\frac{\beta}{2} = 0,$$

$$s_3 = \Sigma \tan\frac{\alpha}{2} \tan\frac{\beta}{2} \tan\frac{\gamma}{2} = \frac{-2(ax_1 - a^2 + b^2)}{by_1}$$

and

$$s_4 = \tan\frac{\alpha}{2} \tan\frac{\beta}{2} \tan\frac{\gamma}{2} \tan\frac{\delta}{2} = -1.$$

Now $$\tan\left(\frac{\alpha}{2}+\frac{\beta}{2}+\frac{\gamma}{2}+\frac{\delta}{2}\right)=\frac{s_1-s_2}{1-s_2+s_4}$$

$$=\frac{s_1-s_3}{0}=\infty$$

$\Rightarrow$ $$\alpha+\beta+\gamma+\delta=\text{an odd multiple of }\pi.$$

Example 3. A perpendicular is drawn from the centre of the ellipse $\frac{x^2}{a^2}+\frac{y^2}{b^2}=1$ to any tangent. Prove that the locus of the foot of the perpendicular is $(x^2+y^2)^2=a^2x^2+b^2y^2$.

Solution. Any tangent to the given ellipse is

$$y=mx+\sqrt{a^2m^2+b^2}\,. \tag{1}$$

The equation of the perpendicular drawn from the centre (0, 0) to (1) is

$$y=-\frac{1}{m}x. \tag{2}$$

The locus of the foot of the perpendicular is obtained by eliminating m between (1) and (2). Hence, the required locus is

$$y=-\frac{x}{y}x+\sqrt{a^2\frac{x^2}{y^2}+b^2}\,,\ \textit{i.e.},\ (x^2+y^2)^2=a^2x^2+b^2y^2.$$

Example 4. Prove that the locus of the pole with respect to the ellipse $\frac{x^2}{a^2}+\frac{y^2}{b^2}=1$, of any tangent to the auxiliary circle is the curve $\frac{x^2}{a^4}+\frac{y^2}{b^4}=\frac{1}{a^2}$.

Solution. Any tangent to the auxiliary circle $x^2+y^2=a^2$ is

$$y=mx+a\sqrt{1+m^2}\,. \tag{1}$$

Let (h, k) be the pole of (1). Then, its polar with respect to the given ellipse is

$$\frac{xh}{a^2}+\frac{yk}{b^2}=1. \tag{2}$$

But the equations (1) and (2) are identical. Therefore, on comparing we get

$$\frac{m}{\frac{h}{a^2}}=\frac{-1}{\frac{k}{b^2}}=\frac{a\sqrt{1+m^2}}{-1}$$

$$\Rightarrow \qquad m = -\frac{b^2 h}{a^2 k} \text{ and } a\sqrt{1+m^2} = \frac{b^2}{k}.$$

Therefore $$a^2\left(1+\frac{b^4h^2}{a^4k^2}\right) = \frac{b^4}{k^2}, \text{ i.e., } \frac{h^2}{a^4}+\frac{k^2}{b^4} = \frac{1}{a^2}.$$

Hence, the locus of (h, k) is $\dfrac{x^2}{a^4}+\dfrac{y^2}{b^4}=\dfrac{1}{a^2}$.

Example 5. Prove that the locus of the middle points of normal chords of the ellipse $\dfrac{x^2}{a^2}+\dfrac{y^2}{b^2}=1$ is the curve

$$\left(\frac{x^2}{a^2}+\frac{y^2}{b^2}\right)^2\left(\frac{a^6}{x^2}+\frac{b^6}{y^2}\right) = (a^2-b^2)^2.$$

Solution. The equation of the normal at the point $(a \cos \theta, b \sin \theta)$ to the given ellipse is

$$ax \sec \theta - by \operatorname{cosec} \theta = a^2 - b^2. \qquad (1)$$

Let (h, k) be the middle point of the line (1). Then, the equation of the chord with middle point (h, k) is

$$\frac{xh}{a^2}+\frac{yk}{b^2} = \frac{h^2}{a^2}+\frac{k^2}{b^2}. \qquad (2)$$

Since (1) and (2) represent the same line, we have

$$-ax \sec \theta + by \operatorname{cosec} \theta + a^2 - b^2 \equiv \frac{xh}{a^2}+\frac{yk}{b^2}-\left(\frac{h^2}{a^2}+\frac{k^2}{b^2}\right)$$

$$\Rightarrow \qquad \frac{-a\sec\theta}{\frac{h}{a^2}} = \frac{b\operatorname{cosec}\theta}{\frac{k}{b^2}} = \frac{a^2-b^2}{-\left(\frac{h^2}{a^2}+\frac{k^2}{b^2}\right)}$$

$$\Rightarrow \qquad \cos\theta\,\frac{a^3\left(\frac{h^2}{a^2}+\frac{k^2}{b^2}\right)}{h(a^2-b^2)}, \sin\theta = \frac{b^3\left(\frac{h^2}{a^2}+\frac{k^2}{b^2}\right)}{k(a^2-b^2)}$$

$$\Rightarrow \qquad \cos^2\theta+\sin^2\theta = \frac{a^6\left(\frac{h^2}{a^2}+\frac{k^2}{b^2}\right)^2}{h^2(a^2-b^2)^2}+\frac{b^6\left(\frac{h^2}{a^2}+\frac{k^2}{b^2}\right)^2}{k^2(a^2-b^2)^2}$$

$$\Rightarrow \qquad \left(\frac{h^2}{a^2}+\frac{k^2}{b^2}\right)^2\left(\frac{a^6}{h^2}+\frac{b^6}{k^2}\right) = (a^2 - b^2)^2.$$

Hence, the locus of (h, k) is

$$\left(\frac{x^2}{a^2}+\frac{y^2}{b^2}\right)^2\left(\frac{a^6}{x^2}+\frac{b^6}{y^2}\right) = (a^2 - b^2)^2.$$

Example 6. If the chord joining two points whose eccentric angles are α and β cut the major axis of an ellipse at a distance d from the centre, prove that

$$\tan\frac{\alpha}{2}\tan\frac{\beta}{2} = \frac{d-a}{d+a}.$$

Solution. Let the equation of the ellipse be $\dfrac{x^2}{a^2}+\dfrac{y^2}{b^2} = 1.$

The equation of the chord joining the points whose eccentric angles are α and β, is

$$\frac{x}{a}\cos\frac{\alpha+\beta}{2}+\frac{y}{b}\sin\frac{\alpha+\beta}{2} = \cos\frac{\alpha-\beta}{2}.$$

This chord passes through $(d, 0)$. Therefore

$$\frac{d}{a}\cos\frac{\alpha+\beta}{2} = \cos\frac{\alpha-\beta}{2}$$

$$\Rightarrow \qquad \frac{\cos\frac{\alpha-\beta}{2}}{\cos\frac{\alpha+\beta}{2}} = \frac{d}{a}$$

$$\Rightarrow \qquad \frac{\cos\frac{\alpha-\beta}{2}-\cos\frac{\alpha+\beta}{2}}{\cos\frac{\alpha-\beta}{2}+\cos\frac{\alpha+\beta}{2}} = \frac{d-a}{d+a}$$

$$\Rightarrow \qquad \tan\frac{\alpha}{2}\tan\frac{\beta}{2} = \frac{d-a}{d+a}.$$

Exercises

1. Prove that the line $lx + my + n = 0$ is a normal to the ellipse $\dfrac{x^2}{a^2}+\dfrac{y^2}{b^2} = 1$, if $\dfrac{a^2}{l^2}+\dfrac{b^2}{m^2} = \dfrac{(a^2-b^2)^2}{n^2}$.

2. Prove that the locus of a point whose polar with respect to the ellipse $\frac{x^2}{a^2}+\frac{y^2}{b^2}=1$ touches its director circle is $\frac{x^2}{a^4}+\frac{y^2}{b^4}=\frac{1}{a^2+b^2}$.

3. Prove that the locus of the points of intersection of the tangents to the ellipse $\frac{x^2}{a^2}+\frac{y^2}{b^2}=1$ drawn at the extremities of a normal chord is

$$\frac{a^6}{x^2}+\frac{b^6}{y^2}=(a^2-b^2)^2.$$

4. If θ and ϕ are the eccentric angles of two points collinear with a focus of an ellipse, prove that $\cos\frac{\theta-\phi}{2}=\pm e\cos\frac{\theta+\phi}{2}$.

5. Prove that the product of the distances of the foci from any tangent to the ellipse $\frac{x^2}{a^2}+\frac{y^2}{b^2}=1$ is b^2.

6. Prove that the line joining the points α and β on the ellipse $\frac{x^2}{a^2}+\frac{y^2}{b^2}=1$ passes through a focus if $\tan\frac{\alpha}{2}\tan\frac{\beta}{2}=\frac{e-1}{e+1}$ or $\frac{e+1}{e-1}$, where e is the eccentricity.

7. P and Q are the points α and β on the ellipse $\frac{x^2}{a^2}+\frac{y^2}{b^2}=1$. If PQ subtends a right angle at $(a, 0)$, prove that $\tan\frac{\alpha}{2}\tan\frac{\beta}{2}=-\frac{b^2}{a^2}$ and deduce that PQ passes through a fixed point. Further, obtain the coordinates of this fixed point.

8. Tangents are drawn to an ellipse at the points whose eccentric angles are α, β, γ. Prove that the area of the triangle formed by them is

$$ab\tan\frac{\beta-\gamma}{2}\tan\frac{\gamma-\alpha}{2}\tan\frac{\alpha-\beta}{2}.$$

9. If the tangent to the ellipse $\frac{x^2}{a^2}+\frac{y^2}{b^2}=1$ at the point $(a\cos\alpha, b\sin\alpha)$ meets the coordinate axes at P and Q, and M is the mid-point of PQ, find the coordinates of M and hence prove that as α varies, the locus of M is the curve $\frac{a^2}{x^2}+\frac{b^2}{y^2}=4$.

10. P is the point $(a \cos \theta, b \sin \theta)$ on the ellipse $\frac{x^2}{a^2}+\frac{y^2}{b^2}=1$ and Q is the corresponding point on the auxiliary circle. C is the centre of the ellipe and the normal to the ellipse at P meets CQ in R. Find the coordinates of R in terms of a, b and θ. Hence, prove that, as θ varies, the locus of R is the circle $x^2 + y^2 = (a + b)^2$.

11. Any tangent to an ellipse is cut by the tangents at the ends of the major axis in the points P, Q. Prove that the circle whose diameter is PQ will pass through the foci.

12. Prove that the locus of the point of intersection of tangents to an ellipse at two points whose eccentric angles differ by a constant is an ellipse.

13. If Q be the point on the auxiliary circle corresponding to the point P on an ellipse, prove that the normals at P and Q meet on a fixed circle.

14. If any two chords be drawn through two points on the major axis of an ellipse equidistant from the centre, prove that $\tan\frac{\alpha}{2}\tan\frac{\beta}{2}\tan\frac{\gamma}{2}\tan\frac{\delta}{2}$ $= 1$, where, α, β, γ, δ are eccentric angles of the extremities of the chords.

15. If the normals be drawn at the extremities of any focal chord of an ellipse, a line through their intersection parallel to the major axis will bisect the chord.

16. If SY and $S'Y'$ be the perpendiculars from the foci upon the tangent at any point P of the ellipse, then Y and Y' lie on the auxiliary circle, and $SY \cdot S'Y' = b^2$. Also CY and $S'P$ are parallel.

17. Two tangents to the ellipse intersect at right angles. Prove that the sum of the squares of the chords which the auxiliary circle intercepts on them is constant, and equal to the square on the line joining the foci.

18. Prove that the circle of any focal distance as diameter touches the auxiliary circle.

19. A tangent to the ellipse $\frac{x^2}{a^2}+\frac{y^2}{b^2}=1$ meets the ellipse $\frac{x^2}{a^2}+\frac{y^2}{b^2}=a+b$ in the points P and Q. Prove that the tangents at P and Q are at right angles.

20. If the normals at four points (x_r, y_r), $r = 1, 2, 3, 4$ on the ellipse $\frac{x^2}{a^2}+\frac{y^2}{b^2}$ $= 1$ are concurrent, prove that

$$(x_1+x_2+x_3+x_4)\left(\frac{1}{x_1}+\frac{1}{x_2}+\frac{1}{x_3}+\frac{1}{x_4}\right)=4.$$

21. If the normals at the points whose eccentric angles are α, β, γ are concurrent, prove that $\sin(\beta+\gamma)+\sin(\gamma+\alpha)+\sin(\alpha+\beta)=0$.

8. Diameter

A line through the centre of an ellipse is called a **diameter** *of the ellipse.*

Note. The definitions of a diameter in respect of a circle, parabola and ellipse are consistent. In each case it is a line through the centre of the corresponding curve. Though, at first sight it appears that a diameter of a parabola is quite different from that of a circle or an ellipse, yet it is not so actually, since, a line drawn through a point P on a parabola parallel to the axis is the limiting case of a line joining P to a point on the axis of the parabola at a very great distance, and so that lines parallel to the axis of the parabola may be regarded as lines through its centre at infinity.

8.1 Equation of diameter. *To find the equation of a diameter of an ellipse.*

Let equation of the ellipse be $\dfrac{x^2}{a^2}+\dfrac{y^2}{b^2}=1$.

Let $P(a\cos\theta, b\sin\theta)$ and $Q(a\cos\phi, b\sin\phi)$ be any two points on the ellipse. Then equation of the chord PQ is

$$\frac{x}{a}\cos\frac{\theta+\phi}{2}+\frac{y}{b}\sin\frac{\theta+\phi}{2}=\cos\frac{\theta-\phi}{2}$$

This chord is parallel to the line $y=mx$ if

$$m=-\frac{b}{a}\cot\frac{\theta+\phi}{2}. \tag{1}$$

Let the coordinates of the middle point of the chord PQ be (h, k). Then

$$h=\frac{a\cos\theta+a\cos\phi}{2}=a\cos\frac{\theta+\phi}{2}\cos\frac{\theta-\phi}{2}$$

and
$$k=\frac{b\sin\theta+b\sin\phi}{2}=b\sin\frac{\theta+\phi}{2}\cos\frac{\theta-\phi}{2}$$

Therefore
$$\frac{h}{k}=\frac{a}{b}\cot\frac{\theta+\phi}{2} \tag{2}$$

Thus, from (1) and (2), we have

$$k = -\frac{b^2}{a^2 m} h.$$

Hence, locus of (h, k) is

$$y = -\frac{b^2}{a^2 m} x.$$

This is the equation of the diameter. Clearly all the diameters of the ellipse pass through its centre.

8.2 Alternative definition of a diameter. Following the arguments as in case of parabola (see § 9 of Chapter VII) one can give an alternative definition to a diameter of an ellipse as

The locus of the middle points of a system of parallel chords of an ellipse is called a diameter of the ellipse.

Let $y = m'x$ be the diameter of the ellipse which bisects all the chords parallel to the lime $y = mx$. Then

$$m' = -\frac{b^2}{a^2 m}, \text{i.e., } mm' = -\frac{b^2}{a^2}.$$

The symmetry of the result shows that the diameter $y = mx$ bisects all the chords parallel to the diameter $y = m'x$. Hence, if one diameter of an ellipse bisects chords parallel to a second, the second diameter will bisect all chords parallel to the first.

8.3 Conjugate diameters.

Two diameters of an ellipse are said to be **conjugate** *when each bisects chords parallel to the other.*

Thus, two dimeters $y = mx$ and $y = m'x$ are conjugate if $mm' = -\frac{b^2}{a^2}$.

Ex. Prove that the tangents drawn at the extremities of any diameter of an ellipse are parallel.

8.4. Relation between conjugate diameters and conjugate lines. We know that the polar of a point and the chord of contact of tangents drawn to an ellipse are the same. Two conjugate lines are such that each contains the pole of the other. Now the pole of the diameter PCP' is the point of intersection of the tangents at P and P'. But these tangents are parallel and thus their point of intersection is at infinity and must lie on the line CD, which is parallel to them. Similarly, the pole of the diameter DCD' lies on PCP'. Thus, CP and CD are conjugate lines in view of the definition given in § 5.3 of Chapter 5. Hence, the conjugate diameters are the particular case of conjugate lines.

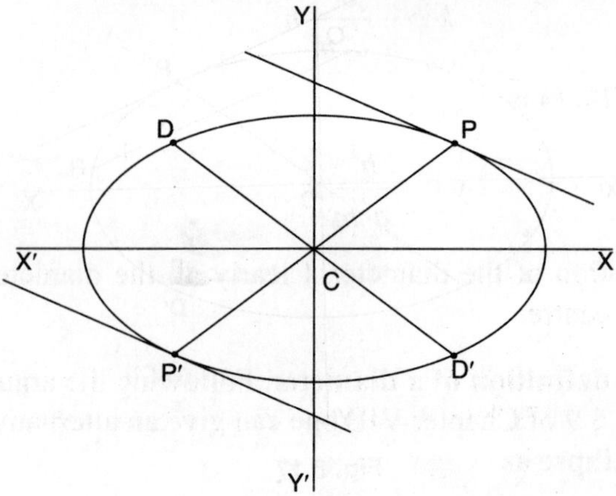

Fig. 8.16

8.5 Propositions on conjugate diameters. Throughout this section we shall consider the equation of the ellipse in the form

$$\frac{x^2}{a^2}+\frac{y^2}{b^2}=1.$$

Proposition I. *The tangent at an extremity of any diameter is parallel to the chords bisected by that diameter.*

Let $P(h, k)$ be any point on the ellipse and the tangent at P be parallel to the chord QR whose equation is $y = mx + c$.

But the equation of the tangent at P is $\dfrac{xh}{a^2}+\dfrac{yk}{b^2}=1.$

Therefore $$m=-\frac{b^2h}{a^2k}, \text{i.e., } k=-\frac{b^2}{a^2m}h.$$

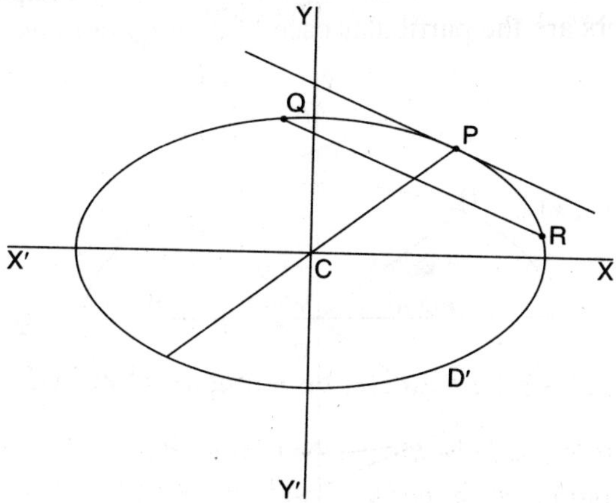

Fig. 8.17

Hence, the locus of the point $P(h, k)$ is $y = -\dfrac{b^2}{a^2 m}x$.

This is the equation of the diameter which bisects the chord QR and all chords parallel to it.

Proposition II. *The tangents at the ends of any chord meet on the diameter which bisects the chord.*

Let Q and R be the ends of a given chord $y = mx + c$.

Let $P(h, k)$ be the point of intersection of the tangents at Q and R. Then QR is the chord of contact of $P(h, k)$ and so its equation is $\dfrac{xh}{a^2} + \dfrac{yk}{b^2} = 1$.

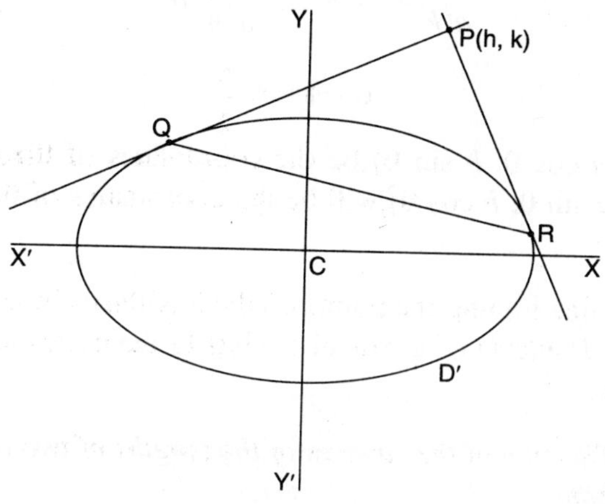

Fig. 8.18

But it must be the same as $y = mx + c$. Therefore

$$m = -\frac{b^2 h}{a^2 k}.$$

Hence, locus of $P(h, k)$ is

$$y = -\frac{b^2}{a^2 m} x,$$

which is the equation of the diameter bisecting the chord QR.

Proposition III. *If P and D be the extremities of a pair of conjugate diameters whose eccentric angles are θ and ϕ; then θ and ϕ differ by a right angle.*

The coordinates of the points P and D are $(a \cos \theta, b \sin \theta)$ and $(a \cos \phi, b \sin \phi)$, respectively. Therefore

$$m = \text{slope of } CP = \frac{b \sin \theta}{a \cos \theta}$$

and $$m' = \text{slope of } CD = \frac{b \sin \phi}{a \cos \phi}.$$

The diameters CP and CD are conjugate if $mm' = -\frac{b^2}{a^2}$, *i.e.*, if

$$\frac{b \sin \theta}{a \cos \theta} \cdot \frac{b \sin \phi}{a \cos \phi} = -\frac{b^2}{a^2}$$

$$\Rightarrow \quad \tan \theta = -\cot \phi = \tan\left(\frac{\pi}{2} + \phi\right)$$

$$\Rightarrow \quad \theta - \phi = \pm \frac{\pi}{2}.$$

Corollary 1. If $(a \cos \theta, b \sin \theta)$ be the coordinates of the extremity of a diameter, then $(-a \sin \theta, b \cos \theta)$ will be the coordinates of the extremity of its conjugate.

Corollary 2. The line joining the points on the auxiliary circle corresponding to the extremities P and D of a pair of conjugate diameters subtends a right angle at the centre.

Proposition IV. *The sum of the squares of the lengths of two conjugate semi-diameters is constant.*

Let P and D be the extremities of two conjugate diameters of the ellipse. Let $(a \cos \theta, b \sin \theta)$ be the coordinates of P. Then, the coordinates of D are $(-a \sin \theta, b \cos \theta)$. Now

$$CP^2 = a^2 \cos^2 \theta + b^2 \sin^2 \theta \text{ and } CD^2 = a^2 \sin^2 \theta + b^2 \cos^2 \theta.$$

Therefore $$CP^2 + CD^2 = a^2 + b^2 = \text{constant}.$$

Proposition V. *The area of the parallelogram which touches an ellipse at the ends of conjugate diameters is constant.*

Let PCP' and DCD' be the conjugate diameters of the ellipse. Let $(a \cos \theta, b \sin \theta)$ be the coordinates of the point P.

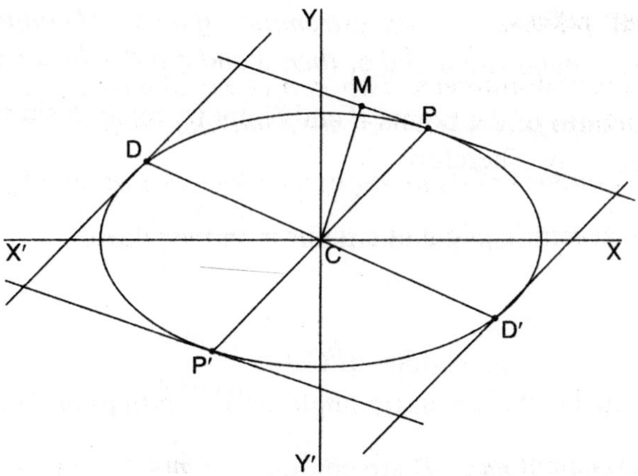

Fig. 8.19

The area of the parallelogram which touches the ellipse at the points P, P', D, D' is $4CD.CM$, where CM is the perpendicular distance of the tangent at P from C. But the equation of the tangent at $P(a \cos \theta, b \sin \theta)$ is

$$\frac{x}{a} \cos \theta + \frac{y}{b} \sin \theta = 1.$$

Therefore $$CM = \frac{1}{\sqrt{\dfrac{\cos^2 \theta}{a^2} + \dfrac{\sin^2 \theta}{b^2}}} = \frac{ab}{\sqrt{b^2 \cos^2 \theta + a^2 \sin^2 \theta}}.$$

Since, the coordinates of D are $(-a \sin \theta, b \cos \theta)$, we have

$$CD = \sqrt{a^2 \sin^2 \theta + b^2 \cos^2 \theta}.$$

Thus $$4CD \cdot CM = 4\sqrt{a^2 \sin^2 \theta + b^2 \cos^2 \theta} \cdot \frac{ab}{\sqrt{b^2 \cos^2 \theta + a^2 \sin^2 \theta}}$$

$$= 4ab,$$

which is constant.

Proposition VI. *The product of the focal distances of a point P on an ellipse is equal to the square of the length of the semi-diameter parallel to the tangent at P.*

Let $(a \cos \theta, b \sin \theta)$ be the coordinates of P. Then

$$SP = a - ae \cos \theta \text{ and } S'P = a + ae \cos \theta.$$

Therefore $\quad SP \cdot S'P = (a - ae \cos \theta)(a + ae \cos \theta) = a^2 \sin^2 \theta + b^2 \cos^2 \theta.$

The semi-diameter CD parallel to the tangent at P is conjugate to CP. Therefore, the coordinates of D are $(-a \sin \theta, b \cos \theta)$. Thus

$$CD^2 = a^2 \sin^2 \theta + b^2 \cos^2 \theta.$$

Hence, the result follows.

8.6 Equi-conjugate diameters. *Two conjugate diameters of an ellipse are said to be* **equi-conjugate** *when they are equal in length.*

To find the equations of equi-conjugate diameters of an ellipse.

Let ϕ be the eccentric angle of a point P on the ellipse

$$\frac{x^2}{a^2} + \frac{y^2}{b^2} = 1.$$

Then $\phi + \frac{\pi}{2}$ will be the eccentric angle of the extremity D of a diameter conjugate to CD. Therefore

$$CP^2 = a^2 \cos^2 \phi + b^2 \sin^2 \phi \text{ and } CD^2 = a^2 \sin^2 \phi + b^2 \cos^2 \phi.$$

Now, CP and CD are equi-conjugate diameters if $CP = CD$, *i.e.*,

$$a^2 \cos^2 \phi + b^2 \sin^2 \phi = a^2 \sin^2 \phi + b^2 \cos^2 \phi$$

$$\Rightarrow \quad (a^2 - b^2) \cos 2\phi = 0$$

$$\Rightarrow \quad \phi = \frac{\pi}{4} \text{ or } \frac{3\pi}{4}.$$

The equation of CP is

$$y = \frac{b \sin \phi}{a \cos \phi} x, \text{ i.e., } y = \pm \frac{b}{a} x.$$

and equation to CD is

$$y = \frac{b\cos\phi}{-a\sin\phi}x, \text{ i.e., } y = \mp\frac{b}{a}x.$$

Note. The equi-conjugate diameters of an ellipse are coincident in direction with the diagonals of the rectangle formed by the tangents at the ends of its axes.

8.7 Supplemental chords. *The two lines drawn from any point on an ellipse to the extremities of any diameter are called* **supplemental chords.**

Any two supplemental chords of an ellipse are parallel to a pair of conjugate diameters.

Let the chords be formed by joining the point Q to the extremities P, P' of the diameter PCP'.

Let θ be the eccentric angle of P. Then eccentric angle of P' will be $\pi + \theta$. Let ϕ be the eccentric angle of Q. Therefore

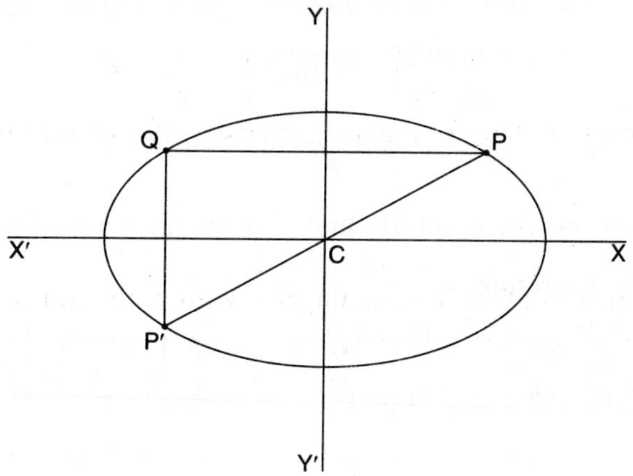

Fig. 8.20

$$m = \text{slope of } QP = \frac{b\sin\theta - b\sin\phi}{a\cos\theta - a\cos\phi} = -\frac{b}{a}\cot\frac{\theta+\phi}{2}$$

and $$m' = \text{slope of } QP' = \frac{b\sin\theta + b\sin\phi}{a\cos\theta + a\cos\phi} = \frac{b}{a}\tan\frac{\theta+\phi}{2}$$

Thus $mm' = -\frac{b^2}{a^2}$. This shows that QP and P' are parallel to a pair of conjugate diameters.

8.8 Examples

Example 1. If the circles be described on two conjugate semi-diameters of the ellipse $\frac{x^2}{a^2}+\frac{y^2}{b^2}=1$ as diameters, prove that the locus of their second points of intersection is the curve $2(x^2+y^2)^2 = a^2x^2+b^2y^2$.

Solution. Let CP and CD be the conjugate semi-diameters of the ellipse. Let ϕ be the eccentric angle of P. Then the coordinates of P and D are, respectively $(a \cos \phi, b \sin \phi)$ and $(-a \sin \phi, b \cos \phi)$.

The equations of the circles on CP and CD as diameters are

$$x(x-a\cos\phi)+y(y-b\sin\phi)=0, \text{ i.e., } x^2+y^2=ax\cos\phi+by\sin\phi$$

and $$x(x+a\sin\phi)+y(y-b\cos\phi)=0, \text{ i.e., } x^2+y^2=-ax\sin\phi+by\cos\phi$$

The required locus is obtained by eliminating ϕ between the above equations, *i.e.*, $2(x^2+y^2)^2 = a^2x^2+b^2y^2$.

Example 2. If the points of intersection of the ellipses

$$\frac{x^2}{a^2}+\frac{y^2}{b^2}=1 \text{ and } \frac{x^2}{\alpha^2}+\frac{y^2}{\beta^2}=1$$

be the end points of the conjugate diameters of the former, prove that $\frac{a^2}{\alpha^2}+\frac{b^2}{\beta^2}=2.$

Solution. The equation of the lines joining the centre to the points of intersection of the ellipses is $\frac{x^2}{a^2}+\frac{y^2}{b^2} = \frac{x^2}{\alpha^2}+\frac{y^2}{\beta^2}$, *i.e.*, $\left(\frac{1}{a^2}-\frac{1}{\alpha^2}\right)x^2+\left(\frac{1}{b^2}-\frac{1}{\beta^2}\right)y^2$ $=0.(1)$

Let m_1 and m_2 be the slopes of the lines represented by (1). Then

$$m_1m_2 = \frac{\frac{1}{a^2}-\frac{1}{\alpha^2}}{\frac{1}{b^2}-\frac{1}{\beta^2}}.$$

The lines represented by (1) will be the conjugate diameters of the former ellipse if $m_1m_2 = -\frac{b^2}{a^2}$, *i.e.*, $\frac{a^2}{\alpha^2}+\frac{b^2}{\beta^2}=2.$

Example 3. CP and CD are conjugate diameters of the ellipse $\frac{x^2}{a^2}+\frac{y^2}{b^2}=1.$ Prove that the locus of the orthocentre of the triangle CPD is the curve

$$2(b^2y^2+a^2x^2)^3 = (a^2-b^2)^2(b^2y^2-a^2x^2)^2.$$

Solution. Let ϕ be the eccentric angle of P. Then, the coordinates of P and D are, respectively, $(a \cos \phi, b \sin \phi)$ and $(-a \sin \phi, b \cos \phi)$.

Since, the tangent at P is parallel to CD, the perpendicular from P on CD is the normal at P and its equation is

$$ax \sec \phi - by \operatorname{cosec} \phi = a^2 - b^2. \quad (1)$$

Similarly, the equation of the perpendicular from D on CP is

$$-ax \operatorname{cosec} \phi - by \sec \phi = a^2 - b^2. \quad (2)$$

The locus of the orthocentre of the triangle CPD is obtained by eliminating ϕ between (1) and (2). Now, from equations (1) and (2), we have

$$\frac{\sec \phi}{(by - ax)(a^2 - b^2)} = \frac{\operatorname{cosec} \phi}{(by + ax)(a^2 - b^2)} = \frac{1}{-(b^2y^2 + a^2x^2)}$$

$$\Rightarrow \quad \cos \phi = \frac{-(b^2y^2 + a^2x^2)}{(by - ax)(a^2 - b^2)} \text{ and } \sin \phi = \frac{-(b^2y^2 + a^2x^2)}{(by + ax)(a^2 - b^2)}$$

$$\text{Now } \cos^2 \phi + \sin^2 \phi = \frac{(b^2y^2 + a^2x^2)^2}{(by - ax)^2(a^2 - b^2)^2} + \frac{(b^2y^2 + a^2x^2)^2}{(by + ax)^2(a^2 - b^2)^2}$$

$$\Rightarrow \quad 1 = \frac{(b^2y^2 + a^2x^2)^2}{(a^2 - b^2)^2}\left\{\frac{1}{(by - ax)^2} + \frac{1}{(by + ax)^2}\right\}$$

$$\Rightarrow \quad 2(b^2y^2 + a^2x^2)^3 = (a^2 - b^2)^2(b^2y^2 - a^2x^2)^2.$$

Example 4. If P and D be the ends of conjugate diameters of the ellipse $x^2/a^2 + y^2/b^2 = 1$, prove that the locus of the foot of the perpendicular from the centre of the ellipse on PD is $2(x^2 + y^2)^2 = a^2x^2 + b^2y^2$.

Solution. Let ϕ be the eccentric angle of P. Then the coordinates of P and D are, respectively, $(a \cos \phi, b \sin \phi)$ and $\left(a \cos\left(\frac{\pi}{2} + \phi\right), b \sin\left(\frac{\pi}{2} + \phi\right)\right)$ and the equation of PD is

$$\frac{x}{a} \cos\left(\frac{\pi}{4} + \phi\right) + \frac{y}{b} \sin\left(\frac{\pi}{4} + \phi\right) = \frac{1}{\sqrt{2}}. \quad (1)$$

The equation of the perpendicular drawn from the centre (0, 0) to (1) is

$$\frac{x}{b} \sin\left(\frac{\pi}{4} + \phi\right) - \frac{y}{a} \cos\left(\frac{\pi}{4} + \phi\right) = 0. \quad (2)$$

The locus of the foot of the perpendicular is obtained by eliminating ϕ between (1) and (2). Now, from equation (2), we have

$$\tan\left(\frac{\pi}{4}+\phi\right)=\frac{by}{ax}$$

$$\Rightarrow \quad \sin\left(\frac{\pi}{4}+\phi\right)=\frac{by}{\sqrt{a^2x^2+b^2y^2}} \text{ and } \cos\left(\frac{\pi}{4}+\phi\right)=\frac{ax}{\sqrt{a^2x^2+b^2y^2}}.$$

Substituting these values in (1), we get

$$\frac{x}{a}\cdot\frac{ax}{\sqrt{a^2x^2+b^2y^2}}+\frac{y}{b}\cdot\frac{by}{\sqrt{a^2x^2+b^2y^2}}=\frac{1}{\sqrt{2}}$$

$$\Rightarrow \quad 2(x^2+y^2)^2=a^2x^2+b^2y^2.$$

Example 5. *CP* and *CD* are conjugate diameters of the ellipse $\frac{x^2}{a^2}+\frac{y^2}{b^2}=1.$ Prove that the locus of the point of intersection of normal at *P* with *CD* is

$$\frac{a^2}{x^2}+\frac{b^2}{y^2}=\left(\frac{a^2-b^2}{x^2+y^2}\right)^2.$$

Solution. Let ϕ be the eccentric angle of *P*. Then, the coordinates of *P* and *D* are, respectively,

$$(a\cos\phi,\ b\sin\phi) \text{ and } \left(a\cos\left(\frac{\pi}{2}+\phi\right), b\sin\left(\frac{\pi}{2}+\phi\right)\right).$$

The equation of the normal at *P* is

$$ax\sec\phi-by\operatorname{cosec}\phi=a^2-b^2. \tag{1}$$

and the equation of *CD* is

$$y=\frac{b\sin\left(\frac{\pi}{2}+\phi\right)}{\cos a\left(\frac{\pi}{2}+\phi\right)}x, \text{ i.e., } ay\sec\phi+bx\operatorname{cosec}\phi=0. \tag{2}$$

The required locus is obtained by eliminating ϕ between (1) and (2). Now, from equations (1) and (2), we get

$$\frac{\sec\phi}{bx(a^2-b^2)}=\frac{\operatorname{cosec}\phi}{-ay(a^2-b^2)}=\frac{1}{abx^2+aby^2}$$

$$\Rightarrow \qquad \cos\phi = \frac{a(x^2+y^2)}{x(a^2-b^2)} \text{ and } \sin\phi = -\frac{b(x^2+y^2)}{y(a^2-b^2)}.$$

Thus

$$\cos^2\phi + \sin^2\phi = \frac{a^2(x^2+y^2)^2}{x^2(a^2-b^2)^2} + \frac{b^2(x^2+y^2)^2}{y^2(a^2-b^2)^2}$$

$$\Rightarrow \qquad \frac{a^2}{x^2} + \frac{b^2}{y^2} = \frac{(a^2-b^2)^2}{(x^2+y^2)^2}.$$

Example 6. *CP* and *CD* are conjugate diameters of the ellipse $\frac{x^2}{a^2}+\frac{y^2}{b^2}=1$. A tangent parallel to *PD* meets *CP* and *CD* is *L* and *M* respectively. Prove that *L* and *M* lie on the ellipse $\frac{x^2}{a^2}+\frac{y^2}{b^2}=2$.

Solution. Let ϕ be the eccentric angle of *P*. Then the coordinates of *P* and *D* are, respectively $(a\cos\phi, b\sin\phi)$ and $(-a\sin\phi, b\cos\phi)$. Now

$$\text{slope of } PD = \frac{b\sin\phi - b\cos\phi}{a\cos\phi + a\sin\phi} = \frac{b(\tan\phi - 1)}{a(1+\tan\phi)}$$

The equation of a tangent parallel to *PD* is

$$y = mx + \sqrt{a^2m^2+b^2}, \qquad (1)$$

where $m = \frac{b(\tan\phi-1)}{a(1+\tan\phi)}$.

Further, the equation of *CP* is

$$y = \frac{b\sin\phi}{a\cos\phi}x, \text{ i.e., } \tan\phi = \frac{ay}{bx}.$$

Therefore

$$m = \frac{b\left(\frac{ay}{bx}-1\right)}{a\left(1+\frac{ay}{bx}\right)} = \frac{b(ay-bx)}{a(bx+ay)}.$$

Substituting this value of *m* in (1), we get

$$y = \frac{b(ay-bx)}{a(ay+bx)}.x + \sqrt{a^2.\frac{b^2(ay-bx)^2}{a^2(ay+bx)^2}+b^2}$$

$$\Rightarrow \quad \frac{x^2}{a^2}+\frac{y^2}{b^2}=2.$$

This is the lócus of the point L. Similarly, the locus of M can be obtained as the same ellipse. Hence, the result follows.

Exercises

1. P and D are the extremities of a pair of conjugate radii of the ellipse $\frac{x^2}{a^2}+\frac{y^2}{b^2}=1$. Prove that

   (*i*) PD touches the ellipse $\frac{x^2}{a^2}+\frac{y^2}{b^2}=\frac{1}{2}$.

   (*ii*) The locus of the middle point of PD is $\frac{x^2}{a^2}+\frac{y^2}{b^2}=\frac{1}{2}$.

   (*iii*) The tangents at P and D intersect on the ellipse $\frac{x^2}{a^2}+\frac{y^2}{b^2}=2$.

2. PCP', DCD' are conjugate diameters of an ellipse; PD is produced to R so the $DR = 2PD$. Find the locus of R.

3. Prove that the locus of the intersections of the normals at the ends of conjugate diameters of the ellipse $\frac{x^2}{a^2}+\frac{y^2}{b^2}=1$ is the curve

$$2(a^2x^2+b^2y^2)^3=(a^2-b^2)^2(a^2x^2-b^2y^2)^2.$$

4. In an ellipse, a pair of conjugate diameters is produced to meet the directrix. Prove that the orthocentre of the triangle so formed is the focus of the ellipse.

5. If $x\cos\alpha + y\sin\alpha = p$ is a chord joining the ends of conjugate semi-diameters of the ellipse $\frac{x^2}{a^2}+\frac{y^2}{b^2}=1$, prove that $a^2\cos^2\alpha + b^2\sin^2\alpha = 2p^2$.

6. Lines are drawn through the foci of an ellipse perpendicular respectively to a pair of conjugate diameters and intersect in R. Prove that the locus of R is a concentric ellipse.

7. Lines are drawn through the origin perpendicular to the tangents from a point P to the ellipse $\frac{x^2}{a^2}+\frac{y^2}{b^2}=1$. If the lines are conjugate diameters of the ellipse, prove that P lies on the curve $a^2x^2+b^2y^2=a^4+b^4$.

8. If P, D be the extremities of conjugate diameters of an ellipse, and PP', DD' be chords parallel to an axis of the ellipse; prove that PD' and $P'D$ are parallel to the equi-conjugate diameters.

9. CP and CD are conjugate semi-diameters of an ellipse, and the tangent at P meets any other pair of conjugate diameters in T and T'. Prove that $TP \cdot PT' = CD^2$.

10. If P, D are extremities of conjugae diameters, and the tangent at P cut the major axis in T, and the tangent at D cut the minor axis in T'; prove that TT' will be parallel to one of the equi-conjugates.

11. Two fixed conjugate diameters of an ellipse are met in the points P, Q respectively by two lines OP, OQ which pass through a fixed point O and are parallel to any other pair of conjugate diameters. Prove that the locus of the middle point of PQ is a line.

12. CP and CD are conjugate semi-diameters of an ellipse whose foci are S and S'. The tangents to the ellipse at P and D meet in T. Prove that SP, $S'P$, SD, $S'D$ touch a circle whose centre is T. Find the radius of the circle.

13. A variable line meets and ellipse $\frac{x^2}{a^2}+\frac{y^2}{b^2}=1$ at A and B. P is a fixed point on the ellipse. If PA and PB are parallel to conjugate diameters of the fixed ellipse $\frac{x^2}{a^2+c^2}+\frac{y^2}{b^2+c^2}=1$, prove that AB passes through a fixed point on the normal at P to the first ellipse.

14. Perpendiculars PM, PN are drawn from any point P on the equi-conjugate diameters of an ellipse. Prove that the perpendicular from P on its polar line bisects MN.

15. Two conjugate diameters of an ellipse are drawn, and their four extremities are joined to any point on a given circle whose centre is at the centre of the ellipse. Prove that the sum of the squares of the lengths of these four lines is constant.

16. Prove that the sum of the products of the perpendiculars from the two extremities of each of two conjugate diameters on any tangent to an ellipse is equal to the square of the perpendicular from the centre on that tangent.

17. Q is a point on the normal at any point P of an ellipse whose centre is C such that the lines CP, CQ make equal angles with the axis of the ellipse. Prove that PQ is proportional to the diameter conjugate to CP.

18. Normals to an ellipse are drawn at the extremities of a chord parallel to one of the equi-conjugate diameters. Prove that the locus of the intersection of the normals is a diameter which is perpendicular to the other equi-conjugate diameter.

19. If θ and φ be the angles subtended by the major axis of an ellipse at the extremities of a pair of conjugate diameters, prove that $\cot^2 \theta + \cot^2 \phi$ is constant.

Miscellaneous Exercises

1. If S be the focus of an ellipse and BB' the minor axis, prove that $SB = SB' = a$, where $2a$ is the length of the major axis.

2. Find the equations of the tangents and normals at the ends of the latera recta of $2x^2 + 3y^2 = 6$.

3. Find the equations of the tangents to the ellipse

$$\frac{x^2}{a^2} + \frac{y^2}{b^2} = 1$$

which make equal intercepts on the axes.

4. From a point P on a directrix of an ellipse a pair of tangents PQ and PR are drawn to an ellipse. Prove that Q and R are collinear with the corresponding focus of the ellipse.

5. The pole of the normal at a point P on an ellipse is at Q. Prove that the product of the perpendiculars from Q and the centre of the ellipse to this normal is equal to the product of the focal distances from P.

6. The tangent to an ellipse cuts the directrix corresponding to the focus S at Q. Prove that the angle PSQ is a right angle.

7. Prove that the lines $lx + my + n = 0$, and $l'x + m'y + n' = 0$, are conjugate lines of the ellipse $\frac{x^2}{a^2} + \frac{y^2}{b^2} = 1$ if $a^2ll' + b^2mm' = nn'$. If two conjugate lines cut at right angles at (x_1, y_1), and one of them make an angle θ with the x-axis, then

$$\tan 2\theta = \frac{2x_1y_1}{(x_1^2 - a^2) - (y_1^2 - b^2)}.$$

8. Prove that the area of the triangle with vertices at the points α, β and γ of the ellipse $\frac{x^2}{a^2} + \frac{y^2}{b^2} = 1$ is $2ab \sin \frac{\beta - \gamma}{2} \sin \frac{\gamma - \alpha}{2} \sin \frac{\alpha - \beta}{2}$.

9. Tangents are drawn to an ellipse at the points whose eccentric angles are α, β, γ. Find the condition that the circle circumscribing the triangle formed by them should pass through the centre of the ellipse.

10. The point S is a focus of the ellipse $\frac{x^2}{a^2} + \frac{y^2}{b^2} = 1$, and P is a point on the ellipse such that SP is perpendicular to the x-axis. The tangent and normal to the ellipse at P meet the y-axis in Q and R, respectively. If S' is the other focus of the ellipse, prove that $QR = S'P$.

11. A circle $x^2 + y^2 + 2gx + 2fy + c = 0$ meets the ellipse $\frac{x^2}{a^2} + \frac{y^2}{b^2} = 1$ in the points $(a \cos \phi, b \sin \phi)$, where $\phi = \alpha, \beta, \gamma, \delta$. Prove that $\alpha + \beta + \gamma + \delta = 2n\pi$, where n is an integer.

12. If tangents at P and Q to an ellipse meet in T, and PQ meets the directrix in K, prove that SK bisects the angle PSQ externally. Deduce that the angle KST is a right angle.

13. A system of ellipses is described with a given focus and corresponding directrix. Prove that the locus of the extremities of their minor axes is a parabola.

14. PNP' is a double ordinate of an ellipse whose centre is C, and the normal at P meets the CP' in R. Prove that the locus of R is an ellipse.

15. If the perpendicular from the centre C of an ellipse to the tangent at any point P meet the focal distance, produced if necessary, in R, prove that the locus of R is a circle.

16. Prove that the locus of the point of intersection of normals at the ends of chords drawn through any fixed point on the major axis of an ellipse is an ellipse, which cuts the major axis in two points the product of whose distances from the centre is a^2e^4.

17. Prove that the locus of the middle points of chords of the conic $\frac{x^2}{a^2+\lambda} + \frac{y^2}{b^2+\lambda} = 1$ which touch the ellipse $\frac{x^2}{a^2} + \frac{y^2}{b^2} = 1$ is

$$\left(\frac{x^2}{a^2+\lambda} + \frac{y^2}{b^2+\lambda}\right)^2 = \frac{a^2x^2}{(a^2+\lambda)^2} + \frac{b^2y^2}{(b^2+\lambda)^2}.$$

18. Prove that if (x', y') be the middle point of a chord of the ellipse $\frac{x^2}{a^2} + \frac{y^2}{b^2} = 1$, (x_1, y_1) the point of intersection of the normals and (x_2, y_2) that of the tangents at its extremities, then

$$\frac{a^2x_1}{x_2} + \frac{b^2y_1}{y_2} = (a^2 - b^2)\left(\frac{x_2x'}{a^2} - \frac{y_2y'}{b^2}\right).$$

19. From any point of eccentric angle ϕ on the ellipse $\frac{x^2}{a^2} + \frac{y^2}{b^2} = 1$, three normals are drawn to the curve. Prove that the equation of the circum circle of the triangle formed by their feet is

$$x^2 + y^2 - \frac{b^2}{a} x \cos \phi - \frac{a^2}{b} y \sin \phi = a^2 + b^2.$$

20. A point P moves so that the chord of contact of the tangents from P to the ellipse $\frac{x^2}{a^2} + \frac{y^2}{b^2} = 1$ touches the ellipse $\frac{x^2}{a^2} + \frac{y^2}{b^2} = \frac{1}{4}$. Find the locus of P.

21. P and Q are variable points on the ellipse $\frac{x^2}{a^2} + \frac{y^2}{b^2} = 1$ such that the lines joining them to its centre are perpendicular, and the tangents at P and Q meet in T. Find the locus of T.

22. P, Q, R are points on the ellipse $\frac{x^2}{a^2} + \frac{y^2}{b^2} = 1$ with eccentric angles θ, α, β. PQ goes through $(ae, 0)$ and PR through $(-ae, 0)$. Prove that

$$\tan \frac{\alpha}{2} \tan \frac{\beta}{2} = \frac{(1-e)^2}{(1+e)^2}.$$

23. The area of quadrilateral formed by joining the points on the ellipse $\frac{x^2}{a^2} + \frac{y^2}{b^2} = 1$ whose eccentric angles are $\alpha, \beta, \gamma, \delta$ is

$$\tfrac{1}{2} ab \{\sin (\beta - \alpha) + \sin (\gamma - \beta) + \sin (\delta - \gamma) + \sin (\alpha - \delta)\}.$$

24. From any point on the ellipse $\frac{x^2}{a^2} + \frac{y^2}{b^2} = 1$ normals are drawn to the curve. Prove that the locus of the orthocentre of the triangle formed by joining their feet is the ellipse

$$a^2x^2 + b^2y^2 = \left(\frac{a^4 + b^4}{a^2 - b^2}\right)^2.$$

25. If PSQ, $PS'R$ be focal chords of the ellipse $\frac{x^2}{a^2} + \frac{y^2}{b^2} = 1$, where S and S' are the foci, prove that the equation of the chord QR is

$$\frac{x}{a} \cos \theta + \frac{y}{b} \frac{(1+e^2)}{(1-e^2)} \sin \theta + 1 = 0,$$

where θ is the eccentric angle of P.

26. If the pole of the normal at P lies on the normal at Q, then prove that the pole of the normal at Q lies on the normal at P.

27. Find the locus of the point of intersection of the tangent at one end of a focal chord of an ellipse with the normal at the other end.

28. If a pair of tangents to a conic be at right angles to one another, prove that the product of the perpendiculars from the centre and the intersection of the tangents to the chord of contact is constant.

29. The lines PS, PS' joining any point P on the ellipse $\frac{x^2}{a^2}+\frac{y^2}{b^2}=1$ to the foci S and S' meet the curve again in Q, Q'. The tangents at Q and Q' meet in T. Prove that the locus of T as P moves round the curve is

$$(1+e^2)^2\frac{x^2}{a^2}+(1-e^2)^2\frac{y^2}{b^2}=(1+e^2)^2.$$

30. If a point P on an ellipse be such that all chords drawn through a given point Q subtend a right angle at P, prove that P must be the foot of one of the normals that pass through Q.

31. Focal chords are drawn through any point on an ellipse, prove that the radical axis of the two circles described on them as diameters is normal to a coaxal and concentric ellipse.

32. A point P of an ellipse $\frac{x^2}{a^2}+\frac{y^2}{b^2}=1$ is joined to the points $(\pm c, 0)$, and the joints meet the ellipse again in Q and R. Prove that the locus of the point of intersection of the tangents to the ellipse at Q and R is the ellipse

$$\frac{x^2}{a^2}+\frac{(a^2-c^2)^2}{(a^2+c^2)^2}\cdot\frac{y^2}{b^2}=1.$$

33. The lines joining the point P to the foci S and S' of an ellipse cut it again at A and B. Prove that the tangents at A and B and the normal at P are concurrent.

34. SP and SQ are radii vectros from a focus of an ellipse, parallel to two conjugate diameters. Prove that the tangents at P and Q intersect in an ellipse having the same eccentricity as the original curve with the centre at a distance $\frac{ae}{1-e^2}$ from its centre.

35. If PL and PM be the perpendiculars from the point P to two fixed conjugate diameters of an ellipse, prove that the perpendicular from P to its polar divides LM in a constant ratio.

36. PQ is any chord of an ellipse parallel to one of the equi-conjugates and the tangents at P and Q meet in T. Prove the circle PTQ passes through the centre.

37. The line joining two extremities of any two diameters of an ellipse is either parallel or conjugate to the line joining two extremities of their conjugate diameters.

38. If a length PQ be taken in the normal at any point P of an ellipse whose centre is C, equal in length to the semi-diameter which is conjugate to CP, prove that Q lies on one, or the other of two circles.

39. Prove that the locus of the centre of a circle which cuts the ellipse $\frac{x^2}{a^2}+\frac{y^2}{b^2}=1$ in the fixed point (α, β) and in two other points at the extremities of a diameter is the ellipse $2a^2x^2 + 2b^2y^2 = (a^2 - b^2)(\alpha x - \beta y)$.

40. Prove that a chord which joins the ends of a pair of conjugate diameters of an ellipse always touches a similar ellipse.

41. A pair of conjugate diameters is produced to meet the directrix. Prove that the orthocentre of the triangle so formed is at the focus.

The Hyperbola

9 CHAPTER

1. Definition

The **hyperbola** *is the locus of a point which moves so that its distance from a fixed point bears a constant ratio (> 1) to its distance from a fixed line.*

The fixed point is called the **focus**, constant ratio the **eccentricity** and the fixed line the **directrix**.

1.1 Standard equation of a hyperbola. *To find the equation of a hyperbola.*

Let S be the focus and ZK the directrix. Draw SZ perpendicular to the directrix. Let e be the eccentricity of the hyperbola. Let A divide ZS such that $AS \cdot ZA = e : 1$. Then A lies on the curve.

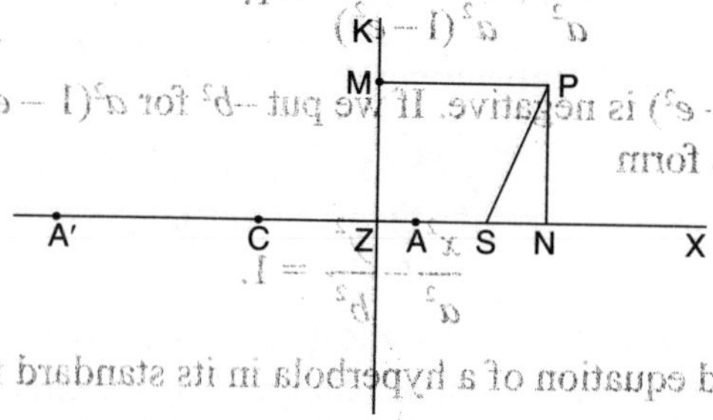

Fig. 9.1

Since, $e > 1$, there is another point A' on SZ produced such that $A'S : A'Z = e : 1$. Thus, $AS = e \cdot ZA$ and $A'S = e \cdot A'Z$. Let C be the middle point of AA' and let $AA' = 2a$.

Now $$AS + A'S = e(ZA + A'Z)$$

$$\Rightarrow \quad 2CS = e \cdot AA' = 2ae$$

$$\Rightarrow \quad CS = ae.$$

Further $$A'S - AS = e\,(A'Z - ZA)$$

$$\Rightarrow \quad AA' = e \cdot 2CZ$$

$$\Rightarrow \quad 2a = e \cdot 2CZ$$

$$\Rightarrow \quad CZ = \frac{a}{e}.$$

Let us take C to be the origin, CA the x-axis and a line through C perpendicular to CA the y-axis.

Let $P(h, k)$ be any point on the locus (hyperbola). Draw PM perpendicular from P to the directrix ZK. Then, by definition of hyperbola, we have

$$SP = e \cdot MP.$$

But $\quad SP = \sqrt{(h-ae)^2 + k^2}$ and $MP = ZN = CN - CZ = h - \frac{a}{e}$

Therefore $\quad (h - ae)^2 + k^2 = e^2 \left(h - \frac{a}{e}\right)^2$

$$\Rightarrow \quad h^2(1 - e^2) + k^2 = a^2(1 - e^2)$$

$$\Rightarrow \quad \frac{h^2}{a^2} + \frac{k^2}{a^2(1-e^2)} = 1.$$

Hence, locus of (h, k) is

$$\frac{x^2}{a^2} + \frac{y^2}{a^2(1-e^2)} = 1.$$

Since, $e > 1$, $a^2(1 - e^2)$ is negative. If we put $-b^2$ for $a^2(1 - e^2)$, the equation of the curve takes the form

$$\frac{x^2}{a^2} - \frac{y^2}{b^2} = 1.$$

This is the required equation of a hyperbola in its standard form.

Ex. Find the equation of a hyperbola whose focus is the point (2, 2), directrix is the line $x + 2y = 3$ and eccentricity is $\frac{3}{2}$.

1.2 Shape of a hyperbola. *To trace the hyperbola*

$$\frac{x^2}{a^2} - \frac{y^2}{b^2} = 1.$$

In order to sketch the graph of the given hyperbola, we notice the following:

(*i*) The curve is symmetrical about both axes and the origin, since, if (x, y) lies on the curve, the points $(-x, y)$, $(x, -y)$ and $(-x, -y)$ also lie on the curve.

(*ii*) The curve meets the x-axis in the points $A(a, 0)$ and $A'(-a, 0)$.

(*iii*) The curve meets the y-axis in imaginary points.

(*iv*) Writing the equation of the curve in either of the forms

$$y = \pm \frac{b}{a} \sqrt{x^2 - a^2}, x = \pm \frac{a}{b} \sqrt{b^2 + y^2}.$$

we see that there are no points on the curve if $-a < x < a$ and the points $(\pm a, 0)$ lie on it. There is no limitation on the possible values of y.

Further, from the second equation we note that y may have any real value and from the first that x may have any real value except those for which $x^2 < a^2$. Hence, the hyperbola extends indefinitely far from the axis in each quadrants. Accordingly, the hyperbola consists of two separate parts or branches as shown in Fig. 9.2.

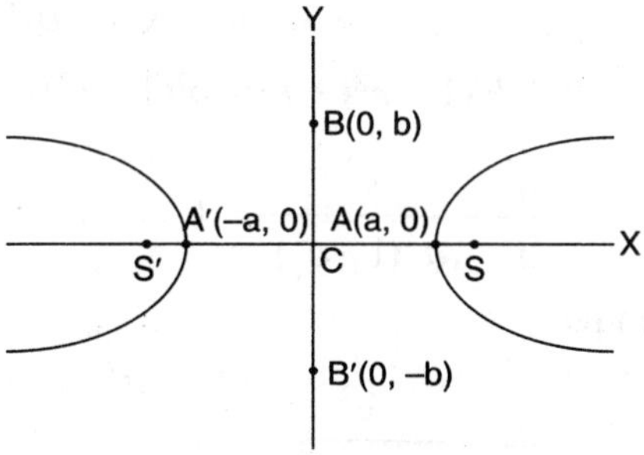

Fig. 9.2

1.3 Pair of lines as a limiting case of a hyperbola. The equation of a hyperbola referred to its axes is

$$\frac{x^2}{a^2} - \frac{y^2}{a^2(e^2-1)} = 1, \; i.e., \; x^2 - \frac{y^2}{e^2-1} = a^2.$$

If we keep e constant and make a smaller and smaller, the equation of the hyperbola approximates to

$$x^2 - \frac{y^2}{e^2-1} = 0.$$

This represents a pair of lines.

Hence, *a pair of lines may be regarded as a limiting case of a hyperbola whose axes are infinitely small, while their ratio is finite.*

Note. An application of the hyperbola is that of locating the place from which a sound, such as gun fire, emanates. From the difference in the times at which the sound reaches two listening posts, the difference between distances of the posts from the gun can be determined. Then the gun is known to be located on a branch of a hyperbola of which the posts are foci.

The position of the gun on this curve can be found by the use of a third listening post. Either of the two posts and the third are foci of a branch of another hyperbola on which the gun is located. Hence, the gun is at the intersection of the two branches.

The principle used in finding the location of a gun is also employed by a reader-equipped airplane to determine its location. In this case, the plane receives singnals from three stations of known locations. □

The points $A(a, 0)$, $A'(-a, 0)$ are called the vertices of the hyperbola. The points $B(0, b)$, $B'(0, -b)$ do not lie on the hyperbola.

The two lines $A'A$ and BB' with respect to which the curve is symmetrical are called the axes of the hyperbola. The axis $A'A$ is called the transverse axis whereas the axis BB' is called the conjugate axis.

Since, the curve is symmetrical about the origin, the point C (the origin) bisects every chord through it. Therefore, the point C is referred to the **centre** of the hyperbola.

Further, the symmetry of the curve about its conjugate axis (y-axis) verifies that there must be a second focus S' situated on the transverse axis at the same distance from C as S, and a second directrix corresponding to S' and parallel to the original directrix, which would be cutting the transverse axis in Z' such that $CZ = CZ'$.

Thus, the coordinates of the foci S and S' are $(ae, 0)$ and $(-ae, 0)$, respectively.

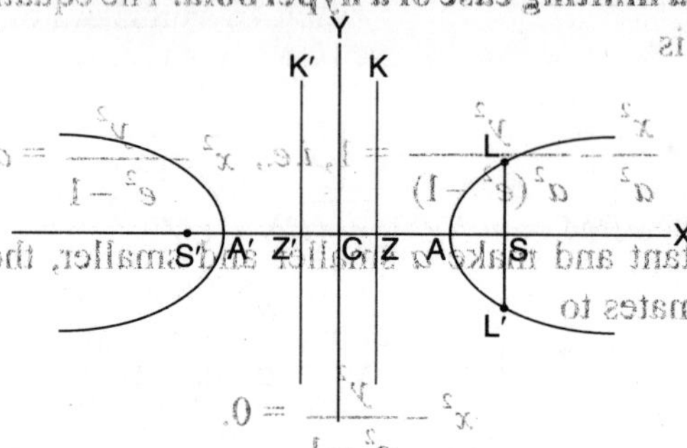

Fig. 9.3

The equations of the directrices ZK and $Z'K'$ are $x = \dfrac{a}{e}$ and $x = -\dfrac{a}{e}$, respectively.

The chord through a focus perpendicular to the transverse axis is called the **latus rectum**.

To find the length of the semi-latus rectum.

Let LSL' be latus rectum through the focus $(ae, 0)$ and let $SL = l$. Then the coordinates of the point L are (ae, l). But L lies on the hyperbola. Therefore

$$\frac{a^2e^2}{a^2} - \frac{l^2}{b^2} = 1, \text{ i.e., } l^2 = b^2(e^2 - 1) = \frac{b^4}{a^2}.$$

Hence, the length of the semi-latus rectum is $\frac{b^2}{a}$.

1.4 Focal distance of a point. *To prove that the difference of the focal distances of any point P on the hyperbola*

$$\frac{x^2}{a^2} - \frac{y^2}{b^2} = 1,$$

from foci S and S′, is constant and is equal to 2a.

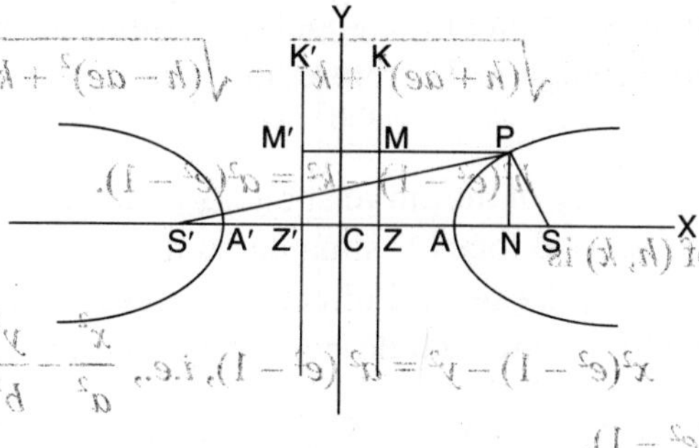

Fig. 9.4

Let (x, y) be the coordinates of the point P. We know that

$$SP = e{\cdot}PM = e{\cdot}NZ = e{\cdot}(CN - CZ)$$

$$= e\cdot\left(x - \frac{a}{e}\right) = ex - a$$

and $$S'P = e{\cdot}PM' = e{\cdot}NZ' = e{\cdot}(CN + CZ')$$

$$= e\left(x + \frac{a}{e}\right) = ex + a.$$

Thus $$S'P - SP = 2a$$

A hyperbola may also be defined as *the locus of a point which moves so that the diffenrence of its distances from two fixed points is constant.*

To find the equation of the curve from the above definition.

Let the distance between the two fixed points S and S' be $2a$.

Let C be the middle point of SS'. Taking CS as x-axis and a line perpendicular to it through C as y-axis. Let $P(h, k)$ be the moving point so that

$$S'P - SP = \text{constant} = 2a \text{ (say)}.$$

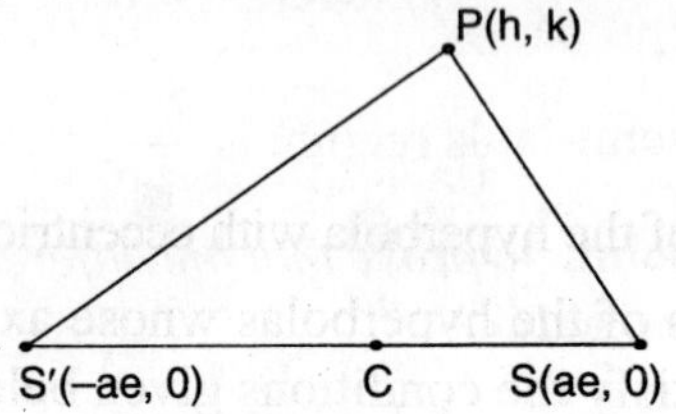

Fig. 9.5

But $\quad SP = \sqrt{(h-ae)^2 + k^2}$ and $S'P = \sqrt{(h+ae)^2 + k^2}$.

Therefore $\quad \sqrt{(h+ae)^2 + k^2} - \sqrt{(h-ae)^2 + k^2} = 2a$

$\Rightarrow \quad h^2(e^2 - 1) - k^2 = a^2(e^2 - 1).$

Hence, the locus of (h, k) is

$$x^2(e^2 - 1) - y^2 = a^2(e^2 - 1), \text{ i.e., } \frac{x^2}{a^2} - \frac{y^2}{b^2} = 1,$$

where $b^2 = a^2(e^2 - 1)$.

1.5 Polar equation of the hyperbola. *To find the polar equation of the hyperbola referred to the centre as pole.*

Let the equation of the hyperbola be

$$\frac{x^2}{a^2} - \frac{y^2}{b^2} = 1.$$

Putting $x = r\cos\theta$, $y = r\sin\theta$, we get

$$\frac{1}{r^2} = \frac{\cos^2\theta}{a^2} - \frac{\sin^2\theta}{b^2}, \text{ i.e., } \frac{1}{r^2} = \frac{1}{a^2} - \left(\frac{1}{a^2} + \frac{1}{b^2}\right)\sin^2\theta.$$

From this equation, it is clear that $\frac{1}{r^2}$ is greatest, and therefore r is the least, when θ is zero. As θ increases, $\frac{1}{r^2}$ decreases, and is zero when $\sin^2\theta = \frac{b^2}{a^2 + b^2}$; so that for this value of θ, r is infinite. If $\sin^2\theta$ be greater than $\frac{b^2}{a^2+b^2}$, $\frac{1}{r^2}$ will be negative, so that a radius vector which makes with the axis an angle greater than $\sin^{-1}\dfrac{b}{\sqrt{a^2+b^2}}$ does not meet the curve in real points.

Exercises

1. Find the equation of the hyperbola with eccentricity $\frac{3}{2}$ and foci at $(\pm 2, 0)$.
2. Find the equations of the hyperbolas whose axes are as coordinate axes and which also satisfy the conditions given below:
   (*i*) Focus (6, 0), vertex (4, 0). (*ii*) Conjugate axis 6, vertex (7, 0).
   (*iii*) End of conjugate axis (3, 0), length of latus rectum 10.
3. Find the coordinates of foci, length of the latus rectum and eccentricity in each of the following:
   (*i*) $x^2 - y^2 = 6$. (*ii*) $21x^2 - 4y^2 = 84$. (*iii*) $4x^2 - 2y^2 = 7$.
4. Find the foci and latus rectum of a hyperbola whose transverse and conjugate axes are respectively 6 and 4 and whose centre is the origin.
5. Find the locus of a point which moves so that the difference of its distances from (–4, 3) and (4, 3) is numerically equal to 6.
6. Find the locus of a point which moves so that its distance from (4, 0) is twice its distance from the line $x = 1$.

2. Some Important Results

Most of the results obtained in the preceding chapter hold good for the hyperbola and in the proofs it is only necessary to change the sign of b^2.

Let the equation of the hyperbola be

$$\frac{x^2}{a^2} - \frac{y^2}{b^2} = 1.$$

Then, we have the following:

(*i*) The equation of the tangent at the point (x_1, y_1) is $\dfrac{xx_1}{a^2} - \dfrac{yy_1}{b^2} = 1$.

(*ii*) The line $y = mx + c$ is tangent to the hyperbola if $c = \sqrt{a^2m^2 - b^2}$.

(*iii*) The equation of the normal at (x_1, y_1) to the hyperbola is

$$\frac{x - x_1}{\dfrac{x_1}{a^2}} = \frac{y - y_1}{-\dfrac{y_1}{b^2}}.$$

(*iv*) The equation of the pair of tangents from (x_1, y_1) to the hyperbola is $T^2 = SS_1$, where

$$S \equiv \frac{x^2}{a^2} - \frac{y^2}{b^2} - 1,\ S_1 \equiv \frac{x_1^2}{a^2} - \frac{y_1^2}{b^2} - 1 \text{ and } T \equiv \frac{xx_1}{a^2} - \frac{yy_1}{b^2} - 1.$$

(*v*) The equation of the chord of contact of the tangents drawn from (x_1, y_1) to the hyperbola is $\frac{xx_1}{a^2} - \frac{yy_1}{b^2} = 1$.

(*vi*) The equation of the polar of the point (x_1, y_1) with respect to the hyperbola is $\frac{xx_1}{a^2} - \frac{yy_1}{b^2} = 1$.

(*vii*) The pole of the line $lx + my + n = 0$ with respect to the hyperbola is $\left(-\frac{a^2 l}{n}, \frac{b^2 m}{n}\right)$.

(*viii*) The equation of the chord with given middle point (x_1, y_1) is $T = S_1$, where S_1 and T are as defined in (*iv*) above.

(*ix*) The locus of the point of intersection of the tangents at right angles is the circle $x^2 + y^2 = a^2 - b^2$.

This is known as **director circle** of the hyperbola and is real if $a > b$.

(*x*) The locus of the middle points of all chords parallel to $y = mx$ of the hyperbola is $y = \frac{b^2}{a^2 m} x$, which is a line through the centre of the hyperbola.

(*xi*) The lines $y = mx$ and $y = m'x$ are conjugate if $mm' = \frac{b^2}{a^2}$.

Note. *The theory of conjugate diameters of a hyperbola differs in several respects from that for an ellipse. We note that of two conjugate diameters of a hyperbola one meets the curve in real points and the other in imaginary points.*

Let $y = mx$ and $y = m'x$ be a pair of conjugate diameters of the hyperbola

$$\frac{x^2}{a^2} - \frac{y^2}{b^2} = 1.$$

Then, the abscissae of the points of intersection of the hyperbola with the diameters $y = mx$ and $y = m'x$ are respectively given by

$$x^2\left(\frac{1}{a^2} - \frac{m^2}{b^2}\right) = 1, \text{ and } x^2\left(\frac{1}{a^2} - \frac{m'^2}{b^2}\right) = 1.$$

Each of the above equations gives the real values of x if and only if the corresponding slope m (or m') is numerically less than $\frac{b}{a}$. This is not possible, since, $mm' = \frac{b^2}{a^2}$ and as such if the slope of one diameter is numerically less than $\frac{b}{a}$, the slope of the other is bounded to be numerically greater than $\frac{b}{a}$.

Hence, *of two conjugate diameters of a hyperbola one meets the curve in real points, and the other in imaginary points.*

3. Parametric form of the Hyperbola

Draw the circle, called the **auxiliary circle** in analogy with that of an ellipse on the transverse axis $A'A$ as diameter. Let PN be the perpendicular from $P(x, y)$ to the x-axis and construct the tangent NT to the circle. Denote the angle NCT by θ. Then

$$x = CN = CT \sec \theta = a \sec \theta.$$

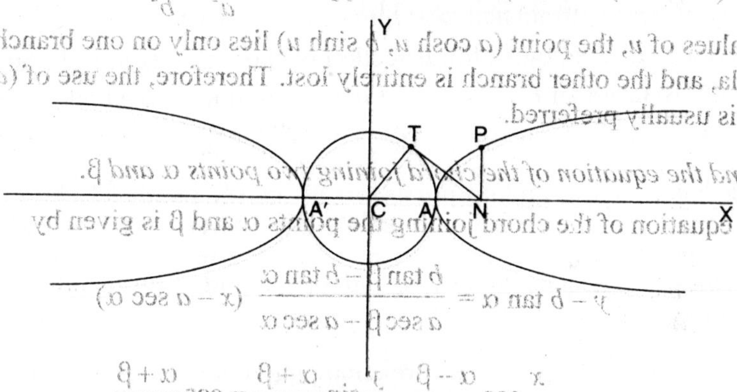

Fig. 9.6

Substituting the value of x in the equation $\frac{x^2}{a^2} - \frac{y^2}{b^2} = 1$ of the hyperbola, we get $y = b \tan \theta$. As θ varies from 0 to $\frac{\pi}{2}$ the point P traces the portion of the hyperbola which lies in the first quadrant starting from A and tending to infinity as θ tends to $\frac{\pi}{2}$. Next, as θ varies from $\frac{\pi}{2}$ to π, the point P traces the portion of the hyperbola in the third quadrant from infinity to A', since, both sec θ and tan θ are negative for $\frac{\pi}{2} < \theta < \pi$. As θ varies from π to $\frac{3\pi}{2}$, P traces out the portion of the hyperbola in the second quadrant from A' to infinity. Finally, as θ varies from $\frac{3\pi}{2}$ to 2π, the point P returns to A from infinity in the fourth quadrant. That is, P traces out every point of the hyperbola as θ ranges from 0 to 2π. Consequently, we may say that the hyperbola has the parametric equations.

$$x = a \sec \theta, y = b \tan \theta.$$

The point $(a \sec \theta, b \tan \theta)$ is also referred to the point θ on the hyperbola.

Note. Instead of using sec θ and tan θ, we may use 'hyperbolic sines and cosines' and write $x = a \cosh u$, $y = b \sinh u$, where $\cosh u = \frac{e^u + e^{-u}}{2}$ and $\sinh u = \frac{e^u - e^{-u}}{2}$ so that $\cosh^2 u - \sinh^2 u = 1$.

It may be pointed out that from the identity $\cosh^2 u - \sinh^2 u = 1$, clearly the point $(a \cosh u, b \sinh u)$ lies on the hyperbola $\frac{x^2}{a^2} - \frac{y^2}{b^2} = 1$, but $\cosh u > 1$ for all values of u, the point $(a \cosh u, b \sinh u)$ lies only on one branch of the hyperbola, and the other branch is entirely lost. Therefore, the use of $(a \sec \theta, b \tan \theta)$ is usually preferred.

3.1 *To find the equation of the chord joining two points α and β.*

The equation of the chord joining the points α and β is given by

$$y - b \tan \alpha = \frac{b \tan \beta - b \tan \alpha}{a \sec \beta - a \sec \alpha} (x - a \sec \alpha)$$

$$\Rightarrow \qquad \frac{x}{a} \cos \frac{\alpha - \beta}{2} - \frac{y}{b} \sin \frac{\alpha + \beta}{2} = \cos \frac{\alpha + \beta}{2}.$$

Corollary 1. The equation of the tangent at $(a \sec \theta, b \tan \theta)$ to the hyperbola $\frac{x^2}{a^2} - \frac{y^2}{b^2} = 1$ is $\frac{x}{a} - \frac{y}{b} \sin \theta = \cos \theta$.

Corollary 2. The equation of the normal at $(a \sec \theta, b \tan \theta)$ to the hyperbola $\frac{x^2}{a^2} - \frac{y^2}{b^2} = 1$ is $by \cot \theta + ax \cos \theta = a^2 + b^2$.

Note. Proceeding as in § 6.2, Chapter VIII, one may easily verify that four normals can be drawn to a hyperbola from a point.

3.2 Examples

Example 1. The chord of contact of the tangents through P to the hyperbola $\frac{x^2}{a^2}-\frac{y^2}{b^2}=1$ subtends a right angle at the origin. Prove that the locus of P is the ellipse $\frac{x^2}{a^4}+\frac{y^2}{b^4}=\frac{1}{a^2}-\frac{1}{b^2}$.

Solution. Let (h, k) be the coordinates of P. Then, the equation of the chord of contact of the given hyperbola is

$$\frac{xh}{a^2}-\frac{yk}{b^2}=1.$$

The equation of the lines joining the origin to the points of intersections of the chord and the hyperbola is

$$\frac{x^2}{a^2}-\frac{y^2}{b^2}=\left(\frac{xh}{a^2}-\frac{yk}{b^2}\right)^2$$

i.e.
$$\left(\frac{h^2}{a^4}-\frac{1}{a^2}\right)x^2-\frac{2hk}{a^2b^2}xy+\left(\frac{k^2}{b^4}+\frac{1}{b^2}\right)y^2=0. \quad (1)$$

But the chord subtends a right at the origin. Therefore, the lines represented by (1) are at right angles. Thus,

coeff. of x^2 + coeff. of $y^2 = 0$, *i.e.*, $\left(\frac{h^2}{a^4}-\frac{1}{a^2}\right)+\left(\frac{k^2}{b^4}+\frac{1}{b^2}\right)=0.$

Hence, the locus of (h, k) is $\frac{x^2}{a^4}+\frac{y^2}{b^4}=\frac{1}{a^2}-\frac{1}{b^2}$.

Example 2. A variable chord of the hyperbola $\frac{x^2}{a^2}-\frac{y^2}{b^2}=1$ is a tangent to the circle $x^2+y^2=c^2$. Prove that the locus of the mid-point of the chord is the curve

$$\left(\frac{x^2}{a^2}-\frac{y^2}{b^2}\right)^2=c^2\left(\frac{x^2}{a^4}+\frac{y^2}{b^4}\right).$$

Solution. Let (h, k) be the middle point of the chord of the given hyperbola. Then, its equation is

$$\frac{xh}{a^2}-\frac{yk}{b^2}=\frac{h^2}{a^2}-\frac{k^2}{b^2}.$$

This chord will be tangent to the circle $x^2 + y^2 = c^2$ if

$$\frac{\frac{h^2}{a^2}-\frac{k^2}{b^2}}{\sqrt{\frac{h^2}{a^4}+\frac{k^2}{b^4}}} = c, \text{ i.e., } \left(\frac{h^2}{a^2}-\frac{k^2}{b^2}\right)^2 = c^2\left(\frac{h^2}{a^4}+\frac{k^2}{b^4}\right).$$

Hence, the locus of (h, k) is

$$\left(\frac{x^2}{a^2}-\frac{y^2}{b^2}\right)^2 = c^2\left(\frac{x^2}{a^4}+\frac{y^2}{b^4}\right).$$

Example 3. Prove that the locus of the poles of normal chords with respect to the hyperbola $\frac{x^2}{a^2}-\frac{y^2}{b^2} = 1$ is the curve $\frac{a^6}{x^2}-\frac{b^6}{y^2} = (a^2+b^2)^2$.

Solution. Any normal chord of the given hyperbola is

$$ax \cos\theta + by \cot\theta = a^2 + b^2. \tag{1}$$

Let (h, k) be the pole of (1). Then, its polar with respect to the given hyperbola is

$$\frac{xh}{a^2}-\frac{yk}{b^2} = 1. \tag{2}$$

But equations (1) and (2) are identical. Therefore, on comparing (1) and (2), we get

$$\frac{a\cos\theta}{\frac{h}{a^2}} = \frac{b\cot\theta}{-\frac{k}{b^2}} = \frac{a^2+b^2}{1}$$

Therefore $$\frac{a^6}{h^2}-\frac{b^6}{k^2} = (a^2+b^2)^2(\sec^2\theta - \tan^2\theta).$$

Hence, the locus of (h, k) is $\frac{a^6}{x^2}-\frac{b^6}{y^2} = (a^2+b^2)^2$.

Example 4. Prove that the locus of the middle points of chords of the hyperbola $\frac{x^2}{a^2}-\frac{y^2}{b^2} = 1$ passing through the fixed point (α, β) is a hyperbola whose centre is $\left(\frac{\alpha}{2}, \frac{\beta}{2}\right)$.

Solution. Let (h, k) be the middle point of the chord of the given hyperbola.

Then, its equation is

$$\frac{xh}{a^2}-\frac{yk}{b^2}=\frac{h^2}{a^2}-\frac{k^2}{b^2}.$$

This chord will pass through (α, β) if $\frac{\alpha h}{a^2}-\frac{\beta k}{b^2}=\frac{h^2}{a^2}-\frac{k^2}{b^2}$.

Hence, the locus of (h, k) is $\frac{x^2}{a^2}-\frac{y^2}{b^2}=\frac{\alpha x}{a^2}-\frac{\beta y}{b^2}$.

This equation can be written as

$$\frac{\left(x-\frac{\alpha}{2}\right)^2}{a^2}-\frac{\left(y-\frac{\beta}{2}\right)^2}{b^2}=\frac{1}{4}\left(\frac{\alpha^2}{a^2}-\frac{\beta^2}{b^2}\right).$$

Clearly it is a hyperbola whose centre is $\left(\frac{\alpha}{2},\frac{\beta}{2}\right)$.

Exercises

1. Prove that the line $x \cos \alpha + y \sin \alpha = p$ touches the hyperbola $\frac{x^2}{a^2}-\frac{y^2}{b^2}=1$ if $p^2 = a^2 \cos^2 \alpha - b^2 \sin^2 \alpha$.
2. Prove that the tangents to a hyperbola at the extremities of a focal chord intersect on the corresponding directrix.
3. Prove that the product of the perpendiculars from the two foci on any tangent to the hyperbola $\frac{x^2}{a^2}-\frac{y^2}{b^2}=1$ is $-b^2$.
4. Find the equation of the locus of middle points of chords of the hyperbola $4x^2 - 9y^2 = 18$ parallel to the line $x + y = 0$.
5. Find the equation of the locus of the foot of the perpendicular from the origin to a variable tangent to the hyperbola.
6. The normal at a variable point on a hyperbola cuts the principal axes in P, Q. Prove that the locus of the mid-points of PQ is a hyperbola. Can it coincide with the original hyperbola?
7. Prove that the locus of the mid-points of the chords of the hyperbola $\frac{x^2}{a^2}-\frac{y^2}{b^2}=1$ which are tangents to the ellipse $\frac{x^2}{a^2}+\frac{y^2}{b^2}=1$ is $a^2b^2(b^2x^2+a^2y^2) = (a^2y^2 - b^2x^2)^2$.

8. If the polars of (x_1, y_1), (x_2, y_2) with respect to the hyperbola $\frac{x^2}{a^2}-\frac{y^2}{b^2}=1$ are at right angles; prove that $\frac{x_1x_2}{y_1y_2}+\frac{a^4}{b^4}=0.$

9. A series of chords of the hyperbola $\frac{x^2}{a^2}-\frac{y^2}{b^2}=1$ are tangents to the circle described on the line joining the foci of the hyperbola as diameter. Prove that the locus of their poles with respect to the hyperbola is

$$\frac{x^2}{a^4}+\frac{y^2}{b^4}=\frac{1}{a^2+b^2}.$$

10. Prove that the line $lx+my+n=0$ is a normal to the hyperbola $\frac{x^2}{a^2}-\frac{y^2}{b^2}=1$, if $\frac{a^2}{l^2}-\frac{b^2}{m^2}=\frac{(a^2+b^2)^2}{n^2}$.

11. P is any point on the hyperbola $\frac{x^2}{a^2}-\frac{y^2}{b^2}=1$. A tangent is drawn at P which cuts the directrix at Q. Prove that PQ subtends a right angle at a focus S.

12. A normal to the hyperbola $\frac{x^2}{a^2}-\frac{y^2}{b^2}=1$ meets the axes in M and N, and lines MP and NP are drawn at right angles to the axes. Prove that the locus of P is the hyperbola $a^2x^2-b^2y^2=(a^2+b^2)^2$.

13. A variable chord of the circle $x^2+y^2=a^2$ is a tangent to the hyperbola $\frac{x^2}{a^2}-\frac{y^2}{b^2}=1$. Prove that the locus of the middle point of the chord is the curve $(x^2+y^2)^2=a^2x^2-b^2y^2$.

14. Prove that the locus of the middle points of chords of the hyperbola $\frac{x^2}{a^2}-\frac{y^2}{b^2}=1$, which subtend a right angle at the centre, is

$$\frac{x^2}{a^4}+\frac{y^2}{b^4}=\left(\frac{1}{a^2}-\frac{1}{b^2}\right)\left(\frac{x^2}{a^2}-\frac{y^2}{b^2}\right)^2.$$

4. Asymptotes

An **asymptote** *is a line which meets a curve in two coincident points at infinity, but which is not altogether at infinity.*

4.1 Equation of an asymptote. *To find the equation of an asymptote of a hyperbola.*

Let equation of the hyperbola be $\frac{x^2}{a^2} - \frac{y^2}{b^2} = 1$.

First Method

Any line $y = mx + c$ meets the hyperbola at points, where

$$\frac{x^2}{a^2} - \frac{(mx+c)^2}{b^2} = 1$$

i.e. $$x^2\left(\frac{1}{a^2} - \frac{m^2}{b^2}\right) - \frac{2mc}{b^2}x - \frac{c^2}{b^2} - 1 = 0. \quad (1)$$

The line $y = mx + c$ meets the hyperbola in two coincident points at infinity if the coefficients of x^2 and x in (1) both are zero, *i.e.*, if

$$\frac{1}{a^2} - \frac{m^2}{b^2} = 0 \text{ and } mc = 0$$

$$\Rightarrow \qquad m = \pm\frac{b}{a} \text{ and } c = 0.$$

Hence, $y = \pm\frac{b}{a}x$ are two real asymptotes of the hyperbola whose combined equation is $\frac{x^2}{a^2} - \frac{y^2}{b^2} = 0$.

Second Method

Consider the point $P(a \sec\theta, b\tan\theta)$ on the hyperbola, and suppose that $0 < \theta < \frac{\pi}{2}$. Then, both x and y coordinates of P are positive and as such P lies in the first quadrant. As $\theta \to \frac{\pi}{2}$, $a\sec\theta \to \infty$ so that $P \to \infty$. Similarly, as $\theta \to \frac{3\pi}{2}$, $P \to \infty$.

The asymptote, if exists, is the limiting form of the tangent as $\theta \to \frac{\pi}{2}$. Therefore, as $\theta \to \frac{\pi}{2}$, the tangent

$$\frac{x}{a}\sec\theta - \frac{y}{b}\tan\theta = 1, \text{ i.e., } \frac{x}{a} - \frac{y}{b}\sin\theta = \cos\theta$$

becomes $\frac{x}{a}-\frac{y}{b}=0.$

Similarly, as $\theta \to \frac{3\pi}{2}$, we have as another asymptote $\frac{x}{a}+\frac{y}{b}=0.$

Hence, the required asymptotes are $\frac{x}{a}\pm\frac{y}{b}=0.$

Remarks. (*i*) If we draw lines through B, B' parallel to the transverse axis and through A, A' parallel to the conjugate axis, then from $\frac{x^2}{a^2}-\frac{y^2}{b^2}=0$, it is clear that the asymptotes are the diagonals of the rectangle so formed. This rectangle is called **associated rectangle** of the hyperbola.

(*ii*) The asymptotes of a hyperbola are helpful in sketching the hyperbola. A rough drawing can made from the associated rectangle and its extended diagonals. The accuracy may be improved considerably, however, by plotting the end points of each latus rectum.

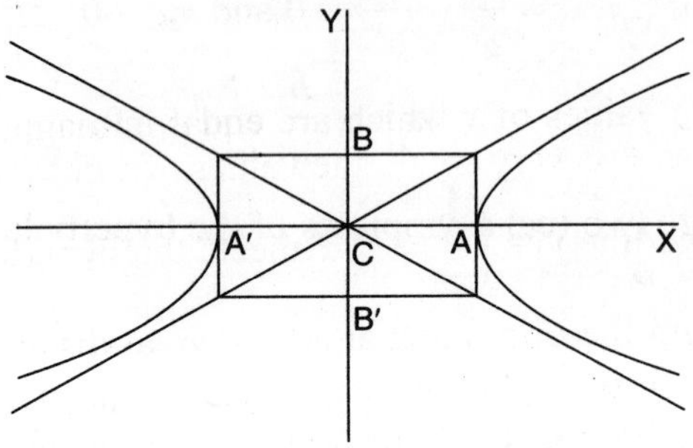

Fig. 9.7

(*iii*) Each asymptote lies along a pair of coincident conjugate diameters.

Corollary. Any line parallel to an asymptote will meet the curve in one point at infinity.

One root of the equation (1) will be infinite, if the coefficient of x^2 is zero.

This will be the case if $m = \pm\frac{b}{a}$. So, the line

$$y=\pm\frac{b}{a}x+c$$

meets the curve in one point at infinity, whatever, the value of c may be.

4.2 Certain important observations about the asymptotes. *The asymptotes of a hyperbola are the pair of tangents from its centre.*

(*I*) The equation of the pair of tangents from the centre of the hyperbola is obtained by putting $x_1 = 0$ and $y_1 = 0$ in the equation $T^2 = SS_1$, *i.e.*,

$$\frac{x^2}{a^2} - \frac{y^2}{b^2} = 0.$$

(*II*) By actually solving simultaneously the equation of the hyperbola and any equation of its asymptotes, it appears that the asymptotes do not meet the hyperbola at all in the points, real or imaginary. But each asymptote being a tangent drawn from the centre to the hyperbola, there must be two coincident points in which each asymptote meets the hyperbola. We would now like to give an explanation as below so as to remove this apparent inconsistency.

Consider the line $y = \left(\frac{b}{a} - \varepsilon\right)x$, where ε is very small. This line is inclined at a very small angle to the asymptote $y = \frac{b}{a}x$. We see that the line meets the hyperbola in two points, where

$$x^2\left\{\frac{1}{a^2} - \left(\frac{1}{a} - \frac{}{h}\right)^2\right\} = 1, \text{ i.e., } x^2 = \frac{1}{\frac{2\in}{ab} - \frac{\in^2}{b^2}}.$$

This gives two values of x which are equal in magnitude but opposite in sign. Also, as $\in$ is very small, the values of x obtained would be very large. Thus, the line $y = \left(\frac{b}{a} - \varepsilon\right)x$ meets the hyperbola in two points equidistant from the origin and at a very great distance from it. Since, the points on a line in two opposite directions and at a sufficiently large distance tend to become the same point, known as the point infinity, this line, as ε becomes sufficiently small, tends to become a tangent to the hyperbola at the point infinity.

The same thing is true for the line $y = -\left(\frac{b}{a} - \varepsilon\right)x$.

Thus, the lines $y = \pm\frac{b}{a}x$, the tangents from the centre, are seen to be the limiting case of the two lines $y = \pm\left(\frac{b}{a} - \varepsilon\right)x$, where ε is efficiently small.

Hence, the tangents from the centre of a hyperbola have their points of contact at an infinite distancc.

(*III*) Any diameter $y = mx$ cuts the hyperbola $\frac{x^2}{a^2} - \frac{y^2}{b^2} = 1$ at the points where

$$x^2\left(\frac{1}{a^2}-\frac{m^2}{b^2}\right)=1.$$

These points are real if and only if $-\frac{b}{a}<m<\frac{b}{a}$.

In case m tends to $\pm\frac{b}{a}$, the points of intersection tend to infinity. Thus, the two lines given by

$$\frac{x^2}{a^2}-\frac{y^2}{b^2}=0$$

are those diameters each of which meets the hyperbola in two coincident points at infinity. Therefore, they represent the asymptotes to the hyperbola and they divide all the diameters into two classes. One class consists of those diameters, of slope numerically less than $\frac{b}{a}$, which intersect both the branches of the hyperbola. The second class consists of those diameters, of slope numerically greater than $\frac{b}{a}$, which do not intersect the hyperbola.

Hence, *the asymptotes can be regarded as the diameters lying on the boundary of the two classes of diameters, intersecting and non-intersecting.*

Note. Since, the equations of an ellipse and a hyperbola are alike in the sense that b^2 is replaced by $-b^2$ in case of hyperbola, one is interested to study the asymptotes for an ellipse also. In fact the equations of the pair of tangents from the centre of the ellipse $\frac{x^2}{a^2}+\frac{y^2}{b^2}=1$ is given by $\frac{x^2}{a^2}+\frac{y^2}{b^2}=0$.

In analogy with the hyperbola case, the lines given by this equation are to be regarded as the asymptotes of the ellipse. But these lines being imaginary having no real point upon them except the origin, they do not have the geometrical importance. However, they possess the same algebraic properties as the asymptotes of the hyperbola.

5. Conjugate Hyperbola

For a given hyperbola the hyperbola, having for, respectively, its transverse and conjugate axes the conjugate and transverse axes of the given hyperbola, is called the **conjugate hyperbola**.

If the hyperbola be $\frac{x^2}{a^2} - \frac{y^2}{b^2} = 1$, then the equation of the conjugate hyperbola will be $\frac{y^2}{b^2} - \frac{x^2}{a^2} = 1$.

Both of these hyperbolas have the same asymptotes, *viz.*,

$$\frac{x^2}{a^2} - \frac{y^2}{b^2} = 0.$$

Note. If the axes of coordinates be changed in any manner, we should, in order to obtain the new equations of a hyperbola, of the asymptotes and of the conjugate hyperbola, have to make the same substitutions in all three cases.

Hence, *for all positions of the axes of coordinates, the equations of a hyperbola and of the conjugate hyperbola will only differ from the equation of the asymptotes by constants, and the two constants will be equal and opposite for the two hyperbolas.*

5.1 Properties of a pair of conjugate hyperbolas. (*i*) *The two hyperbolas have the same asymptotes.*

(*ii*) *If two diameters be conjugate with respect to one of the hyperbolas, they will be conjugate with respect to the other.*

(*iii*) *The equations of the hyperbolas*

$$\frac{x^2}{a^2} - \frac{y^2}{b^2} = 1 \text{ and } \frac{y^2}{b^2} - \frac{x^2}{a^2} = 1$$

can be written in polar forms

$$\frac{1}{r^2} = \frac{\cos^2\theta}{a^2} - \frac{\sin^2\theta}{b^2} \text{ and } -\frac{1}{r^2} = \frac{\cos^2\theta}{a^2} - \frac{\sin^2\theta}{b^2},$$

respectively. It is evident that if, for any value of θ, r^2 is positive for one curve it is negative for the other.

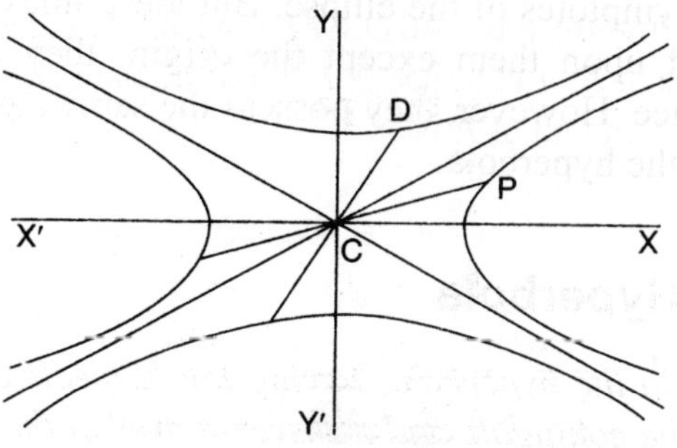

Fig. 9.8

Hence, every diameter meets one curve in real points and the other in imaginary points. Moreover the lengths of semi-diameters of the two curves are connected by the relation $r_1^2 = -r_2^2$ for all values of θ.

(iv) *If two conjugate diameters cut the hyperbola and its conjugate in P and D, respectively, then $CP^2 - CD^2 = a^2 - b^2$.*

Let $P(a \sec \theta, b \tan \theta)$ be any point on the hyperbola

$$\frac{x^2}{a^2} - \frac{y^2}{b^2} = 1. \tag{1}$$

Then m = slope of the diameter $CP = \dfrac{b \tan \theta - 0}{a \sec \theta - 0} = \dfrac{b}{a} \sin \theta.$

Let m' be the slope of the diameter CD. Since, CP and CD are conjugate diameters, we have

$$mm' = \frac{b^2}{a^2}, \text{ i.e., } m' = \frac{b}{a \sin \theta}.$$

Therefore, equation of CD is
$$y = \frac{b}{a \sin \theta} x. \tag{2}$$

The equation of the hyperbola conjugate to (1) is

$$\frac{y^2}{b^2} - \frac{x^2}{a^2} = 1. \tag{3}$$

On solving (2) and (3), we get two points of intersection whose coordinates are $(a \tan \theta, b \sec \theta)$ and $(-a \tan \theta, -b \sec \theta)$. Hence, the coordinates of D are $(a \tan \theta, b \sec \theta)$.

Now $\quad CP^2 - CD^2 = a^2 \sec^2 \theta + b^2 \tan^2 \theta - a^2 \tan^2 \theta - b^2 \sec^2 \theta$

$$= a^2 - b^2.$$

(v) *The parallelogram formed by the tangents at P, P', D, D' is of constant area.*

Area of the parallelogram = $4CD.CM$, where CM is perpendicular from C to the tangent at P. Let $(a \sec \theta, b \tan \theta)$ be the coordinates of P. The equation of the tangent at P is

$$\frac{x}{a} - \frac{y}{b} \sin \theta = \cos \theta.$$

Therefore $\quad CM = \dfrac{\cos \theta}{\sqrt{\dfrac{1}{a^2} + \dfrac{\sin^2 \theta}{b^2}}} = \dfrac{ab \cos \theta}{\sqrt{b^2 + a^2 \sin^2 \theta}}.$

and $CD = \sqrt{a^2 \tan^2 \theta + b^2 \sec^2 \theta} = \dfrac{1}{\cos\theta}\sqrt{b^2 + a^2 \sin^2 \theta}$

Hence, *area of the parallelogram is* $4ab$, *which is constant.*

(*vi*) *The asymptotes bisect PD and PD′.*

The coordinates of P and D are $(a \sec \theta, b \tan \theta)$ and $(a \tan \theta, b \sec \theta)$, respectively. Let (h, k) be the middle point of PD. Then

$$h = \frac{a\sec\theta + a\tan\theta}{2} = \frac{a}{2}(\sec\theta + \tan\theta)$$

and $$k = \frac{b\tan\theta + b\sec\theta}{2} = \frac{b}{2}(\sec\theta + \tan\theta)$$

Therefore $\dfrac{k}{h} = \dfrac{b}{a}$.

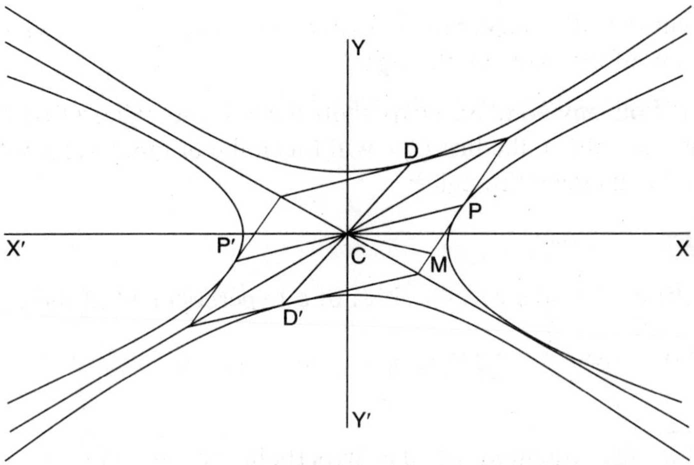

Fig. 9.9

Hence, locus of (h, k) is $y = \dfrac{b}{a}x$, which is the equation of an asymptote.

Similarly, we can prove that $y = -\dfrac{b}{a}x$ is the locus of the middle point of PD'. Hence, the result follows.

(*vii*) *The polars of any point with respect to a hyperbola and its conjugate are parallel to one another and equidistant from the centre.*

The equations of the polars of any point (x_1, y_1) with respect to the hyperbola

$$\frac{x^2}{a^2} - \frac{y^2}{b^2} = 1, \qquad (1)$$

and its conjugate are

$$\frac{xx_1}{a^2} - \frac{yy_1}{b^2} = 1 \text{ and } \frac{yy_1}{b^2} - \frac{xx_1}{a^2} = 1.$$

Clearly these polars are parallel and equidistant from the centre.

Further, let $P(x_1, y_1)$ be any point on the hyperbola (1). Then, polar of $P(x_1, y_1)$ with respect to the conjugate hyperbola

$$\frac{y^2}{b^2} - \frac{x^2}{a^2} = 1 \text{ is } \frac{yy_1}{b^2} - \frac{xx_1}{a^2} = 1.$$

This, can be written as

$$\frac{x(-x_1)}{a^2} - \frac{y(-y_1)}{b^2} = 1.$$

But this equation is the tangent to the hyperbola (1) at $(-x_1, -y_1)$, which is the other extremity of the diameter through P.

Hence, if from any point on a hyperbola the tangents PQ, PQ' be drawn to the conjugate hyperbola, the line QQ' will touch the original hyperbola at the other end of the diameter through P.

5.2 Examples

Example 1. If e, e' be the eccentricities of a hyperbola and of the conjugate hyperbola; prove that $\frac{1}{e^2} + \frac{1}{e'^2} = 1$.

Solution. Let the equation of the hyperbola be $\frac{x^2}{a^2} - \frac{y^2}{b^2} = 1$. Then $b^2 = a^2(e^2 - 1)$, which means that

$$\frac{1}{e^2} = \frac{a^2}{a^2 + b^2}. \qquad (1)$$

Further, equation of the conjugate hyperbola is $\frac{y^2}{b^2} - \frac{x^2}{a^2} = 1$. Therefore $a^2 = b^2(e'^2 - 1)$. This means that

$$\frac{1}{e'^2} = \frac{b^2}{a^2 + b^2}. \qquad (2)$$

On adding equations (1) and (2), we get $\frac{1}{e^2}+\frac{1}{e'^2}=1.$

Example 2. Prove that the tangent to a hyperbola makes with the asymptotes a triangle of constant area.

Solution. Let the equation of the hyperbola be $\frac{x^2}{a^2}-\frac{y^2}{b^2}=1.$

The equation of the tangent to the hyperbola at the point (a sec θ, b tan θ) is

$$\frac{x}{a}\sec\theta-\frac{y}{b}\tan\theta=1.$$

The equations of the asymptotes are $y=\pm\frac{b}{a}x.$

The coordinates of the points of intersection of the tangent and the asymptotes are

$$\left(\frac{a\cos\theta}{1-\sin\theta},\frac{b\cos\theta}{1-\sin\theta}\right) \text{ and } \left(\frac{a\cos\theta}{1-\sin\theta},\frac{-b\cos\theta}{1+\sin\theta}\right).$$

Hence, the area of the triangle is

$$\frac{1}{2}\left(\frac{a\cos\theta}{1+\sin\theta}\cdot\frac{b\cos\theta}{1-\sin\theta}+\frac{a\cos\theta}{1-\sin\theta}\cdot\frac{b\cos\theta}{1+\sin\theta}\right)$$

$$= ab = \text{constant.}$$

Example 3. The polar lines of a point P, with respect to a hyperbola and with respect to its auxiliary circle, are perpendicular. Prove that P lies on one of the asymptotes of the hyperbola.

Solution. Let the equation of the hyperbola be $\frac{x^2}{a^2}-\frac{y^2}{b^2}=1.$

Then, the equation of the auxiliary circle is $x^2+y^2=a^2$. Let (h, k) be the coordinates of P. Then, its polars with respect to the hyperbola and the auxiliary circle are, respectively

$$\frac{xh}{a^2}-\frac{yk}{b^2}=1 \text{ and } xh+yk=a^2.$$

But these polar lines are perpendicular. Therefore

$$\frac{h^2}{a^2} - \frac{k^2}{b^2} = 0.$$

Hence, the locus of (h, k) is $\dfrac{x^2}{a^2} - \dfrac{y^2}{b^2} = 0.$

Exercises

1. Find the asymptotes of $2xy + 7x - 6y - 18 = 0$. What is the equation of the conjugate hyperbola?
2. Prove that the line $x = 0$ is an asymptote of the hyperbola $3x^2 + 2xy + 4x = 9$. What is the equation of the other asymptote?
3. Find the equation of the hyperbola which has $3x - 4y + 7 = 0$ and $4x + 3y + 1 = 0$ for its asymptotes and passes through the origin.
4. Prove that the portion of any tangent to a hyperbola intercepted by the asymptotes is bisected at the point of contact.
5. Prove that the polar of any point on an asymptote of a hyperbola with respect to the hyperbola is parallel to that asymptote.
6. Prove that a line parallel to an asymptote intersects a hyperbola in only one point.
7. P is any point on the hyperbola and C is the centre of the hyperbola $\dfrac{x^2}{a^2} - \dfrac{y^2}{b^2} = 1$. The diameter conjugate to CP intersects the conjugate hyperbola $\dfrac{y^2}{b^2} - \dfrac{x^2}{a^2} = 1$ at Q. Prove that the locus of the point of intersection of the normals at P and Q is a pair of lines represented by $a^2x^2 - b^2y^2 = 0$.
8. Prove that the product of the perpendiculars from any point of the hyperbola $\dfrac{x^2}{a^2} - \dfrac{y^2}{b^2} = 1$ to the two asymptotes is equal to $\dfrac{a^2b^2}{a^2 + b^2}$.
9. The tangent at the point $(a \sec \theta, b \tan \theta)$ to the hyperbola $\dfrac{x^2}{a^2} - \dfrac{y^2}{b^2} = 1$ meets the asymptotes at P and Q. Find the coordinates of P and Q. Prove that the locus of the point which divides PQ in the ratio $\lambda : 1$ is

$$\frac{x^2}{a^2} - \frac{y^2}{b^2} = \frac{4\lambda}{(\lambda + 1)^2}.$$

10. The normal at a point P to a hyperbola meets the transverse axis in G; the line perpendicular to this axis through P meets an asymptote in Q. Prove that GQ is perpendicular to the asymptote.

11. The locus of the centre of a circle which circumscribes the triangle formed by the asymptotes and any tangent to a given hyperbola is another hyperbola whose asymptotes are perpendicular to those of the given hyperbola.

12. Prove that the two lines joining the points in which any two tangents to a hyperbola meet the asymptotes are parallel to the chord of contact of the tangents and are equidistant from it.

13. A variable tangent to the hyperbola $\frac{x^2}{a^2} - \frac{y^2}{b^2} = 1$ cuts the asymptotes in the points P and Q. Prove that the locus of the centre of the circle CPQ, where C is the origin of coordinates, is given by the equation

$$4(a^2x^2 - b^2y^2) = (a^2 + b^2)^2.$$

6. Rectangular Hyperbola

A hyperbola whose asymptotes are perpendicular is called **rectangular hyperbola**.

Let the equation of the hyperbola be

$$\frac{x^2}{a^2} - \frac{y^2}{b^2} = 1.$$

The angle between the asymptotes

$$\frac{x^2}{a^2} - \frac{y^2}{b^2} = 0$$

of this hyperbola is equal to $2 \tan^{-1} \frac{b}{a}$. Thus, when the angle is a right angle, we must have $b = a$. On this account the curve is sometimes referred to an **equilateral hyperbola**.

The equation of the rectangular hyperbola, referred to its axes as coordinate axes, is $x^2 - y^2 = a^2$. From the relation $b^2 = a^2(e^2 - 1)$, it follows that the eccentricity of the rectangular hyperbola will be $\sqrt{2}$.

Note. If $a = b$, the associated rectangle is a square and the asymptotes are perpendicular to each other. For this reason the hyperbola is said to be equilateral becuase its axes are equal or is said to be rectangular because its asymptotes intersect at right angles.

We now proceed to simplify the equation $x^2 - y^2 = a^2$ of a rectangular hyperbola, by selecting the asymptotes as coordinate axes.

In order to reduce the given equation to the more usual form, we rotate the axes through an angle $-\frac{\pi}{4}$ and we get

$$X = x \cos(-45^\circ) + y \sin(-45^\circ) = \frac{x-y}{\sqrt{2}}$$

and $$Y = -x \sin(-45^\circ) + y \cos(-45^\circ) = \frac{x+y}{\sqrt{2}}.$$

Therefore $\quad x^2 - y^2 = 2XY.$

But $\quad x^2 - y^2 = a^2.$

Thus $$XY = \frac{a^2}{2}.$$

Hence, the equation of the rectangular hyperbola can also be written as $xy = c^2$, where $c^2 = \frac{a^2}{2}$.

6.1 Parametric equations. If we substitute $x = ct$ in the equation $xy = c^2$ of the rectangular hyperbola, we get $y = \frac{c}{t}$. That is, as t varies the point $\left(ct, \frac{c}{t}\right)$ always lies on the hyperbola $xy = c^2$. Accordingly, the rectangular hyperbola can be represented by the parametric equations

$$x = ct, y = \frac{c}{t}.$$

Thus, any point on the curve $xy = c^2$ can be represented by $\left(ct, \frac{c}{t}\right)$.

This point is also referred to **the point t** on the curve.

6.2 Equation of a chord. *To find the equation of the chord joining any two points t_1 and t_2 on the rectangular hyperbola $xy = c^2$.*

The equation of the chord joining the points t_1 and t_2 is

$$x + t_1 t_2 y - c(t_1 + t_2) = 0.$$

Corollary 1. The equation of the tangent at the point t to the curve $xy = c^2$ is $x + t^2y - 2ct = 0$.

Corollary 2. The equation of the normal at the point t to $xy = c^2$ is $t^3x - ty + c(1 - t^4) = 0$.

6.3 Chord of contact. *To find the equation of the chord of contact of tangents drawn from (x_1, y_1) to the hyperbola $xy = c^2$.*

Let the tangents drawn from $A(x_1, y_1)$ to the given hyperbola meet it in the points P and Q. Let $\left(ct_1, \frac{c}{t_1}\right)$ and $\left(ct_2, \frac{c}{t_2}\right)$ be the coordinates of P and Q, respectively. The equations of the tangents at P and Q are $x + t_1^2\ y = 2ct_1$ and $x + t_2^2\ y = 2ct_2$, respectively. Since, these tangents pass through $A(x_1, y_1)$, we have

$$x_1 + t_1^2\, y_1 = 2ct_1 \text{ and } x_1 + t_2^2\, y_1 = 2ct_2.$$

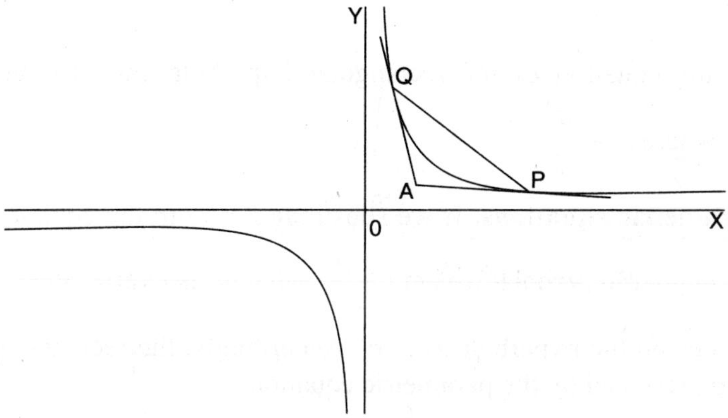

Fig. 9.10

These equations may be written as

$$ct_1y_1 + \frac{c}{t_1}\, x_1 = 2c^2 \text{ and } ct_2y_1 + \frac{c}{t_2}x_1 = 2c^2.$$

Thus, the points $P\left(ct_1, \frac{c}{t_1}\right)$ and $Q\left(ct_2, \frac{c}{t_2}\right)$ both lie on the line

$$xy_1 + yx_1 = 2c^2.$$

This equation represents the chord joining the points of contact of tangents drawn from the point (x_1, y_1) to the hyperbola.

6.4 *To find the locus of the middle points of a system of parallel chords of the hyperbola* $xy = c^2$.

Let AB be one of the parallel chords of the given hyperbola.

Let the coordinates of A and B be $\left(ct_1, \frac{c}{t_1}\right)$ and $\left(ct_2, \frac{c}{t_2}\right)$, respectively. Let $P(h, k)$ be the middle point of AB. Then

$$h = \frac{c(t_1 + t_2)}{2}, \; k = \frac{c(t_1 + t_2)}{2t_1t_2}.$$

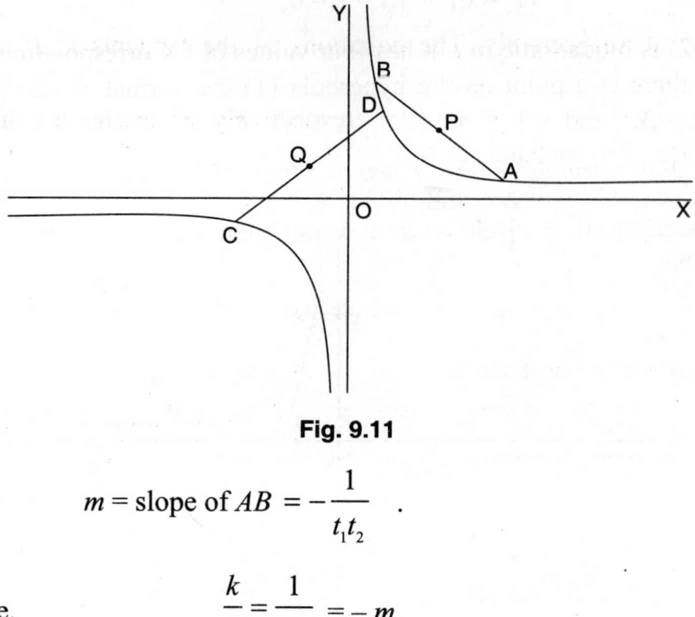

Fig. 9.11

Now $\qquad m = \text{slope of } AB = -\frac{1}{t_1t_2}.$

Therefore. $\qquad \frac{k}{h} = \frac{1}{t_1t_2} = -m.$

Hence, locus of (h, k) is $y + mx = 0$.

But $y + mx = 0$ is the equation of a diameter of the curve. Thus, *the locus of the middle points of a system of parallel chords is a diameter of the hyperbola.*

Let us denote the slope of the above diameter by m'. Then $m' = -m$. Since, $m = -m'$, if follows by symmetry that the locus of the middle points Q of all chords parallel to CD with slope m' is the diameter OP with slope m. That is, each of the diameters $y = mx$ and $y = m'x$ bisects the chords parallel to the other. These are called **conjugate diameters** and the relation between their slopes is $m + m' = 0$.

6.5 *To prove that four normals can be drawn to the hyperbola* $xy = c^2$ *from any given point.*

Let $P(x_1, y_1)$ be any given point. Let $\left(ct, \frac{c}{t}\right)$ be the coordinates of any point on the given hyperbola

$$xy = c^2. \tag{1}$$

The equation of the normal at $\left(ct, \frac{c}{t}\right)$ to the hyperbola (1) is

$$t^3x - ty + c(1 - t^4) = 0.$$

This will pass through $P(x_1, y_1)$ if

$$ct^4 - x_1t^3 + y_1t - c = 0. \tag{2}$$

Equation (2) is biquadratic in t, it has four values of t. Corresponding to each value of t there is a point on the hyperbola (1) the normal at which passes through (x_1, y_1).

Hence, *four normals can be drawn to a hyperbola from a given point.*

6.6 Intersection of a circle and a hyperbola. Let the equation of the hyperbola be

$$xy = c^2 \tag{1}$$

and the equation of the circle be

$$x^2 + y^2 + 2gx + 2fy + k = 0. \tag{2}$$

The equation (1) can also be written as

$$x = ct, y = \frac{c}{t}. \tag{3}$$

On solving (2) and (3), we get

$$c^2t^4 + 2gct^3 + kt^2 + 2fct + c^2 = 0. \tag{4}$$

Equation (4) is biquadratic in t, it has four roots say t_1, t_2, t_3, t_4. Corresponding to each value of t there is a common point of the circle and the hyperbola. Thus, there are four points of intersections of a circle and a hyperbola.

Further, by theory of equations, we have

$$t_1 + t_2 + t_3 + t_4 = -\frac{2g}{c},$$

$$t_1t_2 + t_1t_3 + t_1t_4 + t_2t_3 + t_2t_4 + t_3t_4 = \frac{k}{c^2},$$

$$t_1t_2t_3 + t_1t_2t_4 + t_1t_3t_4 + t_2t_3t_4 = -\frac{2f}{c},$$

and

$$t_1t_2t_3t_4 = 1.$$

The fourth equation yields the necessary condition that four points are **concyclic**. In other words, if the parameters t_1, t_2, t_3, t_4 of four points on the hyperbola satisfy $t_1t_2t_3t_4 = 1$ these points shall be concyclic, since, one condition is enough to ensure that four points should be on a circle.

Conversely, if t_1, t_2, t_3, t_4 be the parameters of any four points on a hyperbola, then for the points to be concyclic the parameters must satisfy equation (4) above which requires that $t_1t_2t_3t_4 = 1$.

6.7 Certain results on the rectangular hyperbola. In this section, we state certain results of the rectangular hyperbola $xy = c^2$ without using parametric forms. The proofs may be obtained on the same lines as in previous chapters.

(*i*) The equation of the tangent at the point (x_1, y_1) to the hyperbola is

$$xy_1 + yx_1 = 2c^2.$$

(*ii*) The equation of the normal at (x_1, y_1) to the hyperbola is

$$xx_1 - yy_1 = x_1^2 - y_1^2.$$

(*iii*) The equation of the pair of tangents from (x_1, y_1) to the hyperbola is

$$(xy - c^2)(x_1y_1 - c^2) = \{\tfrac{1}{2}(xy_1 + yx_1) - c^2\}^2.$$

(*iv*) The equation of the polar of the point (x_1, y_1) with respect to the hyperbola is

$$xy_1 + yx_1 = 2c^2.$$

(*v*) The equation of the chord with given middle point (x_1, y_1) is

$$xy_1 + yx_1 = 2x_1y_1.$$

6.8 Examples

Example 1. A circle cuts the rectangular hyperbola $xy = 1$ in points (x_r, y_r), $r = 1, 2, 3, 4$. Prove that $x_1x_2x_3x_4 = y_1y_2y_3y_4 = 1$.

Solution. Let the equation of the circle be

$$x^2 + y^2 + 2gx + 2fy + c = 0.$$

Solving for x the equation of the circle and the given hyperbola simultaneously, we get

$$x^4 + 2gx^3 + cx^2 + 2fx + 1 = 0.$$

This equation is biquadratic in x, hence, by theory of equations, we have $x_1x_2x_3x_4 = 1$.

Similarly, on solving the equations of the circle and the hyperbola for y, we get $y_1y_2y_3y_4 = 1$. Therefore

$$x_1x_2x_3x_4 = y_1y_2y_3y_4 = 1.$$

Example 2. If the normals at four points (x_r, y_r), $r = 1, 2, 3, 4$ on the rectangular hyperbola $xy = c^2$ meet in the point (α, β): prove that

$$\alpha = x_1 + x_2 + x_3 + x_4, \ \beta = y_1 + y_2 + y_3 + y_4,$$

and $$x_1x_2x_3x_4 = y_1y_2y_3y_4 = -c^4$$

Solution. Let $\left(ct, \frac{c}{t}\right)$ be any point on the hyperbola $xy = c^2$.

Then, equation of the normal at $\left(ct, \frac{c}{t}\right)$ to this hyperbola is

$$ct^4 - xt^3 + yt - c = 0.$$

This normal will pass through (α, β) if $ct^4 - \alpha t^3 + \beta t - c = 0$.

The above equation being biquadratic in t has four values say t_1, t_2, t_3, t_4; hence by theory of equations, we have

$$t_1 + t_2 + t_3 + t_4 = \frac{\alpha}{c}, \tag{1}$$

$$t_1t_2 + t_1t_3 + t_1t_4 + t_2t_3 + t_2t_4 + t_3t_4 = 0, \tag{2}$$

$$t_1t_2t_3 + t_1t_2t_4 + t_1t_3t_4 + t_2t_3t_4 = -\frac{\beta}{c}, \tag{3}$$

$$t_1t_2t_3t_4 = -1. \tag{4}$$

From (1), we get

$$\alpha = ct_1 + ct_2 + ct_3 + ct_4, \ i.e., \ \alpha = x_1 + x_2 + x_3 + x_4.$$

On dividing equation (3) by equation (4), we get

$$\frac{\beta}{c} = \frac{1}{t_1} + \frac{1}{t_2} + \frac{1}{t_3} + \frac{1}{t_4}$$

$$\Rightarrow \qquad \beta = y_1 + y_2 + y_3 + y_4.$$

Further

$$x_1x_2x_3x_4 = ct_1 \cdot ct_2 \cdot ct_3 \cdot ct_4 = c^4t_1t_2t_3t_4 = -c^4 \qquad \text{(from (4))}$$

and $$y_1y_2y_3y_4 = \frac{c}{t_1} \cdot \frac{c}{t_2} \cdot \frac{c}{t_3} \cdot \frac{c}{t_4} = \frac{c^4}{t_1t_2t_3t_4} = -c^4 \qquad \text{(from (4).)}$$

Therefore $$x_1x_2x_3x_4 = y_1y_2y_3y_4 = -c^4.$$

Example 3. If the polar of (h, k) with respect to $y^2 = 4ax$ touches the rectangular hyperbola $x^2 - y^2 = a^2$, prove that the locus of (h, k) is the ellipse $4x^2 + y^2 = 4a^2$.

Solution. The equation of the polar of (h, k) with respect to $y^2 = 4ax$ is

$$ky = 2a(x + h), \text{ i.e., } y = \frac{2a}{k}x + \frac{2ah}{k}.$$

It will touch the hyperbola $x^2 - y^2 = a^2$ if $4h^2 + k^2 = 4a^2$.

Hence, the locus of (h, k) is $4x^2 + y^2 = 4a^2$.

Example 4. Chords of the rectangular hyperbola $x^2 - y^2 = a^2$ touch the parabola $y^2 = 4ax$. Prove that the locus of their middle points is the curve $y^2(x - a) = x^3$.

Solution. Let (h, k) be the middle point of a chord of $x^2 - y^2 = a^2$. Then, its equation is $xh - yk = h^2 - k^2$, *i.e.*, $y = \dfrac{hx}{k} - \dfrac{h^2 - k^2}{k}$.

It will touch the parabola $y^2 = 4ax$ if $k^2(h - a) = h^3$.

Hence, the locus of (h, k) is $y^2(x - a) = x^3$.

Example 5. A line through the origin meets the circle $x^2 + y^2 = a^2$ at P and the hyperbola $x^2 - y^2 = a^2$ at Q. Prove that the locus of the point of intersection of the tangent at P to the circle with the tangent at Q to the hyperbola is the curve

$$(a^4 + 4y^4)\, x^2 = a^6.$$

Solution. Any line through the origin is $y = mx$. This line meets the circle $x^2 + y^2 = a^2$ at $P\left(\dfrac{a}{\sqrt{1+m^2}}, \dfrac{am}{\sqrt{1+m^2}}\right)$ and the hyperbola $x^2 - y^2 = a^2$ at $Q\left(\dfrac{a}{\sqrt{1-m^2}}, \dfrac{am}{\sqrt{1-m^2}}\right)$. Tangents at P and Q are respectively,

$$x + my = a\sqrt{1+m^2} \text{ and } x - my = a\sqrt{1-m^2}.$$

On solving these equations, we get $m = \dfrac{2xy}{a^2}$. Substituting the value of m in either of the equations, we get $(a^4 + 4y^4)\, x^2 = a^6$.

Example 6. Prove that the locus of the poles of normal chords of the rectangular hyperbola $xy = c^2$ is the curve $(x^2 - y^2)^2 + 4c^2xy = 0$.

Solution. Any normal chord of the given rectangular hyperbola is

$$t^3x - ty + c\,(1 - t^4) = 0. \tag{1}$$

Let (h, k) be the pole of (1). Then, its polar with respect to $xy = c^2$ is

$$xk + yh = 2c^2. \tag{2}$$

But equations (1) and (2) represent the same locus. Therefore, on comparing (1) and (2), we get

$$\frac{t^3}{k} = \frac{-t}{h} = \frac{c(1-t^4)}{-2c^2}$$

$$\Rightarrow \qquad t^2 = -\frac{k}{h} \text{ and } t = \frac{h(1-t^4)}{2c}.$$

Therefore $\qquad (h^2 - k^2)^2 + 4c^2hk = 0.$

Hence, the locus of (h, k) is $(x^2 - y^2)^2 + 4c^2xy = 0$.

Exercises

1. P and Q are the points (x_1, y_1) and $(x_1, -y_1)$ on the hyperbola $x^2 - y^2 = a^2$. Prove that the tangent at P is perpendicular to the line joining Q to the centre C of the hyperbola. If the tangent at P meets CQ at R, prove that $CQ.CR = a^2$.

2. If the polar of (h, k) with respect to $y^2 = 4ax$ touches $x^2 + y^2 = 4a^2$; prove that the locus of (h, k) is the rectangular hyperbola $x^2 - y^2 = 4a^2$.

3. If the polar of (h, k) with respect to $y^2 = 4ax$ touches $x^2 = 4by$; prove that the locus of (h, k) is the rectangular hyperbola $xy + 2ab = 0$.

4. In a rectangular hyperbola, prove that $SP{\cdot}S'P = CP^2$,

where S, S' are the foci, C is the centre and P is any point on the hyperbola.

5. From points on the circle $x^2 + y^2 = a^2$ tangents are drawn to the hyperbola $x^2 - y^2 = a^2$. Prove that the locus of the middle points of the chords of contact is the curve $(x^2 - y^2)^2 = a^2(x^2 + y^2)$.

6. A tangent to the parabola $x^2 = 4ay$ meets the hyperbola $xy = c^2$ in two points P and Q. Prove that the middle point of PQ lies on a parabola.

7. The tangents at two points P and Q to the rectangular hyperbola $xy = c^2$ intersect at T. Prove that the line joining T to the centre of the hyperbola bisects the chord PQ.

8. P and Q are the points t_1 and t_2 on the rectangular hyperbola $xy = c^2$. If PQ is normal to the curve at P, prove that $t_1^3\, t_2 + 1 = 0$.

9. A line parallel to the line $y = mx$ cuts the rectangular hyperbola $xy = c^2$ at P and Q. Prove that the locus of the mid-point of PQ is the line $y = -mx$.

10. Find the locus of the mid-point of a variable chord of length $2c$ of the rectangular hyperbola $xy = c^2$.

11. Prove that $y = mx$ and $y = -mx$ are conjugate diameters of $xy = c^2$, for all values of m.

12. Prove that the locus of the middle points of normal chords of the rectangular hyperbola $x^2 - y^2 = a^2$ is $(y^2 - x^2)^3 = 4a^2x^2y^2$.

13. Prove that the locus of the middle points of chords of constant length $2d$ of the rectangular hyperbola $xy = c^2$ is $(x^2 + y^2)(xy - c^2) = d^2xy$.

14. A circle cuts a rectangular hyperbola in four points. Prove that the centroid of these four points bisects the join of the centres of the circle and the hyperbola.

15. If a circle and the rectangular hyperbola $xy = c^2$ cut in the four points t_1, t_2, t_3, t_4; prove that $t_1t_2t_3t_4 = 1$.

16. If t is the point $\left(ct, \frac{c}{t}\right)$ on the rectangular hyperbola $xy = c^2$, find the coordinates of the orthocentre of the triangle formed by the points t_1, t_2, t_3 and prove that it lies on the rectangular hyperbola.

17. A circle with fixed centre $(3h, 3k)$ and of variable radius cuts the rectangular hyperbola $x^2 - y^2 = 9a^2$ at the points A, B, C and D. Prove that the locus of the centroid of the triangle ABC is given by $(x - 2h)^2 - (y - 2k)^2 = a^2$.

Miscellaneous Exercises

1. Find the equation to the chord of the hyperbola $\frac{x^2}{a^2} - \frac{y^2}{b^2} = 1$ whose mid-point is (α, β), and interpret the result when the point (α, β) lies on the hyperbola.

2. Prove that the locus of the poles of a given line with respect to the circles of a coaxal system is a hyperbola.

3. Find the locus of the point of intersection of tangents to the hyperbola $x^2 - y^2 = a^2$ at points with parameters θ, $\theta + \alpha$, as θ varies. What is the locus when $\alpha = \pi$?

4. A circle cuts two fixed perpendicular lines so that each intercept is of given length. Prove that the locus of the centre of the circle is a rectangular hyperbola.

5. If tangents are drawn to a system of coaxal circles parallel to a given line, prove that the locus of their points of contact is a rectangular hyperbola.

6. AOB, COD are two lines which bisect one another at right angles. Prove that the locus of a point which moves so that $PA{\cdot}PB = PC{\cdot}PD$ is a rectangular hyperbola.

7. A line has its extremities on two fixed lines and passes through a fixed point. Find the locus of the middle point of the line.

8. The two lines $x = \alpha$, $y = \beta$ are conjugate with respect to the hyperbola $xy = c^2$. Prove that (α, β) lies on the hyperbola $xy = 2c^2$.

9. Prove that the distance of any point from the centre of a rectangular hyperbola varies inversely as the perpendicular distance of its polar from the centre.

10. If four points be taken on a rectangular hyperbola such that the chord joining any two is perpendicular to the chord joining the other two, and if $\alpha, \beta, \gamma, \delta$ be the inclinations to either asymptote of the lines joining these points respectively to the centre; prove that $\tan\alpha \tan\beta \tan\gamma \tan\delta = 1$.

11. The normals at three points P, Q, R on a rectangular hyperbola intersect at a point S on the curve. Prove that the centre of the hyperbola is the centroid of the triangle PQR.

12. A rectangular hyperbola whose centre is C is cut by any circle of radius r in the four points P, Q, R, S; prove that $CP^2 + CQ^2 + CR^2 + CS^2 = 4r^2$.

13. An ellipse and a hyperbola have the same principal axes. Prove that the polar of any point on either curve with respect to the other touches the first curve.

14. Any tangent to the hyperbola $4xy = ab$ meets the ellipse $\frac{x^2}{a^2} + \frac{y^2}{b^2} = 1$ in points P and Q; prove that the normals to the ellipse at P and Q meet on a fixed diameter of the ellipse.

15. Through any point on the polar of P with respect to a rectangular hyperbola, two chords are drawn each subtending a right angle at P. Prove that the chords are at right angles.

16. A line intersects a hyperbola at the points P, Q and its asymptotes at the points R and S. Prove that $PR = QS$.

17. PA, PB are the perpendiculars from a point P of a rectangular hyperbola on a pair of conjugate diameters. Prove that AB is parallel to the normal at P.

18. The normals at four points P, Q, R, S on a rectangular hyperbola intersect in a point. Prove that the circle PQR passes through the other extremity of the diameter through S.

19. The normal at P_1 to the rectangular hyperbola $xy = c^2$ meets the curve again at P_2 the normal at P_2 meets the curve again in P_3 and so on. Prove that if y_1, y_2, y_3 ... are the ordinates of these points, then

$$y_1^3\, y_2 = y_2^3\, y_3 = \ldots = -c^4.$$

20. The normal at the point t to the rectangular hyperbola $xy = c^2$ meets the curve again at Q. Determine the parameter of Q. If QR is the diameter of the hyperbola through Q; prove that the locus of the mid-point of PR is $4x^3y^3 = c^2(x^2 + y^2)^2$.

21. A line is drawn parallel to the conjugate axis of a hyperbola and meets it and the conjugate hyperbola in the points P and Q. Prove that the tangents at P and Q meet on

$$\frac{y^4}{b^4}\left(\frac{y^2}{b^2} - \frac{x^2}{a^2}\right) = \frac{4x^2}{a^2}$$

and the normals meet on the x-axis.

General Equation of the Second Degree: Tracing of Conics

10

CHAPTER

1. Conic Section

In general, the **conic section** is defined as the locus of a point which moves so that its distance from a fixed point is in a constant ratio to the perpendicular distance of it from a fixed line.

The fixed point is called the focus, the fixed line the directrix and the constant ratio the eccentricity of the conic section. Let us denote the focus by S, the directrix by ZK and eccentricity by e.

We notc the following important particular cases of conic sections which we have already dealt with in the preceding chapters independently.

(I) When the focus does not lie on the directrix, we find that the locus (conic section) is an ellipse, a parabola or a hyperbola according as the eccentricity $e <=$ or > 1.

The circle is further a particular case of an ellipse. In case of a circle $e = 0$, focus is at the centre and directrix is at an infinite distance.

Note. In case of an ellipse it may happen that both the axes be zero, so that the conic section reduces to a point.

(II) When the focus lies on the directrix, the conic section is a pair of lines, real or imaginary.

Since, $SP = e.\ MP$, we have

$$\sin PSM = \frac{MP}{SP} = \frac{1}{e}.$$

There are four cases:

(*i*) When $e > 1$, the point P lies on one or the other of the two lines passing through S inclined to KK' at an angle $\sin^{-1} \frac{1}{e}$.

(*ii*) When $e = 1$, the angle *PSM* is a right angle and the conic section represents a pair of coincident lines coinciding with *SX*.

(*iii*) When $e < 1$, the angle *PSM* is imaginary and the conic section represents a pair of imaginary lines.

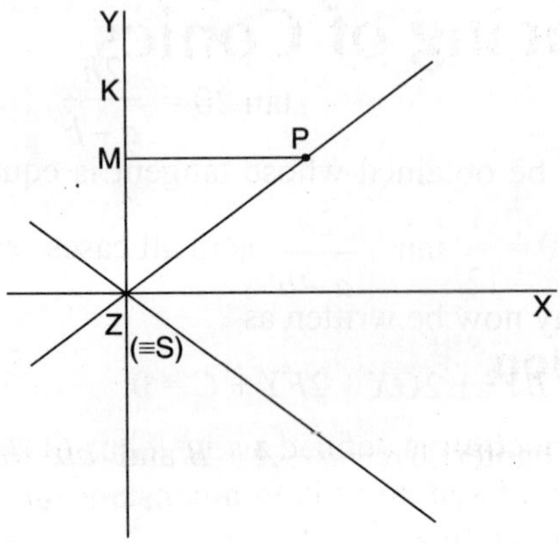

Fig. 10.1

(*iv*) When both *KK′* and *S* lie at infinity and *S* on *KK′*, the lines passing through *S* will meet at infinity and the conic section will represents a pair of parallel lines.

Thus, each of curves, a pair of lines, a circle, a parabola, an ellipse and a hyperbola can be regarded as a particular case of a conic section. We have seen in the preceding chapters that the equations of each of these particular conic sections is always a second degree in *x* and *y*. Now we shall demonstrate the converse that every equation of second degree in x and y represents one of the above conic sections. We shall also show that how to determine from any such equation the nature and the position of the conic section which it represents.

1.1 *To prove that every curve whose equation is of the second degree is a conic.*

Let $$ax^2 + 2hxy + by^2 + 2gx + 2fy + c = 0 \quad (1)$$

be the equation of the curve. Since, this is the most general form of the equation of second degree it will include all possible cases.

In order to eliminate the term containing *xy*, we turn the axes through a certain angle for which we have to substitute.

$$X \cos \theta - Y \sin \theta \text{ and } X \sin \theta + Y \cos \theta$$

for *x* and *y*, respectively. Then, equation (1) will become

$$a(X \cos \theta - Y \sin \theta)^2 + 2h(X \cos \theta - Y \sin \theta)(X \sin \theta + Y \cos \theta)$$

$$+ b(X \sin \theta + Y \cos \theta)^2 + 2g(X \cos \theta - Y \sin \theta)$$

$$+ 2f(X \sin \theta + Y \cos \theta) + c = 0. \quad (2)$$

The coefficient of XY in (2) is $2(b - a) \sin \theta \cos \theta + 2h(\cos^2 \theta - \sin^2 \theta)$; and this will be zero if

$$\tan 2\theta = \frac{2h}{a-b}. \quad (3)$$

Since, an angle can be obtained whose tangent is equal to any real quantity whatever, the angle $\theta = \frac{1}{2} \tan^{-1} \frac{2h}{a-b}$ is in all cases real.

Equation (2) may now be written as

$$AX^2 + BY^2 + 2GX + 2FY + C = 0. \quad (4)$$

By invariants (§5, Chapter 3) $a + b = A + B$ and $ab - h^2 = AB$. There arise two cases:

Case I. If $h^2 = ab$, then either A is zero or B is zero (A and B cannot both be zero). Let $A = 0$. Then, equation (4) can be written as

$$B\left(Y + \frac{F}{B}\right)^2 = -2GX - C + \frac{F^2}{B}. \quad (5)$$

If $G = 0$, this equation represents a pair of parallel lines, which are coincident if $F^2 - BC = 0$. If $G \neq 0$, the equation (5) may be written as

$$\left(Y + \frac{F}{B}\right)^2 = -\frac{2G}{B}\left(X - \frac{F^2}{2BG} + \frac{C}{2G}\right),$$

which represents a parabola, whose axis is parallel to X-axis.

Case II. If $h^2 \neq ab$, then neither A nor B is zero. Hence, equation (4) may be written in the form

$$A\left(X + \frac{G}{A}\right)^2 + B\left(Y + \frac{F}{B}\right)^2 = \frac{G^2}{A} + \frac{F^2}{B} - C = K \text{ (say)}.$$

On taking the origin at $\left(-\frac{G}{A}, -\frac{F}{B}\right)$, this equation takes the form

$$AX^2 + BY^2 = K. \quad (6)$$

If the right side of (6) be zero, this equation will represent a pair of lines. If however the right side of (6) be not zero, then it may be written as

$$\frac{X^2}{\frac{K}{A}} + \frac{Y^2}{\frac{K}{B}} = 1.$$

This is the equation of an ellipse if both the denominators are positive, and a hyperbola if one denominator is positive and the other is negative. If both of the denominators are negative, it is clear that no real values of X and Y will satisfy the equation. In this case, the curve is an imaginary ellipse.

Hence, in all cases the curve represented by the general equation of the second degree is a conic section.

Note. From the above, it is clear that the general equation of second degree $ax^2 + 2hxy + by^2 + 2gx + 2fy + c = 0$ represents

(*i*) a parabola if $h^2 = ab$,

(*ii*) an ellipse if $h^2 < ab$ and

(*iii*) a hyperbola if $h^2 > ab$.

2. Centre of a Conic Section

The **centre** *of a conic section is a point such that all chords of the conic which pass through it are bisected.*

2.1 *To find the coordinates of the centre of a conic section.*

Let $$ax^2 + 2hxy + by^2 + 2gx + 2fy + c = 0 \qquad (1)$$

be the equation of the conic and let (x', y') be its centre. Transferring the origin to the point (x', y') by taking the axes through the point (x', y') parallel to the original axes and substituting $X + x'$ for x and $Y + y'$ for y, equation (1) will become

$$a(X+x')^2 + 2h(X+x')(Y+y') + b(Y+y')^2$$
$$+ 2g(X+x') + 2f(Y+y') + c = 0$$
$$\Rightarrow aX^2 + 2hXY + bY^2 + 2(ax' + hy' + g)X + 2(hx' + by' + f)Y$$
$$+ ax'^2 + 2hx'y' + by'^2 + 2gx' + 2fy' + c = 0 \qquad (2)$$

In this equation, the coefficients of X and Y will both be zero, if x' and y' be so chosen that

$$ax' + hy' + g = 0 \qquad (3)$$

and $$hx' + by' + f = 0. \qquad (4)$$

Thus, equation (2) becomes

$$aX^2 + 2hXY + bY^2 + c' = 0, \qquad (5)$$

where $$c' = ax'^2 + 2hx'y' + by'^2 + 2gx' + 2fy' + c \qquad (6)$$

By solving equations (3) and (4) simultaneously, the centre of the conic can be obtained as

$$\left(\frac{fh-bg}{ab-h^2}, \frac{gh-af}{ab-h^2}\right).$$

Notes. (*i*) If $ab - h^2 \neq 0$, the coordinates of the centre are finite. Thus, the conic sections—circle, ellipse and hyperbola have a finite centre and are referred to **central conics**.

(*ii*) If $ab - h^2 = 0$, the conic section is a parabola and the centre is at infinity. Thus, a parabola has its centre at infinity.

(*iii*) If $ab - h^2 = 0$ and $fh - bg = 0$ *i.e.*, if $\frac{a}{h} = \frac{h}{b} = \frac{g}{f}$, the equations (3) and (4) represent the same line, and any point of that line is a centre. In this case, the locus is a pair of parallel lines. The conic section—pair of lines is referred to **degenerate conic**.

2.2 *To find the equation of a conic referred to the axes through the centre and parallel to the original axes.*

From § 2.1 the equation of a conic referred to the axes through the centre and parallel to the original axes is

$$aX^2 + 2hXY + bY^2 + c' = 0,$$

where

$$\begin{aligned} c' &= ax'^2 + 2hx'y' + by'^2 + 2gx' + 2fy' + c \\ &= x'(ax' + hy' + g) + y'(hx' + by' + f) + gx' + fy' + c \\ &= gx' + fy' + c \qquad \text{by using (3) and (4)} \\ &= g\left(\frac{fh-bg}{ab-h^2}\right) + f\left(\frac{gh-af}{ab-h^2}\right) + c \\ &= \frac{abc + 2fgh - af^2 - bg^2 - ch^2}{ab-h^2} \\ &= \frac{\Delta}{ab-h^2}, \end{aligned}$$

with $\Delta \equiv abc + 2fgh - af^2 - bg^2 - ch^2$.

Now, equation (5) in § 2.1 can be written as

$$aX^2 + 2hXY + bY^2 + \frac{\Delta}{ab-h^2} = 0.$$

This is the required equation of the conic referred to the new axes through the centre.

Ex. Find the centre of the conic $3x^2 - 5xy + 6y^2 + 11x - 17y + 13 = 0$ and its equation when transformed to the centre.

Corollary. The equation of the conic section with its centre at the origin is $Ax^2 + 2Hxy + By^2 = 1$.

From § 2.2, we know that the equation of the conic referred to the axes through the centre parallel to the original axes is

$$aX^2 + 2hXY + bY^2 + c' = 0,$$

where $$c' = \frac{\Delta}{ab - h^2}.$$

The centre of the conic is $\left(\frac{fh - bg}{ab - h^2}, \frac{gh - af}{ab - h^2}\right)$. This centre will lie at the origin if

$$fh - bg = 0 \text{ and } gh - af = 0.$$

This gives $c' = c$. The above equation reduces to

$$ax^2 + 2hxy + by^2 + c = 0,$$

which can be written as $Ax^2 + 2Hxy + By^2 = 1$.

3. Principal Axes and Eccentricity of a Conic

The study of principal axes and eccentricity is in fact relevant for the conic section which is either an ellipse or a hyperbola, and each one of these is a central conic. Further, there is no loss of generality if the equation of a central conic, with origin at the centre which being a finite one, is taken in the form $ax^2 + 2hxy + by^2 = 1$.

3.1 Length and position of the principal axes. *To find the length and position of the principal axes of the conic section whose equation is*

$$ax^2 + 2hxy + by^2 = 1. \tag{1}$$

Let this conic be cut by a concentric circle

$$x^2 + y^2 = r^2. \tag{2}$$

The equation of the lines joining the origin to the points of intersection of (1) and (2) is

$$ax^2 + 2hxy + by^2 = \frac{x^2 + y^2}{r^2}$$

$$\Rightarrow \left(a - \frac{1}{r^2}\right)x^2 + 2hxy + \left(b - \frac{1}{r^2}\right)y^2 = 0. \tag{3}$$

Clearly these lines are the diameters passing through the points of intersection of (1) and (2). Also these diameters will be equally inclined to the axes of the conic and will be coincident if the radius of the circle be equal to either of the semi-axes of the conic. Hence, the lines will be coincident if

$$\left(a-\frac{1}{r^2}\right)\left(b-\frac{1}{r^2}\right) = h^2$$

$$\Rightarrow \qquad \frac{1}{r^4} - (a+b)\frac{1}{r^2} + ab - h^2 = 0, \qquad (4)$$

and then they will coincide with one or the other of the axes of the conic. Thus, the lengths of the semi-axes of the conic are the roots of teh equation (4).

Now, multiplying equation (3) by $a - \dfrac{1}{r^2}$, we get

$$\left(a-\frac{1}{r^2}\right)^2 x^2 + 2h\left(a-\frac{1}{r^2}\right)xy + h^2y^2 = 0$$

$$\Rightarrow \quad \left(a-\frac{1}{r^2}\right)x + hy = 0. \qquad (5)$$

Hence, the axes of the conic will be obtained by substituting the either root of the equation (4) in equation (5).

Ex. Find the principal axes of the conic

$$22x^2 - 12xy + 17y^2 - 112x + 92y + 178 = 0.$$

3.2 Eccentricity. *To find the eccentricity of a conic section whose equation is*

$$ax^2 + 2hxy + by^2 = 1.$$

Let e be an eccentricity of the given conic. On transforming to principal axes, let the given equation become $Ax^2 + By^2 = 1$. Then $A = B(1 - e^2)$.

Also, we know that $A + B = a + b$, and $AB = ab - h^2$.

On eliminating A and B, we get

$$e^4 + \left\{\frac{(a-b)^2 + 4h^2}{ab - h^2}\right\}(e^2 - 1) = 0. \qquad (1)$$

Now there arise two cases:

Case I. When the conic is an ellipse, *i.e.* $ab - h^2 > 0$, the two values of e^2 in equation (1) have opposite sign. The positive value of e^2 gives the real eccentricity.

Case II. When the conic is a hyperbola, *i.e.* $ab - h^2 < 0$, the two values of e^2 in equation (1) are positive, which will give the real eccentricity.

3.3 Asymptotes of a conic. *To find the equation of the asymptotes of a conic.*

Let equation of the conic be

$$ax^2 + 2hxy + by^2 + 2gx + 2fy + c = 0. \tag{1}$$

We know that the equations of a conic and of the asymptotes only differ by a constant [see Chapter 9]. Therefore, the equations of the asymptotes are given by

$$ax^2 + 2hxy + by^2 + 2gx + 2fy + c + \lambda = 0, \tag{2}$$

where λ is so chosen that the equation (2) represents a pair of lines.

The condition that equation (2) may represent a pair of lines is [see Chapter 4 § 3]

$$\begin{vmatrix} a & h & g \\ h & b & f \\ g & f & c+\lambda \end{vmatrix} = 0$$

$$\Rightarrow \qquad \lambda(ab - h^2) + \Delta = 0.$$

Hence, the equations of the asymptotes of the conic (1) are given by

$$ax^2 + 2hxy + by^2 + 2gx + 2fy + c - \frac{\Delta}{ab - h^2} = 0.$$

The equations of the hyperbola and its conjugate differ from the equation of their asymptotes by constants which are equal and opposite to one another [see Chapter IX]. Thus, the equation of the hyperbola conjugate to (1) is

$$ax^2 + 2hxy + by^2 + 2gx + 2fy + c - \frac{2\Delta}{ab - h^2} = 0.$$

Corollary. The lines represetned by the equation $ax^2 + 2hxy + by^2 = 0$ are parallel to the asymptotes of the conic.

Ex. Find the asymptotes of the conic $x^2 - 3xy + y^2 + 10x - 10y + 21 = 0.$

3.4 Condition that the conic represented by the general equation of the second degree may be a rectangular hyperbola.

Let equation of the conic be

$$ax^2 + 2hxy + by^2 + 2gx + 2fy + c = 0. \tag{1}$$

The equation of the lines parallel to the asymptotes of the conic (1) is

$$ax^2 + 2hxy + by^2 = 0. \tag{2}$$

The conic (1) will represent a rectangular hyperbola if the lines given by (2) be at right angles, *i.e.*, if $a + b = 0$.

4. Axis, Latus Rectum, Vertex and Focus of a Parabola

To find the axis, latus rectum, vertex and focus of a parabola.

If the equation

$$ax^2 + 2hxy + by^2 + 2gx + 2fy + c = 0 \tag{1}$$

represents a parabola, then $h^2 = ab$, *i.e.*, second degree terms form a perfect square. Hence, equation (1) can be written as

$$(\alpha x + \beta y)^2 + 2gx + 2fy + c = 0, \tag{2}$$

where $\alpha^2 = a$ and $\beta^2 = b$

By introducing a new constant λ, equation (2) may be rewritten as

$$(\alpha x + \beta y + \lambda)^2 = 2(\lambda\alpha - g)x + 2(\lambda\beta - f)y + \lambda^2 - c. \tag{3}$$

Choose λ such that the lines

$$\alpha x + \beta y + \lambda = 0 \text{ and } 2(\lambda\alpha - g)x + 2(\lambda\beta - f)y + \lambda^2 - c = 0$$

are at right angles for which $\alpha(\lambda\alpha - g) + \beta(\lambda\beta - f) = 0$, *i.e.*,

$$\lambda = \frac{\alpha g + \beta f}{\alpha^2 - \beta^2}.$$

Putting this value of λ in equation (3), we get

$$(\alpha x + \beta y + \lambda)^2 = \frac{2(\alpha f - \beta g)}{\alpha^2 + \beta^2}(\beta x - \alpha y) + \lambda^2 - c$$

$$\Rightarrow \quad \left(\frac{\alpha x + \beta y + \lambda}{\sqrt{\alpha^2 + \beta^2}}\right)^2 = \frac{2(\alpha f - \beta g)}{(\alpha^2 + \beta^2)^{3/2}}\left(\frac{\beta x - \alpha y + c'}{\sqrt{\alpha^2 + \beta^2}}\right).$$

where $\quad c' = \dfrac{(\lambda^2 - c)(\alpha^2 + \beta^2)}{2(\alpha f - \beta g)}.$

The above equation with respect to new axes X and Y can be written as

$$Y^2 = 4AX,$$

where $\quad Y = \dfrac{\alpha x + \beta y + \lambda}{\sqrt{\alpha^2 + \beta^2}}, \quad X = \dfrac{\beta x - \alpha y + c'}{\sqrt{\alpha^2 + \beta^2}}$

and latus rectum $$4A = \frac{2(\alpha f - \beta g)}{(\alpha^2 + \beta^2)^{3/2}}.$$

Hence, the axis of the parabola is $Y = 0$, *i.e.* $\alpha x + \beta y + \lambda = 0$ and tangent at the vertex is $\beta x - \alpha y + c' = 0$.

Vertex. On solving the equations $\alpha x + \beta y + \lambda = 0$ and $\beta x - \alpha y + c' = 0$ simultaneously , we get the coordinates of the vertex.

Equation of latus rectum. The equation of the latus rectum is $X = A$, *i.e.*

$$\frac{\beta x - \alpha y + c'}{\sqrt{\alpha^2 + \beta^2}} = \frac{1}{2} \frac{(\alpha f - \beta g)}{(\alpha^2 + \beta^2)^{3/2}}$$

$$\Rightarrow \qquad \beta x - \alpha y + c' = \frac{1}{2} \frac{(\alpha f - \beta g)}{(\alpha^2 + \beta^2)}.$$

Focus. On solving the equations of the axis of the parabola and the latus rectum, simultaneously, we get the coordinates of the focus.

Ex. Find the axis, latus rectum, vertex and focus of the parabola

$$x^2 + 2xy + y^2 - 2x - 1 = 0.$$

5. Tracing of Conics

Example 1. Trace the conic

$$9x^2 + 24xy + 16y^2 - 2x + 14y + 1 = 0.$$

Solution. Here $a = 9$, $h = 12$, $b = 16$. Since, $h^2 = ab = 144$, the given conic is a parabola. The equation of the parabola can be written as

$$(3x + 4y)^2 = 2x - 14y - 1.$$

Introducing a new constant λ in this equation, we get

$$(3x + 4y + \lambda)^2 = 2(1 + 3\lambda)x + 2(4\lambda - 7)y + \lambda^2 - 1. \qquad (1)$$

Choose λ such that the lines

$$3x + 4y + \lambda = 0$$

and $$2(1 + 3\lambda)x + 2(4\lambda - 7)y + \lambda^2 - 1 = 0$$

are at right angles for which $3(1 + 3\lambda) + 4(4\lambda - 7) = 0$, *i.e.*, $\lambda = 1$.

Hence, equation (1) becomes $(3x + 4y + 1)^2 = 8x - 6y$. This can be written as

$$25 \cdot \left(\frac{3x+4y+1}{\sqrt{3^2+4^2}}\right)^2 = 10 \cdot \frac{(8x-6y)}{\sqrt{8^2+6^2}}$$

$$\Rightarrow \qquad \left(\frac{3x+4y+1}{\sqrt{3^2+4^2}}\right)^2 = \frac{2}{5} \cdot \frac{(4x-3y)}{5}. \qquad (2)$$

This is of the form $Y^2 = 4AX$, where

$$Y = \frac{3x+4y+1}{5},\; X = \frac{4x-3y}{5},\; 4A = \frac{2}{5}.$$

Hence, the axis of the parabola is $Y = 0$, *i.e.*

$$3x + 4y + 1 = 0 \qquad (3)$$

and tangent at the vertex is $X = 0$, *i.e.*

$$4x - 3y = 0. \qquad (4)$$

On solving the equations (3) and (4) simultaneously, the coordinates of the vertex are given by

$$\left(-\frac{3}{25}, \frac{-4}{25}\right).$$

The parabola meets x-axis, where $y = 0$, *i.e.,* in the points given by

$$9x^2 - 2x + 1 = 0$$

which are imaginary.

Further, the parabola meets y-axis, where $x = 0$, *i.e.,* in the points given by

$$16y^2 + 14y + 1 = 0, \text{ i.e., } y = \frac{-7 \pm \sqrt{33}}{16}.$$

Thus, the parabola meets the y-axis in the points

$$\left(0, \frac{-7 \pm \sqrt{33}}{16}\right)$$

and does not meet the x-axis.

Focus. The equation of the latus rectum is $X = A$, *i.e.*

$$4x - 3y = \frac{1}{2}. \qquad (5)$$

On solving (3) and (5), simultaneously, we get the coordinates of focus as

$$\left(-\frac{1}{25}, -\frac{11}{50}\right).$$

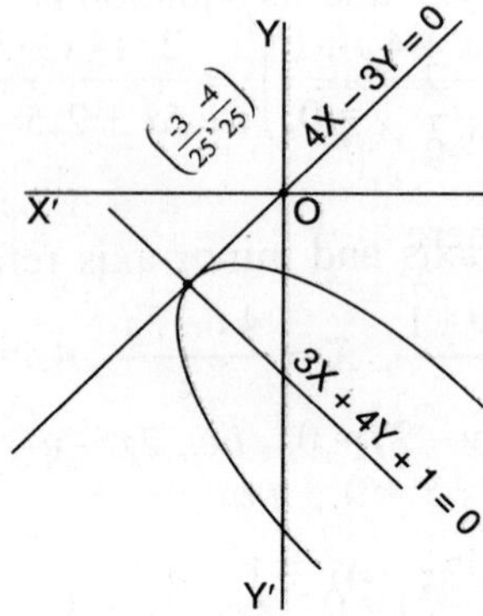

Fig. 10.2

Example 2. Trace the conic $8x^2 - 4xy + 5y^2 - 16x - 14y + 17 = 0$.

Solution. Here $a = 8, h = -2, b = 5$. Since, $h^2 < ab$, the given conic is an ellipse.

Let (x', y') be the centre of the ellipse. Then, from § 2.1, we have

$$8x' - 2y' - 8 = 0, \quad i.e., \quad 4x' - y' - 4 = 0$$

and
$$-2x' + 5y' - 7 = 0, \quad i.e., \quad 2x' - 5y' + 7 = 0.$$

On solving these equations the coordinates of the centre are given by $\left(\frac{3}{2}, 2\right)$.

Let c' be the new constant. Then, from § 2.2, we have

$$c' = -8 \times \frac{3}{2} + (-7) \times 2 + 17 = -9.$$

Hence, the equation of the conic referred to the centre as origin is $8x^2 - 4xy + 5y^2 = 9$, *i.e.*,

$$\frac{8}{9}x^2 - \frac{4}{9}xy + \frac{5}{9}y^2 = 1.$$

The squares of the lengths of the semi-axes of the conic are the roots of the equation

$$\left(\frac{8}{9} - \frac{1}{r^2}\right)\left(\frac{5}{9} - \frac{1}{r^2}\right) = \left(-\frac{2}{9}\right)^2$$

$$\Rightarrow \qquad r^2 = \frac{9}{4}, 1$$

The length of the major axis is 3 and its equation is

$$\left(\frac{8}{9}-\frac{4}{9}\right)x-\frac{2}{9}\,y=0,\quad i.e.,\ 2x-y=0.$$

The length of the minor axis is 2 and its equation is

$$\left(\frac{8}{9}-1\right)x-\frac{2}{9}\,y=0,\quad i.e.,\ x+2y=0.$$

The equations of the major axis and minor axis referred to the original axes are, respectively

$$2\left(x-\frac{3}{2}\right)-(y-2)=0,\quad i.e.,\ 2x-y-1=0$$

and

$$\left(x-\frac{3}{2}\right)+2(y-2)=0,\quad i.e.,\ 2x+4y-11=0.$$

The conic meets the x-axis in the points where $y = 0$, *i.e.*, in the points given by $8x^2 - 16x + 17 = 0$,which are imaginary.

Further, the conic meets the y-axis, where $x = 0$, *i.e.*, in the points given by the equation $5y^2 - 14y + 17 = 0$. This also gives imaginary values. Hence, the ellipse meets neither with x-axis nor with y-axis in real points.

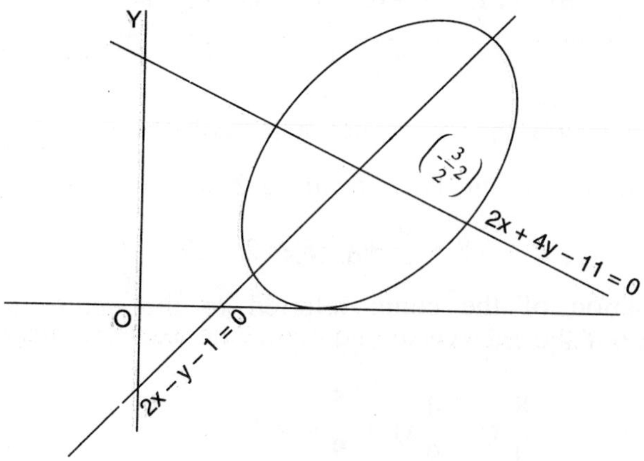

Fig. 10.3

Example 3. Trace the conic $x^2 - 4xy - 2y^2 + 10x + 4y = 0$.

Solution. Here $a = 1$, $h = -2$, $b = -2$. Since, $h^2 > ab$, the given conic is a hyperbola.

Let (x', y') be the centre of the conic. Then

$$x' - 2y' + 5 = 0$$

and $$-2x' - 2y' + 2 = 0 \quad i.e.\ x' + y' - 1 = 0.$$

On solving these equations the coordinates of the centre are given by (–1, 2).

New constant $c' = 5 \times (-1) + 2 \times 2 + 0 = -1.$

Hence, the equation of the conic referred to the centre as origin is

$$x^2 - 4xy - 2y^2 = 1.$$

The squares of the lengths of the semi-axes of the conic are the roots of the equation

$$\left(1 - \frac{1}{r^2}\right)\left(-2 - \frac{1}{r^2}\right) = (-2)^2$$

$$\Rightarrow \qquad r^2 = \frac{1}{2}, -\frac{1}{3},$$

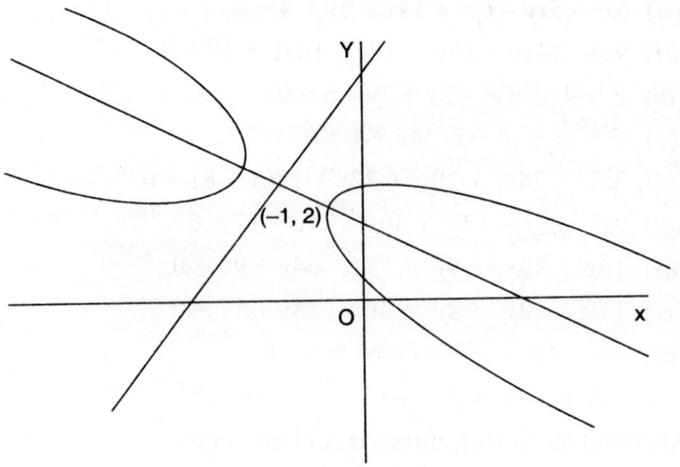

Fig. 10.4

The lengths of the transverse and conjugate axes are, respectively $\frac{2}{\sqrt{2}}$ and $\frac{2}{\sqrt{3}}$.

Equation of the transverse axis referred to centre as origin is

$$(1 - 2)x - 2y = 0, \quad i.e.,\ x + 2y = 0.$$

Equation of the conjugate axis referred to centre as origin is

$$(1 + 3)x - 2y = 0, \quad i.e.,\ 2x - y = 0.$$

The equations of the transverse and conjugate axes referred to original axes are, respectively

$$(x + 1) + 2(y - 2) = 0, \text{ i.e., } x + 2y - 3 = 0$$

and $$2(x + 1) - (y - 2) = 0, \text{ i.e., } 2x - y + 4 = 0.$$

The hyperbola meets x-axis, where $y = 0$, *i.e.*, $x^2 + 10x = 0$, which gives

$$x = 0, -10.$$

The hyperbola meets y-axis, where $x = 0$, *i.e.*, $-2y^2 + 4y = 0$, which gives

$$y = 0, 2.$$

Exercises

1. Trace the following conics:

(*i*) $x^2 + 4xy + y^2 - 2x + 2y + 4 = 0.$

(*ii*) $6x^2 - 5xy - 6y^2 + 14x + 5y + 4 = 0.$

(*iii*) $9x^2 - 24xy + 16y^2 - 18x - 101y + 19 = 0.$

(*iv*) $x^2 + 4xy + y^2 - 2x + 2y - 6 = 0.$

(*v*) $x^2 + xy + y^2 + x - 4y + 1 = 0.$

(*vi*) $36x^2 + 24xy + 29y^2 - 72x + 126y + 81 = 0.$

(*vii*) $9x^2 - 6xy + 17y^2 + 30x - 74y + 17 = 0.$

(*viii*) $16x^2 - 24xy + 9y^2 + 77x - 64y + 95 = 0.$

(*ix*) $17x^2 - 12xy + 8y^2 + 46x - 28y + 17 = 0.$

(*x*) $4x^2 - 4xy + y^2 - 8x - 6y + 5 = 0.$

2. Trace the parabola $x^2 - 4xy + 4y^2 + 10x - 8y + 13 = 0.$
Also find the vertex, directrix and the focus.

3. Trace the conic $2x^2 + 3xy - 2y^2 - 7x + y - 2 = 0.$
Calculate the eccentricity of the conic.

4. Trace the conic $25x^2 + 120xy + 144y^2 - 146x + 89y - 25 = 0,$
and find the coordinates of its foci.

5. Find the equation to the hyperbola which has $3x - 4y + 7 = 0$ and $4x + 3y + 1 = 0$ for asymptotes and passes through the origin. Find its centre and trace the curve.

6. Obtain the foci and directrices of the conic whose equation is

$$x^2 + 12xy - 4y^2 - 6x + 4y + 9 = 0.$$

7. Trace the conic $9x^2 + 6xy + y^2 + 2x + 3y + 4 = 0$, and find its latus rectum.

8. Trace the conic $32x^2 - 52xy - 7y^2 - 64x - 52y - 148 = 0$ and find the coordinates of its foci.

9. Prove that the conic $(a^2 + b^2)(x^2 + y^2) = (bx + ay - ab)^2$ is a parabola of latus rectum $\dfrac{2ab}{\sqrt{a^2 + b^2}}$.

10. Prove that the lengths of the semi-axes of the conic $ax^2 + 2hxy + by^2 = d$ are $\sqrt{\dfrac{d}{a+h}}$ and $\sqrt{\dfrac{d}{a-h}}$, respectively, and that their equations are $x^2 - y^2 = 0$.

11. Find the equation to the directrix and coordinates of focus of the parabola $x^2 - 2xy + y^2 - 2x - 2y + 3 = 0$.

12. Trace the conic $9x^2 - 24xy + 16y^2 - 2x - 39y - 11 = 0$.

13. If r be the length of the semi-axes of the conic $ax^2 + 2hxy + by^2 + 2gx + 2fy + c = 0$, prove that $(ab - h^2)^3 r^4 + \Delta(a + b)(ab - h^2)r^2 + \Delta^2 = 0$, where

$$\Delta = \begin{vmatrix} a & h & g \\ h & b & f \\ g & f & c \end{vmatrix}.$$

Polar Equation of a Conic

11
CHAPTER

1. Introduction

The polar equation of a conic can always be obtained from the corresponding cartesian equation referred to rectangular coordinates by writting $x = r \cos \theta$, $y = r \sin \theta$. Thus, for example, the polar equations of a line, a circle etc. have been obtained in the preceding chapters.

The polar equations of conics are of much use in the case when the pole is at the focus of the conic. In this case, the polar equation of the conic takes very simple form which is usually used in the problems. Now we proceed to obtain the same independently.

2. Polar Equation of a Conic

2.1 *To find the polar equation of a conic, the focus being at the pole.*

Let S be the focus, ZK the directrix and e the eccentricity of the conic.

Draw SZ perpendicular to the diectrix and consider SZ as initial line.

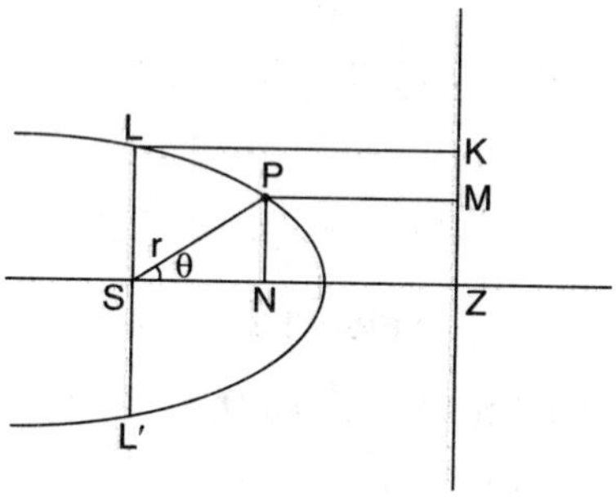

Fig. 11.1

Let $SL(= l)$ be the semi-latus rectum. Then

$$l = SL = e \cdot LK = e \cdot SZ.$$

Let $P(r, \theta)$ be any point on the conic. Draw PM and PN perpendiculars to the directrix and to SZ, respectively. Then, by the definition of the conic, we have

$$SP = e \cdot PM.$$

But $SP = r$, and $PM = NZ = SZ - SN = \dfrac{l}{e} - r\cos\theta$

Therefore $$r = e\left(\frac{l}{e} - r\cos\theta\right) = 1 + e\cos\theta.$$

This is the required equation of the conic in polar form.

Corollary. The equation of the conic whose axis makes an angle α with the initial line is $\dfrac{l}{r} = 1 + e\cos(\theta - \alpha)$.

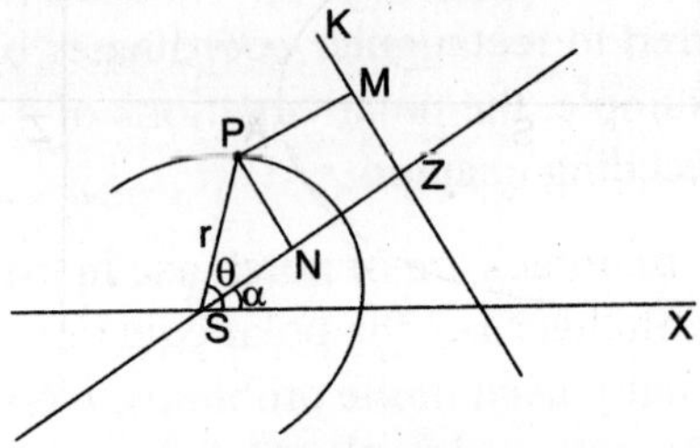

Fig. 11.2

2.2 *To find the equation of a directrix to the conic* $\dfrac{l}{r} = 1 + e\cos\theta$.

Let $P(r, \theta)$ be any point on the directrix ZK. Then

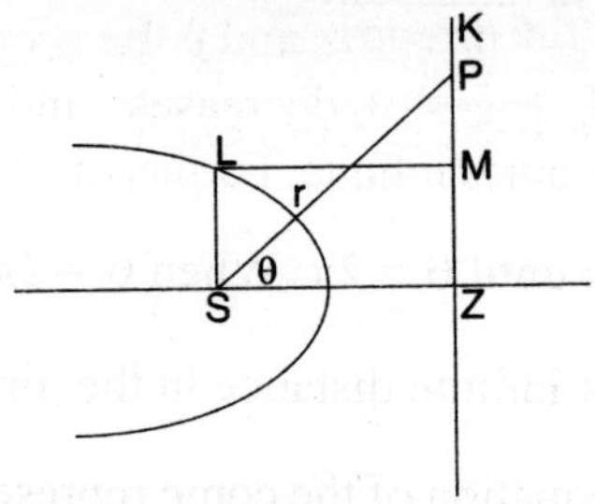

Fig. 11.3

$$SZ = SP\cos\theta = r\cos\theta.$$

But $$SZ = LM = \frac{SL}{e} = \frac{l}{e}. \qquad (\because SL = e \cdot LM)$$

Therefore $$\frac{l}{e} = r\cos\theta, \quad \frac{l}{r} = e\cos\theta.$$

This is the equation of the directrix.

3. Tracing of the Conic $\frac{l}{r} = 1 + e \cos \theta$

Case I. If $e = 0$, the equation of the conic reduces to $r = l$, which represents a circle of radius l and centre at the pole.

Case II. If $e = 1$, the equation of the conic reduces to

$$\frac{l}{r} = 1 + \cos \theta,$$

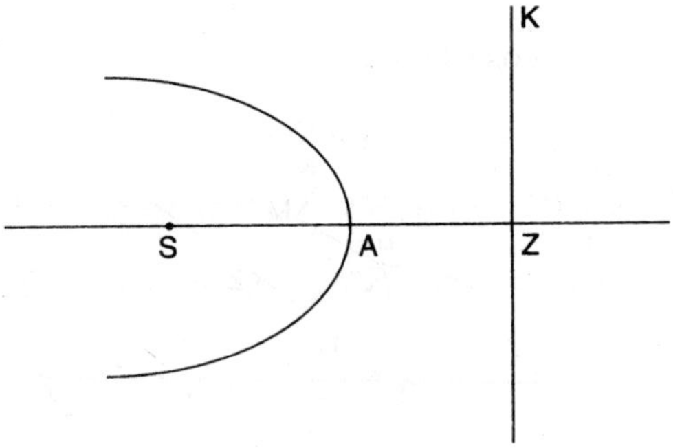

Fig. 11.4

which represents a parabola.

When $\theta = 0$, $r = \frac{l}{2}$, *i.e.*, the curve meets the axis at some point A such that $r = \frac{l}{2}$. This value of r is minimum.

As θ increases, $1 + \cos \theta$ decreases, and therefore r increases until $\theta = \pi$. when $\theta = \pi$, r becomes infinite. Further, as θ increases beyond π, $1 + \cos \theta$ increases continuously until $\theta = 2\pi$. When $\theta = 2\pi$, r takes the value $\frac{l}{2}$. Thus, the curve extends to an infinite distance in the direction of AS, see Fig. 11.4.

Case III. If $e < 1$, the equation of the conic represents an ellipse which can also be written in the form $r = \dfrac{l}{1 + e\cos\theta}$.

When $\theta = 0$, $r = \dfrac{l}{1+e}$, *i.e.*, the curve meets the axis at the point A such that $SA = \dfrac{l}{1+e}$.

As θ increases, $1 + e \cos \theta$ decreases, and therefore r increases until $\theta = \pi$. When $\theta = \pi$, $r = \dfrac{l}{1-e}$ which is positive since, $e < 1$.

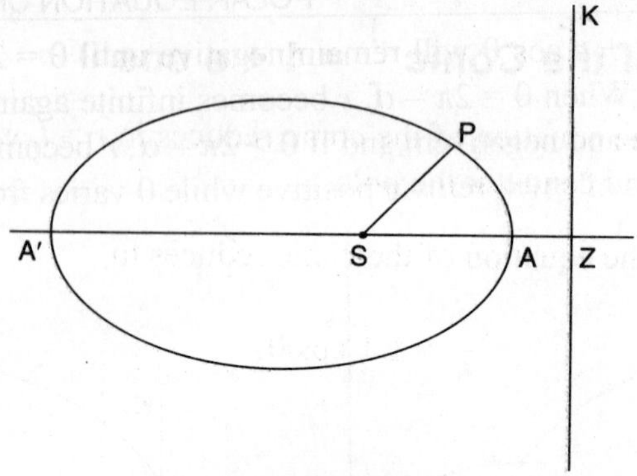

Fig. 11.5

This shows that the curve meets the axis at some point A' other than A such that $SA' = \dfrac{l}{1-e}$. Further, as θ increases from π to 2π, $1 + e\cos\theta$ increases continuously, and hence r decreases continuously from $\dfrac{l}{1-e}$ to $\dfrac{l}{1+e}$. But the curve is symmetrical about the axis, since, for any value of θ, $\cos(2\pi - \theta) = \cos\theta$. Thus, when $e < 1$, the equation $\dfrac{l}{r} = 1 + e\cos\theta$ represents a closed curve, symmetrical about the initial line.

Case IV. If $e > 1$, the equation of the conic represents a hyperbola and can be written as $r = \dfrac{l}{1 + e\cos\theta}$.

When $\theta = 0$, $r = \dfrac{l}{1+e}$, *i.e.*, the curve meets to the axis at the point A such that

$$SA = \frac{l}{1+e}.$$

As θ increases, $1 + e\cos\theta$ decreases, and hence r increases untill $1 + e\cos\theta = 0$. Thus, when $1 + e\cos\theta = 0$ which gives

$$\theta = \cos^{-1}\left(-\frac{1}{e}\right) = \alpha,$$

say (the angle ASK in the Fig. 11.6), the value of r becomes infinite. Further, as θ increases beyond the value α, $1 + e\cos\theta$ becomes negative, and when $\theta = \pi$, $r = \dfrac{-l}{e-1}$ *i.e.* the curve meets the axis at some point A' other than A such that

$$SA' = \frac{-l}{e-1}.$$

The value of $1 + e \cos \theta$ will remain negative until $\theta = 2\pi - \alpha$ (the angle ASK' in Fig. 11.6). When $\theta = 2\pi - \alpha$, r becomes infinite again. If $\theta < 2\pi - \alpha$, r becomes very large and negativem and if $\theta > 2\pi - \alpha$, r becomes very large and positive. The values of r will remain positive while θ varies from $2\pi - \alpha$ to 2π.

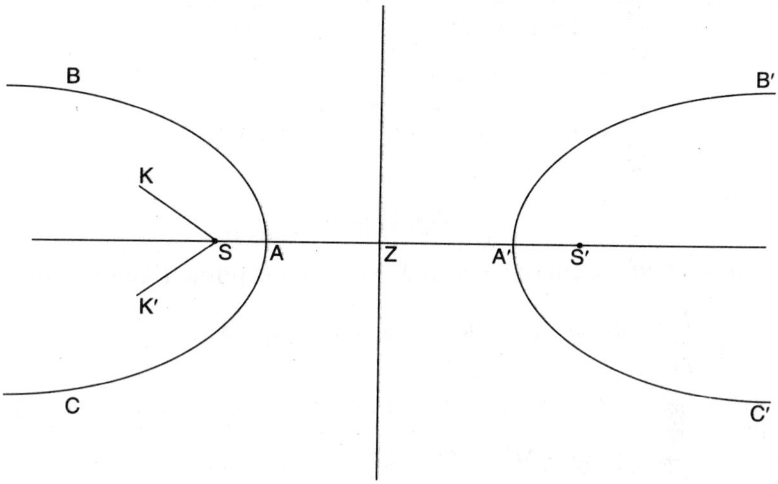

Fig. 11.6

Thus, the portion AB is described first, then the portions $C'A'$, $A'B'$ and CA.

3.1 Examples

Example 1. Prove that the semi-latus rectum of any conic is a harmonic mean between the segments of any focal chord.

Solution. Let the equation of the conic be $\dfrac{l}{r} = 1 + e \cos \theta$.

Let PSQ be the focal chord of the conic. Let α be the vectorial angle of P. Then vectorial angle of Q will be $\pi + \alpha$. Then

$$\frac{l}{SP} = 1 + e \cos \alpha \text{ and } \frac{l}{SQ} = 1 + e \cos (\pi + \alpha) = 1 - e \cos \alpha.$$

Therefore $$\frac{l}{SP} + \frac{l}{SQ} = 2, \textit{ i.e., } \frac{1}{SP} + \frac{1}{SQ} = \frac{2}{l}.$$

Example 2. If PSP' and QSQ' be any two focal chords of a conic which are at right angles to one another; prove that $\dfrac{1}{SP \cdot SP'} + \dfrac{1}{SQ \cdot SQ'}$ is constant.

Solution. Let the equation of the conic be $\dfrac{l}{r} = 1 + e \cos \theta$.

Let α be the vectorial angle of P. Then, the vectorial angles of Q, P' and Q' are respectively, $\frac{\pi}{2}+\alpha$, $\pi+\alpha$ and $\frac{3\pi}{2}+\alpha$. Thus, we have

$$\frac{l}{SP} = 1 + e\cos\alpha,$$

$$\frac{l}{SQ} = 1 + e\cos\left(\frac{\pi}{2}+\alpha\right) = 1 - e\sin\alpha,$$

$$\frac{l}{SP'} = 1 + e\cos(\pi+\alpha) = 1 - e\cos\alpha,$$

and $$\frac{l}{SQ'} = 1 + e\cos\left(\frac{3\pi}{2}+\alpha\right) = 1 + e\sin\alpha.$$

Therefore $$\frac{l}{SP}\cdot\frac{l}{SP'} + \frac{l}{SQ}\cdot\frac{l}{SQ'} = 1 - e^2\cos^2\alpha + 1 - e^2\sin^2\alpha$$

$$\Rightarrow \qquad \frac{1}{SP\cdot SP'} + \frac{1}{SQ\cdot SQ'} = \frac{2-e^2}{l^2} = \text{constant}.$$

Example 3. A circle of given radius passing through the focus S of a given conic intersects it in the points A, B, C and D. Prove that $SA\cdot SB\cdot SC\cdot SD$ is constant.

Solution. Let the equation of the conic be

$$\frac{l}{r} = 1 + e\cos\theta, \tag{1}$$

with pole at the focus S. The equation of the circle (see § 7, Chapter V) passing through the pole is $r = 2a\cos(\theta-\alpha)$, where a is the radius of the circle. This equation may be written as

$$r = 2a(\cos\theta\cos\alpha + \sin\theta\sin\alpha). \tag{2}$$

Eliminating θ between (1) and (2), we get

$$r = 2a\left[\left(\frac{l-r}{er}\right)\cos\alpha + \sqrt{1-\left(\frac{l-r}{er}\right)^2}\cdot\sin\alpha\right]$$

$$\Rightarrow \qquad e^2r^4 + 4ae\cos\alpha\, r^3 + (4a^2 - 4ael\cos\alpha - 4a^2e^2\sin^2\alpha)\, r^2 - 8a^2lr + 4a^2l^2 = 0.$$

This equation being biquadratic in r has four values of r which are given by SA, SB, SC and SD, respectively. Hence, by theory of equations, we have

$$SA \cdot SB \cdot SC \cdot SD = \frac{4a^2l^2}{e^2} = \text{constant}.$$

Exercises

1. If PSQ and PHR be two chords of an ellipse passing through the foci S and H, respectively; prove that $\frac{PS}{SQ} + \frac{PH}{HR}$ is independent of the position of P.

2. Prove that in any conic the sum of the reciprocals of two perpendicular focal chords is constant.

3. Prove that the equations $\frac{l}{r} = 1 + e \cos \theta$ and $\frac{l}{r} = -1 + e \cos \theta$ represent the same conic.

4. A circle which passes through the focus of a conic meets the conic in four points whose distances from the focus are r_1, r_2, r_3 and r_4. Prove that

$$\frac{1}{r_1} + \frac{1}{r_2} + \frac{1}{r_3} + \frac{1}{r_4} = \frac{2}{l},$$

where l is semi-latus rectum of the conic.

5. If PSQ is a focal chord of a conic; prove that the locus of its middle point is another conic of the same nature as the original conic.

6. Prove that the directrix of the conic $\frac{l}{r} = 1 + e \cos \theta$ corresponding to the focus other than the pole is $\frac{l}{r} = -\frac{e(1-e^2)}{(1+e^2)} \cos \theta$.

7. A point moves so that the sum of its distances from two fixed points S, S' is constant and equal to $2a$. Show that P lies on the conic

$$\frac{a(1-e^2)}{r} = 1 - e \cos \theta,$$

referred to S as pole and SS' as the initial line, SS' being equal to $2ae$.

4. Chord Joining Two Points

To find the equation of the chord joining two points on a conic.

Let the equation of the conic be $\frac{l}{r} = 1 + e \cos \theta$.

Let $\alpha - \beta$ and $\alpha + \beta$ be the vectorial angles of any two point P and Q on the conic, respectively. The general equation of any line is

$$\frac{l}{r} = A \cos \theta + B \cos (\theta - \alpha).$$

This will pass through the points P and Q if

$$1 + e \cos (\alpha - \beta) = A \cos (\alpha - \beta) + B \cos \beta$$

and

$$1 + e \cos (\alpha + \beta) = A \cos (\alpha + \beta) + B \cos \beta.$$

Three equations imply that $A = e$ and $B = \sec \beta$. Substituting the values of A and B in the equation of the line, we get

$$\frac{l}{r} = \sec \beta \cos (\theta - \alpha) + e \cos \theta.$$

Corollary. The equation of the chord joining the points whose vectorial angles are $\alpha - \beta$ and $\alpha + \beta$ on the conic $\frac{l}{r} = 1 + e \cos (\theta - \gamma)$ is

$$\frac{l}{r} = \sec \beta \cos (\theta - \alpha) + e \cos (\theta - \gamma).$$

5. Tangent and Normal

5.1 Equation of tangent. *To find the equation of the tangent at a point whose vectorial angle is α to a conic.*

Let the equation of the conic be $\frac{l}{r} = 1 + e \cos \theta$.

The equation of the chord joining the two points whose vectorial angles are $\alpha - \beta$, $\alpha + \beta$ is

$$\frac{l}{r} = \sec \beta \cos (\theta - \alpha) + e \cos \theta.$$

This chord will become the tangent at the point whose vectorial angle is α if we take $\beta = 0$ in its equation, and we get

$$\frac{l}{r} = \cos (\theta - \alpha) + e \cos \theta.$$

Corollary. The equation of the tangent at the point whose vectorial angle is α to the conic $\frac{l}{r} = 1 + e \cos (\theta - \gamma)$ is $\frac{l}{r} = \cos (\theta - \alpha) + e \cos (\theta - \gamma)$.

5.2 Equation of normal. *To find the equation of the normal at a point whose vectorial angle is α to a conic.*

Let the equation of the conic be $\frac{l}{r} = 1 + e\cos\theta$.

The equation of the tangent at the point whose vectorial angle is α, is

$$\frac{l}{r} = \cos(\theta - \alpha) + e\cos\theta.$$

The coordinates of the point of contact of the tangent are $\left(\frac{l}{1+e\cos\alpha}, \alpha\right)$.

The equation of the line perpendicular to the tangent is

$$\frac{A}{r} = \cos\left(\theta + \frac{\pi}{2} - \alpha\right) + e\cos\left(\theta + \frac{\pi}{2}\right), \text{ i.e., } \frac{A}{r} = -\sin(\theta - \alpha) - e\sin\theta.$$

This line will pass through the point $\left(\frac{l}{1+e\cos\alpha}, \alpha\right)$ if

$$A = \frac{-l e\sin\alpha}{1+e\cos\alpha}.$$

Hence, the equation of the normal at the point whose vectorial angle is α, is

$$\frac{l e\sin\alpha}{r(1+e\cos\alpha)} = \sin(\theta - \alpha) + e\sin\theta.$$

5.3 Examples

Example 1. Find the condition that the line $\frac{l}{r} = A\cos\theta + B\sin\theta$ may be a tangent to the conic $\frac{l}{r} = 1 + e\cos\theta$.

Solution. The equation of the tangent at the point whose vectorial angle is α to the given conic is

$$\frac{l}{r} = \cos(\theta - \alpha) + e\cos\theta, \text{ i.e., } \frac{l}{r} = (e + \cos\alpha)\cos\theta + \sin\alpha\sin\theta.$$

The given line will be tangent to the given conic if

$$\frac{l}{r} - A\cos\theta - B\sin\theta \equiv \frac{l}{r} - (e + \cos\alpha)\cos\theta - \sin\alpha\sin\theta$$

$$\Rightarrow \quad A = e + \cos\alpha,\ B = \sin\alpha, \text{ i.e., } (A - e)^2 + B^2 = 1.$$

This is the required condition.

Example 2. Prove that the equation to the locus of the foot of the perpendicular from the focus of the conic $\frac{l}{r} = 1 + e\cos\theta$ or any tangent to it is $r^2(e^2 - 1) - 2ler\cos\theta + l^2 = 0$.

Discuss the particular case when $e = 1$.

Solution. The equation of the tangent at any point on the conic whose vectorial angle is α, is

$$\frac{l}{r} = \cos(\theta - \alpha) + e\cos\theta \qquad (1)$$

Transforming into cartesian coordinates, it becomes $l = x(e + \cos\alpha) + y\sin\alpha$.

Any line perpendicular to it passing through focus (origin) is

$$x\sin\alpha - y(e + \cos\alpha) = 0.$$

Transforming into polar coordinates, it becomes

$$\sin(\theta - \alpha) + e\sin\theta = 0. \qquad (2)$$

In order to obtain the locus of the foot of the perpendicular, we eliminate α from (1) and (2). On squaring and adding, we get

$$r^2(e^2 - 1) - 2ler\cos\theta + l^2 = 0.$$

When $e = 1$, the locus becomes

$$-2lr\cos\theta + l^2 = 0, \text{ i.e., } \frac{l}{r} = 2\cos\theta$$

Thus, when $e = 1$, the locus is tangent at the vertex to the parabola

$$\frac{l}{r} = 1 + \cos\theta.$$

Example 3. The normal to the conic $\frac{l}{r} = 1 + e\cos\theta$ at the point whose vectorial angle is α meets the curve again at the point whose vectorial angle is β. Prove that

$$\tan\frac{\alpha}{2}\tan\frac{\beta}{2} = -\frac{1 + 2e\cos^2\frac{\alpha}{2} + e^2}{1 - 2e\sin^2\frac{\alpha}{2} + e^2}.$$

Solution. The equation of the normal to the given conic at the point whose vectorial angle is α, is

$$\frac{le\sin\alpha}{r(1+e\cos\alpha)} = \sin(\theta-\alpha)+e\sin\theta.$$

This normal will pass through the point $\left(\frac{l}{1+e\cos\beta}, \beta\right)$ if

$$\frac{e\sin\alpha(1+e\cos\beta)}{1+e\cos\alpha} = \sin(\beta-\alpha)+e\sin\beta$$

$$\Rightarrow \quad (1+e\cos\alpha+e^2)\sin(\alpha-\beta)+e(\sin\alpha-\sin\beta)=0$$

$$\Rightarrow \quad (1+e\cos\alpha+e^2)\cdot 2\sin\frac{\alpha-\beta}{2}\cos\frac{\alpha-\beta}{2} + 2e\cos\frac{\alpha+\beta}{2}\sin\frac{\alpha-\beta}{2}=0$$

$$\Rightarrow \quad (1+e\cos\alpha+e^2)\left(\cos\frac{\alpha}{2}\cos\frac{\beta}{2}+\sin\frac{\alpha}{2}\sin\frac{\beta}{2}\right) + e\left(\cos\frac{\alpha}{2}\cos\frac{\beta}{2}-\sin\frac{\alpha}{2}\sin\frac{\beta}{2}\right)=0$$

$$\Rightarrow \quad (1+e\cos\alpha+e^2)\left(1+\tan\frac{\alpha}{2}\tan\frac{\beta}{2}\right) + e\left(1-\tan\frac{\alpha}{2}\tan\frac{\beta}{2}\right)=0.$$

$$\Rightarrow \quad (1+e\cos\alpha+e^2-e)\tan\frac{\alpha}{2}\tan\frac{\beta}{2} + (1+e\cos\alpha+e^2+e)=0$$

$$\Rightarrow \quad \tan\frac{\alpha}{2}\tan\frac{\beta}{2} = -\frac{1+2e\cos^2\frac{\alpha}{2}+e^2}{1-2e\sin^2\frac{\alpha}{2}+e^2}.$$

Exercises

1. Prove that the line $\frac{l}{r} = A\cos\theta + B\sin\theta$ may be a tangent to the conic $\frac{l}{r} = 1 + e\cos(\theta - \gamma)$ if $A^2 + B^2 - 2e(A\cos\gamma + B\sin\gamma) + e^2 - 1 = 0$.

2. Prove that the two conics $\frac{l_1}{r} = 1 + e_1\cos\theta$ and $\frac{l_2}{r} = 1 + e_2\cos(\theta - \alpha)$ will touch one another if $l_1^2(1 - e_2^2) + l_2^2(1 - e_1^2) = 2l_1l_2(1 - e_1e_2\cos\alpha)$.

3. If a chord of the conic $\frac{l}{r} = 1 + e\cos\theta$ subtends a constant angle 2α at the focus; prove that the locus of the point where it meets the internal bisector of the angle 2α is the conic $\frac{l\cos\alpha}{r} = 1 + e\cos\alpha\cos\theta$.

4. PQ is a variable chord of a conic having a focus at S and the angle PSQ is constant. Prove that the locus of the point of intersection of the tangents at P and Q is a conic having S for a focus, and the corresponding directrix in common with the given conic.

5. In any conic prove the the portion of the tangent intercepted between a directrix and the point of contact subtends a right angle at the corresponding focus.

6. In a conic prove that the locus of the point of intersection of the tangents at the ends of a focal chord is the corresponding directrix.

7. Chords of a conic which subtend a constant angle at a focus touch a fixed conic.

8. If the normal at one of the extremities of the latus rectum of the conic $\frac{l}{r} = 1 + e\cos\theta$, meets the curve again at Q; prove that the distance of Q from the focus is $\frac{l(1+3e^2+e^4)}{1+e^2-e^4}$.

9. If the normals at the points whose vectorial angles are α, β, γ on the parabola $\frac{l}{r} = 1 + \cos\theta$ meet in the point (ρ, ϕ); prove that $2\phi = \alpha + \beta + \gamma$.

10. If the normals at the points whose vectorial angles are α, β, γ, δ on the conic $\frac{l}{r} = 1 + e\cos\theta$ meet in the point (ρ, ϕ); prove that

    (*i*) $\alpha + \beta + \gamma + \delta - 2\phi = (2n + 1)\pi$.

(*ii*) $\tan\frac{\alpha}{2}\tan\frac{\beta}{2}\tan\frac{\gamma}{2}\tan\frac{\delta}{2}+\left(\frac{1+e}{1-e}\right)^2=0.$

6. Polar

To find the equation of the polar of a point with respect to a conic.

Let the equation of the conic be

$$\frac{l}{r}=1+e\cos\theta \tag{1}$$

and let (r_1, θ_1) be the coordinates of the point.

Let $\alpha-\beta$ and $\alpha+\beta$ be the vectorial angles of the points P and Q, respectively the tangents at which pass through the point (r_1, θ_1). The equation of the chord of contact PQ is

$$\frac{l}{r}=\sec\beta\cos(\theta-\alpha)+e\cos\theta. \tag{2}$$

The equations of the tangents at P and Q are

$$\frac{l}{r}=\cos(\theta-\overline{\alpha-\beta})+e\cos\theta \tag{3}$$

and $$\frac{l}{r}=\cos(\theta-\overline{\alpha+\beta})+e\cos\theta. \tag{4}$$

These tangents will pass through (r_1, θ_1) if

$$\frac{l}{r_1}=\cos(\theta_1-\overline{\alpha-\beta})+e\cos\theta_1 \tag{5}$$

and $$\frac{l}{r_1}=\cos(\theta_1-\overline{\alpha+\beta})+e\cos\theta_1. \tag{6}$$

From equations (5) and (6), we get

$$\cos(\theta_1-\overline{\alpha-\beta})=\cos(\theta_1-\overline{\alpha+\beta})$$

$$\Rightarrow \quad \theta_1-(\alpha-\beta)=2n\pi\pm\{\theta_1-(\alpha+\beta)\}.$$

We consider here lower sign, since upper sign would give a special value of β so that $\theta_1=n\pi+\alpha$. Substituting this value of α in (5), we get

$$\frac{l}{r_1}=\cos(n\pi+\beta)+e\cos\theta_1$$

$$\Rightarrow \quad \frac{l}{r_1} - e\cos\theta_1 = (-1)^n \cos\beta.$$

Thus, equation (2) becomes

$$\left(\frac{l}{r} - e\cos\theta\right)\left(\frac{l}{r_1} - e\cos\theta_1\right) = (-1)^n \cos(\theta - \overline{\theta_1 - n\pi})$$

$$\Rightarrow \quad \left(\frac{l}{r} - e\cos\theta\right)\left(\frac{l}{r_1} - e\cos\theta_1\right) = \cos(\theta - \theta_1).$$

7. Director Circle

To find the polar equation of the director circle of a conic.

Let the equation of the conic be

$$\frac{l}{r} = 1 + e\cos\theta.$$

The equations of the tangents to the conic at the points whose vectorial angles are α and β are

$$\frac{l}{r} = \cos(\theta - \alpha) + e\cos\theta \text{ and } \frac{l}{r} = \cos(\theta - \beta) + e\cos\theta.$$

The vectorial angle of the point of intersection of these tangents is given by the equation

$$\cos(\theta - \alpha) = \cos(\theta - \beta), \ i.e., \ \theta - \alpha = \pm(\theta - \beta).$$

We consider here lower sign, since, upper sign would give $\alpha = \beta$ which is a contradiction. Therefore

$$\theta = \frac{\alpha + \beta}{2}. \tag{1}$$

Substituting this value of θ in either equation of the tangent and we obtain

$$\frac{l}{r} = \cos\frac{\alpha - \beta}{2} + e\cos\frac{\alpha + \beta}{2}. \tag{2}$$

Transforming the equations of the tangents into cartesian coordinates and taking that the tangents are at right angles, then

$$(e + \cos\alpha)(e + \cos\beta) + \sin\alpha\sin\beta = 0$$

$$\Rightarrow \quad e^2 + 2e\cos\frac{\alpha+\beta}{2}\cos\frac{\alpha-\beta}{2} + 2\cos^2\frac{\alpha-\beta}{2} - 1 = 0. \tag{3}$$

Eliminating α and β from (3) with the help of (1) and (2), we get

$$r^2(1-e^2)+2elr\cos\theta-2l^2=0.$$

This is the required equation of the director circle.

8. Asymptotes

To find the polar equations of the asymptotes of a conic.

Let the equation of the conic be

$$\frac{l}{r}=1+e\cos\theta.$$

Let $P(R, \alpha)$ be any point on the conic. Then equation of the tangent to the conic at $P(R, \alpha)$ is

$$\frac{l}{r}=\cos(\theta-\alpha)+e\cos\theta$$

$$\Rightarrow \quad \frac{l}{r}=(e+\cos\alpha)\cos\theta+\sin\theta\sin\alpha. \qquad (1)$$

Since, the point $P(R, \alpha)$ lies on the conic, we have

$$\frac{l}{R}=1+e\cos\alpha.$$

The tangent at $P(R, \alpha)$ will become the asymptote if P lies at infinity for which $R\to\infty$, *i.e.*, if $0=1+e\cos\alpha$, *i.e.*,

$$\cos\alpha=-\frac{1}{e}. \qquad (2)$$

Eliminating α from (1) with the help of (2), we get

$$\frac{l}{r}=\left(e-\frac{1}{e}\right)\cos\theta\pm\sin\theta\sqrt{1-\frac{1}{e^2}}$$

$$\Rightarrow \quad \left\{\frac{le}{r}-(e^2-1)\cos\theta\right\}^2=(e^2-1)\sin^2\theta.$$

These are the equations of the asymptotes.

Miscellaneous Exercises

1. Prove that the equations $\frac{l}{r} = 1 - e\cos\theta$ and $\frac{l}{r} = -1 - e\cos\theta$ represent the same conic.

2. If a chord of an ellipse makes an angle α with the axis; prove that the angle between the tangents at its extremities is $\tan^{-1}\left(\frac{2e\sin\alpha}{1-e^2}\right)$.

3. Prove that the perpendicular focal chords of a rectangular hyperbola are equal.

4. Prove that the locus of the point of intersection of two tangents to a parabola which cut one another at a constant angle is a hyperbola having the same focus and the directrix as the original parabola.

5. A focal chord PSP' of an ellipse is inclined at an angle α to the major axis. Prove that the perpendicular from the focus on the tangent at P makes an angle $\tan^{-1}\left(\frac{\sin\alpha}{e+\cos\alpha}\right)$ with the axis.

6. Prove that the locus of the pole of a chord of the conic $\frac{l}{r} = 1 + e\cos\theta$ which subtends a constant angle 2α at the focus is

$$\frac{l\sec\alpha}{r} = 1 + e\sec\alpha\cos\theta.$$

Also distinguish between the cases for which $\cos\alpha > = < e$.

7. The tangents at the points P and Q to a parabola meet in T. Prove that $ST^2 = SP\cdot SQ$, where S is the focus of the parabola.

8. An ellipse and a parabola have a common focus S and intersect in two real points P and Q, of which P is the vertex of the parabola. If e be the eccentricity of the ellipse and α the angle made by SP with the major axis; prove that

$$\frac{SQ}{SP} = 1 + \frac{4e^2\sin^2\alpha}{(1-e\cos\alpha)^2}.$$

9. If A, B, C be any three points on a parabola, and the tangents at these points form a triangle $A'B'C'$; prove that $SA\cdot SB\cdot SC = SA'\cdot SB'\cdot SC'$, where S is the focus of the parabola.

10. A conic is described having the same focus and eccentricity as the conic $\frac{l}{r} = 1 + e\cos\theta$, and the two conics touch at the point θ = α; prove that the length of its latus rectum is $\frac{2l(1-e^2)}{e^2+2e\cos\alpha+1}$.

11. If the tangent at any point of an ellipse makes an angle α with its major axis and an angle β with the focal radius to the point of contact; prove that $e \cos \alpha = \cos \beta$.

12. If the tangents from P subtend a fixed angle β at the focus S of the conic $\frac{l}{r} = 1 + e \cos \theta$, prove that the locus of the middle point of SP is a conic eccentricity $e \sec \beta$. Also find its latus rectum.

13. If two conics have a common focus; prove that two of their common chords will pass through the point of intersection of their directrices.

14. Two equal ellipses of, eccentricity e, are placed with their axes at right angles and have a common focus S. If PQ be a common tangent to the ellipses, prove that the angle PSQ is $2\sin^{-1}\left(\frac{e}{\sqrt{2}}\right)$.

15. If PQ is the chord of contact of tangents drawn from a point T to an ellipse $\frac{l}{r} = 1 + e \cos \theta$, whose focus is S; prove that

$$\frac{1}{SP \cdot SQ} - \frac{1}{ST^2} = \frac{1}{b^2} \sin^2 \tfrac{1}{2} PSQ,$$

where b is the semi-minor axis of the ellipse.

16. The eccentric angle of any point P on an ellipse is α, measured from the semi-major axis CA, S is the focus nearest to A. If $\angle ASP = \theta$, prove that

$$\tan \frac{\theta}{2} = \sqrt{\frac{1+e}{1-e}} \tan \frac{\alpha}{2}.$$

17. Conics with latus rectum of given length are described with a fixed point as a focus and touching a fixed line. Prove that the locus of their centres is a conic.

18. Prove that the exterior angle between any two tangents to a parabola is half the difference of the vectorial angles of their point of contact.

19. P, Q, R are three points on the conic $\frac{l}{r} = 1 + e \cos \theta$, the focus S being the pole. SP and SR meet the tangent at Q in M and N so that $SM = SN = l$. Prove that PR touches the conic $\frac{l}{r} = 1 + 2e \cos \theta$.

20. Given the focus and directrix of a conic, prove that the polar of a given point with respect to it passes through a fixed point.

21. Prove that the equation of the pair of tangents which can be drawn to the conic $\frac{l}{r} = 1 + e\cos\theta$ from the point (r_1, θ_1) is

$$\left\{\left(\frac{l}{r} - e\cos\theta\right)^2 - 1\right\}\left\{\left(\frac{l}{r_1} - e\cos\theta_1\right)^2 - 1\right\}$$

$$= \left\{\left(\frac{l}{r} - e\cos\theta\right)\left(\frac{l}{r_1} - e\cos\theta_1\right) - \cos(\theta - \theta_1)\right\}^2.$$

22. If the normals at three points whose vectorial angles are α, β, γ to the parabola $r = a\,\text{cosec}^2\,\frac{\theta}{2}$ meet in the point (R, δ); prove that $2\delta = \alpha + \beta + \gamma - \pi$.

23. Find the equation to the circle circumscribing the triangle formed by the tangents to a parabola and prove that it passes through the focus.

24. If the normals to the parabola $\frac{l}{r} = 1 + \cos\theta$ at the points *P*, *Q*, *R* whose vectorial angles are α, β, γ meet in the point $T(\rho, \delta)$; prove that the diameter of the circumcircle of the triangle formed by the tangents at *P*, *Q*, *R* is equal to *ST*, where *S* is the focus of the parabola.

25. From the focus *S* of the conic $\frac{l}{r} = 1 + e\cos\theta$, radii *SP* and *SQ* are drawn at right angles to one another. Prove that the locus of the pole of *PQ* is a conic having the same focus which is hyperbola, parabola or ellipse according as $e >=< \frac{1}{\sqrt{2}}$.

Oblique Axes

APPENDIX I

1. Introduction

In Chapters 1 to 9, we have studied different loci, viz., straight line, circle, parabola etc. in different forms by taking rectangular axes. In the present chapter we study the above loci by taking oblique axes, *i.e.*, the axes inclined at an angle other than right angle.

2. Distance between Two Points

To express the distance between two points in terms of their coordinates.

Let the axes be inclined at an angle ω and let $P(x_1, y_1)$, $Q(x_2, y_2)$ be any two given points. Draw PM, QN parallel to y-axis and PL parallel to x-axis. Then

$$OM = x_1,\ MP = y_1,\ ON = x_2 \text{ and } NQ = y_2.$$

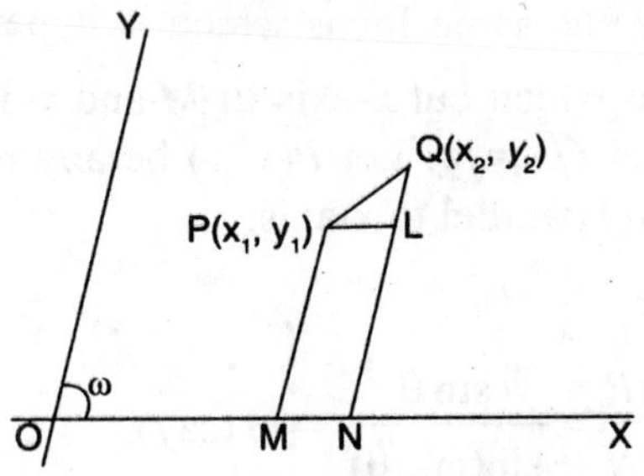

Fig. 1

In $\Delta\, PLQ$, we have

$$PQ^2 = PL^2 + LQ^2 - 2PL.LQ \cos QLP.$$

But $$PL = MN = ON - OM = x_2 - x_1,$$

$$LQ = NQ - NL = NQ - MP = y_2 - y_1$$

and $$\angle QLP = \pi - \omega.$$

$$\therefore \quad PQ^2 = (x_2 - x_1)^2 + (y_2 - y_1)^2 - 2(x_2 - x_1)(y_2 - y_1)\cos(\pi - \omega)$$

$$\Rightarrow \qquad PQ = \sqrt{(x_2 - x_1)^2 + (y_2 - y_1)^2 + 2(x_2 - x_1)(y_2 - y_1)\cos\omega}\,.$$

Corollary. The distance of the point $P(x_1, y_1)$ from the origin is

$$\sqrt{x_1^2 + y_1^2 + 2x_1y_1\cos\omega}\,.$$

Remarks. One may easily verify the following when the axes are inclined at an angle ω:

(*i*) The coordinates of a point which divides a line joining two points $P(x_1, y_1)$ and $Q(x_2, y_2)$ in a given ration $m_1 : m_2$ are

$$\left(\frac{m_1x_2 + m_2x_1}{m_1 + m_2}, \frac{m_1y_2 + m_2y_1}{m_1 + m_2}\right).$$

(*ii*) The equation of the line parallel to *y*-axis is $x = a$ whereas the equation of the line parallel to *x*-axis is $y = b$.

(*iii*) The area of the triangle whose vertices are (x_1, y_1), (x_2, y_2) and (x_3, y_3) is

$$\frac{1}{2}\begin{vmatrix} x_1 & y_1 & 1 \\ x_2 & y_2 & 1 \\ x_3 & y_3 & 1 \end{vmatrix} \sin\omega.$$

3. Equation of a Line

To find the equation of a line in slope-intercept form referred to the axes inclined at an angle ω.

Let *AB* be any line which cut *x*-axis in *M* and *y*-axis in *N* and make an angle θ with *x*-axis. Let $ON = c$. Let $P(x, y)$ be any point on *AB*. Draw *PL* parallel to *y*-axis and *NQ* parallel to *x*-axis.

In Δ *NQP*, we have

$$\frac{QP}{NQ} = \frac{\sin QNP}{\sin QPN} = \frac{\sin\theta}{\sin(\omega - \theta)} = m \text{ (say)}.$$

But $\qquad QP = LP - LQ = LP - ON = y - c,$

and $\qquad NQ = OL = x.$

Therefore $\qquad \dfrac{y - c}{x} = m$, *i.e.*, $y = mx + c$.

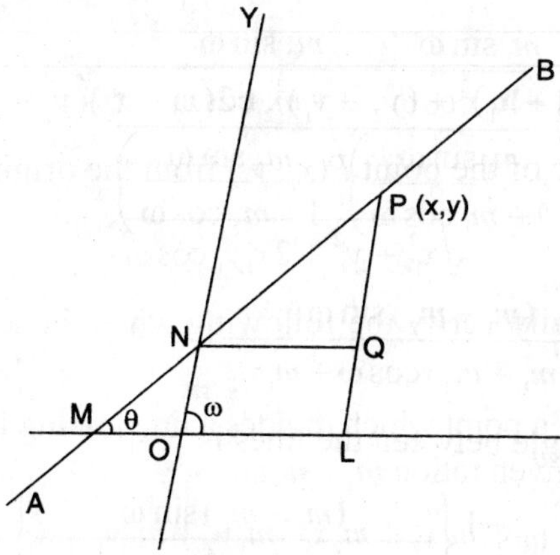

Fig. 2

Now $\quad m = \dfrac{\sin\theta}{\sin(\omega-\theta)} = \dfrac{\sin\theta}{\sin\omega\cos\theta - \cos\omega\sin\theta}$

$$= \frac{\tan\theta}{\sin\omega - \cos\omega\tan\theta}$$

which gives

$$\tan\theta = \frac{m\sin\omega}{1+m\cos\omega}.$$

Thus, $y = mx + c$ is the equation of the line which cuts off an intercept c on y-axis and makes an angle $\tan^{-1}\left(\dfrac{m\sin\omega}{1+m\cos\omega}\right)$ with x-axis.

4. Angle between Two Lines

To find the angle between two lines referred to the axes inclined at an angle ω.

Let $\quad y = m_1x + c_1$ and $y = m_2x + c_2$

be the equations of the two lines which make angles θ_1, θ_2, respectively with x-axis. Then from § 3, we have

$$\tan\theta_1 = \frac{m_1\sin\omega}{1+m_1\cos\omega} \text{ and } \tan\theta_2 = \frac{m_2\sin\omega}{1+m_2\cos\omega}.$$

Now

$$\tan(\theta_1-\theta_2) = \frac{\tan\theta_1 - \tan\theta_2}{1+\tan\theta_1\tan\theta_2}$$

$$= \frac{\dfrac{m_1 \sin\omega}{1+m_1\cos\omega} - \dfrac{m_2 \sin\omega}{1+m_2\cos\omega}}{1+\left(\dfrac{m_1 \sin\omega}{1+m_1\cos\omega}\right)\left(\dfrac{m_2 \sin\omega}{1+m_2\cos\omega}\right)}$$

$$= \frac{(m_1 - m_2)\sin\omega}{1+(m_1+m_2)\cos\omega + m_1 m_2}.$$

Hence, the required angle between the lines is

$$\tan^{-1}\left\{\frac{(m_1 - m_2)\sin\omega}{1+(m_1+m_2)\cos\omega + m_1 m_2}\right\}.$$

Corollary 1. The lines $y = m_1 x + c_1$ and $y = m_2 x + c_2$ are parallel if $m_1 = m_2$ and perpendicular if $1 + (m_1 + m_2)\cos\omega + m_1 m_2 = 0$.

Corollary 2. The angle between the lines $a_1 x + b_1 y + c_1 = 0$ and $a_2 x + b_2 y + c_2 = 0$ is

$$\tan^{-1}\left\{\frac{(a_2 b_1 - a_1 b_2)\sin\omega}{a_1 a_2 + b_1 b_2 - (a_1 b_2 + a_2 b_1)\cos\omega}\right\}.$$

The lines will be parallel if $\dfrac{a_1}{b_1} = \dfrac{a_2}{b_2}$ and perpendicular if

$$a_1 a_2 + b_1 b_2 - (a_1 b_2 + a_2 b_1)\cos\omega = 0.$$

Corollary 3. The equation of the line perpendicular to the line $ax + by + c = 0$ is

$$(b - a\cos\omega)\,x - (a - b\cos\omega)\,y = \lambda.$$

where λ is an arbitrary constant.

Corollary 4. The lines $x + y\cos\omega = 0$ and $y + x\cos\omega = 0$ are perpendicular to the axes $y = 0$ and $x = 0$, respectively.

Corollary 5. The equation of the line parallel to $ax + by + c = 0$ is $ax + by + \lambda = 0$, where λ is an arbitrary constant.

5. Length of Perpendicular

To find the perpendicular distance of any point (x_1, y_1) from the line $ax + by + c = 0$ referred to the axes inclined at an angle ω.

The equation of the line passing through $P(x_1, y_1)$ and perpendicular to the line $ax + by + c = 0$ is

$$(b - a\cos\omega)(x - x_1) - (a - b\cos\omega)(y - y_1) = 0. \quad (1)$$

Let M be the foot of the perpendicular drawn from $P(x_1, y_1)$ on the given line and let (x_2, y_2) the coordinates of M. But M lies on the perpendicular as well as on the given line. Therefore

$$(b - a\cos\omega)(x_2 - x_1) - (a - b\cos\omega)(y_2 - y_1) = 0 \quad (2)$$

and
$$ax_2 + by_2 + c = 0. \quad (3)$$

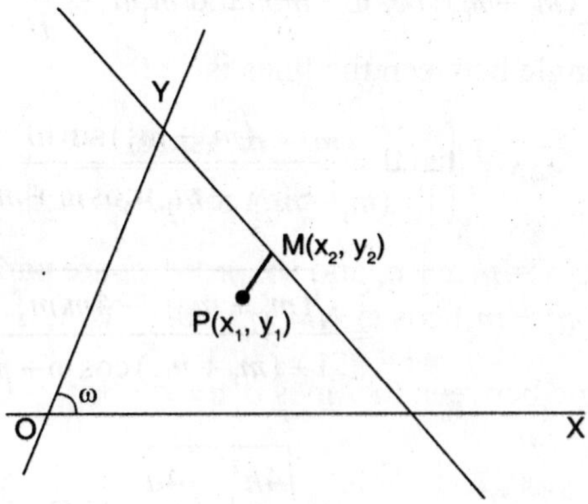

Fig. 3

Equation (3) can also be written as

$$a(x_2 - x_1)\sin\omega + b(y_2 - y_1)\sin\omega = -(ax_1 + by_1 + c)\sin\omega. \quad (4)$$

On squaring and adding (2) and (4), we get

$$(a^2 + b^2 - 2ab\cos\omega)\{(x_2 - x_1)^2 + (y_2 - y_1)^2 + 2(x_2 - x_1)(y_2 - y_1)\cos\omega\}$$
$$= (ax_1 + by_1 + c)^2 \sin^2\omega.$$

But
$$PM = \sqrt{(x_2 - x_1)^2 + (y_2 - y_1)^2 + 2(x_2 - x_1)(y_2 - y_1)\cos\omega}.$$

Hence,
$$PM = \frac{(ax_1 + by_1 + c)\sin\omega}{\sqrt{a^2 + b^2 - 2ab\cos\omega}}$$

Corollary. The equations of the bisectors of the angles between the lines

$$a_1x + b_1y + c_1 = 0 \text{ and } a_2x + b_2y + c_2 = 0$$

are
$$\frac{a_1x + b_1y + c_1}{\sqrt{a_1^2 + b_1^2 - 2a_1b_1\cos\omega}} = \pm\frac{a_2x + b_2y + c_2}{\sqrt{a_2^2 + b_2^2 - 2a_2b_2\cos\omega}}.$$

6. Angle between the Pair of Lines

To find the angle between the lines $ax^2 + 2hxy + by^2 = 0$, referred to the axes inclined at an angle ω.

Let $y = m_1x$ and $y = m_2x$ be the lines represented by the given equation and let θ be angle between them. Then

$$m_1 + m_2 = -\frac{2h}{b} \text{ and } m_1m_2 = \frac{a}{b}.$$

By § 4, we have

$$\tan\theta = \frac{(m_1 - m_2)\sin\omega}{1 + (m_1 + m_2)\cos\omega + m_1m_2}$$

$$= \frac{\sqrt{(m_1 + m_2)^2 - 4m_1m_2}\,.\sin\omega}{1 + (m_1 + m_2)\cos\omega + m_1m_2}$$

$$= \frac{\sqrt{\frac{4h^2}{b^2} - \frac{4a}{b}}\,.\sin\omega}{1 + \left(-\frac{2h}{b}\right)\cos\omega + \frac{a}{b}}$$

$$= \frac{2\sqrt{h^2 - ab}\,.\sin\omega}{a + b - 2h\cos\omega}.$$

Corollary 1. The lines represented by $ax^2 + 2hxy + by^2 = 0$ will be parallel if $h^2 = ab$ and perpendicular if $a + b - 2h \cos \omega = 0$.

Corollary 2. The equations of the bisectors of the angles between the lines $ax^2 + 2hxy + by^2 = 0$ are given by

$$h(x^2 - y^2) - (a - b)\,xy + (ax^2 - by^2)\cos\omega = 0$$

i.e. $$(-a\cos\omega - h)\,x^2 + (b\cos\omega + h)\,y^2 - (b - a)\,xy = 0.$$

7. Oblique Axes from Rectangular Axes

To transform rectangular axes into oblique axes.

Let (x', y') be the coordinates of a point P referred to rectangular axes OX, OY' and (x, y) the coordinates of the same point P referred to oblique axes OX. OY.

Draw PM and PN parallel to OY' and OY, respectively. Then

$$\begin{cases} x = ON = OM - NM = x' - y' \cot \omega \\ y = NP = y' \operatorname{cosec} \omega \end{cases}$$

and
$$\begin{cases} x' = OM = ON + NM = x + y \cos \omega \\ y' = MP = y \sin \omega. \end{cases}$$

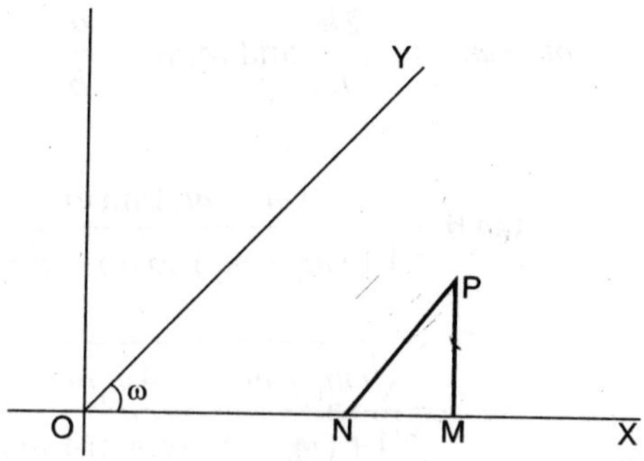

Fig. 4

8. Transformation from Set of Oblique Axes to Another without Changing the Origin

Let OX' OY' be a set of oblique axes containing an angle ω and OX, OY be another set of oblique axes containing an angle ω. Let (x', y') be the coordinates of a point P with respect to first set of axes and (x, y) the coordinates of the same point with respect to second set of axes. Let OX make an angle α with OX'.

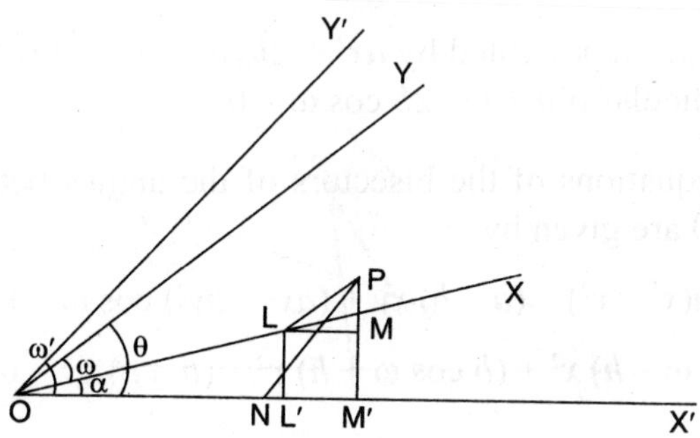

Fig. 5

Drawn PN parallel to OY' and PL parallel to OY.

Then $ON = x'$, $NP = y'$, $OL = x$ and $LP = y$.

Further, draw PM' perpendicular to OX', LL' perpendicular to OX' and LM perpendicular to PM'. Then

$$y' \sin \omega' = M'P = L'L + MP = x \sin \alpha + y \sin \theta.$$

Also OX and OY make angles $\omega' - \alpha$ and $\omega' - \theta$ with OY'. Therefore,

$$x' \sin \omega' = x \sin (\omega' - \alpha) + y \sin (\omega' - \theta).$$

Thus $$x' = \frac{x \sin(\omega' - \alpha)}{\sin \omega'} + \frac{y \sin(\omega' - \theta).}{\sin \omega'},$$

$$y' = \frac{x \sin \alpha}{\sin \omega'} + \frac{y \sin \theta}{\sin \omega'},$$

where $\theta - \alpha = \omega$.

9. Invariants

If by any change of axes the expression $ax^2 + 2hxy + by^2$ be changed into $a'x^2 + 2h'xy + b'y^2$, then

$$\frac{a + b - 2h \cos \omega}{\sin^2 \omega} = \frac{a' + b' - 2h' \cos \omega'}{\sin^2 \omega'}$$

and $$\frac{ab - h^2}{\sin^2 \omega} = \frac{a'b' - h'^2}{\sin^2 \omega'},$$

where ω and ω' are the angles of inclination of the two sets of axes.

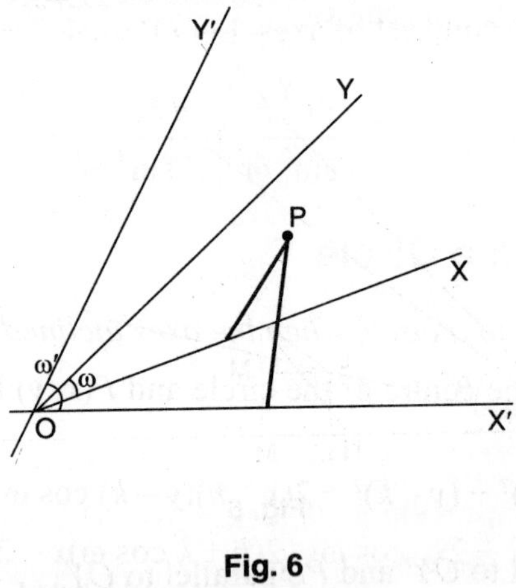

Fig. 6

Let (x', y') be the coordinates of a point P referred to the axes OX' and OY' and (x, y) the coordinates of the same point referred to the axes OX and OY.

Then $$OP^2 = x^2 + y^2 + 2xy \cos \omega.$$

Also $$OP^2 = x'^2 + y'^2 + 2x'y' \cos \omega'.$$

This shows that the expression $x^2 + y^2 + 2xy \cos \omega$ is changed into $x'^2 + y'^2 + 2x'y' \cos \omega'$. But it is given that $ax^2 + 2hxy + by^2$ is changed into $a'x'^2 + 2h'x'y' + b'y'^2$. Therefore, if λ be any constant, then

$$ax^2 + 2hxy + by^2 + \lambda(x^2 + y^2 + 2xy \cos \omega)$$

will be changed into

$$a'x'^2 + 2h'x'y' + b'y'^2 + \lambda(x'^2 + y'^2 + 2x'y' \cos \omega').$$

Thus, if λ be so chosen that one of these expressions is a perfect square, the other will also be a perfect square for the same value of λ. The expressions will be perfect square if

$$(a + \lambda)(b + \lambda) - (h + \lambda \cos \omega)^2 = 0$$

and $$(a' + \lambda)(b' + \lambda) - (h' + \lambda \cos \omega')^2 = 0,$$

respectively, *i.e.*,

$$\lambda^2 \sin^2\omega + (a + b - 2h \cos \omega) \lambda + ab - h^2 = 0$$

and $$\lambda^2 \sin^2\omega' + (a' + b' - 2h' \cos \omega') \lambda + a'b' - h'^2 = 0.$$

For finding the value of λ, these equations must have the same roots. Thus, we get

$$\frac{a + b - 2h \cos \omega}{\sin^2 \omega} = \frac{a' + b' - 2h' \cos \omega'}{\sin^2 \omega'}$$

and $$\frac{ab - h^2}{\sin^2 \omega} = \frac{a'b' - h'^2}{\sin^2 \omega'}.$$

10. Equation to a Circle

To find the equation to a circle when the axes inclined at an angle ω.

Let $C(h, k)$ be the centre of the circle and $P(x, y)$ be any point on it. Then from § 2, we have

$$(x - h)^2 + (y - k)^2 + 2(x - h)(y - k) \cos \omega = a^2$$

$$\Rightarrow \quad x^2 + y^2 + 2xy \cos \omega - 2(h + k \cos \omega)x - 2(k + h \cos \omega)y + h^2 + k^2 + 2hk \cos \omega - a^2 = 0.$$

This is the required equation of the circle.

Corollary. The equation of the circle with centre at origin is

$$x^2 + y^2 + 2xy \cos \omega = a^2.$$

Remark. The general equation of the circle may be considered as

$$x^2 + y^2 + 2xy \cos \omega + 2gx + 2fy + c = 0.$$

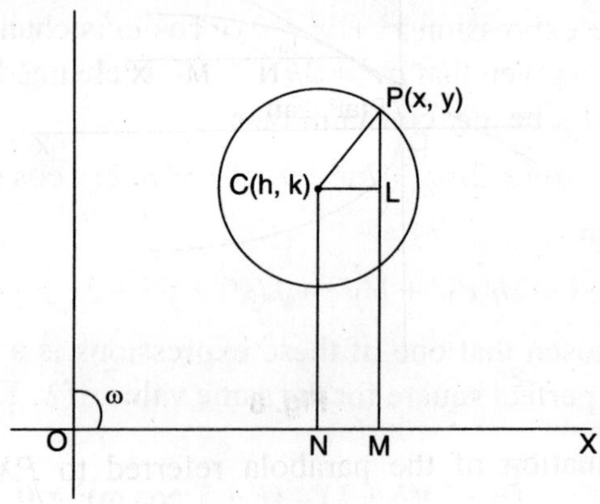

Fig. 7

11. Equation to a Parabola

To find the equation to a parabola when the axes are any diameter and the tangent at the extremity of that diameter.

Let $P(at^2, 2at)$ be any point on the parabola $y^2 = 4ax$. Transferring the origin to the point P by taking axes parallel to the original axes. Let (x', y') be the coordinates of a point Q with respect to these axes. Then equation of the parabola becomes

$$(y' + 2at)^2 = 4a(x' + at^2).$$

Now, consider the new axes as PX' and the tangent at P(*i.e.* PY') so that $\angle X'PY' = \omega$. Let (x'', y'') be the coordinates of Q with respect to the axes PX' and PY'. Then

$$x' = x'' + y'' \cos \omega,$$

$$y' = y'' \sin \omega.$$

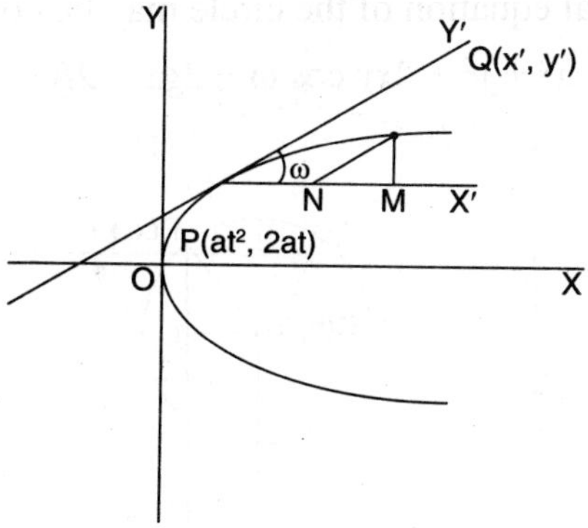

Fig. 8

Therefore, the equation of the parabola referred to PX' and PY' as axes becomes

$$(y'' \sin \omega + 2at)^2 = 4a(x'' + y'' \cos \omega + at^2)$$

$$\Rightarrow \qquad y''^2 \sin^2\omega + 4ay'' (t \sin \omega - \cos \omega) = 4ax''. \qquad (1)$$

But $\tan \omega$ = slope of the tangent at $P = \dfrac{1}{t}$. Hence, the equation (1) becomes

$$y''^2 = \frac{4ax''}{\sin^2 \omega}.$$

This equation may be written as $y^2 = 4bx$, where $b = \dfrac{4a}{\sin^2 \omega}$.

This shows that the equation of the parabola $y^2 = 4ax$ is a particular case of the equation of the parabola referred to the diameter and the tangent at the extremity of that diameter.

12. Equation of an Ellipse

To find the equation of an ellipse referred to any pair of conjugate diameters as axes.

Let the equation of the ellipse be

$$\frac{x^2}{a^2} + \frac{y^2}{b^2} = 1. \qquad (1)$$

Let (x, y) be the coordinates of a point Q on (1) referred to CA and CB as axes.

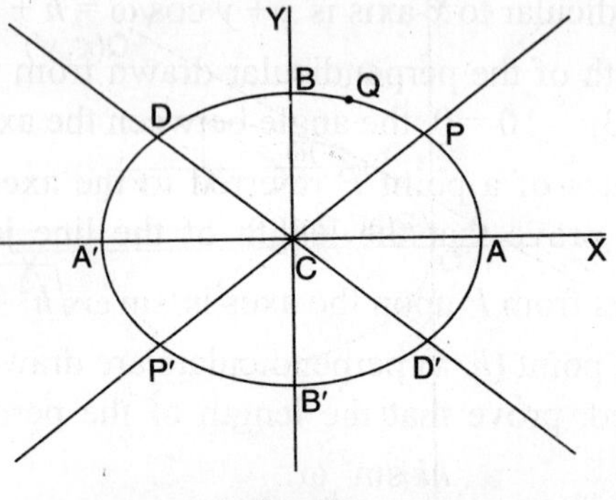

Fig. 9

Let (x', y') be the coordinates of the same point Q referred to the conjugate diameters CP and CD as axes. Then, by § 8, we have

$$x = lx' + my', y = l'x' + m'y',$$

where l, m, l', m' are functions of the angles ACP and PCD. Writing these values of x and y in (1), we get

$$\frac{(lx' + my')^2}{a^2} + \frac{(l'x' + m'y')^2}{b^2} = 1.$$

This equation can be written as

$$Ax'^2 + 2Hx'y' + By'^2 = 1. \tag{2}$$

Let $CP = a'$ and $CD = b'$. Then, the points $(a', 0)$ and $(0, b')$ will satisfy equation (2) if

$$Aa'^2 = 1, \text{ i.e., } A = \frac{1}{a'^2} \text{ and } Bb'^2 = 1, \text{ i.e., } B = \frac{1}{b'^2}.$$

Also, the line $x' = a'$ is the tangent at P. Substiuting $x' = a'$ in (2), we get

$$Aa'^2 + 2Ha'y' + By'^2 = 1, \text{ i.e., } 2Ha'y' + By'^2 = 0.$$

The two roots of this equation have to be zero, therefore, $H = 0$. Hence, the equation of the ellipse is $\dfrac{x'^2}{a'^2} + \dfrac{y'^2}{b'^2} = 1.$

13. Equation of a Hyperbola

To find the equation of a hyperbola referred to any pair of conjugate diameters as axes.

Proof is similar to above.

Exercises

1. Prove that the equation of the line which passes through the point (h, k) and is perpendicular to x-axis is $x + y \cos \omega = h + k \cos \omega$.
2. Find the length of the perpendicular drawn from the point $(4, -3)$ upon the line $6x + 3y - 10 = 0$, the angle between the axes being 60°.
3. The coordinates of a point P referred to the axes meeting at an angle ω are (h, k), prove that the length of the line joining the feet of the perpendiculars from P upon the axes is $\sin \omega \sqrt{h^2 + k^2 + 2hk \cos \omega}$.
4. From a given point (h, k) perpendiculars are drawn to the axes and their feet are joined, prove that the length of the perpendicular from (h, k) upon this line is $\dfrac{hk \sin^2 \omega}{\sqrt{h^2 + k^2 + 2hk \cos \omega}}$.

   and that its equation is $hx - ky = h^2 - k^2$, ω being the angle between the axes.
5. Prove that the lines $y = m_1 x + c_1$ and $y = m_2 x + c_2$ make equal angles with x-axis if $m_1 + m_2 + 2m_1 m_2 \cos \omega = 0$.
6. A line is drawn through a fixed point A cutting two given lines OX, OY in the points P, Q respectively, and the parallelogram $OPRQ$ is completed. Find the locus of R.
7. If the axes be inclined at an angle ω, prove that the equation
   $$x^2 + 2xy \cos \omega + y^2 \cos 2\omega = 0.$$
   represents a pair of perpendicular lines.
8. If the lines $ax^2 + 2hxy + by^2 = 0$ make equal angles with x-axis, prove that $h = a \cos \omega$.
9. Prove that the equation to the bisectors of the angle between the lines $ax^2 + 2hxy + by^2 = 0$ is $h(x^2 - y^2) + (b - a)\, xy = (ax^2 - by^2) \cos \omega$, the axes being inclined at an angle ω.
10. If one system of oblique axes be transformed to another system according to the formula $x = mx' + ny'$, $y = m'x' + n'y'$, the origin being the same, then prove that $mm'(n^2 + n'^2 - 1) = nn'(m^2 + m'^2 - 1)$.
11. If (x, y) and (x', y') be the coordinates of the same point referred to two sets of axes with the same origin and $ux + vy$ be transformed to $u'x' + v'y'$, then prove that
   $$\frac{u^2 - 2uv \cos \omega + v^2}{\sin^2 \omega} - \frac{u'^2 - 2u'v' \cos \omega' + v'^2}{\sin^2 \omega'},$$
   where ω and ω' are the angles between the axes in the two sets of axes.

12. The axes being inclined at an angle 60°, find the equation to the circle whose centre is (–3, –5) and radius is 6.

13. The axes being inclined at an angle ω, find the centre and radius of the circle $x^2 + 2xy \cos \omega + y^2 - 2gx - 2fy = 0$.

14. If the axes be inclined at an angle ω the directrix of the parabola $y^2 = 4ax$ is $x + y \cos \omega + a = 0$.

15. Parabolas are drawn to touch the axes, which are inclined at an angle ω, and their directrices all pass through a fixed point (*h*, *k*). Prove that all the parabolas touch the line

$$\frac{x}{h + k \sec \omega} + \frac{y}{k + h \sec \omega} = 1.$$

16. Prove that the equation to the director circle of the conic $xy = c^2$ is

$$x^2 + 2xy \cos \omega + y^2 = 4c^2 \cos \omega.$$

PART B

Three Dimensions

Preliminary

12
CHAPTER

1. Introduction

The purpose of this chapter is to explain the basic concepts of the subject like coordinate systems, ratio formula, locus, projection and direction cosines in three dimensional space which are necessary for the development of the subject in the subsequent chapters.

2. Cartesian Coordinates

The non-coplanar lines $X'OX$, $Y'OY$ and $Z'OZ$ are called the x-axis, y-axis and z-axis, respectively. These lines are also referred to **coordinate axes** or simply the **axes**. The planes XOY, YOZ and ZOX (respectively, xy-, yz- and zx-plane) are called **coordinate planes**. These planes divide the space into eight parts $OXYZ$, $OX'YZ$, $OX'Y'Z$, $OXY'Z$, $OXYZ'$, $OX'YZ'$, $OXY'Z'$ and $OX'Y'Z'$ which are called the **octants**. The point O is called the origin. Here too we have the same rule for the sign convention as in trigonometry i.e., the lengths measured along OX, OY and OZ are positive while the lengths measured along OX', OY' and OZ' are negative.

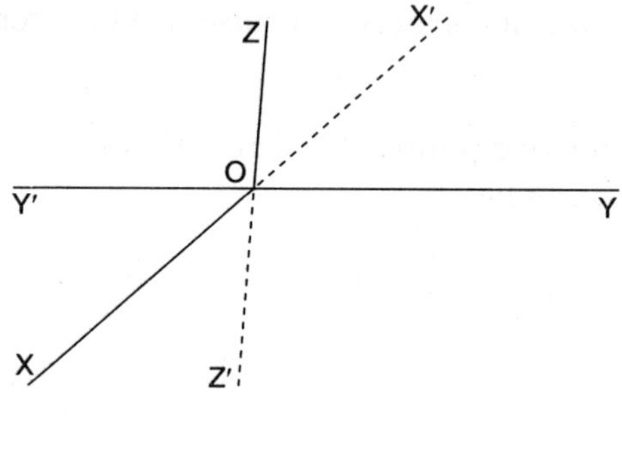

Fig. 12.1

Let P be any point in the space. Draw perpendicular PM from P on xy-plane. Draw the lines MB and AM parallel to x-axis and y-axis, respectively.

Let $OA = x$, $OB = y$ and $OC = z$. Then x, y and z are, respectively, called the x-coordinate, y-coordinate and z-coordinate of the point P. The triad (x, y, z) is referred to the **cartesian coordinates** or simply coordinates of the point P with reference to the coordinate axes $X'OX$, $Y'OY$ and $Z'OZ$ and is usually denoted by $P(x, y, z)$.

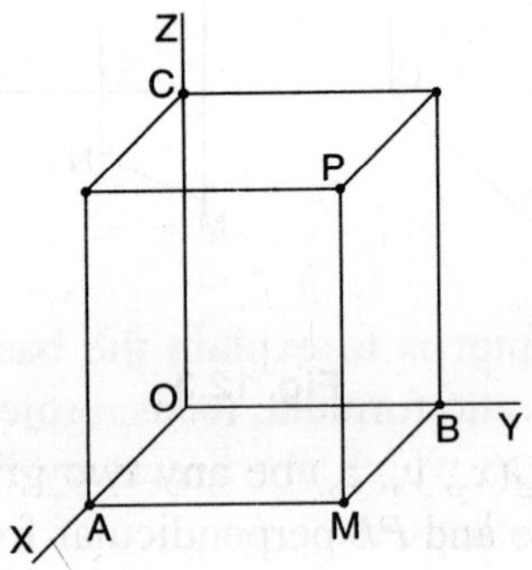

Fig. 12.2

Note. When the axes $X'OX$, $Y'OY$ and $Z'OZ$ are mutually perpendicular to each other, the coordinate axes are called **rectangular** otherwise they are called **oblique axes**. Throughout we shall be dealing with rectangular coordinate axes unless specified otherwise, and therefore by coordinate axes or simply axes we shall mean rectangular coordinate axes.

Remark. With reference to an arbitrary but fixed system of coordinate axes, for a given point P in the space, there correspond unique coordinates (x, y, z) and conversely, for a given ordered triad (x, y, z) of real numbers, there corresponds a unique point in the space with (x, y, z) as its coordinates. Thus, there is a one-to-one correspondence between the set of all points in the space and the set of all ordered triads of real numbers, of course, with reference to a fixed system of coordinate axes.

2.1 Distance between two points. *To express the distance between two points in terms of their coordinates.*

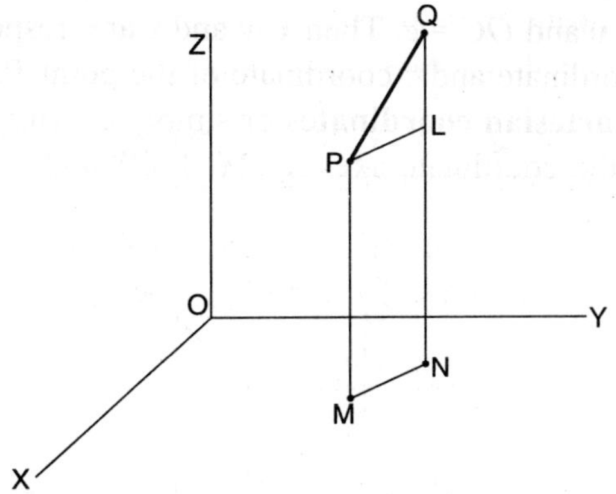

Fig. 12.3

Let $P(x_1, y_1, z_1)$ and $Q(x_2, y_2, z_2)$ be any two given points. Draw PM, QN perpendiculars to xy-plane and PL perpendicular from P to QN. Since, M and N lie in xy-plane, their coordinates with reference to OX and OY as coordinate axes are given by (x_1, y_1) and (x_2, y_2), respectively. By analytical geometry of two dimensions, we have

$$MN^2 = (x_2 - x_1)^2 + (y_2 - y_1)^2.$$

Further, from right angled triangle PLQ, we have

$$PQ^2 = PL^2 + LQ^2.$$

But $$PL = MN = \sqrt{(x_2 - x_1)^2 + (y_2 - y_1)^2},$$

and $$LQ = NQ - NL = NQ - MP = z_2 - z_1.$$

Therefore $$PQ^2 = (x_2 - x_1)^2 + (y_2 - y_1)^2 + (z_2 - z_1)^2.$$

$\Rightarrow$ $$PQ = \sqrt{(x_2 - x_1)^2 + (y_2 - y_1)^2 + (z_2 - z_1)^2}.$$

Thus, we have obtained the distance formula between two points P and Q when both of these are lying in the octant $OXYZ$. But this formula will be found to be true for all positions of P and Q in the space when due regard is paid to the signs of the coordinates (x_1, y_1, z_1) and (x_2, y_2, z_2).

Corollary. The distance of the point $P(x, y, z)$ from origin is $\sqrt{x^2 + y^2 + z^2}$.

Ex. 1. Find the distance between the points (2, 5, –4) and (8, 2, –6).

Ex. 2. Prove that the points (–2, 4, –3), (4, –3, –2), (–3, –2, 4) form an equilateral triangle.

2.2 The ratio formula. *To find the coordinates of a point which divides the line joining two given points in a given ratio.*

Let $P(x_1, y_1, z_1)$ and $Q(x_2, y_2, z_2)$ be any two given points and let the point $R(x, y, z)$ divide PQ internally in the ratio $m_1 : m_2$.

Draw PM, RL and QN parallel to z-axis and $PL'N'$, RN'' parallel to MLN. Then from similar triangles $PL'R$ and $RN''Q$, we have

$$\frac{L'R}{N''Q} = \frac{PR}{RQ} = \frac{m_1}{m_2}.$$

But $$L'R = LR - LL' = LR - MP = z - z_1,$$

and $$N''Q = NQ - NN'' = NQ - LR = z_2 - z.$$

Therefore $$\frac{z - z_1}{z_2 - z} = \frac{m_1}{m_2}, \textit{ i.e., } z = \frac{m_1 z_2 + m_2 z_1}{m_1 + m_2}.$$

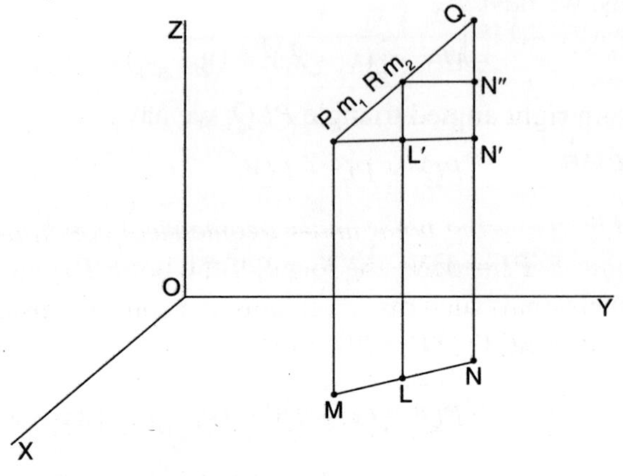

Fig. 12.4

Similarly $$x = \frac{m_1 x_2 + m_2 x_1}{m_1 + m_2} \text{ and } y = \frac{m_1 y_2 + m_2 y_1}{m_1 + m_2}.$$

Corollary 1. The coordinates of the middle point of the line joining the points (x_1, y_1, z_1) and (x_2, y_2, z_2) are

$$\left(\frac{x_1 + x_2}{2}, \frac{y_1 + y_2}{2}, \frac{z_1 + z_2}{2}\right).$$

Corollary 2. If R divides PQ externally in the ratio $m_1 : m_2$, then the coordinates of R are

$$\left(\frac{m_1 x_2 - m_2 x_1}{m_1 - m_2}, \frac{m_1 y_2 - m_2 y_1}{m_1 - m_2}, \frac{m_1 z_2 - m_2 z_1}{m_1 - m_2}\right), m_1 \neq m_2.$$

Since, in this case the distances PR and RQ are measured in opposite sense, we have

$$\frac{PR}{RQ} = -\frac{m_1}{m_2}.$$

which gives

$$z = \frac{m_1 z_2 - m_2 z_1}{m_1 - m_2}, \; m_1 \neq m_2.$$

Similarly
$$x = \frac{m_1 x_2 - m_2 x_1}{m_1 - m_2} \text{ and } y = \frac{m_1 y_2 - m_2 y_1}{m_1 - m_2}, \; m_1 \neq m_2.$$

Remark. The above result can be extended to other octants provided the coordinates of the points are taken with proper signs.

Ex. 1. Find the coordinates of the point which divides the line joining the points (2, –4, 3) and (–4, 5, –6) in the ratio 2 : 1.

Ex. 2. Prove that the centroid of the triangle whose vertices are (x_1, y_1, z_1), (x_2, y_2, z_2) and (x_3, y_3, z_3) is $\left(\frac{x_1 + x_2 + x_3}{3}, \frac{y_1 + y_2 + y_3}{3}, \frac{z_1 + z_2 + z_3}{3}\right)$.

3. The Locus

The path traced by a moving point under geometrical conditions is called the **locus** *of that point.* For instance, the locus of the point P which moves in the space (three dimensional) such that it remains at a constant distance 3 from the fixed point A is a sphere, Fig. 12.5.

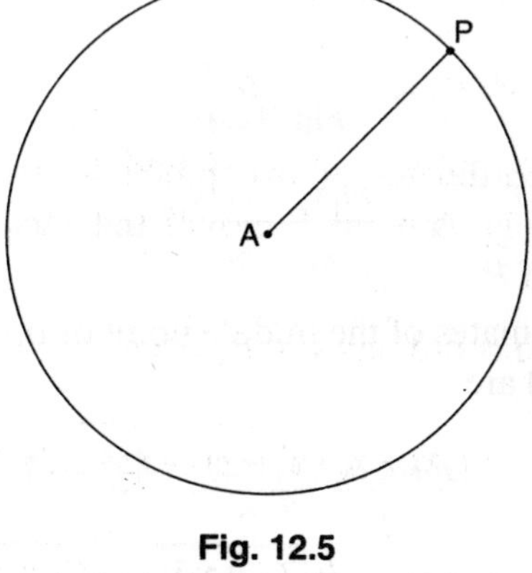

Fig. 12.5

The sciences of algebra and geometry correlated by means of the coordinate systems leads to a study of analytical geometry. In analytical geometry of two dimensions, the underlying feature is the correspondence between an equation in x and y, the coordinates of the variable point and a geometric figure of the locus (path) of the variable point. Similarly, in analytical geometry of three dimensions, the underlying feature is the correspondence between an equation in x, y and z, the coordinates of the variable point and a geometric figure of the locus (path) of the variable point. In regard to this there are two problems to be studied. One is that we are given some geometric facts which limits the positions of the moving point P and asked to find what relation is then satisfied by x, y and z, the coordinates of P. Conversely, an equation in x, y and z is given, and the problem is to find out on what geometrical locus P lies.

3.1 Equation of the locus

An equation of a locus is a relation among x, y and z which is satisfied by the coordinates of all points of the locus and by no others.

Illustration 1. Find the equation of the locus of the points which are equidistant from the points $A(1, 2, 3)$ and $B(3, 2, 1)$.

Let (x', y', z') be the coordinates of the point P which is equidistant from A and B. Then

$$PA = \sqrt{(x'-1)^2 + (y'-2)^2 + (z'-3)^2},$$

and
$$PB = \sqrt{(x'-3)^2 + (y'-2)^2 + (z'-1)^2}$$

But $PA = PB$. Therefore

$$\sqrt{(x'-1)^2 + (y'-2)^2 + (z'-3)^2} = \sqrt{(x'-3)^2 + (y'-2)^2 + (z'-1)^2}$$

$$\Rightarrow \qquad x' - z' = 0.$$

Hence, the locus of (x', y', z') is $x - z = 0$.

Illustration 2. The coordinates of two points A and B are $(-2, 2, 3)$ and $(13, -3, 13)$, respectively. A point P moves such that $3PA = 2PB$. Find the equation of the locus of P.

Let (x', y', z') be the coordinates of P. Then

$$PA = \sqrt{(x'+2)^2 + (y'-2)^2 + (z'-3)^2},$$

and
$$PB = \sqrt{(x'-13)^2 + (y'+3)^2 + (z'-13)^2}.$$

We are given that $3PA = 2PB$. Therefore

$$3\sqrt{(x'+2)^2+(y'-2)^2+(z'-3)^2} = 2\sqrt{(x'-13)^2+(y'+3)^2+(z'-13)^2}$$

$$\Rightarrow \qquad x'^2 + y'^2 + z'^2 + 28x' - 12y' - 247 = 0.$$

Hence, the locus of (x', y', z') is $x^2 + y^2 + z^2 + 28x - 12y - 247 = 0$.

3.2 Locus of certain equations

I. Locus of f(x) = 0. By the locus of the equation $f(x) = 0$ in the space, we mean the path traced by the variable point (x, y, z) for which x is restricted by the relation $f(x) = 0$ whereas y and z are free to take any value whatsoever.

Suppose $f(x)$ be a polynomial of degree n. Then, it can be factorised into n factors as

$$f(x) \equiv (x - \alpha_1)(x - \alpha_2) \ldots (x - \alpha_n) = 0,$$

where $\alpha_1, \alpha_2, \ldots, \alpha_n$ are the roots of the equation $f(x) = 0$. Thus, by theory of equations, it gives

$$x = \alpha_1, x = \alpha_2, \ldots, x = \alpha_n. \qquad (1)$$

But each of these equations represents a plane parallel to yz-plane. Thus, the coordinates of any point on either of the planes in (1) satisfy the equation $f(x) = 0$. Hence, the locus of the equation $f(x) = 0$ is a system of planes parallel to yz- plane.

Similarly, the equations $f(y) = 0$ and $f(z) = 0$ represent systems of planes parallel to zx- and xy-planes, respectively.

II. Locus of f(x, y) = 0. By the locus of the equation $f(x, y) = 0$ in the space, we mean the locus of the variable point (x, y, z) for which x and y are restricted by the relation $f(x, y) = 0$ where z is free to take any value whatsoever.

We know that $f(x, y) = 0$ being an equation in x and y represents a conic in xy-plane. Suppose a and b are the values of x and y, respectively, satisfying the equation $f(x, y) = 0$. As such (a, b) is a point lying on two dimensional curve $f(x, y) = 0$ in xy-plane.

Any point (a, b, z), z a variable, lies on the line through $(a, b, 0)$ and parallel to z-axis. Thus, the point (a, b, z) satisfies the equation $f(x, y) = 0$ and hence, the equation $f(x, y) = 0$ represents a cylinder which is generated by the line through $(a, b, 0)$ and parallel to z-axis.

Similarly, the equations $f(y, z) = 0$ and $f(z, x) = 0$, represent cylinders having generators parallel to x-axis and y-axis, respectively.

III. Locus of $f(x, y, z) = 0$. It is easy to prove that the equation $f(x, y, z) = 0$ represents a surface.

Exercises

1. The points $A(3, 2, -4)$, $B(5, 4, -6)$, $C(9, 8, 10)$ are collinear, find the ratio in which B divides AC.
2. If A and B be the points (3, 4, 5) and (–1, 3, –7), respectively, find the locus of a point P such that $PA^2 + PB^2 = k^2$, where k is constant.
3. Prove that the points (0, 7, 10), (–1, 6, 6), (–4, 9, 6) form an isoscele triangle which is right angled.
4. Find the coordinates of the point which is equidistant from the four points (0, 0, 0), (2, 0, 0), (0, 3, 0) and (0, 0, 4).
5. Prove that the centroid of the tetrahedron whose vertices are (x_1, y_1, z_1), (x_2, y_2, z_2), (x_3, y_3, z_3) and $(x_4, y_4. z_4)$ is
$$\left(\frac{x_1+x_2+x_3+x_4}{4}, \frac{y_1+y_2+y_3+y_4}{4}, \frac{z_1+z_2+z_3+z_4}{4}\right).$$
6. Prove that the lines drawn from the vertices of a tetrahedron to the centroids of the opposite faces meet in a point which divides them in the ratio 3 : 1.
7. Find the ratios in which the coordinate planes divide the line joining the points (–2, 4, 7) and (3, –5, 8).
8. Find the coordinates of the circumcentre of the triangle formed by the points whose vertices are (1, 1, 0), (1, 2, 1) and (–2, 2, –1).
9. Prove that the locus of a point, the sum of whose distances from the points $(a, 0, 0)$, $(-a, 0, 0)$ is cosntant $2k$, is the curve $\dfrac{x^2}{k^2} + \dfrac{y^2+z^2}{k^2-a^2} = 1.$
10. Find the ratio in which the join of the points (2, –1, –3) and (3, 5, –4) is divided by the plane YOZ.
11. Prove that the coordinates of any point on the plane determined by the three points (x_1, y_1, z_1), (x_2, y_2, z_2) and (x_3, y_3, z_3) may be expressed in the form
$$\left(\frac{lx_1+mx_2+nx_3}{l+m+n}, \frac{ly_1+my_2+ny_3}{l+m+n}, \frac{lz_1+mz_2+nz_3}{l+m+n}\right).$$

4. Projection

4.1 Projection of a point on a line. *The foot of the perpendicular drawn from a point P to a line is called the* **projection** *of the point on that line. In other words, the projection of a point on a line is the point in which the plane through that point and perpendicular to the line meets it.*

4.2 Projection of a line segment on a given line

Let AB be any line segment and CD the given line. Let A', B' be the projections of the points A, B, respectively, on the line CD. Then the line segment $A'B'$ is the projection of the line segment AB on the line CD.

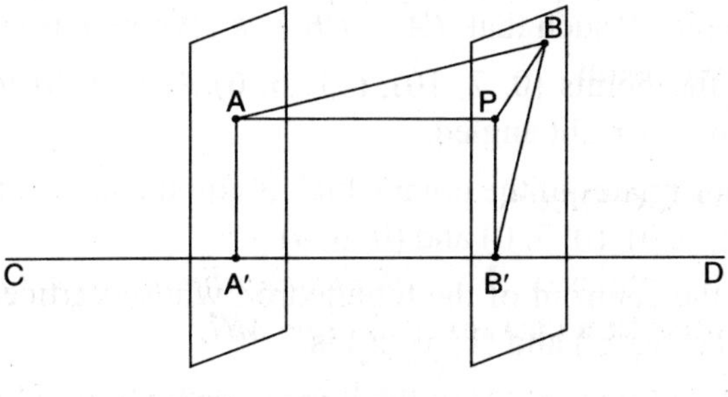

Fig. 12.6

Draw AP parallel and equal to $A'B'$ and join PB. Then PB is perpendicular to AP. Therefore

$$A'B' = AP = AB \cos \theta,$$

where θ is the angle between AP and AB. Hence, the length of the projection of the line segment AB on any line CD is $AB \cos \theta$, where θ is the angle between AB and CD.

4.3 *If $P_1, P_2, \ldots, P_n$ be n arbitrary points in the space, then the sum of the projections of the line segments $P_1P_2, P_2P_3, \ldots, P_{n-1}P_n$ on any line is equal to the projection of the line segment P_1P_n on the same.*

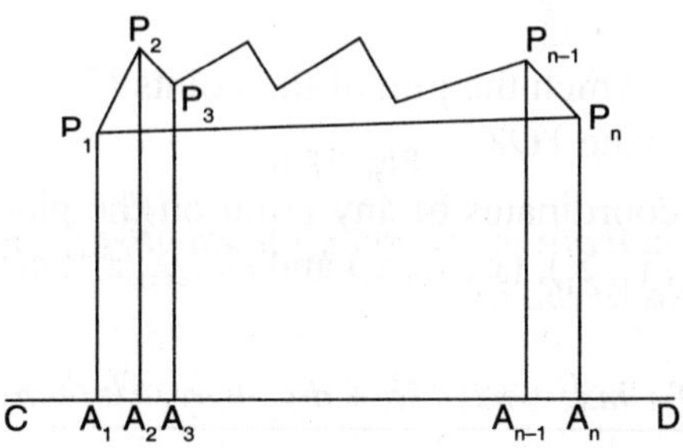

Fig. 12.7

Let CD be any given line. Suppose $A_1, A_2, \ldots, A_n$ be the projections of the points $P_1, P_2, \ldots, P_n$, respectively, on the line CD. Then

$$A_1A_2 = \text{projection of } P_1P_2 \text{ on } CD,$$

$$A_2A_3 = \text{projection of } P_2P_3 \text{ on } CD,$$

....

....

$$A_{n-1}A_n = \text{projection of } P_{n-1}P_n \text{ on } CD.$$

Also $\qquad A_1A_n = \text{projection of } P_1P_n$ on CD.

But $A_1, A_2, \ldots, A_n$ being points on the same line CD, we have

$$A_1A_n = A_1A_2 + A_2A_3 + \ldots + A_{n-1}A_n.$$

This proves the result.

5. Direction Cosines

If a line MN makes angles α, β, γ *with the coordinate axes, then* $\cos\alpha$, $\cos\beta$, $\cos\gamma$ *are called* **direction cosines** *of the line MN.*

In general, direction cosines of the line are denoted by l, m, n. Thus

$$l = \cos\alpha, \quad m = \cos\beta, \quad n = \cos\gamma.$$

Corollary. The direction cosines of the coordinate axes, viz. x-axis, y-axis and z-axis, are respectively, 1, 0, 0 ; 0, 1, 0 and 0, 0, 1.

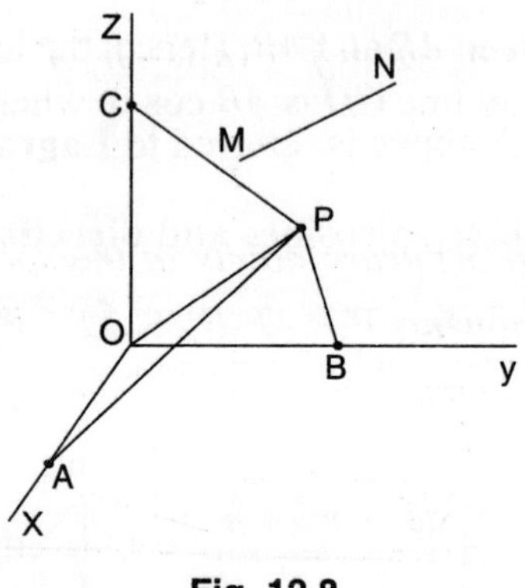

Fig. 12.8

The quantities which are proportional to the direction cosines of a line are called **direction ratios** *of the line.*

5.1 A relation in l, m, n. *To prove that if* l, m, n *are the direction cosines of a line, then* $l^2 + m^2 + n^2 = 1$.

Let MN be a given line and l, m, n be its direction cosines. Draw a line OP through origin O parallel to MN. Then l, m, n are also the direction cosines of OP. Let OP makes angles α, β, γ with the axes. Then

$$l = \cos\alpha, \ m = \cos\beta, \ n = \cos\gamma.$$

Draw perpendiculars PA, PB and PC on the axes. Let (x, y, z) be the coordinates of P and let $OP = r$. Then, $OA = x$, $OB = y$, $OC = z$. Therefore

$$l = \cos\alpha = \frac{OA}{OP} = \frac{x}{r} \Rightarrow x = lr,$$

$$m = \cos\beta = \frac{OB}{OP} = \frac{y}{r} \Rightarrow y = mr,$$

and $$n = \cos\gamma = \frac{OC}{OP} = \frac{z}{r} \Rightarrow z = nr.$$

Now $\quad x^2 + y^2 + z^2 = (l^2 + m^2 + n^2)r^2.$

But $\quad x^2 + y^2 + z^2 = r^2.$

Therefore $\quad l^2 + m^2 + n^2 = 1.$

Ex. 1. A line makes angles α, β, γ with the axes. Prove that $\sin^2\alpha + \sin^2\beta + \sin^2\gamma = 2$.

Ex. 2. If l_1, m_1, n_1 and l_2, m_2, n_2 be the direction cosines of any two lines, prove that

$$(l_1^2 + m_1^2 + n_1^2)(l_2^2 + m_2^2 + n_2^2) - (l_1l_2 + m_1m_2 + n_1n_2)^2$$
$$= (m_1n_2 - m_2n_1)^2 + (n_1l_2 - n_2l_1)^2 + (l_1m_2 - l_2m_1)^2.$$

Note. The relation in Ex. 2 above is referred to **Lagrange's identity**.

5.2 Relation between direction cosines and direction ratios of a line.

Let l, m, n, be the direction cosines of a line AB and a, b, c be its direction ratios. Since, direction ratios of a line are proportional to its direction cosines, we have

$$\frac{l}{a} = \frac{m}{b} = \frac{n}{c} = \pm\frac{\sqrt{l^2 + m^2 + n^2}}{\sqrt{a^2 + b^2 + c^2}} = \pm\frac{1}{\sqrt{a^2 + b^2 + c^2}}$$

$$\Rightarrow \; l = \pm\frac{a}{\sqrt{a^2 + b^2 + c^2}},\; m = \pm\frac{b}{\sqrt{a^2 + b^2 + c^2}},\; n = \pm\frac{c}{\sqrt{a^2 + b^2 + c^2}}.$$

Note. The direction cosines of the line AB are taken with positive signs whereas for BA with negative signs in the above values of l, m, n.

Remark. The direction ratios of a line are also referred to **direction numbers** of the line.

Ex. The direction ratios of a line are 1, 2, 3. Find its direction cosines.

5.3 Direction cosines of a line joining two points. *To find the direction cosines of the line joining the points* (x_1, y_1, z_1) *and* (x_2, y_2, z_2).

Let l, m, n be the direction cosines of the line AB joining the points $A(x_1, y_1, z_1)$ and $B(x_2, y_2, z_2)$ and let $AB = r$. Draw perpendiculars AM and BN on z-axis. Then

$$MN = \text{projection of } AB \text{ on } z\text{-axis} = AB \cos \gamma,$$

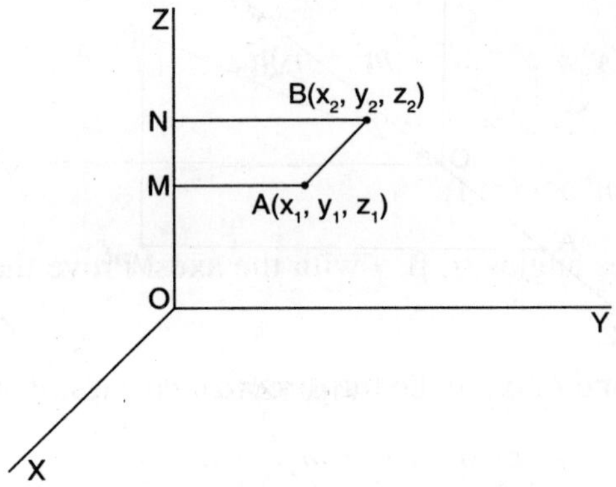

Fig. 12.9

where γ is the angle between AB and z-axis. But

$$MN = ON - OM = z_2 - z_1, \text{ and } \cos \gamma = n.$$

Therefore $$z_2 - z_1 = nr \Rightarrow n = \frac{z_2 - z_1}{r}.$$

Similarly, $$l = \frac{x_2 - x_1}{r}, \; m = \frac{y_2 - y_1}{r}.$$

Hence $\dfrac{x_2 - x_1}{r}, \dfrac{y_2 - y_1}{r}, \dfrac{z_2 - z_1}{r}$ are the direction cosines of line AB.

Remark. The direction ratios of the line joining the points (x_1, y_1, z_1) and (x_2, y_2, z_2) are $x_2 - x_1, y_2 - y_1, z_2 - z_1$.

Ex. If P is the point (x_1, y_1, z_1), prove that the projection of OP on a line whose direction cosines are l_1, m_1, n_1 is $l_1x_1 + m_1y_1 + n_1z_1$.

5.4 Angle between two lines. *To find the angle between two lines whose direction cosines are* l_1, m_1, n_1 *and* l_2, m_2, n_2.

Let AB and CD be the two lines whose direction cosines are respectively, l_1, m_1, n_1 and l_2, m_2, n_2. Draw OP parallel to AB and OQ parallel to CD. Let (x_1, y_1, z_1) and (x_2, y_2, z_2) be the coordinates of P and Q, respectively, and let $OP = r_1$, $OQ = r_2$. Further, let θ be the angle between OP and OQ. Then, θ is also the angle between AB and CD. Now projection of OP on OQ = projection of OA' on OQ

+ projection of $A'M$ on OQ

+ projection of MP on OQ

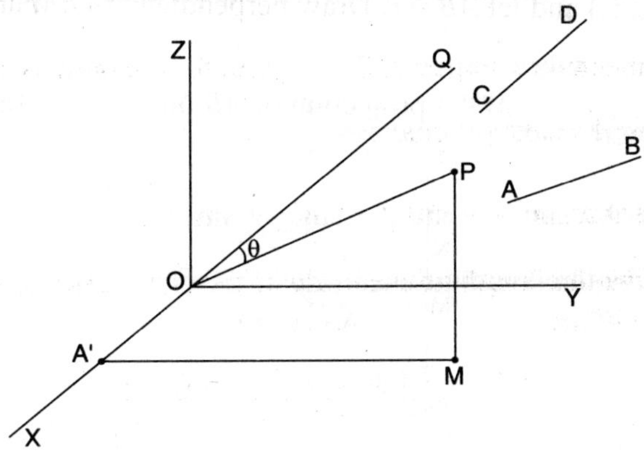

Fig. 12.10

i.e. $$r_1 \cos\theta = l_2 x_1 + m_2 y_1 + n_2 z_1$$

$$\Rightarrow \qquad \cos\theta = l_2 \frac{x_1}{r_1} + m_2 \frac{y_1}{r_1} + n_2 \frac{z_1}{r_1}.$$

But $\frac{x_1}{r_1} = l_1, \frac{y_1}{r_1} = m_1, \frac{z_1}{r_1} = n_1$. Hence, the angle θ between the lines is given by $\cos\theta = l_1 l_2 + m_1 m_2 + n_1 n_2$.

Corollary 1. $\sin\theta = \sqrt{(m_1 n_2 - m_2 n_1)^2 + (n_1 l_2 - n_2 l_1)^2 + (l_1 m_2 - l_2 m_1)^2}$. This follows by making use of Lagrange's identity.

Corollary 2. The lines with direction cosines l_1, m_1, n_1 and l_2, m_2, n_2 are parallel if $\frac{l_1}{l_2} = \frac{m_1}{m_2} = \frac{n_1}{n_2}$ and perpendicular if $l_1 l_2 + m_1 m_2 + n_1 n_2 = 0$.

Corollary 3. The angle between the lines whose direction ratios are a_1, b_1, c_1 and a_2, b_2, c_2 is given by

$$\cos^{-1} \frac{a_1 a_2 + b_1 b_2 + c_1 c_2}{\sqrt{a_1^2 + b_1^2 + c_1^2} \cdot \sqrt{a_2^2 + b_2^2 + c_2^2}}.$$

5.5 Examples

Example 1. Find the direction cosines of the line joining the points (4, 3, –5) and (–2, 1, –8).

Solution. Let r be the distance between the given points. Then

$$r = \sqrt{(4+2)^2 + (3-1)^2 + (-5+8)^2} = 7.$$

Hence, the direction cosines of the line joining the given points are $\frac{6}{7}, \frac{2}{7}, \frac{3}{7}$.

Example 2. A line makes angles $\alpha, \beta, \gamma, \delta$ with four diagonals of a cube. Prove that $\cos^2\alpha + \cos^2\beta + \cos^2\gamma + \cos^2\delta = \frac{4}{3}$.

Also, prove that $\sin^2\alpha + \sin^2\beta + \sin^2\gamma + \sin^2\delta = \frac{8}{3}$.

Solution. Let a be the length of each side of the cube. Then, direction cosines of the diagonal OP are

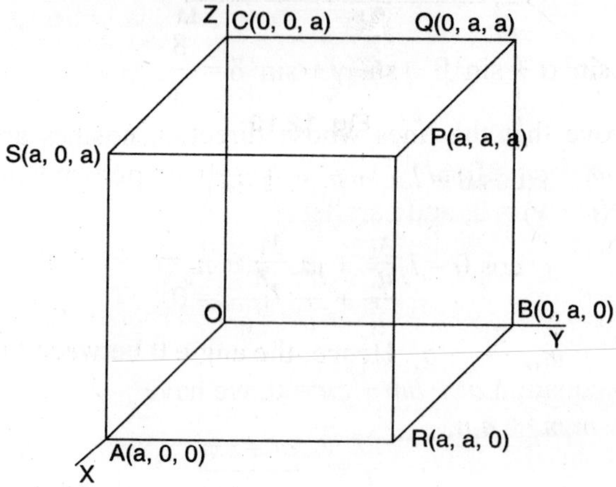

Fig. 12.11

$$\frac{a-0}{\sqrt{a^2+a^2+a^2}}, \frac{a-0}{\sqrt{a^2+a^2+a^2}}, \frac{a-0}{\sqrt{a^2+a^2+a^2}}$$

i.e. $$\frac{1}{\sqrt{3}}, \frac{1}{\sqrt{3}}, \frac{1}{\sqrt{3}}.$$

Similarly, direction cosines of the diagonals AQ, BS and CR are, respectively,

$$-\frac{1}{\sqrt{3}}, \frac{1}{\sqrt{3}}, \frac{1}{\sqrt{3}}; \frac{1}{\sqrt{3}}, -\frac{1}{\sqrt{3}}, \frac{1}{\sqrt{3}}; \text{ and } \frac{1}{\sqrt{3}}, \frac{1}{\sqrt{3}}, -\frac{1}{\sqrt{3}}.$$

Let l, m, n be the direction cosines of the line which make angles $\alpha, \beta, \gamma, \delta$ with the diagonals of the cube. Then

$$\cos\alpha = l.\frac{1}{\sqrt{3}} + m.\frac{1}{\sqrt{3}} + n.\frac{1}{\sqrt{3}} = \frac{l+m+n}{\sqrt{3}},$$

$$\cos\beta = \frac{-l+m+n}{\sqrt{3}}, \cos\gamma = \frac{l-m+n}{\sqrt{3}}, \cos\delta = \frac{l+m-n}{\sqrt{3}}.$$

Now

$$\cos^2\alpha + \cos^2\beta + \cos^2\gamma + \cos^2\delta = \frac{1}{3}\{(l+m+n)^2 + (-l+m+n)^2 + (l-m+n)^2 + (l+m-n)^2\}$$

$$= \frac{4}{3} \because l^2 + m^2 + n^2 = 1.$$

This can also be written as

$$\sin^2\alpha + \sin^2\beta + \sin^2\gamma + \sin^2\delta = \frac{8}{3}.$$

Example 3. Prove that the lines whose direction cosines are given by the equations $al + bm + cn = 0$, $ul^2 + vm^2 + wn^2 = 0$ are perpendicular if $a^2(v + w) + b^2(w + u) + c^2(u + v) = 0$; and parallel if

$$\frac{a^2}{u} + \frac{b^2}{v} + \frac{c^2}{w} = 0.$$

Solution. From equation $al + bm + cn = 0$, we have

$$n = -\frac{al + bm}{c}.$$

Substituting this value of n in equation $ul^2 + vm^2 + wn^2 = 0$, we get

$$(a^2w + c^2u)\frac{l^2}{m^2} + 2abw\frac{l}{m} + b^2w + c^2v = 0. \quad (1)$$

Let l_1, m_1, n_1 and $l_2, m_2\ n_2$ be the direction cosines of the lines. Then, $\frac{l_1}{m_1}$ and $\frac{l_2}{m_2}$ are the roots of the equation (1). Now

$$\frac{l_1}{m_1}.\frac{l_2}{m_2} = \frac{b^2w + c^2v}{a^2w + c^2u}, \text{ i.e., } \frac{l_1l_2}{b^2w + c^2v} = \frac{m_1m_2}{a^2w + c^2u}.$$

Similarly, we obtain

$$\frac{m_1m_2}{a^2w + c^2u} = \frac{n_1n_2}{a^2v + b^2u}.$$

Therefore $$\frac{l_1 l_2}{b^2 w + c^2 v} = \frac{m_1 m_2}{c^2 u + a^2 w} = \frac{n_1 n_2}{a^2 v + b^2 u} = \lambda(\text{say}).$$

The lines will be perpendicular if $l_1 l_2 + m_1 m_2 + n_1 n_2 = 0$, *i.e.*, if $\lambda(b^2 w + c^2 v) + \lambda(c^2 u + a^2 w) + \lambda(a^2 v + b^2 u) = 0$

$$a^2(v + w) + b^2(w + u) + c^2(u + v) = 0.$$

Further, the lines will be parallel if

$$\frac{l_1}{l_2} = \frac{m_1}{m_2} \text{ i.e. } \frac{l_1}{m_1} = \frac{l_2}{m_2}.$$

For this the roots of (1) should be coincident, i.e.,

$$4a^2b^2w^2 = 4(a^2w + c^2u)(b^2w + c^2v)$$

$$\Rightarrow \qquad \frac{a^2}{u} + \frac{b^2}{v} + \frac{c^2}{w} = 0.$$

Example 4. Prove that if two pairs of opposite edges of a tetrahedron are perpendicular, then the third pair is also perpendicular.

Solution. Let *OABC* be a tetrahedron. Let *O* be the origin and the coordinates of the vertices *A*, *B* and *C* be (x_1, y_1, z_1), (x_2, y_2, z_2) and (x_3, y_3, z_3), respectively.

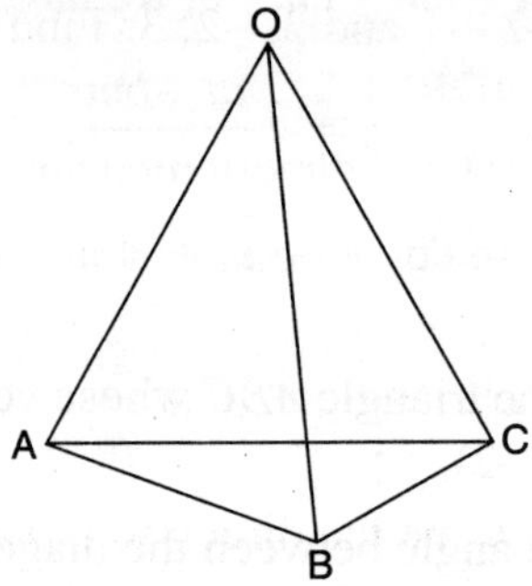

Fig. 12.12

The direction cosines of *OA*, *OB*, *OC*, *AB*, *BC* and *CA* are, respectively, proportional to

$$x_1, y_1, z_1; x_2, y_2, z_2; x_3, y_3, z_3; x_2 - x_1, y_2 - y_1, z_2 - z_1;$$
$$x_3 - x_2, y_3 - y_2, z_3 - z_2 \text{ and } x_1 - x_3, y_1 - y_3, z_1 - z_3.$$

Let *OA* be perpendicular to *BC* and *OB* perpendicular to *CA*. Then

$$x_1(x_3 - x_2) + y_1(y_3 - y_2) + z_1(z_3 - z_2) = 0, \qquad (1)$$

and $$x_2(x_1 - x_3) + y_2(y_1 - y_3) + z_2(z_1 - z_3) = 0. \qquad (2)$$

On adding (1) and (2), we get

$$x_3(x_1 - x_2) + y_3(y_1 - y_2) + z_3(z_1 - z_2) = 0.$$

This equation shows that OC is perpendicular to AB. Hence, the result.

Exercises

1. If P, Q are the points (x_1, y_1, z_1), (x_2, y_2, z_2), prove that the projection of PQ on a line whose direction cosines are l_1, m_1, n_1 is

$$l_1(x_2 - x_1) + m_1(y_2 - y_1) + n_1(z_2 - z_1).$$

2. Find the direction cosines of the line which is perpendicular to the lines whose direction ratios are 2, –1, 1; 3, –3, 4.

3. Prove that the points (–2, 3, 5), (1, 2, 3) and (7, 0, –1) are collinear.

4. Prove that the line joining the points (1, 2, 3), (–1, –2, –3) is perpendicular to the line joining the points (–2, 1, 5), (3, 3, 2).

5. Find the direction cosines of the line which is equally inclined to the axes.

6. If A, B, C, D are the points (6, –6, 0), (–1, –7, 6), (3, –4, 4), (2, –9, 2), respectively, prove that AB is perpendicular to CD.

7. Lines OA and OB are drawn from the origin O with direction cosines proportional to 1, –2, –1 and 3, –2, 3. Find the direction cosines of the normal to the plane AOB.

8. Prove that the lines whose direction cosines are proportional to 1, 1, 2; $\sqrt{3}-1, -\sqrt{3}-1, 4$; $-\sqrt{3}-1, \sqrt{3}-1, 4$ are inclined to one another at an angle $\pi/3$.

9. Find the angles of the triangle ABC whose vertices are the points (2, 3,5), (–1, 3, 2) and (3, 5, –2).

10. Prove that the acute angle between the diagonals of a cube is $\cos^{-1}\frac{1}{3}$.

11. The coordinates of the angular points A, B, C, D of a tetrahedron are (–2, 1, 3), (3, –1, 2), (2, 4, –1) and (1, 2, 3), respectively. Find the angle between the edges AC and BD.

12. P, Q, R, S are the points (6, 3, 2), (5, 1, 4), (3, –4, 7), (0, 2, 5), respectively. Find the projection of PQ on RS.

13. Find the angle between the two lines whose direction cosines are given by the equations $l + 2m + 3n = 0$, $mn - 4nl + 3lm = 0$.

14. Prove that the lines whose direction cosines are given by the equations

$$ul + vm + wn = 0, fmn + gnl + hlm = 0$$

are parallel if $\sqrt{uf} + \sqrt{vg} + \sqrt{wh} = 0$ and perpendicular if $\frac{f}{u} + \frac{g}{v} + \frac{h}{w} = 0$.

6. Distance of a Point from a Line

Let $P(x_1, y_1, z_1)$ be the given point and l, m, n be the direction cosines of the given line AB. Draw perpendicular $PM(= p)$ on AB. Then the projection of AP on AB is AM. Therefore

$$AM = (x_1 - \alpha)l + (y_1 - \beta)m + (z_1 - \gamma)n,$$

where (α, β, γ), are the coordinates of A.

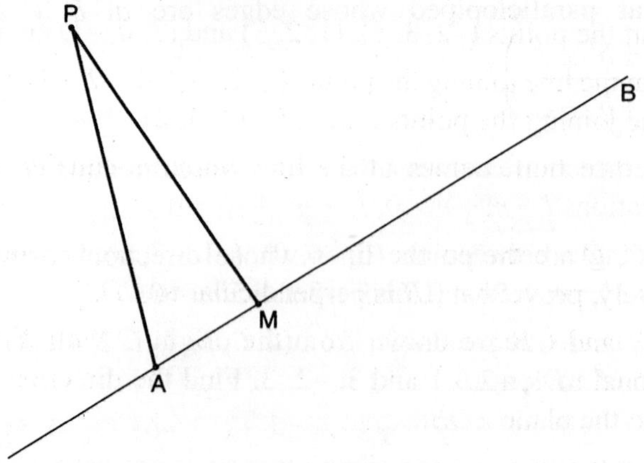

Fig. 12.13

Now, in right angled triangle AMP, we have $PM^2 = AP^2 - AM^2$, *i.e.*,

$$\begin{aligned} p^2 &= (x_1 - \alpha)^2 + (y_1 - \beta)^2 + (z_1 - \gamma)^2 - \\ &\qquad \{(x_1 - \alpha)\, l + (y_1 - \beta)\, m + (z_1 - \gamma)\, n\}^2 \\ &= \{(x_1 - \alpha)^2 + (y_1 - \beta)^2 + (z_1 - \gamma)^2\}\, (l^2 + m^2 + n^2) - \\ &\qquad \{(x_1 - \alpha)\, l + (y_1 - \beta)\, m + (z_1 - \gamma) n\}^2 \\ &= \{(y_1 - \beta)\, n - (z_1 - \gamma)\, m\}^2 + \{(z_1 - \gamma)\, l - (x_1 - \alpha) n\}^2 \\ &\qquad + \{(x_1 - \alpha)\, m - (y_1 - \beta)\, l\}^2, \end{aligned}$$

(on using Lagrange's identity.)

Hence, *the distance of a point* (x_1, y_1, z_1) *from the line AB is*

$$\sqrt{\{(y_1 - \beta)n - (z_1 - \gamma)m\}^2 + \{(z_1 - \gamma)l - (x_1 - \alpha)n\}^2 + \{(x_1 - \alpha)m - (y_1 - \beta)}$$

Miscellaneous Exercises

1. Prove that the points (3, –2, 4), (1, 1, 1) and (–1, 4, 2) are collinear.
2. Prove that the three lines joining the mid-points of opposite edges of a tetrahedron meet in a point which bisects them.
3. A variable line in two adjacent positions has direction cosines l, m, n; $l + \delta l, m + \delta m, n + \delta n$.

   Prove that the small angle $\delta\theta$ between the two position is given by $\delta\theta^2 = \delta l^2 + \delta m^2 + \delta n^2$.
4. Find the angle between the lines whose direction cosines are proportional to 1, 2, 4 and –2, 1, 5.
5. Prove that the angles between the four diagonals of a rectangular parallelopiped whose edges are a, b, c are given by $\cos^{-1}\left(\dfrac{a^2 \pm b^2 \pm c^2}{a^2 + b^2 + c^2}\right)$.
6. Find the acute angle between the lines whose direction cosines are given by the relations $l + m + n = 0, l^2 + m^2 + n^2 = 0$.
7. Find the angle between the lines whose direction cosines satisfy the equations $3l + m + 5n = 0, 6mn - 2nl + 5lm = 0$.
8. If A, B, C, D are the points (3, 4, 5), (4, 6, 3), (–1, 2, 4), (1, 0, 5); find the projection of CD on AB.
9. If O is the origin and P, Q are the points (2, 3, –6), (3, –4, 5), respectively; find the angle that OP makes with OQ.
10. Prove by direction cosines that the points (1, 2, 3), (4, 0, 4), (–2, 4, 2), (7, –2, 5) are collinear.
11. Prove that the points (4, 7, 8), (2, 3, 4), (–1, –2, 1), (1, 2, 5) are the vertices of a parallelogram.
12. Prove that the points (0, 4, 1), (2, 3, –1), (4, 5, 0) (2, 6, 2) are the vertices of a square.
13. The direction cosines of two concurrent lines are l_1, m_1, n_1; l_2, m_2, n_2. Prove that the direction cosines of the lines bisecting the angles between them are proportional to $l_1 \pm l_2, m_1 \pm m_2, n_1 \pm n_2$.
14. Prove that the three concurrent lines with direction ratios l_1, m_1, n_1; l_2, m_2, n_2; l_3, m_3, n_3 are coplanar if

$$\begin{vmatrix} l_1 & m_1 & n_1 \\ l_2 & m_2 & n_2 \\ l_3 & m_3 & n_3 \end{vmatrix} = 0.$$

15. Find the direction cosines of the line which is perpendicular to the lines with direction cosines proportional to 3, –1, 1 and –3, 2, 4.

16. There are three lines through the origin with direction cosines proportional to 1, 2, 2; 2, 3, 6; 3, 4, 12. Find the direction cosines of a line which is equally inclined to the three given lines.

17. If l_1, m_1, n_1 and l_2, m_2, n_2 be the direction ratios of two intersecting lines, prove that the line through their point of intersection with direction ratios $l_1 + \lambda l_2, m_1 + \lambda m_2, n_1 + \lambda n_2$ is coplanar with them for all values of λ.

18. Prove that the direction cosines of a line equally inclined to three mutually perpendicular lines whose direction cosines are l_1, m_1, n_1; l_2, m_2, n_2 and l_3, m_3, n_3 are $\dfrac{l_1 + l_2 + l_3}{\sqrt{3}}, \dfrac{m_1 + m_2 + m_3}{\sqrt{3}}, \dfrac{n_1 + n_2 + n_3}{\sqrt{3}}$.

The Plane

13 CHAPTER

1. Introduction

A **plane** *is a surface such that if any two points be taken on it, the line* joining them lies wholly in the surface.*

2. Particular Planes

Before we come to a general discussion of the equations of a plane, we examine some particular cases.

To find the equation of a plane parallel to one of the coordinate planes.

Let $ABCD$ be a plane parallel to yz-plane. Clearly, the plane $ABCD$ is the locus of a point which moves such that its distance from yz-plane remains fixed. Let $P(x', y', z')$ be the moving point.

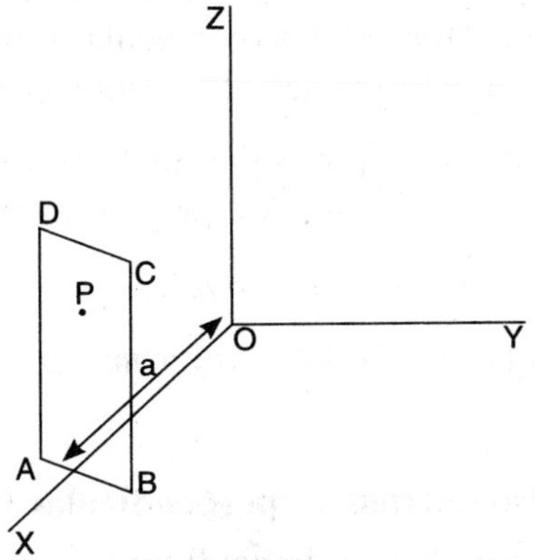

Fig. 13.1

Then, by definition, P is at a distance a (say) from yz-plane. Therefore, $x' = a$. Thus, the locus of (x', y', z') is $x = a$. This is the required equation of the plane $ABCD$.

Similarly, the equations of the planes parallel to zx-plane and xy-plane are,

* Throughout, by line we mean straight line unless specified otherwise.

respectively, $y = b$ and $z = c$.

Corollaries.

(i) The equation of yz-plane is $x = 0$.
(ii) The equation of zx-plane is $y = 0$.
(iii) The equation of xy-plane is $z = 0$.

3. Various forms of the Equation of a Plane

3.1 Normal form. *To find the equation of a plane in terms of the length of the normal drawn from the origin to the plane and its direction cosines.*

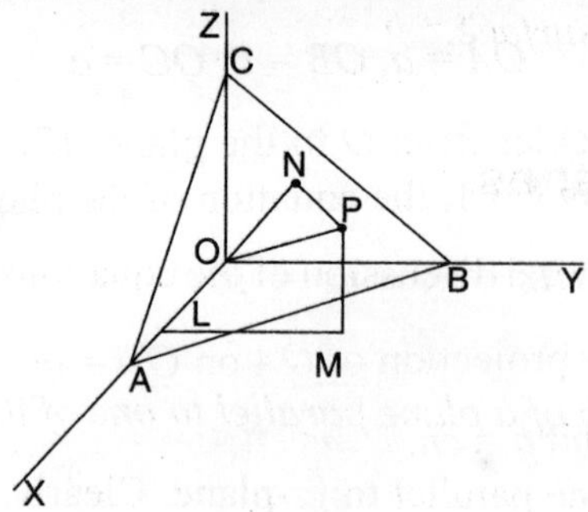

Fig. 13.2

Let ABC be a plane and ON $(= p)$ the normal to the plane. Let $P(x', y', z')$ be any point on the plane. Draw PM perpendicular from P to xy-plane and ML parallel to y-axis. Let l, m, n be the actual direction cosines of ON. Then

projection of OP on ON = projection of OL on ON + projection of LM on ON + projection of MP on ON

i.e. $$p = lx' + my' + nz'.$$

Hence, the locus of (x', y', z') is $lx + my + nz = p$.

Remarks.

(i) Normal form is also referred to **'perpendicular form'**.
(ii) The foot of the perpendicular drawn from the origin to the plane $lx + my + nz = p$ is (lp, mp, np).

3.2 Intercept form. *To find the equation of the plane in terms of the intercepts which it makes on the axes.*

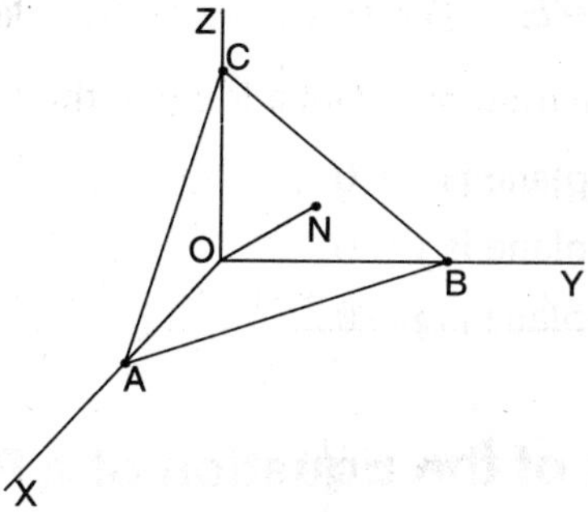

Fig. 13.3

Let ABC be a plane which cuts intercepts a, b, c on the axes, so that

$$OA = a,\ OB = b,\ OC = c.$$

Draw ON $(= p)$ perpendicular from O to the plane ABC and let l, m, n be its direction cosines. Then, by § 3.1, the equation of the plane ABC is

$$lx + my + nz = p. \tag{1}$$

But $\quad p = ON =$ projection of OA on $ON = al.$

Similarly, $\quad p = bm$ and $p = cn.$

Therefore, equation (1) becomes

$$\frac{x}{a} + \frac{y}{b} + \frac{z}{c} = 1.$$

This is the required equation of the plane.

4. General Equation of First Degree

From the various forms of the equations of a plane obtained so for in § 3, we observe that the equation of a plane is a first degree (or linear) equation in x, y and z, the coordinates of the moving point on the plane. It is a natural temptation to examine the converse problem, viz., whether a first degree equation in x, y and z always represents a plane.

4.1 *To prove that the general equation of first degree in x, y and z represents a plane.*

The general equation of first degree in x, y and z is

$$Ax + By + Cz + D = 0. \tag{1}$$

Let $P(x_1, y_1, z_1)$ and $Q(x_2, y_2, z_2)$ be any two points on the locus of (1). Then

$$Ax_1 + By_1 + Cz_1 + D = 0, \text{ and } Ax_2 + By_2 + Cz_2 + D = 0,$$

Multiplying the first equation by λ and adding in the second, we get

$$A(x_1 + \lambda x_2) + B(y_1 + \lambda y_2) + C(z_1 + \lambda z_2) + D(1 + \lambda) = 0$$

$$\Rightarrow \quad A\left(\frac{x_1 + \lambda x_2}{1+\lambda}\right) + B\left(\frac{y_1 + \lambda y_2}{1+\lambda}\right) + C\left(\frac{z_1 + \lambda z_2}{1+\lambda}\right) + D = 0.$$

This equation shows that the point

$$\left(\frac{x_1 + \lambda x_2}{1+\lambda}, \frac{y_1 + \lambda y_2}{1+\lambda}, \frac{z_1 + \lambda z_2}{1+\lambda}\right)$$

which divides the line PQ in the ratio $\lambda : 1$ also lies on the locus. Since, λ is arbitrary, all the points lying on the line PQ also lie on the locus. Moreover, the line PQ is arbitrary, hence, all the lines joining any two points on the locus lie wholly on the locus. This will be true only if the locus is a plane. Thus, any first degree equation in x, y and z always represents a plane.

Remark. We have the following observations about the equation of the plane:*

(i) When $A = 0$, the plane is parallel to x-axis.

(ii) When $B = 0$, the plane is parallel to y-axis.

(iii) When $C = 0$, the plane is parallel to z-axis.

(iv) When $D = 0$, the plane passes through the origin.

(v) When $A = B = 0$, the plane is parallel to xy-plane.

(vi) When $B = C = 0$, the plane is parallel to yz-plane.

(vii) When $A = C = 0$, the plane is parallel to zx-plane.

(viii) When $A = B = C = 0$, and D finite, the plane is at an infinite distance.

Note. Suppose $D \neq 0$. Then, the equation of the plane can be written as

$$\frac{A}{D}x + \frac{B}{D}y + \frac{C}{D}z + 1 = 0.$$

In case $D = 0$, then divide by A, B or C which is different from zero. All the constants A, B, C and D cannot be simultaneously zero. We note from either case that there are three unknown constants in the equation of a plane. As such

* A plane is said to be parallel to a line if any perpendicular to the line is a normal to the plane. For further details see § 3, in Chapter 3.

A plane is said to be parallel to another plane if normals of one plane are also the normals of the other.

to specify completely the equation of a plane, we need three conditions.

4.2 *To find the equation of a plane which passes through a given point.*

Let (x_1, y_1, z_1) be any given point. Let the equaiton of a plane be

$$Ax + By + Cz + D = 0. \tag{1}$$

This plane will pass through the point (x_1, y_1, z_1) if $D = -(Ax_1 + By_1 + Cz_1)$. Substituting this value of D in (1), we get

$$A(x - x_1) + B(y - y_1) + C(z - z_1) = 0. \tag{2}$$

Note. The equation (2) is called **'one-point form'** of the plane. To determine the plane completely in this case, we need two more conditions.

Corollary. The equation of the plane passing through origin is

$$Ax + By + Cz = 0.$$

4.3 *To find the equation of a plane which passes through three points* (x_1, y_1, z_1), (x_2, y_2, z_2) *and* (x_3, y_3, z_3).

Let the equation of a plane be

$$Ax + By + Cz + D = 0. \tag{1}$$

It will pass through the given points if

$$Ax_1 + By_1 + Cz_1 + D = 0, \tag{2}$$

$$Ax_2 + By_2 + Cz_2 + D = 0, \tag{3}$$

and

$$Ax_3 + By_3 + Cz_3 + D = 0. \tag{4}$$

Eliminating A, B, C and D from equations (1), (2), (3) and (4), we get

$$\begin{vmatrix} x & y & z & 1 \\ x_1 & y_1 & z_1 & 1 \\ x_2 & y_2 & z_2 & 1 \\ x_3 & y_3 & z_3 & 1 \end{vmatrix} = 0. \tag{5}$$

This is the required equation of the plane.

Note. The equaiton (5) of the plane is called **'three-point form'**.

Since, a plane is completely specified if it passes through three given points, a fourth point may or may not lie on it. However, we have the following.

Corollary. The four points (x_1, y_1, z_1), (x_2, y_2, z_2), (x_3, y_3, z_3) and (x_4, y_4, z_4) lie in a plane if and only if

$$\begin{vmatrix} x_1 & y_1 & z_1 & 1 \\ x_2 & y_2 & z_2 & 1 \\ x_3 & y_3 & z_3 & 1 \\ x_4 & y_4 & z_4 & 1 \end{vmatrix} = 0.$$

4.4 Reduction of general equation. *To reduce the general equation of the plane into standard forms.*

Let $$Ax + By + Cz + D = 0 \quad (1)$$
be the general equation of the plane.

I. Normal form. If the equation is of the form $lx + my + nz = p$, then on comparing, we get

$$\frac{l}{A} = \frac{m}{B} = \frac{n}{C} = \frac{-p}{D}.$$

Now $$\frac{l}{A} = \frac{m}{B} = \frac{n}{C} = \pm\frac{\sqrt{l^2 + m^2 + n^2}}{\sqrt{A^2 + B^2 + C^2}}$$

$$= \pm\frac{1}{\sqrt{A^2 + B^2 + C^2}},$$

which gives

$$l = \pm\frac{A}{\sqrt{A^2 + B^2 + C^2}}, \quad m = \pm\frac{B}{\sqrt{A^2 + B^2 + C^2}}, \quad n = \pm\frac{C}{\sqrt{A^2 + B^2 + C^2}}$$

and $$p = \pm\frac{-D}{\sqrt{A^2 + B^2 + C^2}}.$$

Since, p is always positive, there arise two cases:

Case I. When D is positive,

$$p = \frac{D}{\sqrt{A^2 + B^2 + C^2}},$$

$$l = -\frac{A}{\sqrt{A^2 + B^2 + C^2}}, \quad m = -\frac{B}{\sqrt{A^2 + B^2 + C^2}}, \quad n = -\frac{C}{\sqrt{A^2 + B^2 + C^2}}.$$

Case II. When D is negative, $p = \dfrac{-D}{\sqrt{A^2 + B^2 + C^2}}$,

$$l = \frac{A}{\sqrt{A^2 + B^2 + C^2}}, \quad m = \frac{B}{\sqrt{A^2 + B^2 + C^2}}, \quad n = \frac{C}{\sqrt{A^2 + B^2 + C^2}}.$$

Note. The direction cosines of the normal to the plane $Ax + By + Cz + D = 0$ are proportional to A, B, C.

II. Intercept form. On dividing equation (1) by $-D$ $(D \neq 0)$, we get

$$\Rightarrow \qquad \frac{x}{-\frac{D}{A}} + \frac{y}{-\frac{D}{B}} + \frac{z}{-\frac{D}{C}} = 1.$$

This is the intercept form of the plane. Clearly, $-\frac{D}{A}$, $-\frac{D}{B}$ and $-\frac{D}{C}$ are the intercepts on the axes.

4.5 Examples

Example 1. If P is the point (a, b, c), find the equation of the plane through the point P and at right angles to OP, where O is the origin.

Solution. The direction cosines of OP are proportional to a, b, c. Since, the plane is at right angles to OP, the normal to the plane will be parallel to OP and its direction cosines will be proportional to a, b, c. Thus, the equaiton of the plane is

$$ax + by + cz + d = 0. \qquad (1)$$

But it passes through the point (a, b, c). Therefore

$$a^2 + b^2 + c^2 + d = 0$$

$$\Rightarrow \qquad d = -(a^2 + b^2 + c^2).$$

Substituting this value of d in (1), we get

$$ax + by + cz = a^2 + b^2 + c^2.$$

This is the required equation of the plane.

Example 2. A variable plane passes through a fixed point (a, b, c) and meets the coordinates axes in A, B, C. Prove that the locus of the point of intersection of the planes through A, B, C parallel to the coordinate planes is

$$\frac{a}{x} + \frac{b}{y} + \frac{c}{z} = 1.$$

Solution. Let $OA = \alpha$, $OB = \beta$, $OC = \gamma$, where O is the origin. Then, the equation of the plane ABC is

$$\frac{x}{\alpha} + \frac{y}{\beta} + \frac{z}{\gamma} = 1.$$

But it passes through the point (a, b, c). Therefore

$$\frac{a}{\alpha}+\frac{b}{\beta}+\frac{c}{\gamma}=1. \quad (1)$$

The equations of the planes through A, B, C parallel to the coordinate planes are

$$x=\alpha, y=\beta \text{ and } z=\gamma\} \quad (2)$$

The required locus is obtained by eliminating α, β, γ from equations (1) and (2), i.e., $\frac{a}{x}+\frac{b}{y}+\frac{c}{z}=1.$

Exercises

1. If P is the point $(2, 3, -1)$; find the equation of the plane through the point P and at right angles to OP, where O is the origin.
2. Find the equation of the plane through the three points $(0, 1, 1)$, $(1, 1, 2)$ and $(-1, 2, -2)$.
3. Find the equation of the plane passing through the points $(2, 3, -4)$ and $(1, -1, 3)$ and parallel to x-axis.
4. Find the equation of the plane passing through the point $(3, -3, 1)$ and normal to the join of the points $(3, 4, -1)$ and $(2, -1, 5)$.
5. Prove that the points $(0, -1, 0)$, $(2, 1, -1)$, $(1, 1, 1)$ and $(3, 3, 0)$ are coplanar.
6. For what value of λ, the points $(0, -1, \lambda)$, $(4, 5, 1)$, $(3, 9, 4)$ and $(-4, 4, 4)$ are coplanar?
7. A plane meets the coordinate axes at A, B, C such that the centroid of the triangle ABC is the point (a, b, c). Prove that the equation of the plane is
$$\frac{x}{a}+\frac{y}{b}+\frac{z}{c}=3.$$
8. A variable plane makes intercepts on the coordinate axes the sum of whose squares is constant and is equal to k^2. Prove that the locus of the foot of the perpendicular from the origin to the plane is
$$(x^2+y^2+z^2)^2(x^{-2}+y^{-2}+z^{-2})=k^2.$$

5. Angle between Two Planes

The **angle between two planes** *is equal to the angle which the positive direction* *of a normal to one plane makes with the positive direction of a normal to the other plane.*

5.1 *To find the angle between any two given planes.*

Let the equations of the given planes be

$$Ax + By + Cz + D = 0 \text{ and } A'x + B'y + C'z + D' = 0.$$

The direction cosines of the normals to these planes are, respectively, proportional to A, B, C and A', B', C'. Let θ be the angle between them. Then

$$\cos\theta = \frac{AA' + BB' + CC'}{A^2 + B^2 + C^2 \cdot \sqrt{A'^2 + B'^2 + C'^2}}$$

$$\Rightarrow \qquad \theta = \cos^{-1} \frac{AA' + BB' + CC'}{A^2 + B^2 + C^2 \cdot \sqrt{A'^2 + B'^2 + C'^2}}.$$

Remark. Sometimes the value of $\cos\theta$ may be negative which, in fact, is due to the factors $\sqrt{A^2 + B^2 + C^2}$ and $\sqrt{A'^2 + B'^2 + C'^2}$, and it corresponds to the obtuse angle between the planes.

Corollary 1. The planes $Ax + By + Cz + D = 0$ and $A'x + B'y + C'z + D' = 0$ are parallel if and only if $\frac{A}{A'} = \frac{B}{B'} = \frac{C}{C'}$.

Corollary 2. The planes $Ax + By + Cz + D = 0$ and $A'x + B'y + C'z + D' = 0$ are perpendicular if and only if $AA' + BB' + CC' = 0$.

5.2 Examples

Example 1. Find the equation of the plane which passes through the point $(2, -3, 4)$ and is parallel to the plane $2x - 5y - 7z = 6$.

Solution. Any plane parallel to the given plane is

$$2x - 5y - 7z + \lambda = 0. \qquad (1)$$

This plane will pass through the point $(2, -3, 4)$ if $\lambda = 9$. Substituting this value of λ in (1), we get

$$2x - 5y - 7z + 9 = 0.$$

This is the required equation of the plane.

Example 2. Find the equation of the plane through the point $(-1, 3, 2)$ and perpendicular to the planes $x + 2y + 2z = 5$ and $3x + 3y + 2z = 8$.

Solution. Let the equation of any plane be

$$Ax + By + Cz + D = 0. \tag{1}$$

This plane will pass through (–1, 3, 2) if

$$-A + 3B + 2C + D = 0. \tag{2}$$

The plane (1) will be perpendicular to the given planes if

$$A + 2B + 2C = 0 \text{ and } 3A + 3B + 2C = 0.$$

Therefore, we get

$$\frac{A}{2} = \frac{B}{-4} = \frac{C}{3}. \tag{3}$$

From equations (2) and (3), we get $D = 8$. Hence, the required equation of the plane becomes $2x - 4y + 3z + 8 = 0$.

Exercises

1. Find the angle between the planes:
   (i) $2x - y + z = 6, x + y + 2z = 3;$
   (ii) $3x - 2y + z + 17 = 0, 4x + 3y - 6z + 25 = 0.$
2. Prove that the following planes are at right angles:
   (i) $x + 2y + 2z = 0, 2x + y - 2z = 0;$
   (ii) $3x + 4y - 5z = 9, 2x + 6y + 6z = 7.$
3. Find the equation of the plane through the points (2, 2, 1) and (9, 3, 6), and perpendicular to the plane $2x + 6y + 6z = 9$.
4. Find the equaiton of the plane through the points (–1, 1, 1) and (1, –1, 1), and perpendicular to the plane $x + 2y + 2z = 5$.
5. Find the equation of the plane which passes through the origin and is perpendicular to each of the planes $x + 2y + 2z = 0$ and $2x + y - 2z = 0$.
6. Find the equation of the plane which passes through the point (1, 2, 3) and is parallel to $3x + 4y - 5z = 0$.
7. Find the equation of the plane which passes through the point (2, 5, –8) and is perpendicular to each of the planes $2x - 3y + 4z + 1 = 0$ and $4x + y - 2z + 6 = 0$.
8. Find the equation of the planes which pass through the points (2, 3, 1), (4, –5, 3) and are parallel to the coordinate axes.

6. Perpendicular Distance of a Point from a Plane

6.1 *To find the perpendicular distance of a given point from a given plane.*

Let us first consider the equation of the plane ABC in the normal form

$$lx + my + nz = p. \tag{1}$$

Let $P(x_1, y_1, z_1)$ be the point from which the perpendicular distance of the plane ABC is to be calculated. Draw PM perpendicular from P to the plane (1). Also draw a plane parallel to (1) passing through $P(x_1, y_1, z_1)$. Its equation is given by

$$lx + my + nz = p_1, \tag{2}$$

where p_1 is the length of the normal drawn from the origin to the plane (2). But the plane (2) passes through $P(x_1, y_1, z_1)$. Therefore

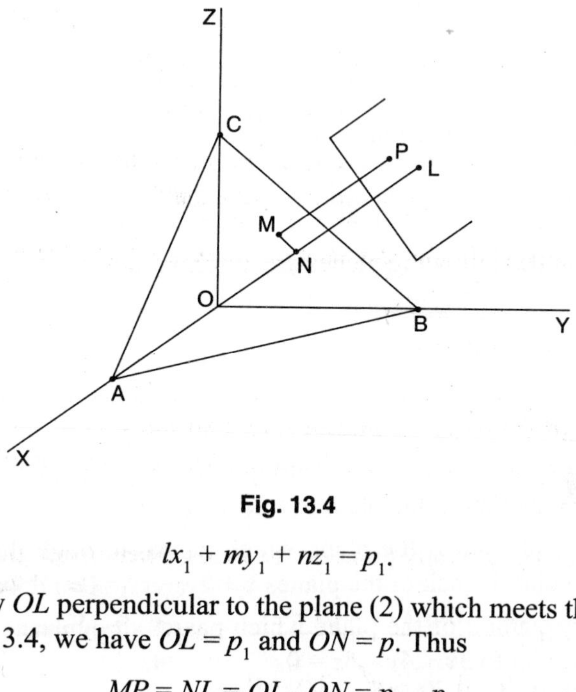

Fig. 13.4

$$lx_1 + my_1 + nz_1 = p_1. \tag{3}$$

Now, draw OL perpendicular to the plane (2) which meets the plane (1) in N. From Fig. 13.4, we have $OL = p_1$ and $ON = p$. Thus

$$MP = NL = OL \cdot ON = p_1 - p.$$

Hence, *the perpendicular distance of the point* $P(x_1, y_1, z_1)$ *from the plane* $lx + my + nz = p$ is $lx_1 + my_1 + nz_1 - p$. ▮

Consider now the general equation of the plane

$$Ax + By + Cz + D = 0. \tag{4}$$

This can be written in the normal form as

$$\frac{A}{\sqrt{A^2+B^2+C^2}}x+\frac{B}{\sqrt{A^2+B^2+C^2}}y+\frac{C}{\sqrt{A^2+B^2+C^2}}z + \frac{D}{\sqrt{A^2+B^2+C^2}}=0.$$

Hence, the perpendicular distance of the point $P(x_1, y_1, z_1)$ from the plane (4) is

$$\frac{A}{\sqrt{A^2+B^2+C^2}}x_1+\frac{B}{\sqrt{A^2+B^2+C^2}}y_1+\frac{C}{\sqrt{A^2+B^2+C^2}}z_1 + \frac{D}{\sqrt{A^2+B^2+C^2}}$$

i.e. $$\pm\frac{Ax_1+By_1+Cz_1+D}{\sqrt{A^2+B^2+C^2}}.$$

Positive or negative sign is to be taken according as the expression $Ax_1 + By_1 + Cz_1 + D$ is positive or negative in order to have the distance always positive.

Corollary. The perpendicular distance of the origin from the plane $Ax + By + Cz + D = 0$ is $\pm\frac{D}{\sqrt{A^2+B^2+C^2}}$.

6.2 Examples

Example 1. Find the locus of a point such that the sum of the squares of its distance from the planes $x + y + z = 0$ and $x - 2y + z = 0$ is equal to its distance from the plane $x - z = 0$.

Solution. Let (α, β, γ) be any point on the locus. Then, from given, we have

$$\frac{(\alpha+\beta+\gamma)^2}{3}+\frac{(\alpha-2\beta+\gamma)^2}{6}=\frac{(\alpha-\gamma)^2}{2}, \text{ i.e.,} \beta^2+2\gamma\alpha=0.$$

Hence, the locus of (α, β, γ) is $y^2 + 2zx = 0$.

Example 2. A variable plane is at a constant distance p from the origin and cuts the axes at A, B, C. Prove that the locus of the centroid of the triangle ABC is

$$\frac{1}{x^2}+\frac{1}{y^2}+\frac{1}{z^2}=\frac{9}{p^2}.$$

Solution. Let a, b, c be the distances A, B and C from the origin, respectively. Then the coordinates of A, B, C are $A(a, 0, 0)$, $B(0, b, 0)$ and $C(0, 0, c)$. Also the equation of the plane ABC is

$$\frac{x}{a}+\frac{y}{b}+\frac{z}{c}=1. \tag{1}$$

But p is the distance of the origin from the plane (1). Therefore

$$\frac{1}{a^2}+\frac{1}{b^2}+\frac{1}{c^2}=\frac{1}{p^2}. \tag{2}$$

Further, let (α, β, γ) be the centroid of the triangle ABC. Then

$$\left.\begin{aligned}\alpha &= \frac{a+0+0}{3}=\frac{a}{3},\ i.e., a=3\alpha\\ \beta &= \frac{0+b+0}{3}=\frac{b}{3},\ i.e., b=3\beta.\\ \gamma &= \frac{0+0+c}{3}=\frac{c}{3},\ i.e., c=3\gamma\end{aligned}\right\} \tag{3}$$

Eliminating a, b, c from (2) and (3), we get

$$\frac{1}{9\alpha^2}+\frac{1}{9\beta^2}+\frac{1}{9\gamma^2}=\frac{1}{p^2}.$$

Hence, the locus of (α, β, γ) is

$$\frac{1}{x^2}+\frac{1}{y^2}+\frac{1}{z^2}=\frac{9}{p^2}.$$

Exercises

1. Find the distance of the origin from the plane $6x-3y+2z-14=0$.
2. Find the locus of a point the sum of the squares of whose distances from the planes $x+y+z=0$, $x-z=0$ and $x-2y+z=0$ is 9.
3. Find the locus of a point whose distance from the origin is 7 times its distance from the plane $2x+3y-6z=2$.
4. A variable plane is at a constant distance p from the origin and meets the axes at A, B and C. Through A, B, C the planes are drawn parallel to the coordinate planes. Prove that the locus of their points of intersection is given by $\frac{1}{x^2}+\frac{1}{y^2}+\frac{1}{z^2}=\frac{1}{p^2}$.
5. A variable plane is at a constant distance p from the origin and meets the

axes at A, B, C. Prove that the locus of the centroid of the tetrahedron $OABC$ is $\frac{1}{x^2}+\frac{1}{y^2}+\frac{1}{z^2}=\frac{16}{p^2}$.

7. Positive and Negative Sides of a Plane

We know that the xy-plane ($z = 0$) divides the space into two parts, one above the plane where $z > 0$ and one below the plane where $z < 0$. The boundary between the two spaces is the plane $z = 0$. In general, any plane $Ax + By + Cz + D = 0$ also divides the space into two parts. It is natural curiousity to know if the expression $L \equiv Ax + By + Cz + D$ has the same sign, positive or negative, for all the points lying on one side of the plane.

7.1 *To prove that the expression $L \equiv Ax + By + Cz + D$ has the same sign for all the points lying on one side of the plane $L = 0$.*

Let $P(x_1, y_1, z_1)$ and $Q(x_2, y_2, z_2)$ be any two points in the space and let their join meets the plane $L = 0$ in R, where

$$\frac{PR}{RQ}=\frac{\lambda}{\mu}.$$

Then, the coordinates of the point R are $\left(\frac{\lambda x_2+\mu x_1}{\lambda+\mu},\frac{\lambda y_2+\mu y_1}{\lambda+\mu},\frac{\lambda z_2+\mu z_1}{\lambda+\mu}\right)$

Since R lies on the plane $L = 0$, we have

$$A\left(\frac{\lambda x_2+\mu x_1}{\lambda+\mu}\right)+B\left(\frac{\lambda y_2+\mu y_1}{\lambda+\mu}\right)+C\left(\frac{\lambda z_2+\mu z_1}{\lambda+\mu}\right)+D=0$$

$$\Rightarrow \qquad \frac{\lambda}{\mu}=-\frac{Ax_1+By_1+Cz_1+D}{Ax_2+By_2+Cz_2+D}.$$

If $Ax_1 + By_1 + Cz_1 + D$ and $Ax_2 + By_2 + Cz_2 + D$ have the same sign, the ratio $\frac{\lambda}{\mu}$ is negative and so R divides PQ externally. Hence, P and Q lie on the same side of the plane $L = 0$. If $Ax_1 + By_1 + Cz_1 + D$ and $Ax_2 + By_2 + Cz_2 + D$ have opposite signs, the ratio $\frac{\lambda}{\mu}$ is positive and so R divides PQ internally. Hence, P and Q lie on the opposite sides of the plane $L = 0$.

Thus, the expression $L \equiv Ax + By + Cz + D$ has the same sign for all the points lying on one side of the plane $L = 0$.

Corollary. The point (x_1, y_1, z_1) and the origin are on the same side of the plane $Ax + By + Cz + D = 0$ if and only if $Ax_1 + By_1 + Cz_1 + D$ and D have the same sign.

A side of the plane $Ax + By + Cz + D = 0$ is said to be **positive** *or the* **negative side** *according as the expression $Ax + By + Cz + D$ has the positive or negative sign.*

Remark. It may be noted that the sign of $Ax_1 + By_2 + Cz_1 + D$ has in itself no particular significance for deciding the positive or negative side of the plane

$$Ax + By + Cz + D = 0$$

since $$Ax + By + Cz + D = 0 \text{ and } -Ax - By - Cz - D = 0$$

represent the same plane and as such the positive and negative regions of the space would be interchanged. Therefore, we cannot attach the term positive or negative infallibly to the two sides of the plane. However, to fix up the ideas in a particular situation one may adopt the above definition for the positive or negative side of a plane.

Exercises

1. Find the distances of the points (2, 3, –5), (3, 4, 7) from the plane $x + 2y - 2z = 9$ and prove that these points lie on the opposite sides of the plane.
2. Prove that the points (1, 1, 1) and (–3, 0, 1) lie on the opposite sides of the plane $3x + 4y - 12z + 13 = 0$.
3. Prove that the two points (2, 3, –5) and (3, 4, 7) lie on the opposite sides of the plane which meets the axes in A, B, C such that the centroid of the triangle ABC is the point (1, 2, 4).
4. Prove that the origin lies in the acute angle between the planes $x + 2y + 2z - 9 = 0$ and $4x - 3y + 12z + 13 = 0$.

8. Bisectors of Angles

Bisector of angles *between the two planes is the locus of a point which moves such that its distances from the two planes are equal in magnitude.*

The locus so obtained is, in fact, a plane

8.1 *To find the equations of the planes which bisect the angles between the two given planes.*

Let

$$A_1x + B_1y + C_1z + D_1 = 0 \tag{1}$$

and $$A_2x + B_2y + C_2z + D_2 = 0 \quad (2)$$

be any two given intersecting planes.

Let $P(x', y', z')$ be any point on either of the bisectors. Then, the perpendicular distances of the point $P(x', y', z')$ from the planes (1) and (2) must be equal in magnitude, i.e.

$$\frac{A_1x' + B_1y' + C_1z' + D_1}{\sqrt{A_1^2 + B_1^2 + C_1^2}} = \pm\frac{A_2x' + B_2y' + C_2z' + D_2}{\sqrt{A_2^2 + B_2^2 + C_2^2}}$$

Hence, the locus of (x', y', z') is

$$\frac{A_1x + B_1y + C_1z + D_1}{\sqrt{A_1^2 + B_1^2 + C_1^2}} = \pm\frac{A_2x + B_2y + C_2z + D_2}{\sqrt{A_2^2 + B_2^2 + C_2^2}}. \quad (3)$$

The two planes given by (3) are therefore the required bisectors.

Note. We can distinguish between the two bisectors given by (3) with the help of origin. Let us assume,without any loss of generality that D_1 and D_2 both be positive for if it is not so it could be made on multiplying the equations by -1. Let (x', y', z') be any point on a bisector. Now, we have the following different cases:

Case I. When both the expressions $A_1x' + B_1y' + C_1z' + D_1$ and $A_2x' + B_2y' + C_2z' + D_2$ are positive. Then, the point (x', y', z') and the origin lie on the same side of each of the planes

$$A_1x + B_1y + C_1z + D_1 = 0 \text{ and } A_2x + B_2y + C_2z + D_2 = 0.$$

Thus, the point (x', y', z') lies on the bisector of the angle which contains the origin. We call it **internal bisector** and its equation is given by

$$\frac{A_1x + B_1y + C_1z + D_1}{\sqrt{A_1^2 + B_1^2 + C_1^2}} = \frac{A_2x + B_2y + C_2z + D_2}{\sqrt{A_2^2 + B_2^2 + C_2^2}}.$$

Case II. When the expressions $A_1x' + B_1y' + C_1z' + D_1$ and $A_2x' + B_2y' + C_2z' + D_2$ have opposite signs. Then the point (x', y', z') and the origin lie on the same side of one plane and on opposite sides of the other. Thus, the point (x', y', z') lies on the bisector of the angle which does not contain the origin. We call it **external bisector** and its equation is given by

$$\frac{A_1x + B_1y + C_1z + D_1}{\sqrt{A_1^2 + B_1^2 + C_1^2}} = -\frac{A_2x + B_2y + C_2z + D_2}{\sqrt{A_2^2 + B_2^2 + C_2^2}}.$$

Remark. The planes representing the internal and external bisectors of the angles between any two planes, of course intersecting, are perpendicular to each other.

Working rule. In order to obtain the bisectors of the angles between the two planes we use the above formula with a remark that the equations of the planes should be so written that the constant terms in the equations have the same sign.

8.2 Examples

Example. Find the equations of the planes bisecting the angles between the planes $3x + 4y - 5z + 1 = 0$ and $5x + 12y - 13z = 0$.

Solution. The equation of the internal bisector is

$$\frac{3x+4y-5z+1}{\sqrt{3^2+4^2+(-5)^2}} = \frac{5x+12y-13z}{\sqrt{5^2+12^2+(-13)^2}}$$

$$\Rightarrow \qquad 14x - 8y + 13 = 0.$$

Further, the equation of the other bisector is

$$\frac{3x+4y-5z+1}{\sqrt{3^2+4^2+(-5)^2}} = -\frac{5x+12y-13z}{\sqrt{5^2+12^2+(-13)^2}}$$

$$\Rightarrow \qquad 64x + 112y - 130z + 13 = 0.$$

Exercises

1. Find the equations of the planes bisecting the angles between the planes $3x - 4y + 12z = 26$ and $x + 2y - 2z = 9$.
2. Find the equations of the planes bisecting the angles between the planes $x + 2y + 2z - 9 = 0$ and $4x - 3y + 12z + 13 = 0$.

9. Systems of Planes

We have expressed the equations of the planes in various forms. Among these we consider the equation

$$\frac{x}{a}+\frac{y}{b}+\frac{z}{c} = 1.$$

This equation has three constants a, b, c which have geometrical significance. When definite values are assigned to these letters, a plane is completely determined. Other values of these, of course, determine other planes. Thus, the quantities a, b and c are fixed for any particular plane but change from plane to plane. These letters are called **parameters**.

If we fix the value of one of the parameters (say) a and allow the others two, b and c, to vary, we shall obtain numerous planes each of which would make an intercept a on the x-axis. This, in fact, may be regarded as a family (or system) of planes with a particular property that each member of the family is making an intercept a on the x-axis. Further, if we fix the values of two parameters (say) a and b, and allow the third one, c, to vary, we shall again obtain numerous planes each of which would make intercept a on x-axis and b on y-axis. This represents another family of planes. Of course, the later family is included in the former one.

The collection of all planes defined by a linear equation with one or two parameters is called a **family** *or* **system** *of planes.*

Note. The collection of all the planes defined by a linear equation with all the three parameters may be considered a system of planes. But in that situation the system would contain the totality of all the planes in the space.

A system is said to be a **one-parameter** *or* **two-parameters system** *according as its equation has, respectively, one or two parameters.*

9.1 System of planes parallel to a given plane.

Let the equation of the given plane be

$$Ax + By + Cz + D = 0. \tag{1}$$

Then, the system of planes parallel to the plane (1) is $Ax + By + Cz + \lambda = 0$, where λ is a parameter.

Note. This is one-parameter system of planes.

9.2 System of planes through the intersection of two planes. *To find the system of planes passing through the line of intersection of two given planes.*

Let the two distinct intersecting planes be

$$A_1x + B_1y + C_1z + D_1 = 0, \tag{1}$$

and $$A_2x + B_2y + C_2z + D_2 = 0. \tag{2}$$

Then, the equation

$$A_1x + B_1y + C_1z + D_1 + \lambda(A_2x + B_2y + C_2z + D_2) = 0, \tag{3}$$

where λ is a parameter, represents a system of planes, through the line of intersection of the planes (1) and (2). To verify this statement we observe that

(i) Equation (3) being linear in x, y and z for any value of the parameter λ, represents a plane;

(ii) If (x_1, y_1, z_1) is the point lying on (1) and (2), then

$$A_1x_1 + B_1y_1 + C_1z_1 + D_1 = 0 \text{ and } A_2x_1 + B_2y_1 + C_2z_1 + D_2 = 0,$$

and thus, equation (3) is also satisfied by the coordinates (x_1, y_1, z_1). Therefore, the point (x_1, y_1, z_1) lies on the plane (3). Hence, the equation (3) represents a system of planes through the line of intersection of the given planes.

Note. This is also one-parameter system.

9.3 System of planes through a given point. *To find the equation of the system of planes through the point* (x_1, y_1, z_1).

Any plane through the point (x_1, y_1, z_1) is

$$A(x - x_1) + B(y - y_1) + C(z - z_1) = 0.$$

If $A \neq 0$, dividing by A throughout, we get

$$(x - x_1) + \frac{B}{A}(y - y_1) + \frac{C}{A}(z - z_1) = 0.$$

Putting $\lambda = \frac{B}{A}$, $\mu = \frac{C}{A}$, we obtain

$$(x - x_1) + \lambda(y - y_1) + \mu(z - z_1) = 0.$$

This is the equation of the system of planes each member of which passes through (x_1, y_1, z_1), where λ and μ are parameters.

Note. This is a two-parameters system of planes.

9.4 Examples

Example 1. Find the equation of the plane passing through the line of intersection of the planes $x + y + z = 6$ and $2x + 3y + 4z + 5 = 0$ and perpendicular to the plane $4x + 5y - 3z = 8$.

Solution. The equation of the plane passing through the line of intersection of the planes

$$x + y + z = 6 \text{ and } 2x + 3y + 4z + 5 = 0$$

is $$x + y + z - 6 + \lambda(2x + 3y + 4z + 5) = 0$$

i.e. $$(1 + 2\lambda)x + (1 + 3\lambda)y + (1 + 4\lambda)z - 6 + 5\lambda = 0. \quad (1)$$

This plane will be perpendicular to the plane $4x + 5y - 3z = 8$

if $$4(1 + 2\lambda) + 5(1 + 3\lambda) - 3(1 + 4\lambda) = 0,$$

which gives $\lambda = -\dfrac{6}{11}$. Substituting this value of λ in (1), we get

$$x + 7y + 13z + 96 = 0.$$

This is the required equation of the plane.

Example 2. The plane $lx + my = 0$ is rotated about its line of intersection with the plane $z = 0$ through an angle α. Prove that the equation of the plane in its new position is $lx + my \pm z\sqrt{l^2 + m^2}\ \tan\alpha = 0$.

Solution. Any plane through the line of intersection of the planes $lx + my = 0$ and $z = 0$ is

$$lx + my + \lambda z = 0. \tag{1}$$

If the plane $lx + my = 0$ after rotation through an angle α takes the position of (1), then

$$\cos\alpha = \frac{l.l + m.m + 0.\lambda}{\sqrt{l^2 + m^2}\cdot\sqrt{l^2 + m^2 + \lambda^2}}, \text{ i.e., } \lambda = \pm\sqrt{l^2 + m^2}\ \tan\alpha.$$

Substituting this value of λ in (1), we get

$$lx + my \pm z\sqrt{l^2 + m^2}\ \tan\alpha = 0.$$

Exercises

1. Prove that the equation of the plane through the point (α, β, γ) and parallel to the plane $ax + by + cz = 0$ is $ax + by + cz = a\alpha + b\beta + c\gamma$.
2. Find the equation of the plane containing the line of intersection of the planes $x + y + z = 6$, $2x + 3y + 4z + 5 = 0$ and the point $(1, 1, 1)$.
3. Find the equation of the plane which passes through the intersection of the planes $2x + 3y + 10z = 8$, $2x - 3y + 7z = 2$ and is perpendicular to the plane $3x - 2y + 4z = 5$.
4. Find the equation of the plane passing through the line of intersection of the planes $ax + by + cz + d = 0$ and $a'x + b'y + c'z + d' = 0$ and perpendicular to xy-plane.
5. Find the equation of the plane which is perpendicular to the plane $5x + 3y + 6z + 8 = 0$ and contains the line of intersection of the planes $x + 2y + 3z - 4 = 0$, $2x + y - z + 5 = 0$.
6. Find the equation of the plane passing through the line of intersection of the planes $2x - y = 0$ and $3z - y = 0$ and perpendicular to the plane $4x + 3y - 3z = 8$.

10. Pair of Planes

Consider the equation

$$(lx + my + nz + p)(l'x + m'y + n'z + p') = 0. \quad (1)$$

This can be satisfied by the points whose coordinates satisfy the equation

$$lx + my + nz + p = 0 \text{ or } l'x + m'y + n'z + p' = 0. \quad (2)$$

Thus, all the points which satisfy (1) lie on one or the other of the equations in (2), and conversely, any point whose coordinates satisfy any of the two equations in (2) satisfies the equation (1). The equations in (2) being linear, the equation (1), therefore, represents a pair of planes.

If we multiply the two factors in (1) together, then it takes the form

$$ll'x^2 + mm'y^2 + nn'z^2 + (mn' + m'n)yz$$
$$+ (nl' + n'l)zx + (lm' + l'm)xy + (lp' + l'p)x$$
$$+ (mp' + m'p)y + (np' + n'p)z + pp' = 0.$$

This is a second degree equation in x, y and z. Thus, we note that a pair of planes is given by a second degree equation in x, y and z.

Conversely, one is interested to know whether the most general second degree equation in x, y and z, i.e.

$$ax^2 + by^2 + cz^2 + 2fyz + 2gzx + 2hxy + 2ux + 2vy + 2wz + d = 0 \quad (3)$$

represents a pair of planes. However, it would only be possible when the left hand expression in equation (3) can be decomposed into two real linear factors for which some relation among a, b, c, d, f, g, h, u, v, w is required to be satisfied. Before, we deal with this problem in a later section, we come to a particular case when both the plane pass through the origin. In this case, the equation (1) takes the form $(lx + my + nz)(l'x + m'y + n'z) = 0$, *i.e.*

$$ll'x^2 + mm'y^2 + nn'z^2 + (mn' + m'n)yz + (nl' + n'l)zx + (lm' + l'm)xy = 0.$$

This is a homogeneous* equation of degree two in x, y and z. The general equation of this form (homogeneous form) is

$$ax^2 + by^2 + cz^2 + 2fyz + 2gzx + 2hxy = 0. \quad (4)$$

Note. In case of two dimensions, though the general equation of second degree in x and y, i.e., $ax^2 + 2hxy + by^2 + 2gx + 2fy + c = 0$

does not always represent a pair of lines, yet its homogeneous part i.e. $ax^2 + 2hxy + by^2 = 0$ always represents a pair of lines passing through the origin

* An equation in x, y and z in which the sum of the powers of x, y and z in every term is the same (say, n) is called a homogeneous equation (of degree n) in x, y and z.

and the two lines are real and distinct, coincident or imaginary according as $h^2 - ab >, =$ or < 0. In analogy with this one may think in case of three dimensions that the homogeneous equation of second degree in x, y and z i.e. equation (4) should represent a pair of planes passing through the origin. But, this is not the case always, see for instance § 1.2, Chapter 16.

10.1 *To find the necessary condition that the homogeneous equation of second degree*

$$ax^2 + by^2 + cz^2 + 2fyz + 2gzx + 2hxy = 0 \tag{1}$$

may represent a pair of planes.

First Method. Multiplying the given equation by $a(a \neq 0)$ and rearranging the terms, we get

$$a^2x^2 + 2a(hy + gz)x = -aby^2 - acz^2 - 2afyz$$

i.e.
$$a^2x^2 + 2a(hy + gz)x + (hy + gz)^2 = h^2y^2 + 2ghyz + g^2z^2 - aby^2 - acz^2 - 2afyz$$

$$\Rightarrow \quad (ax + hy + gz)^2 = (h^2 - ab)y^2 + 2(gh - af)yz + (g^2 - ac)z^2$$

$$\Rightarrow \quad ax + hy + gz = \pm\sqrt{(h^2 - ab)y^2 + 2(gh - af)yz + (g^2 - ac)z^2}. \tag{2}$$

The equation (1) will represent a pair of planes if the quantity under the radical sign in (2) becomes a perfect square for which

$$4(gh - af)^2 = 4(h^2 - ab)(g^2 - ac)$$

$$\Rightarrow \quad abc + 2fgh - af^2 - bg^2 - ch^2 = 0.$$

If $a = 0$ and $b \neq 0$, multiply equation (1) by b and arrange the terms in descending powers of y. Proceeding similarly as in the previous case, we get the required condition. Further, if $a = 0$, $b = 0$ and $c \neq 0$. Proceeding as above, we get the required condition.

Second Method. Suppose equation (1) represents a pair of planes. Let the planes be

$$lx + my + nz = 0 \text{ and } l'x + m'y + n'z = 0.$$

Then

$$ax^2 + by^2 + cz^2 + 2fyz + 2gzx + 2hxy \equiv (lx + my + nz)(l'x + m'y + n'z).$$

Comparing the coefficients of various terms on both the sides, we get

$$ll' = a,\ mm' = b,\ nn' = c,\ mn' + m'n = 2f,\ nl' + n'l = 2g,\ lm' + l'm = 2h.$$

Now

$$0 = \begin{vmatrix} l & l' & 0 \\ m & m' & 0 \\ n & n' & 0 \end{vmatrix} \times \begin{vmatrix} l' & l & 0 \\ m' & m & 0 \\ n' & n & 0 \end{vmatrix}$$

$$\Rightarrow \quad \begin{vmatrix} ll' + l'l & lm' + l'm & ln' + l'n \\ lm' + l'm & mm' + m'm & mn' + m'n \\ nl' + n'l & mn' + m'n & nn' + n'n \end{vmatrix} = 0$$

$$\Rightarrow \quad \begin{vmatrix} 2a & 2h & 2g \\ 2h & 2b & 2f \\ 2g & 2f & 2c \end{vmatrix} = 0$$

$$\Rightarrow \quad \begin{vmatrix} a & h & g \\ h & b & f \\ g & f & c \end{vmatrix} = 0, \text{i.e., } abc + 2fgh - af^2 - bg^2 - ch^2 = 0.$$

Note. The above condition is also sufficient.

10.2 *Let the second degree homogeneous equation*

$$ax^2 + by^2 + cz^2 + 2fyz + 2gzx + 2hxy = 0$$

represent a pair of planes. Find the angle between them.

Let $lx + my + nz = 0$ and $l'x + m'y + n'z = 0$ be the two planes represented by the given homogeneous equation. Then, as above, we have

$ll' = a$, $mm' = b$, $nn' = c$, $mn' + m'n = 2f$, $nl' + n'l = 2g$, $lm' + l'm = 2h$.

Let θ be the angle between the planes. Then

$$\tan\theta = \frac{\sqrt{(mn' - m'n)^2 + (nl' - n'l)^2 + (lm' - l'm)^2}}{ll' + mm' + nn'}$$

$$= \frac{\sqrt{(mn' + m'n)^2 - 4mm'nn' + (nl' + n'l)^2 - 4nn'll' + (lm' + l'm)^2 - 4ll'mm'}}{ll' + mm' + nn'}$$

$$= \frac{\sqrt{4f^2 - 4bc + 4g^2 - 4ca + 4h^2 - 4ab}}{a + b + c}$$

$$= \frac{2\sqrt{f^2 + g^2 + h^2 - bc - ca - ab}}{a + b + c}.$$

Corollary. The planes represented by $ax^2 + by^2 + cz^2 + 2fyz + 2gzx + 2hxy = 0$ will be at right angles if $a + b + c = 0$.

10.3 *To find the necessary condition that the general equation of the second degree*

$$ax^2 + by^2 + cz^2 + 2fyz + 2gzx + 2hxy + 2ux + 2vy + 2wz + d = 0 \qquad (1)$$

may represent a pair of planes.

Suppose the equation (1) represents a pair of planes. Let the planes be

$$lx + my + nz + p = 0 \text{ and } l'x + m'y + n'z + p' = 0.$$

Then

$$ax^2 + by^2 + cz^2 + 2fyz + 2gzx + 2hxy + 2ux + 2vy + 2wz + d$$
$$\equiv (lx + my + nz + p)(l'x + m'y + n'z + p').$$

Comparing the coefficients of various terms on both the sides, we get

$$ll' = a,\ mm' = b,\ nn' = c,\ mn' + m'n = 2f,$$
$$nl' + n'l = 2g,\ lm' + l'm = 2h,\ lp' + l'p = 2u,$$
$$mp' + m'p = 2v,\ np' + n'p = 2w,\ pp' = d.$$

Now

$$\begin{vmatrix} l & l' & 0 \\ m & m' & 0 \\ n & n' & 0 \end{vmatrix} \times \begin{vmatrix} l' & l & 0 \\ m' & m & 0 \\ n' & n & 0 \end{vmatrix} = 0$$

$$\Rightarrow \quad \begin{vmatrix} 2ll' & lm' + l'm & nl' + n'l \\ lm' + l'm & 2mm' & mn' + m'n \\ nl' + n'l & mn' + m'n & 2nn' \end{vmatrix} = 0$$

$$\Rightarrow \quad \begin{vmatrix} 2a & 2h & 2g \\ 2h & 2b & 2f \\ 2g & 2f & 2c \end{vmatrix} = 0 \Rightarrow \begin{vmatrix} a & h & g \\ h & b & f \\ g & f & c \end{vmatrix} = 0.$$

Similarly,

$$\begin{vmatrix} l & l' & 0 \\ m & m' & 0 \\ p & p' & 0 \end{vmatrix} \times \begin{vmatrix} l' & l & 0 \\ m' & m & 0 \\ p' & p & 0 \end{vmatrix} = 0$$

and $$\begin{vmatrix} m & m' & 0 \\ n & n' & 0 \\ p & p' & 0 \end{vmatrix} \times \begin{vmatrix} m' & m & 0 \\ n' & n & 0 \\ p' & p & 0 \end{vmatrix} = 0$$

give $$\begin{vmatrix} a & h & u \\ h & b & v \\ u & v & d \end{vmatrix} = 0 \text{ and } \begin{vmatrix} b & f & v \\ f & c & w \\ v & w & d \end{vmatrix} = 0$$

Thus $$\begin{vmatrix} a & h & g \\ h & b & f \\ g & f & c \end{vmatrix} = 0, \begin{vmatrix} a & h & u \\ h & b & v \\ u & v & d \end{vmatrix} = 0 \text{ and } \begin{vmatrix} b & f & v \\ f & c & w \\ v & w & d \end{vmatrix} = 0.$$

are the necessary conditions for the equation (1) to represent a pair of planes.

Note. The above conditions are also sufficient.

10.4 Examples

Example. Prove that the equation $2x^2 - 2y^2 + 4z^2 + 2yz + 6zx + 3xy = 0$ represents a pair of planes and the angle between them is $\cos^{-1}\dfrac{4}{9}$.

Solution. Here $a = 2, b = -2, c = 4, f = 1, g = 3, h = \dfrac{3}{2}$.

Therefore $\quad abc + 2fgh - af^2 - bg^2 - ch^2 = 0$

Hence, the given equation represents a pair of planes.

Further, let θ be the angle between the planes. Then

$$\tan\theta = \frac{2\sqrt{f^2 + g^2 + h^2 - bc - ca - ab}}{a + b + c} = \frac{\sqrt{65}}{4}.$$

Therefore $\quad \cos\theta = \dfrac{4}{9} \Rightarrow \theta = \cos^{-1}\dfrac{4}{9}$.

Exercises

1. Prove that the equation $2x^2 - 6y^2 - 12z^2 + 18yz + 2zx + xy = 0$ represents a pair of planes and the angle between them is $\cos^{-1}\dfrac{16}{21}$.
2. Prove that the equation $2x^2 + 6y^2 - 12z^2 + yz + 2zx + 7xy = 0$ represents a pair of planes passing through the origin and find the angle between them.

3. Prove that the equation $\dfrac{3}{y-z}+\dfrac{4}{z-x}+\dfrac{5}{x-y}=0$ represents a pair of planes.

4. Let the equation $\phi(x, y, z) = 0$, where

$$\phi(x, y, z) \equiv ax^2 + by^2 + cz^2 + 2fyz + 2gzx + 2hxy$$

represent a pair of planes. Prove that the product of the distances of the planes from (α, β, γ) is $\dfrac{\phi(\alpha,\beta,\gamma)}{\sqrt{\Sigma a^2 + 4\Sigma f^2 - 2\Sigma bc}}$.

11. Area of a Triangle and Volume of a Tetrahedron

11.1 The area of a triangle

Let the area of the triangle ABC be represented by Δ. Let l, m, n be the direction cosines of the normal to the plane of the triangle.

The orthogonal projections of the vertices $A(x_1, y_1, z_1)$, $B(x_2, y_2, z_2)$, $C(x_3, y_3, z_3)$ on the yz-plane are the points $A'(0, y_1, z_1)$, $B'(0, y_2, z_2)$, $C'(0, y_3, z_3)$. Thus, the triangle $A'B'C'$ is the orthogonal projection of the triangle ABC on the yz-plane. If the area of the triangle $A'B'C'$ is denoted by Δx, we find that

$$\Delta_x = l\Delta.$$

Similarly, if Δ_y and Δ_z denote the areas of the orthogonal projections of the triangle ABC on the zx- and xy-planes, respectively, we find that

$$\Delta_y = m\Delta \text{ and } \Delta_z = n\Delta.$$

On squaring and adding these, we get

$$\Delta^2 = \Delta_x^2 + \Delta_y^2 + \Delta_z^2. \quad (1)$$

Also, from analytical geometry of two dimensions, we know that

$$\Delta_x = \frac{1}{2}\begin{vmatrix} y_1 & z_1 & 1 \\ y_2 & z_2 & 1 \\ y_3 & z_3 & 1 \end{vmatrix}, \Delta_y = \frac{1}{2}\begin{vmatrix} z_1 & x_1 & 1 \\ z_2 & x_2 & 1 \\ z_3 & x_3 & 1 \end{vmatrix}, \quad \Delta_z = \frac{1}{2}\begin{vmatrix} x_1 & y_1 & 1 \\ x_2 & y_2 & 1 \\ x_3 & y_3 & 1 \end{vmatrix}.$$

Substituting the values of Δ_x, Δ_y and Δ_z in (1), we obtain an expression for the area of the triangle ABC.

Ex. 1. Find the area of the triangle whose vertices are the points (1, 2, 3), (–2, 1, –4) and (3, 4, –2).

Ex. 2. Find the area of the triangle which is included between the plane $2x - 3y + 4z = 12$ and the coordinate planes.

11.2 The volume of a tetrahedron

Let V denote the volume of the tetrahedron whose vertices are the points $A(x_1, y_1, z_1)$, $B(x_2, y_2, z_2)$, $C(x_3, y_3, z_3)$, $D(x_4, y_4, z_4)$. Then

$$V = \frac{1}{3} p \times \Delta,$$

where Δ is the area of the triangle BCD and p the perpendicular distance of A from the plane BCD.

Further, the equation of the plane BCD is

$$\begin{vmatrix} x & y & z & 1 \\ x_2 & y_2 & z_2 & 1 \\ x_3 & y_3 & z_3 & 1 \\ x_4 & y_4 & z_4 & 1 \end{vmatrix} = 0. \qquad (1)$$

Let Δ_x, Δ_y, Δ_z be the orthogonal projections of Δ on the coordinate planes. Then, we have

$$2\Delta_x = \begin{vmatrix} y_2 & z_2 & 1 \\ y_3 & z_3 & 1 \\ y_4 & z_4 & 1 \end{vmatrix} = \text{coefficient of } x \text{ in (1)},$$

$$2\Delta_y = \begin{vmatrix} z_2 & x_2 & 1 \\ z_3 & x_3 & 1 \\ z_4 & x_4 & 1 \end{vmatrix} = \text{coefficient of } y \text{ in (1)},$$

$$2\Delta_z = \begin{vmatrix} x_2 & y_2 & 1 \\ x_3 & y_3 & 1 \\ x_4 & y_4 & 1 \end{vmatrix} = \text{coefficient of } z \text{ in (1)}.$$

Also, we note that

$$p = \frac{2}{2\sqrt{\Delta_{x^2} + \Delta_{y^2} + \Delta_{z^2}}} \begin{vmatrix} x_1 & y_1 & z_1 & 1 \\ x_2 & y_2 & z_2 & 1 \\ x_3 & y_3 & z_3 & 1 \\ x_4 & y_4 & z_4 & 1 \end{vmatrix}$$

$$= \frac{1}{2\Delta} \begin{vmatrix} x_1 & y_1 & z_1 & 1 \\ x_2 & y_2 & z_2 & 1 \\ x_3 & y_3 & z_3 & 1 \\ x_4 & y_4 & z_4 & 1 \end{vmatrix}$$

Hence, $$V = \frac{1}{3} p\Delta = \frac{1}{6} \begin{vmatrix} x_1 & y_1 & z_1 & 1 \\ x_2 & y_2 & z_2 & 1 \\ x_3 & y_3 & z_3 & 1 \\ x_4 & y_4 & z_4 & 1 \end{vmatrix}.$$

Ex. 1. *A*, *B*, *C* **are the points (3, 2, 1), (–2, 0, –3), (0, 0, –2). Find the locus of** *P* **if the volume of the tetrahedron** *PABC* **is 5.**

Ex. 2. Prove that the volume of the tetrahedron formed by the planes $y + z = 0$, $z + x = 0$, $x + y = 0$, $x + y + z = 1$ is $\frac{2}{3}$.

Ex. 3. Prove that the volume of the tetrahedron formed by the planes. $my + nz = 0$, $nz + lx = 0$, $lx + my = 0$, and $lx + my + nz = p$ is $\frac{2}{3} \cdot \frac{p^3}{lmn}$.

Miscellaneous Exercises

1. Prove that the point $(-\frac{1}{2}, 2, 0)$ is the circumcentre of the triangle formed by the points (1, 1, 0), (1, 2, 1), (–2, 2, –1).
2. Two systems of rectangular axes have the same origin. If a plane cuts them at distances *a*, *b*, *c* and *a′*, *b′*, *c′*, respectively, from the origin, prove that
$$\frac{1}{a^2} + \frac{1}{b^2} + \frac{1}{c^2} = \frac{1}{a'^2} + \frac{1}{b'^2} + \frac{1}{c'^2}.$$
3. Prove that the planes $7x + 4y - 4z + 30 = 0$, $36x - 51y + 12z + 17 = 0$, $14x + 8y - 8z - 12 = 0$, and $12x - 17y + 4z - 3 = 0$ form the four faces of a reactangular parallelopiped.
4. Prove that the equation of the two planes inclined at an angle α to the *xy*-plane and containing the line of intersection of the planes $y = 0$ and $z \cos\beta = x \sin\beta$ is $z^2 + (x^2 + y^2)\tan^2\beta - 2zx \tan\beta = y^2 \tan^2\alpha$.
5. Prove that $(\frac{1}{4}, \frac{1}{8}, \frac{1}{8})$ is the centre of the tetrahedron formed by the planes $x = 0$, $y = 0$, $z = 0$ and $x + 2y + 2z = 1$.

6. A triangle, the lengths of whose sides are a, b, c is placed so that the middle points of the sides are on the axes. Prove that the equation to its plane is $\frac{x}{\alpha}+\frac{y}{\beta}+\frac{z}{\gamma}=1$, where $8\alpha^2 = b^2 + c^2 - a^2$, $8\beta^2 = c^2 + a^2 - b^2$, and $8\gamma^2 = a^2 + b^2 - c^2$.

7. OA, OB, OC are three mutually perpendicular lines through the origin and their direction cosines are $l_1, m_1, n_1; l_2, m_2, n_2; l_3, m_3, n_3$. If $OA = OB = OC = a$, prove that the equation of the plane ABC is

$$(l_1 + l_2 + l_3)x + (m_1 + m_2 + m_3)y + (n_1 + n_2 + n_3)z = a.$$

8. Prove that the equation $\frac{a}{y-z}+\frac{b}{z-x}+\frac{c}{x-y}=0$ represents a pair of planes.

9. A point P moves on the fixed plane $\frac{x}{a}+\frac{y}{b}+\frac{z}{c}=1$, and a plane through P, perpendicular to OP meets the axes in A, B, C. If the planes through A, B, C parallel to the coordinate planes meet at Q, prove that the locus of Q is

$$\frac{1}{x^2}+\frac{1}{y^2}+\frac{1}{z^2}=\frac{1}{ax}+\frac{1}{by}+\frac{1}{cz}.$$

10. Through a point $P(\alpha, \beta, \gamma)$, a plane is drawn at right angles to OP to meet the axes in A, B, C. Prove that the area of the triangle ABC is $\frac{r^5}{2\alpha\beta\gamma}$, where $OP = r$.

11. A plane makes the intercepts $OA = a$, $OB = b$, $OC = c$ on the axes. Prove that the area of the triangle ABC is $\frac{1}{2}\sqrt{b^2c^2 + c^2a^2 + a^2b^2}$.

12. A variable plane makes with the coordinate planes a tetrahedron of constant volume $64k^3$, prove that the locus of the centroid of the tetrahedron is $xyz = 6k^3$.

13. A variable plane makes with the coordinate planes a tetrahedron of constant volume k^3. Prove that the locus of the foot of the perpendicular from the origin to the plane is $(x^2 + y^2 + z^2)^3 = 6k^3xyz$.

14. Find the volume of a tetrahedron in terms of the edges which meet in a point and of the angles which they make with each other.

The Straight Line

14
CHAPTER

1. Introduction

The intersection of two surfaces S_1 and S_2 is the locus of all the common points on both the surfaces S_1 and S_2. This locus, in general, would be a surface which in particular case could be a plane, a plane curve, a line, a single point or even no point. In most of the cases of common use in analytical geometry the intersection of two surfaces S_1 and S_2 is a space curve. The curve of intersection of a plane and a surface is called a **plane section** of the surface, and it is an ordinary plane curve lying in the plane of intersection. Furthermore, if both the surfaces are taken as, planes, their intersection, in general, would be a line. Conversely, any given line can be uniquely determined by any of the two planes containing the line. Thus, *a line can be regarded as the locus of the common points of two intersecting planes.*

Consider now the equations of the two planes

$$ax + by + cz + d = 0 \text{ and } a'x + b'y + c'z + d' = 0\}.$$

These two equations taken together would represent the equation of a line (the locus of the intersection of two planes). The form of the equation given above is called a **non-symmetrical form** of the equation of a line.

Note. There may be a case when the two planes are non-intersecting, in that case, the line represented by the above equations would be an imaginary one called the **virtual line** of intersection of the two planes.

2. Various forms of the Equations of a Line

2.1 One-point form. *To find the equations of the line passing through a point $A(\alpha, \beta, \gamma)$ and having direction cosines l, m, n.*

Let $P(x, y, z)$ be any point on the line such that $AP = r$. Then

$$\frac{x-\alpha}{r} = l, \frac{y-\beta}{r} = m, \frac{z-\gamma}{r} = n.$$

These equations may also be written as

$$\frac{x-\alpha}{l}=\frac{y-\beta}{m}=\frac{z-\gamma}{n}\ (=r). \tag{1}$$

This is also called the **symmetrical form** of the equations of a line.

Conversely, it can be easily verified that any equations of the form (1) represent a line through (α, β, γ) having direction cosines l, m, n.

Note 1. If l, m, n are taken as direction ratios instead of the direction cosines, the form of the equations of a line remain the same. However, in case of direction ratios r would not represent the distance of the point on the line through (α, β, γ), rather, r would simply be acting as a parameter.

Note 2. The form of the equations (1) of a line suggests us to express the coordinates of any point on the line in terms of a single parameter as

$$x=\alpha+lr,\ y=\beta+mr,\ z=\gamma+nr.$$

Ex. Find the equations of a line passing through (1, 2, 3) and having direction cosines 2, 4, 6.

2.2 Two-point form. *To find the equations of a line passing through two points* $P(x_1, y_1, z_1)$ *and* $Q(x_2, y_2, z_2)$.

The direction cosines of the line joining the points $P(x_1, y_1, z_1)$ and $Q(x_2, y_2, z_2)$ are proportional to $x_2-x_1, y_2-y_1, z_2-z_1$. Hence, the equations of the line PQ are given by

$$\frac{x-x_1}{x_2-x_1}=\frac{y-y_1}{y_2-y_1}=\frac{z-z_1}{z_2-z_1}\left(\text{or}\ \frac{x-x_2}{x_2-x_1}=\frac{y-y_2}{y_2-y_1}=\frac{z-z_2}{z_2-z_1}\right).$$

Ex. Find the equations of the line joining the points (1, 2, 3) and (2, 4, 6).

2.3 *To transform the equations*

$$\left.\begin{aligned}ax+by+cz+d&=0\\a'x+b'y+c'z+d'&=0\end{aligned}\right\} \tag{1}$$

of a line into symmetrical form.

The line parallel to (1) passing through origin is given by

$$\left.\begin{aligned}ax+by+cz&=0\\a'x+b'y+c'z&=0\end{aligned}\right\}. \tag{2}$$

The equations (2) may be written as

$$\frac{x}{bc'-b'c}=\frac{y}{ca'-c'a}=\frac{z}{ab'-a'b}.$$

Thus, the direction cosines of the line (2) and hence of (1) are proportional to

$$bc' - b'c,\ ca' - c'a,\ ab' - a'b.$$

Now, we need one point on the line (1). In fact, there are infinite number of points on the line. For the sake of simplicity we choose the point of intersection of the line (1) with the plane $z = 0$ whose coordinates are given by the equations

$$\left.\begin{aligned} ax + by + d &= 0 \\ a'x + b'y + d' &= 0 \\ z &= 0 \end{aligned}\right\}.$$

On solving the equations, we get the point

$$\left(\frac{bd' - b'd}{ab' - b'c}, \frac{da' - d'a}{ab' - a'b}, 0\right).$$

Thus, the equations of the line in symmetrical form are

$$\frac{x - \dfrac{bd' - b'd}{ab' - a'b}}{bc' - b'c} = \frac{y - \dfrac{da' - d'a}{ab' - a'b}}{ca' - c'a} = \frac{z}{ab' - a'b}.$$

Note. It is not necessary to take $z = 0$ always for finding a point on the line. This depends upon the convenience. In some cases, there may be need to take either $x = 0$ or $y = 0$. It is also possible to take some suitable value of x, y or z different from zero.

2.4 Examples

Example 1. Transform the equations

$$\left.\begin{aligned} 4x + 4y - 5z - 12 &= 0 \\ 8x + 12y - 13z - 32 &= 0 \end{aligned}\right\} \qquad (1)$$

of a line into symmetrical form.

Solution. The line parallel to (1) passing through the origin is given by

$$\left.\begin{aligned} 4x + 4y - 5z &= 0 \\ 8x + 12y - 13z &= 0 \end{aligned}\right\}. \qquad (2)$$

These equations may be written as

$$\frac{x}{-52 + 60} = \frac{y}{-40 + 52} = \frac{z}{48 - 32}$$

i.e. $$\frac{x}{2}=\frac{y}{3}=\frac{z}{4}.$$

Thus, the direction cosines of the line (2) and hence of (1) are proportional to 2, 3, 4.

Now, we find the point of intersection of the line (1) with the plane $z = 0$ whose coordinates are given by

$$\left.\begin{aligned}4x+4y-12=0\\8x+12y-32=0\\z=0\end{aligned}\right\}.$$

On solving these equations, we get $x = 1, y = 2, z = 0$.

Hence, the equations of the line in symmetrical form are

$$\frac{x-1}{2}=\frac{y-2}{3}=\frac{z}{4}.$$

Example 2. Find where the line

$$\frac{x-1}{2}=\frac{y-2}{-3}=\frac{z+3}{4}$$

meets the plane $2x + 4y - z + 1 = 0$.

Solution. Let $\frac{x-1}{2}=\frac{y-2}{-3}=\frac{z+3}{4}=r.$

Then, any point on the line is $(1 + 2r, 2 - 3r, -3 + 4r)$. If this point lies on the plane $2x+4y-z+1=0$, then $r=\frac{7}{6}$. Thus, the point of intersection of the line with the plane is $\left(\frac{10}{3},-\frac{3}{2},\frac{5}{3}\right)$.

Example 3. Find the distance of the point $(1, -2, 3)$ from the plane $x-y+z=5$ measured parallel to the line $\frac{x}{2}=\frac{y}{3}=\frac{z}{-6}$.

Solution. The equations of the line through the point $(1, -2, 3)$ and parallel to the given line are

$$\frac{x-1}{2}=\frac{y+2}{3}=\frac{z-3}{-6}=r \text{ (say)}.$$

Any point on this line is $(1 + 2r, -2 + 3r, 3 - 6r)$. This point will lie on the given plane if $r = \frac{1}{7}$. Thus, the point of intersection of the line and the plane is $\left(\frac{9}{7}, -\frac{11}{7}, \frac{15}{7}\right)$.

Hence, the required distance

$$= \sqrt{\left(\frac{9}{7}-1\right)^2 + \left(-\frac{11}{7}+2\right)^2 + \left(\frac{15}{7}-3\right)^2} = 1.$$

Example 4. Prove that the equations of the line through the point (α, β, γ) and at right angles to the lines

$$\frac{x}{l_1} = \frac{y}{m_1} = \frac{z}{n_1}; \frac{x}{l_2} = \frac{y}{m_2} = \frac{z}{n_2}$$

are
$$\frac{x-\alpha}{m_1n_2 - m_2n_1} = \frac{y-\beta}{n_1l_2 - n_2l_1} = \frac{z-\gamma}{l_1m_2 - l_2m_1}.$$

Solution. The equations of the line through (α, β, γ) having direction cosines l, m, n are

$$\frac{x-\alpha}{l} = \frac{y-\beta}{m} = \frac{z-\gamma}{n}. \tag{1}$$

This line will be at right angles to the given lines if

$$ll_1 + mm_1 + nn_1 = 0, \tag{2}$$
$$ll_2 + mm_2 + nn_2 = 0. \tag{3}$$

On solving the equations (2) and (3), we get

$$\frac{l}{m_1n_2 - m_2n_1} = \frac{m}{n_1l_2 - n_2l_1} = \frac{n}{l_1m_2 - l_2m_1}. \tag{4}$$

Eliminating l, m, n from (1) and (4), we get

$$\frac{x-\alpha}{m_1n_2 - m_2n_1} = \frac{y-\beta}{n_1l_2 - n_2l_1} = \frac{z-\gamma}{l_1m_2 - l_2m_1}.$$

Exercises

1. Find the distance of the point $(-1, -5, -10)$ from the point of intersection of the line $\frac{x-2}{3} = \frac{y+1}{4} = \frac{z-2}{12}$ and the plane $x - y + z = 5$.

2. Find the coordinates of the point in which the line $\frac{x}{1}=\frac{y-1}{2}=\frac{z+2}{3}$ meets the plane $2x+3y+z=0$.

3. Find the coordinates of the point in which the line $\frac{x-\alpha}{l}=\frac{y-\beta}{m}=\frac{z-\gamma}{n}$ meets the plane $ax+by+cz+d=0$.

4. Prove that the equations of the perpendicular from the point (α, β, γ) to the plane $ax+by+cz+d=0$ are $\frac{x-\alpha}{a}=\frac{y-\beta}{b}=\frac{z-\gamma}{c}$.

5. Transform the equations $x+y+z+1=0$, $4x+y-2z+2=0$ of the line into symmetrical form.

6. Transform the equations of the line determined by $x+2y+z=3$, $6x+8y+3z=13$ into symmetrical form.

7. Prove that the lines $3x+2y+z-5=0$, $x+y-2z-3=0$ and $8x-4y-4z=0$, $7x+10y-8z=0$ are at right angles.

8. Find the point in which the perpendicular from the origin on the line joining the points $(-9, 4, 5)$ and $(11, 0, -1)$ meets it.

9. Find the equations of the line through the point $(1, 2, 3)$ and parallel to the line $x-y+2z=5$, $3x+y+z=6$.

10. Find the perpendicular distance from the point $(-1, 3, 9)$ to the line $\frac{x-13}{5}=\frac{y+8}{-8}=\frac{z-31}{1}$.

11. Prove that the lines $2x+3y-4z=0$, $3x-4y+z=0$ and $5x-y-3z+12=0$, $x-7y+5z-6=0$ are parallel.

3. Plane and a Line

3.1 Angle between a plane and a line: *Suppose a line L meets the plane P in the point A. Then the acute angle between the line L and its projection on the plane P is called the angle between the line L and plane P.*

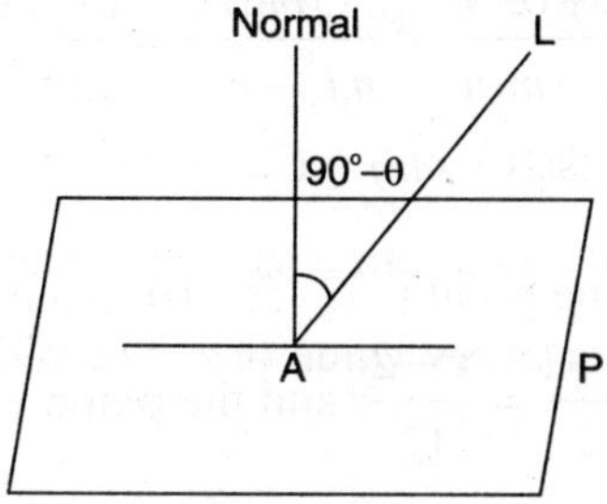

Fig. 14.1

To find the angle between a given plane and a given line.

Let the equations of the given plane and the line be

$$Ax + By + Cz + D = 0 \quad (1)$$

and $$\frac{x-\alpha}{l} = \frac{y-\beta}{m} = \frac{z-\gamma}{n}. \quad (2)$$

Let θ be the angle between the plane (1) and the line (2). Then, the angle between the normal to the plane (1) and the line (2) is 90° – θ. Also ,the direction cosines of the normal to the plane are proportional to *A*, *B*, *C* whereas those of the line are proportional to *l*, *m*, *n*. Thus

$$\cos(90^\circ - \theta) = \frac{Al + Bm + Cn}{\sqrt{A^2 + B^2 + C^2} \cdot \sqrt{l^2 + m^2 + n^2}}$$

$$\Rightarrow \qquad \sin\theta = \frac{Al + Bm + Cn}{\sqrt{A^2 + B^2 + C^2} \cdot \sqrt{l^2 + m^2 + n^2}}.$$

Corollary 1. The line is parallel to the plane if $Al + Bm + Cn = 0$ which is consistent with the condition that the normal to the plane is perpendicular to the line.

Corollary 2. The line is perpendicular to the plane if $\frac{A}{l} = \frac{B}{m} = \frac{C}{n}$ which is consistent with the condition that the normal to the plane is parallel to the line.

3.2 Conditions for a line to lie in a plane. *To find the conditions that a line may lie in a plane.*

Let the equations of the line and the plane be

$$\frac{x-\alpha}{l} = \frac{y-\beta}{m} = \frac{z-\gamma}{n} \ (= r), \quad (1)$$

and $$Ax + By + Cz + D = 0. \quad (2)$$

Any point on the line is $(\alpha + lr, \beta + mr, \gamma + nr)$. This point will lie on the plane if

$$A(\alpha + lr) + B(\beta + mr) + C(\gamma + nr) + D = 0$$

i.e. $$(Al + Bm + Cn)r + A\alpha + B\beta + C\gamma + D = 0.$$

But this equation is true for every value of *r*. Therefore, it must be identity.

Hence

$$Al + Bm + Cn = 0, \text{ and } A\alpha + B\beta + C\gamma + D = 0.$$

These are the conditions for the line (1) to lie in the plane (2).

Remark. The condition $Al + Bm + Cn = 0$ shows that the line is perpendicular to the normal to the plane and hence, parallel to the plane whereas the condition $A\alpha + B\beta + C\gamma + D = 0$ shows that the point (α, β, γ) lies on the plane. Thus, for a line to lie in a plane it is necessary and sufficient that the line should be parallel to the plane and one point of the line must lie on the plane.

3.3 Plane containing a line. *To find the equation of the plane which contains the line*

$$\frac{x-\alpha}{l} = \frac{y-\beta}{m} = \frac{z-\gamma}{n}. \tag{1}$$

The equation of the plane passing through the point (α, β, γ) is

$$A(x-\alpha) + B(y-\beta) + C(z-\gamma) = 0.$$

This plane will contain the line (1) if

$$Al + Bm + Cn = 0.$$

Hence, the equation of the plane containing the line (1) is

$$A(x-\alpha) + B(y-\beta) + C(z-\gamma) = 0,$$

where $$Al + Bm + Cn = 0.$$

3.4 Coplanar lines. *To find the necessary condition that the two lines*

$$\frac{x-\alpha}{l} = \frac{y-\beta}{m} = \frac{z-\gamma}{n}, \tag{1}$$

$$\frac{x-\alpha'}{l'} = \frac{y-\beta'}{m'} = \frac{z-\gamma'}{n'} \tag{2}$$

may lie in a plane.

The equation of the plane containing the line (1) is

$$A(x-\alpha) + B(y-\beta) + C(z-\gamma) = 0, \tag{3}$$

where $$Al + Bm + Cn = 0. \tag{4}$$

The plane (3) will also contain the line (2) if

$$A(\alpha'-\alpha) + B(\beta'-\beta) + C(\gamma'-\gamma) = 0, \tag{5}$$

where $$Al' + Bm' + Cn' = 0. \tag{6}$$

Eliminating A, B, C from equations (4), (5) and (6), we get

$$\begin{vmatrix} \alpha' - \alpha & \beta' - \beta & \gamma' - \gamma \\ l & m & n \\ l' & m' & n' \end{vmatrix} = 0. \tag{7}$$

This is the required condition.

Note. The above condition is also sufficient.

Further, if we eliminate A, B, C from (3), (4) and (6), we get

$$\begin{vmatrix} x - \alpha & y - \beta & z - \gamma \\ l & m & n \\ l' & m' & n' \end{vmatrix} = 0.$$

This, in fact, is the equation of the plane containing the lines (1) and (2).

The lines lying in the same plane are called **coplanar lines.**

Ex. Find the equation of the plane containing the lines

$$\frac{x-1}{1} = \frac{y-2}{2} = \frac{z+3}{3}, \frac{x-3}{2} = \frac{y-4}{3} = \frac{z-5}{4}.$$

3.5 Plane containing a line and parallel to other line. *To find the equation of the plane containing the line*

$$\frac{x-\alpha}{l} = \frac{y-\beta}{m} = \frac{z-\gamma}{n} \tag{1}$$

and parallel to the line

$$\frac{x-\alpha'}{l'} = \frac{y-\beta'}{m'} = \frac{z-\gamma'}{n'}. \tag{2}$$

The equation of the plane containing the line (1) is

$$A(x - \alpha) + B(y - \beta) + C(z - \gamma) = 0, \tag{3}$$

where $$Al + Bm + Cn = 0. \tag{4}$$

The plane (3) will be parallel to the line (2) if

$$Al' + Bm' + Cn' = 0. \tag{5}$$

Eliminating A, B, C from equations (3), (4) and (5), we get

$$\begin{vmatrix} x - \alpha & y - \beta & z - \gamma \\ l & m & n \\ l' & m' & n' \end{vmatrix} = 0.$$

This is the required equation of the plane.

3.6 Examples

Example 1. Find the equation of the plane through the line

$$\frac{x+1}{2} = \frac{y+2}{3} = \frac{z+3}{4}$$

and perpendicular to the plane $x - 2y + 3z - 4 = 0$.

Solution. The equation of the plane containing the given line is

$$A(x + 1) + B(y + 2) + C(z + 3) = 0, \quad (1)$$

where $$2A + 3B + 4C = 0. \quad (2)$$

The plane (1) will be perpendicular to the plane $x - 2y + 3z - 4 = 0$ if

$$A - 2B + 3C = 0. \quad (3)$$

Eliminating A, B, C from (1), (2) and (3), we get

$$\begin{vmatrix} x+1 & y+2 & z+3 \\ 2 & 3 & 4 \\ 1 & -2 & 3 \end{vmatrix} = 0, \text{ i.e., } 17x - 2y - 7z = 8.$$

This is the required equation of the plane.

Example 2. Find the equation of the plane which passes through the line

$$ax + by + cz + d = 0, a'x + b'y + c'z + d' = 0$$

and is parallel to the line $\dfrac{x}{l} = \dfrac{y}{m} = \dfrac{z}{n}$.

Solution. The equation of the plane through the line

$$ax + by + cz + d = 0, a'x + b'y + c'z + d' = 0$$

is $$ax + by + cz + d + \lambda(a'x + b'y + c'z + d') = 0 \quad (1)$$

i.e. $$(a + \lambda a')x + (b + \lambda b')y + (c + \lambda c')z + d + \lambda d' = 0.$$

This plane will be parallel to $\dfrac{x}{l} = \dfrac{y}{m} = \dfrac{z}{n}$ if

$$l(a + \lambda a') + m(b + \lambda b') + n(c + \lambda c') = 0$$

$$\Rightarrow \qquad \lambda = -\frac{al + bm + cn}{a'l + b'm + c'n}.$$

Substituting this value of λ in (1), we get

$$(ax + by + cz + d)(a'l + b'm + c'n) = (a'x + b'y + c'z + d')(al + bm + cn).$$

This is the required equation of the plane.

Example 3. Prove that the lines

$$\frac{x-\alpha}{l} = \frac{y-\beta}{m} = \frac{z-\gamma}{n}$$

and $ax + by + cz + d = 0$, $a'x + b'y + c'z + d' = 0$ are coplanar if

$$\frac{a\alpha + b\beta + c\gamma + d}{al + bm + cn} = \frac{a'\alpha + b'\beta + c'\gamma + d'}{a'l + b'm + c'n}.$$

Solution. The equation of the plane containing the line

$$ax + by + cz + d = 0, a'x + b'y + c'z + d' = 0$$

is $$ax + by + cz + d + \lambda(a'x + b'y + c'z + d') = 0$$

i.e. $$(a + \lambda a')x + (b + \lambda b')y + (c + \lambda c')z + d + \lambda d' = 0.$$

This plane will contain the line $\frac{x-\alpha}{l} = \frac{y-\beta}{m} = \frac{z-\gamma}{n}$ if

$$\begin{cases} (a+\lambda a')\alpha + (b+\lambda b')\beta + (c+\lambda c')\gamma + d + \lambda d' = 0 \\ (a+\lambda a')l + (b+\lambda b')m + (c+\lambda c')n = 0 \end{cases}$$

$$\Rightarrow \qquad \lambda = -\frac{a\alpha + b\beta + c\gamma + d}{a'\alpha + b'\beta + c'\gamma + d'}, \lambda = -\frac{al + bm + cn}{a'l + b'm + c'n}.$$

$$\Rightarrow \qquad \frac{a\alpha + b\beta + c\gamma + d}{al + bm + cn} = \frac{a'\alpha + b'\beta + c'\gamma + d'}{a'l + b'm + c'n}.$$

Example 4. Prove that the lines

$$\frac{x}{1} = \frac{y-2}{2} = \frac{z+3}{3} \text{ and } \frac{x-2}{2} = \frac{y-6}{3} = \frac{z+3}{4}$$

are coplanar and lie in the plane $x - 2y + z + 7 = 0$.

Solution. The equation of the plane containing the first line and parallel to the second is

$$\begin{vmatrix} x & y-2 & z+3 \\ 1 & 2 & 3 \\ 2 & 3 & 4 \end{vmatrix} = 0, \text{ i.e., } x - 2y + z + 7 = 0.$$

Clearly, this plane is satisfied by the point (2, 6, 3) which lies on the second line. Thus, the given lines are coplanar and the plane containing them is

$$x - 2y + z + 7 = 0.$$

Example 5. Prove that the lines

$$ax + by + cz + d = 0,\ a'x + b'y + c'z + d' = 0$$

and $$\alpha x + \beta y + \gamma z + \delta = 0,\ \alpha' x + \beta' y + \gamma' z + \delta' = 0$$

are coplanar if

$$\begin{vmatrix} a & a' & \alpha & \alpha' \\ b & b' & \beta & \beta' \\ c & c' & \gamma & \gamma' \\ d & d' & \delta & \delta' \end{vmatrix} = 0.$$

Solution. The equation of the plane containing first line is

$$ax + by + cz + d + \lambda(a'x + b'y + c'z + d') = 0$$

i.e. $$(a + \lambda a')x + (b + \lambda b')y + (c + \lambda c')z + d + \lambda d' = 0, \qquad (1)$$

and the equation of the plane containing the second line is

$$\alpha x + \beta y + \gamma z + \delta + \mu(\alpha' x + \beta' y + \gamma' z + \delta') = 0$$

i.e. $$(\alpha + \mu\alpha')x + (\beta + \mu\beta')y + (\gamma + \mu\gamma')z + \delta + \mu\delta' = 0. \qquad (2)$$

The given lines will be coplanar if the planes (1) and (2) are identical for some values of λ and μ. Hence, on comparing the equations (1) and (2), we get

$$\frac{a + \lambda a'}{\alpha + \mu\alpha'} = \frac{b + \lambda b'}{\beta + \mu\beta'} = \frac{c + \lambda c'}{\gamma + \mu\gamma'} = \frac{d + \lambda d'}{\delta + \mu\delta'} = \nu \text{ (say)}.$$

Then

$$a + \lambda a' - \nu\alpha - \mu\nu\alpha' = 0, \qquad (3)$$

$$b + \lambda b' - \nu\beta - \mu\nu\beta' = 0, \qquad (4)$$

$$c + \lambda c' - \nu\gamma - \mu\nu\gamma' = 0, \qquad (5)$$

and $$d + \lambda d' - \nu\delta - \mu\nu\delta' = 0. \qquad (6)$$

Eliminating $\lambda, -\nu, -\mu\nu$ from equations (3), (4), (5) and (6), we get

$$\begin{vmatrix} a & a' & \alpha & \alpha' \\ b & b' & \beta & \beta' \\ c & c' & \gamma & \gamma' \\ d & d' & \delta & \delta' \end{vmatrix} = 0.$$

Exercises

1. Prove that the line $\frac{x-3}{2}=\frac{y-4}{3}=\frac{z-5}{4}$ lies on the plane $4x+4y-5z-3=0$.

2. Find the equation of the plane through the points (2, –1, 0) and (3, –4, 5) and parallel to the line $2x = 3y = 4z$.

3. Find the equation of the plane containing the line $\frac{x-1}{2}=\frac{y+1}{-1}=\frac{z-3}{4}$ and perpendicular to the plane $x + 2y + z = 12$.

4. Find the equation of the plane containing the line $\frac{x+1}{-3}=\frac{y-3}{2}=\frac{z+2}{1}$ and the point (0, 7, –7). Prove that the line $\frac{x}{1}=\frac{y-7}{-3}=\frac{z+7}{2}$ also lies on the same plane.

5. Prove that the lines $\frac{x+5}{3}=\frac{y+4}{1}=\frac{z-7}{-2}$ and $3x+2y+z-2=0,\ x-3y+2z-13=0$ are coplanar. Find the equation of the plane in which they lie.

6. Prove that the lines $\frac{x+1}{-3}=\frac{y-3}{2}=\frac{z+2}{1}$ and $\frac{x}{1}=\frac{y-7}{-3}=\frac{z+7}{2}$ intersect. Find the coordinates of the point of intersection and the equation of the plane containing them.

7. Prove that the lines $\frac{x-a}{a'}=\frac{y-b}{b'}=\frac{z-c}{c'}$ and $\frac{x-a'}{a}=\frac{y-b'}{b}=\frac{z-c'}{c}$ are coplanar. Also find the point of intersection of these lines.

8. Prove that the lines $\frac{x-a+d}{\alpha-\delta}=\frac{y-a}{\alpha}=\frac{z-a-d}{\alpha+\delta}$ and $\frac{x-b+c}{\beta-\gamma}=\frac{y-b}{\beta}=\frac{z-b-c}{\beta+\gamma}$ are coplanar. Also find the equation of the plane containing them.

9. Prove that the equation of the plane through the line $\frac{x}{l}=\frac{y}{m}=\frac{z}{n}$ and perpendicular to the plane containing the lines $\frac{x}{m}=\frac{y}{n}=\frac{z}{l}$ and $\frac{x}{n}=\frac{y}{l}=\frac{z}{m}$ is $(m-n)x+(n-l)y+(l-m)z=0$.

10. Prove that the lines $x + y + z - 3 = 0$, $2x + 3y + 4z - 5 = 0$ and $4x - y + 5z - 7 = 0$, $2x - 5y - z - 3 = 0$ are coplanar. Find the equation of the plane in which they lie.

4. Shortest Distance

Two non-intersecting and non-parallel lines are called **skew lines**. There are infinite number of lines connecting the points on the two skew lines. Out of such lines we shall be interested in a line which has the shortest line segment between the two skew lines.

4.1 *To prove that the line joining the points on two skew lines which is at right angles to both of them has the shortest line segment between the two skew lines.*

Let AB and CD be any two skew lines. Let MN be the line segment of the line which is perpendicular to both AB and CD.

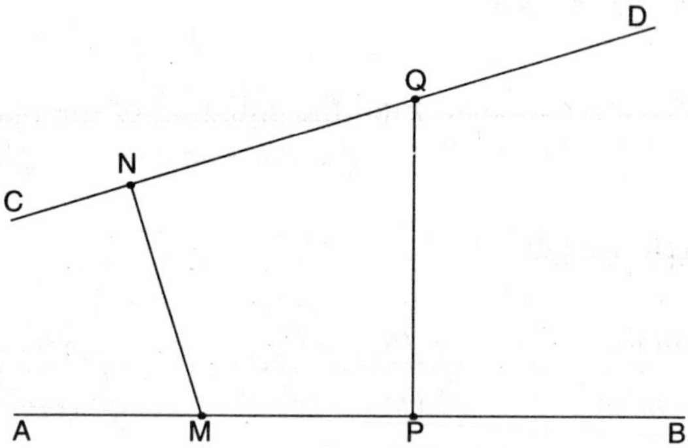

Fig. 14.2

Let PQ be the line segment of any line joining the points P and Q on the lines AB and CD, respectively. Then projection of PQ on MN is MN itself. If θ be the angle between MN and PQ. Then by § 4.2, Chapter 12, we have

$$MN = PQ \cos \theta.$$

But $|\cos \theta| \leqslant 1$. Therefore $MN \leqslant PQ$. Hence, the length of MN is the shortest.

The **shortest distance** *between two skew lines is the length of the line segment which is at right angles to both of them.*

Note. The concept of the shortest distance between two skew lines in the space is analogous to that of the distance between two parallel lines in a plane.

4.2 Length and equations of the shortest distance. *To find the shortest distance between the lines*

$$\frac{x-\alpha}{l}=\frac{y-\beta}{m}=\frac{z-\gamma}{n}, \quad (1)$$

and

$$\frac{x-\alpha'}{l'}=\frac{y-\beta'}{m'}=\frac{z-\gamma'}{n'}. \quad (2)$$

Also to obtain the equations of the line of shortest distance.

First Method. Let MN be the shortest distance between AB and CD given by (1) and (2), respectively. Let λ, μ, ν be the direction cosines of MN. Since, MN is perpendicular to both AB and CD, we have

$$\lambda l+\mu m+\nu n=0, \quad (3)$$

and

$$\lambda l'+\mu m'+\nu n'=0. \quad (4)$$

On solving (3) and (4), we get

$$\frac{\lambda}{mn'-m'n}=\frac{\mu}{nl'-n'l}=\frac{\nu}{lm'-l'm}=\pm\frac{\sqrt{\lambda^2+\mu^2+\nu^2}}{\sqrt{\Sigma(mn'-m'n)^2}}=\pm\frac{1}{\sqrt{\Sigma(mn'-m'n)^2}}.$$

Taking positive sign, we get

$$\lambda=\frac{mn'-m'n}{\sqrt{\Sigma(mn'-m'n)^2}},\ \mu=\frac{nl'-n'l}{\sqrt{\Sigma(mn'-m'n)^2}},\ \nu=\frac{lm'-l'm}{\sqrt{\Sigma(mn'-m'n)^2}}.$$

Also, $P(\alpha, \beta, \gamma)$ and $Q(\alpha', \beta', \gamma')$ are the points on AB and CD, respectively. Then

$$\begin{aligned} MN &= \text{projection of } PQ \text{ on } MN \\ &= (\alpha-\alpha')\lambda+(\beta-\beta')\mu+(\gamma-\gamma')\nu \\ &= \frac{(\alpha-\alpha')(mn'-m'n)+(\beta-\beta')(nl'-n'l)+(\gamma-\gamma')(lm'-l'm)}{\sqrt{\Sigma(mn'-m'n)^2}} \\ &= \frac{1}{\sqrt{\Sigma(mn'-m'n)^2}}\begin{vmatrix}\alpha-\alpha' & \beta-\beta' & \gamma-\gamma \\ l & m & n \\ l' & m' & n'\end{vmatrix}, \end{aligned}$$

This is the magnitude of the shortest distance.

Further, the equation of the plane containing the lines AB and MN is

$$\begin{vmatrix} x-\alpha & y-\beta & z-\gamma \\ l & m & n \\ \lambda & \mu & \nu \end{vmatrix} = 0, \tag{5}$$

and the equation of the plane containing the lines *CD* and *MN* is

$$\begin{vmatrix} x-\alpha' & y-\beta' & z-\gamma' \\ l' & m' & n' \\ \lambda & \mu & \nu \end{vmatrix} = 0. \tag{6}$$

The planes (5) and (6) together represent the equations of the line of shortest distance.

Second Method. Let $MP = r$ and $NQ = r'$. Then, the coordinates of M and N are, respectively, $(\alpha + lr, \beta + mr, \gamma + nr)$ and $(\alpha' + l'r', \beta' + m'r', \gamma' + n'r')$. The direction cosines of MN are proportional to

$$\alpha + lr - \alpha' - l'r',\ \beta + mr - \beta' - m'r',\ \gamma + nr - \gamma' - n'r'.$$

The line segment MN will be the shortest distance between AB and CD if it is perpendicular to both. For this

$$l(\alpha + lr - \alpha' - l'r') + m(\beta + mr - \beta' - m'r') + n(\gamma + nr - \gamma' - n'r') = 0, \tag{7}$$

and $$l'(\alpha + lr - \alpha' - l'r') + m'(\beta + mr - \beta' - m'r') + n'(\gamma + nr - \gamma' - n'r') = 0. \tag{8}$$

On solving the equations (7) and (8) simultaneously, we get the values of r and r'. Substituting these values of r and r' in the coordinates of M and N, and then the distance between the points M and N will give the shortest distance. Also, we can write the equations of the line of shortest distance in symmetrical form.

Third Method. The equation of the plane containing the line (1) and parallel to the line (2) is (see §3.5)

$$\begin{vmatrix} x-\alpha & y-\beta & z-\gamma \\ l & m & n \\ l' & m' & n' \end{vmatrix} = 0$$

i.e. $$(x-\alpha)(mn' - m'n) + (y-\beta)(nl' - n'l) + (z-\gamma)(lm' - l'm) = 0. \tag{9}$$

Now, the shortest distance between the lines (1) and (2) is the distance of any point on the line (2) from the plane (9). But $(\alpha', \beta', \gamma')$ is a point on the line (2). Therefore

$$\text{shortest distance} = \frac{(\alpha' - \alpha)(mn' - m'n) + (\beta' - \beta)(nl' - n'l) + (\gamma' - \gamma)(lm' - l'm)}{\sqrt{\Sigma(mn' - m'n)^2}}$$

4.3 Shortest distance between the lines in non-symmetrical form.

To find the shortest distance and equations of its line between the lines

$$ax + by + cz + d = 0, \quad a'x + b'y + c'z + d' = 0; \tag{1}$$

$$Ax + By + Cz + D = 0, \quad A'x + B'y + C'z + D' = 0. \tag{2}$$

The equation of the plane containing the line (1) is

$$ax + by + cz + d + \lambda(a'x + b'y + c'z + d') = 0, \tag{3}$$

and the plane containing the line (2) is

$$Ax + By + Cz + D + \mu(A'x + B'y + C'z + D') = 0, \tag{4}$$

where λ and μ are constants. We choose λ and μ such that the planes (3) and (4) are parallel. Then the shortest distance between the lines (1) and (2) is the distance between the planes (3) and (4).

Further, the equations of the line of shortest distance are given by the plane containing the line (1) and perpendicular to (3) or (4) and the plane containing the line (2) and perpendicular to (3) or (4). These two planes together represent the line of shortest distance.

4.4 Examples

Example 1. Find the shortest distance between the lines

$$\frac{x-1}{2} = \frac{y-2}{3} = \frac{z-3}{4}, \quad \frac{x-2}{3} = \frac{y-4}{4} = \frac{z-5}{5}$$

and the equations of the line of shortest distance.

Solution. The given lines are

$$\frac{x-1}{2} = \frac{y-2}{3} = \frac{z-3}{4}, \tag{1}$$

and

$$\frac{x-2}{3} = \frac{y-4}{4} = \frac{z-5}{5}. \tag{2}$$

Let, l, m, n be the direction cosines of the line of shortest distance. Since, the shortest distance is perpendicular to both the lines, we have

$$2l + 3m + 4n = 0, \tag{3}$$

and

$$3l + 4m + 5n = 0. \tag{4}$$

On solving equations (3) and (4), we get

$$l = -\frac{1}{\sqrt{6}}, m = \frac{2}{\sqrt{6}}, n = -\frac{1}{\sqrt{6}}.$$

The points on the lines (1) and (2) are, respectively, (1, 2, 3) and (2, 4, 5). Hence, the shortest distance is $(2-1)l+(4-2)m+(5-3)n$

$$=1\cdot\frac{(-1)}{\sqrt{6}}+2\cdot\frac{2}{\sqrt{6}}+2\cdot\frac{(-1)}{\sqrt{6}}=\frac{1}{\sqrt{6}}.$$

The equation of the plane containing the line (1) and the line of shortest distance is

$$\begin{vmatrix} x-1 & y-2 & z-3 \\ 2 & 3 & 4 \\ -1 & 2 & -1 \end{vmatrix}=0. \tag{5}$$

The equation of the plane containing the line (2) and the line of shortest distance is

$$\begin{vmatrix} x-2 & y-4 & z-5 \\ 3 & 4 & 5 \\ -1 & 2 & -1 \end{vmatrix}=0. \tag{6}$$

On simplifying the equations (5) and (6), we get

$$11x+2y-7z+6=0, \quad 7x+y-5z+7=0.$$

These two planes together represent the equations of the line of shortest distance.

Example 2. Find the shortest distance between the lines

$$\frac{x-3}{1}=\frac{y-5}{-2}=\frac{z-7}{1}, \frac{x+1}{7}=\frac{y+1}{-6}=\frac{z+1}{1}.$$

Also, find its equations.

Solution. The given lines are

$$\frac{x-3}{1}=\frac{y-5}{-2}=\frac{z-7}{1} \ (=r), \tag{1}$$

$$\frac{x+1}{7}=\frac{y+1}{-6}=\frac{z+1}{1} \ (=r'). \tag{2}$$

Any point on the line (1) is $P(3+r, 5-2r, 7+r)$ and on (2) is $Q(-1+7r', -1-6r', -1+r')$.

The direction cosines of the line PQ are proportional to

$$4+r-7r', 6-2r+6r', 8+r-r'.$$

Let PQ be the shortest distance. Then, it must be perpendicular to both (1) and (2), i.e.

$$1(4+r-7r')-2(6-2r+6r')+1(8+r-r')=0, \quad (3)$$

and $$7(4+r-7r')-6(6-2r+6r')+1(8+r-r')=0. \quad (4)$$

On solving (3) and (4) simultaneously, we get $r=0$, $r'=0$.

Thus, the coordinates of P and Q become (3, 5, 7) and (–1, –1, –1). Therefore

$$\text{shortest distance} = PQ = \sqrt{(3+1)^2+(5+1)^2+(7+1)^2} = 2\sqrt{29}.$$

The equations of the line of shortest distance are

$$\frac{x-3}{2}=\frac{y-5}{3}=\frac{z-7}{4} \text{ or } \frac{x-1}{2}=\frac{y-2}{3}=\frac{z-3}{4}.$$

Example 3. Prove that the equation of the plane containing the line $\frac{y}{b}+\frac{z}{c}=1$, $x=0$ and parallel to the line $\frac{x}{a}-\frac{z}{c}=1$, $y=0$ is $\frac{x}{a}-\frac{y}{b}-\frac{z}{c}+1=0$. If $2d$ is the shortest distance between the given lines, prove that $\frac{1}{a^2}+\frac{1}{b^2}+\frac{1}{c^2}=\frac{1}{d^2}$.

Solution. The given lines in symmetrical form are

$$\frac{x}{0}=\frac{y}{b}=\frac{z-c}{-c}, \quad (1)$$

and $$\frac{x}{a}=\frac{y}{0}=\frac{z+c}{c}. \quad (2)$$

The equation of the plane containing the line (1) and parallel to the line (2) is

$$\begin{vmatrix} x & y & z-c \\ 0 & b & -c \\ a & 0 & c \end{vmatrix} = 0, \text{ i.e. } bcx-cay-ab(z-c)=0$$

$$\Rightarrow \quad \frac{x}{a}-\frac{y}{b}-\frac{z}{c}+1=0. \quad (3)$$

Any point on the line (2) is (0, 0, $-c$). Then, the shortest distance between the lines (1) and (2) is the distance of the point (0, 0, $-c$) from the plane (3). Thus

$$2d=\frac{2}{\sqrt{\frac{1}{a^2}+\frac{1}{b^2}+\frac{1}{c^2}}}, \text{ i.e., } \frac{1}{a^2}+\frac{1}{b^2}+\frac{1}{c^2}=\frac{1}{d^2}.$$

Example 4. Find the shortest distance between the z-axis and the line

$$ax + by + cz + d = 0,\ a'x + b'y + c'z + d' = 0.$$

Solution. The equation of the plane containing the given line is

$$ax + by + cz + d + \lambda(a'x + b'y + c'z + d') = 0$$

i.e. $$(a + \lambda a')x + (b + \lambda b')y + (c + \lambda c')z + d + \lambda d' = 0. \quad (1)$$

This plane will be parallel to z-axis whose direction cosines are 0, 0, 1 if

$$(a + \lambda a')\cdot 0 + (b + \lambda b')\cdot 0 + (c + \lambda c')\cdot 1 = 0,\ i.e.,\ \lambda = -\frac{c}{c'}.$$

Substituting this value of λ in (1), we get

$$(ac' - a'c)x + (bc' - b'c)y + dc' - d'c = 0. \quad (2)$$

This is the plane parallel to z-axis. Any point on z-axis is $(0, 0, k)$. Then

shortest distance = distance of $(0, 0, k)$ from the plane (2)

$$= \frac{dc' - d'c}{\sqrt{(ac' - a'c)^2 + (bc' - b'c)^2}}.$$

Example 5. Find the length and the equations of the line of shortest distance between the lines

$$3x - 9y + 5z = 0. \quad x + y - z = 0; \quad (1)$$

$$6x + 8y + 3z - 13 = 0, \quad x + 2y + z - 3 = 0. \quad (2)$$

Solution. The equations of the plane containing the line (1) and the plane containing the line (2) are, respectively,

$$3x - 9y + 5z + \lambda(x + y - z) = 0,$$

and $$6x + 8y + 3z - 13 + \mu(x + 2y + z - 3) = 0$$

i.e. $$(3 + \lambda)x + (-9 + \lambda)y + (5 - \lambda)z = 0, \quad (3)$$

and $$(6 + \mu)x + (8 + 2\mu)y + (3 + \mu)z - 13 - 3\mu = 0. \quad (4)$$

We choose λ and μ such that the planes (3) and (4) are parallel for which

$$\frac{3+\lambda}{6+\mu} = \frac{-9+\lambda}{8+2\mu} = \frac{5-\lambda}{3+\mu} = \nu \text{ (say)}.$$

Then, we have

$$3 + \lambda - 6\nu - \mu\nu = 0, -9 + \lambda - 8\nu - 2\mu\nu = 0, \quad 5 - \lambda - 3\nu - \mu\nu = 0.$$

Solving these equations simultaneously, we get

$$\lambda = \frac{53}{5}, \mu = -\frac{31}{8}.$$

Thus, the planes (3) and (4) become

$$17x + 2y - 7z = 0, \tag{5}$$

and $$17x + 2y - 7z - 11 = 0. \tag{6}$$

$\therefore$ shortest distance = perpendicular distance between the planes (5) and (6)

= perpendicular distance of a point (0, 0, 0) on (5) from the plane (6)

$$= \frac{11}{\sqrt{17^2 + 2^2 + (-7)^2}} = \frac{11}{\sqrt{342}}.$$

Further, the plane (3) will be perpendicular to the plane (5) or (6) if

$$17(3 + \lambda) + 2(-9 + \lambda) - 7(5 - \lambda) = 0 \Rightarrow \lambda = \frac{1}{13}.$$

Similarly, the plane (4) will be perpendicular to the plane (5) or (6) if

$$17(6 + \mu) + 2(8 + 2\mu) - 7(3 + \mu) = 0 \Rightarrow \mu = -\frac{97}{14}.$$

Thus, the equations (3) and (4) become

$$10x - 29y + 16z = 0,$$

$$13x + 82y + 55z - 109 = 0.$$

These equations together represent the equations of the line of shortest distance.

Exercises

1. Find the length and the equations of the line of shortest distance between the lines

$$\frac{x-3}{3} = \frac{y-8}{-1} = \frac{z-3}{1} \text{ and } \frac{x+3}{-3} = \frac{y+7}{2} = \frac{z-6}{4}.$$

2. Find the length and the equations of the line of shortest distance between the lines

$$\frac{x-8}{3}=\frac{y+9}{-16}=\frac{z-10}{7} \text{ and } \frac{x-15}{3}=\frac{y-29}{8}=\frac{z-5}{-5}.$$

3. Find the length and the equations of the line of shortest distance between the lines

$$\frac{x+3}{-4}=\frac{y-6}{6}=\frac{z}{2} \text{ and } \frac{x+2}{-4}=\frac{y}{1}=\frac{z-7}{1}.$$

4. Find the shortest distance between the lines

$$\frac{x-1}{2}=\frac{y-2}{3}=\frac{z-3}{4} \text{ and } \frac{x-2}{3}=\frac{y-3}{4}=\frac{z-4}{5}$$

and prove that the lines are coplanar.

5. Find the length and the equations of the common perpendicular to the two lines

$$\frac{x+3}{-4}=\frac{y-6}{3}=\frac{z}{2} \text{ and } \frac{x+2}{-4}=\frac{y}{1}=\frac{z-7}{1}.$$

6. Find the length and the equations of the line of shortest distance between the lines

$$\frac{x-3}{-1}=\frac{y-4}{2}=\frac{z+2}{1} \text{ and } \frac{x-1}{1}=\frac{y+7}{3}=\frac{z+2}{2}.$$

7. Find the shortest distance between the lines

$$\frac{x-7}{2}=\frac{y+4}{3}=\frac{z-2}{1}$$

and $2x+5y-8z-52=0,\ 3x-3y+2z+27=0.$

8. Prove that the shortest distance between the lines

$\frac{x-2}{2}=\frac{y+1}{3}=\frac{z}{4}$ and $2x+3y-5z-6=0,\ \ 3x-2y-z+3=0$ is $\frac{97}{13\sqrt{6}}$.

9. Find the shortest distance between the lines

$$5x-y-z=0,\quad x-2y+z+3=0$$

and

$$7x-4y-2z=0,\quad x-y+z-3=0.$$

10. Prove that the shortest distance between the lines

$$y = az + b, \quad z = \alpha x + \beta$$

and

$$y = a'z + b', z = \alpha' x + \beta'$$

is

$$\frac{(\alpha - \alpha')(b - b') - (\alpha'\beta - \alpha\beta')(a - a')}{\sqrt{\alpha^2\alpha'^2(a - a')^2 + (\alpha - \alpha')^2 + (a'\alpha - a\alpha')^2}}.$$

5. Intersection of Three Planes

Any three planes (no two of them are parallel) may intersect in a point, in a line or form a triangular prism.

5.1 *To find the condition that the three planes may intersect in a common line.*

Let the three planes be

$$a_1x + b_1y + c_1z + d_1 = 0, \tag{1}$$

$$a_2x + b_2y + c_2z + d_2 = 0, \tag{2}$$

$$a_3x + b_3y + c_3z + d_3 = 0. \tag{3}$$

Any plane through the line of intersection of (1) and (2) is given by

$$a_1x + b_1y + c_1z + d_1 + \lambda(a_2x + b_2y + c_2z + d_2) = 0, \tag{4}$$

where λ is arbitrary constant.

In order that the three planes intersect in a common line the planes (3) and (4) must be identical. Hence, on comparing (3) and (4), we get

$$\frac{a_1 + \lambda a_2}{a_3} = \frac{b_1 + \lambda b_2}{b_3} = \frac{c_1 + \lambda c_2}{c_3} = \frac{d_1 + \lambda d_2}{d_3} = \mu \text{ (say)}.$$

Then

$$a_1 + \lambda a_2 - \mu a_3 = 0, b_1 + \lambda b_2 - \mu b_3 = 0,$$

$$c_1 + \lambda c_2 - \mu c_3 = 0, d_1 + \lambda d_2 - \mu d_3 = 0.$$

Eliminating λ and μ from these equations, we get

$$\Delta \equiv \begin{bmatrix} a_1 & b_1 & c_1 & d_1 \\ a_2 & b_2 & c_2 & d_2 \\ a_3 & b_3 & c_3 & d_3 \end{bmatrix} = 0.$$

This is the required condition.

Note. The above matrix notation may also be expressed as

$$\Delta_1 \equiv \begin{vmatrix} b_1 & c_1 & d_1 \\ b_2 & c_2 & d_2 \\ b_3 & c_3 & d_3 \end{vmatrix} = 0, \Delta_2 \equiv \begin{vmatrix} c_1 & d_1 & a_1 \\ c_2 & d_2 & a_2 \\ c_3 & d_3 & a_3 \end{vmatrix} = 0,$$

$$\Delta_3 \equiv \begin{vmatrix} d_1 & a_1 & b_1 \\ d_2 & a_2 & b_2 \\ d_3 & a_3 & b_3 \end{vmatrix} = 0 \text{ and } \Delta_4 \equiv \begin{vmatrix} a_1 & b_1 & c_1 \\ a_2 & b_2 & c_2 \\ a_3 & b_3 & c_3 \end{vmatrix} = 0.$$

If $\Delta_3 = \Delta_4 = 0$, then $\Delta_1 = \Delta_2 = 0$.

5.2 *To find the condition that three planes may form a triangular prism.*

The three planes will form a triangular prism if the lines of intersection of the planes taken in pairs are parallel.

The direction cosines of the line of intersection of the planes (1) and (2) are proportional to $b_1c_2 - b_2c_1$, $c_1a_2 - c_2a_1$, $a_1b_2 - a_2b_1$. This line will be parallel to the plane (3) if

$$a_3(b_1c_2 - b_2c_1) + b_3(c_1a_2 - c_2a_1) + c_3(a_1b_2 - a_2b_1) = 0$$

$$\text{i.e. } \begin{vmatrix} a_1 & b_1 & c_1 \\ a_2 & b_2 & c_2 \\ a_3 & b_3 & c_3 \end{vmatrix} = 0.$$

This is the required condition.

Note. By symmetry it is clear that the above condition can be obtained by taking the line of intersection of any two planes parallel to the third.

5.3 *To find the condition that the three planes may intersect in a point.*

Solving the equations (1), (2) and (3), we get

$$\frac{x}{\begin{vmatrix} b_1 & c_1 & d_1 \\ b_2 & c_2 & d_2 \\ b_3 & c_3 & d_3 \end{vmatrix}} = \frac{y}{\begin{vmatrix} c_1 & a_1 & d_1 \\ c_2 & a_2 & d_2 \\ c_3 & a_3 & d_3 \end{vmatrix}} = \frac{z}{\begin{vmatrix} a_1 & b_1 & d_1 \\ a_2 & b_2 & d_2 \\ a_3 & b_3 & d_3 \end{vmatrix}} = \frac{-1}{\begin{vmatrix} a_1 & b_1 & c_1 \\ a_2 & b_2 & c_2 \\ a_3 & b_3 & c_3 \end{vmatrix}}.$$

Thus, the planes intersect in a finite point if

$$\begin{vmatrix} a_1 & b_1 & c_1 \\ a_2 & b_2 & c_2 \\ a_3 & b_3 & c_3 \end{vmatrix} \neq 0.$$

Working rule. The three planes (1), (2) and (3)

(i) intersect in a point if $\Delta_4 \neq 0$;

(ii) from a triangular prism if $\Delta_4 = 0$ but $\Delta_3 \neq 0$;

(iii) intersect in a line if $\Delta_3 = 0 = \Delta_4$, and hence $\Delta_1 = 0 = \Delta_2$

5.4 Examples

Example 1. Prove that the planes $2x + 4y + 2z = 7$, $5x + y - z = 9$, $x - y - z = 6$ form a triangular prism.

Solution. Here

$$\Delta_4 = \begin{vmatrix} 2 & 4 & 2 \\ 5 & 1 & -1 \\ 1 & -1 & -1 \end{vmatrix} = 0,\ \Delta_3 = \begin{vmatrix} 7 & 2 & 4 \\ 9 & 5 & 1 \\ 6 & 1 & -1 \end{vmatrix} = -96 \neq 0.$$

Hence, the given planes form a triangular prism.

Example 2. Prove that the planes $x + ay + (b + c)z + d = 0$, $x + by + (c + a)z + d = 0$, and $x + cy + (a + b)z + d = 0$ intersect in a line.

Solution. Here

$$\Delta_4 = \begin{vmatrix} 1 & a & b+c \\ 1 & b & c+a \\ 1 & c & a+b \end{vmatrix} = \begin{vmatrix} 1 & a & a+b+c \\ 1 & b & a+b+c \\ 1 & c & a+b+c \end{vmatrix} = (a+b+c)\begin{vmatrix} 1 & a & 1 \\ 1 & b & 1 \\ 1 & c & 1 \end{vmatrix} = 0,$$

$$\Delta_3 = \begin{vmatrix} d & 1 & a \\ d & 1 & b \\ d & 1 & c \end{vmatrix} = d\begin{vmatrix} 1 & 1 & a \\ 1 & 1 & b \\ 1 & 1 & c \end{vmatrix} = 0.$$

Hence, the given planes intersect in a line.

Exercises

1. Prove that the planes $2x + 3y + 4z = 6$, $3x + 4y + 5z = 20$, $x + 2y + 3z = 2$ form a triangular prism. Find the equations of the edges and the equation of the right section passing through $(4, 5, -1)$.

2. Prove that the planes $2x-3y-7z=0$, $3x-14y-13z=0$, $8x-31y-33z=0$ intersect in a line, and find its equations.
3. Prove that the planes $bx - ay = n$, $cy - bz = l$, $az - cx = m$ will intersect in a line if $al + bm + cn = 0$ and the direction cosines of the common line are proportional to a, b, c.
4. Prove that the planes $3x + 3y - z - 2 = 0$, $3x + 3y + z - 4 = 0$, $x - y + 2z - 5 = 0$ intersect in a point. Find the point of intersection.
5. Prove that the planes $x = cy + bz$, $y = az + cx$, $z = bx + ay$ intersect in a line if $a^2 + b^2 + c^2 + 2abc = 1$, and the line of intersection then is $\dfrac{x}{\sqrt{1-a^2}}=\dfrac{y}{\sqrt{1-b^2}}=\dfrac{z}{\sqrt{1-c^2}}$.

6. Intersection of Lines

6.1 Line intersecting two given lines. If

$$u_1 = 0,\ v_1 = 0,\ u_2 = 0,\ v_2 = 0$$

represent planes, then the two lines may be taken as

$$u_1 = 0,\ v_1 = 0; \tag{1}$$

and

$$u_2 = 0,\ v_2 = 0. \tag{2}$$

Planes through these lines are, respectively, given by

$$u_1 + \lambda v_1 = 0, \tag{3}$$

and

$$u_2 + \mu v_2 = 0, \tag{4}$$

where λ and μ are arbitrary constant.

The line of intersection, say L, of the planes (3) and (4) lies in the plane (3). Therefore, it is coplanar with the line (1). Thus, the line L intersects the line (1) if it is not parallel to (1). Similarly, the line L intersects the line (2) if it is not parallel to (2). Hence, the planes (3) and (4) together represent the line which intersects the given two lines.

6.2 Line intersecting three given lines. Let any three lines be

$$u_1 = 0 = v_1;\ u_2 = 0 = v_2;\ u_3 = 0 = v_3.$$

If the three planes

$$u_1 + \lambda v_1 = 0, \tag{1}$$

$$u_2 + \mu v_2 = 0, \tag{2}$$

$$u_3 + \nu v_3 = 0 \qquad (3)$$

intersect in a common line which is coplanar with each of the three given lines, then it must intersect all three given lines. The condition obtained in §5.1 shows that there are two independent conditions for the three planes to have a common line of intersection. Let the two conditions be represented by

$$f(\lambda, \mu, \nu) = 0, \qquad (4)$$

$$g(\lambda, \mu, \nu) = 0. \qquad (5)$$

If λ, μ, ν be so chosen to satisfy the equations (4) and (5), then any two of the equations (1), (2), (3) will represent a line which will intersect the three given lines.

Let us consider the planes (1) and (2). Now, eliminating ν from (4) and (5), we obtain an equation of the form

$$\phi(\lambda, \mu) = 0. \qquad (6)$$

Since, equation (6) can be satisfied by an infinite number of values of λ and μ, there are infinite number of lines which intersect the given three lines.

Now, one is interested to find the locus of such lines. Eliminating λ and μ from (1), (2) and (6), we get

$$\phi\left(\frac{u_1}{v_1}, \frac{u_2}{v_2}\right) = 0. \qquad (7)$$

This equation is satisfied by the coordinates of the points which lie on the lines intersecting the three given lines. Thus, equation (7) represents a surface generated by the lines which intersect the given three lines. Eliminating λ, μ, ν from equations (1), (2), (3), (4) and (5), we get

$$f\left(\frac{u_1}{v_1}, \frac{u_2}{v_2}, \frac{u_3}{v_3}\right) = 0, \; g\left(\frac{u_1}{v_1}, \frac{u_2}{v_2}, \frac{u_3}{v_3}\right) = 0.$$

These equations together represent the required surface.

6.3 Examples

Example 1. Prove that the locus of the lines which intersect the three lines

$$y - z = 1, x = 0; \; z - x = 1, y = 0; \; x - y = 1, z = 0$$

is

$$x^2 + y^2 + z^2 - 2yz - 2zx - 2xy = 1.$$

Solution. The planes through the given lines are, respectively, given by

$$\left.\begin{aligned} y-z-1+\lambda x=0 \\ z-x-1+\mu y=0 \\ x-y-1+\nu z=0 \end{aligned}\right\}. \tag{1}$$

These planes will intersect in a line if

$$\begin{bmatrix} \lambda & 1 & -1 & -1 \\ -1 & \mu & 1 & -1 \\ 1 & -1 & \nu & -1 \end{bmatrix} = 0$$

i.e. if

$$\begin{vmatrix} \lambda & 1 & -1 \\ -1 & \mu & 1 \\ 1 & -1 & \nu \end{vmatrix} = 0$$

$$\Rightarrow \qquad \lambda\mu\nu + \lambda + \mu + \nu = 0. \tag{2}$$

Eliminating λ, μ, ν from (1) and (2), we get

$$\left(\frac{y-z-1}{-x}\right)\left(\frac{z-x-1}{-y}\right)\left(\frac{x-y-1}{-z}\right) - \frac{y-z-1}{x} - \frac{z-x-1}{y} - \frac{x-y-1}{z} = 0$$

$$\Rightarrow \qquad x^2 + y^2 + z^2 - 2yz - 2zx - 2xy = 1.$$

Example 2. Prove that the locus of a line which meets the lines $y = \pm\, mx$, $z = \pm\, c$ and the circle $x^2 + y^2 = a^2$, $z = 0$ is $c^2m^2(cy - mzx)^2 + c^2(yz - cmx)^2 = a^2m^2(z^2 - c^2)^2$.

Solution. Any line intersecting the given lines is

$$y - mx + \lambda(z - c) = 0,\; y + mx + \mu(z + c) = 0. \tag{1}$$

This line will meet the given circle if

$$y - mx - \lambda c = 0,\; y + mx + \mu c = 0 \text{ and } x^2 + y^2 = a^2$$

are satisfied simultaneously.

Now, eliminating x, y and z from these equations, we get

$$c^2(\lambda + \mu)^2 + c^2m^2(\lambda - \mu)^2 = 4a^2m^2. \tag{2}$$

Substituting the values of λ and μ from equation (1) in (2), we get

$$c^2\left(-\frac{y-mx}{z-c} - \frac{y+mx}{z+c}\right)^2 + c^2m^2\left(-\frac{y-mx}{z-c} + \frac{y+mx}{z+c}\right)^2 = 4a^2m^2$$

$$\Rightarrow \qquad c^2(yz - cmx)^2 + c^2m^2(cy - mzx)^2 = a^2m^2(z^2 - c^2)^2.$$

Exercises

1. Find the surface generated by a line which intersects the lines $y = a = z$, $x + 3z = a = y + z$ and is parallel to the plane $x + y = 0$.
2. Prove that all lines which intersect the lines $y = mx, z = c, y = -mx, z = -c$ and the x-axis, lie on the surface $mzx = cy$.
3. Find the equation of the surface generated by a line which intersects the three lines $y - 2z = 0, x = a; z - 2x = 0, y = a$ and $x - 2y = 0, z = a$.
4. A variable line intersects x-axis and the curve $x = y, y^2 = cz$ and is parallel to yz-plane. Prove that it generates the paraboloid.
5. Prove that the locus of a variable line which intersects the three lines $y = mx, z = c; y = -mx, z = -c; y = z, mx = -c$ is the surface $y^2 - m^2x^2 = z^2 - c^2$.
6. A line is drawn through a vairable point on the ellipse $\frac{x^2}{a^2} + \frac{y^2}{b^2} = 1, z = 0$ to meet two fixed lines $y = mx, z = c; y = -mx, z = -c$. Find the locus of the line.

Miscellaneous Exercises

1. Find the symmetrical form of the equations $x + y + z + 1 = 0, 4x + y - 2z + 2 = 0$ and hence find the equation of the plane through the point (1, 1, 1) and perpendicular to the given line.
2. Find the distance from the point (2, –3/2, –1) to the point where the line $\frac{x-1}{2} = \frac{y-2}{-3} = \frac{z+3}{4}$ meets the plane $2x + 4y + z + 1 = 0$.
3. Find the equation of the plane through the point (3, 1, –1) perpendicular to the line given by $3x + 4y + 7z + 4 = 0, x - y + 2z + 3 = 0$.
4. Prove that the lines $x = ay + b, z = cy + d; x = a'y + b', z = c'y + d'$ are perpendicular if $aa' + cc' = -1$.
5. Find the foot and the length of the perpendicular from the point (5, 7, 3) to the line $\frac{x-15}{3} = \frac{y-29}{8} = \frac{z-5}{-5}$. Find also the equation of the plane in which the perpendicular and the given line lie.
6. Prove that the line joining the points (2, 3, 4) and (3, 4, 5) is at right angles to the plane passing through the points (–2, –3, 6), (4, 0, –3) and (0, –1, 2).

7. Find the equations of the perpendicular from the origin to the line $x + 2y + 3z + 4 = 0, 2x + 3y + 4z + 5 = 0$. Find also the coordinates of the foot of the perpendicular.

8. Find the equations of the planes through the point $(-1, 0, 1)$ and the lines $4x - 3y + 1 = 0, \quad y - 4z + 13 = 0; 2x - y - 2 = 0, \quad z - 5 = 0$, and prove that the equations of the line through the given point which intersects the two given lines can be written as $\frac{x}{1} = \frac{y-1}{1} = \frac{z-2}{1}$.

9. Prove that the line joining the points (a, b, c) and (a', b', c') passes through the origin if $aa' + bb' + cc' = pp'$, where p and p' are the distances of the points from the origin.

10. Prove that the line $\frac{x-\alpha}{l} = \frac{y-\beta}{m} = \frac{z-\gamma}{n}$ will intersect the locus $ax^2 + by^2 = 1, z = 0$, if $a(\alpha n - l\gamma)^2 + b(\beta n - m\gamma)^2 = n^2$.

11. Prove that the lines

$$\frac{x-\alpha}{\alpha'} = \frac{y-\beta}{\beta'} = \frac{z-\gamma}{\gamma'}; \quad \frac{x-\alpha'}{\alpha} = \frac{y-\beta'}{\beta} = \frac{z-\gamma'}{\gamma}$$

intersect. Find their point of intersection and the plane containing them.

12. Prove that the lines $\frac{x}{\alpha} = \frac{y}{\beta} = \frac{z}{\gamma}; \frac{x}{a\alpha} = \frac{y}{b\beta} = \frac{z}{c\gamma}; \frac{x}{l} = \frac{y}{m} = \frac{z}{n}$ are coplanar if $\frac{l(b-c)}{\alpha} + \frac{m(c-a)}{\beta} + \frac{n(a-b)}{\gamma} = 0$.

13. Find the equations of the line which intersects the lines

$$x + y + z - 1 = 0, \quad 2x - y - z - 2 = 0;$$
$$x - y - z - 3 = 0, \quad 2x + 4y - z - 4 = 0$$

and passes through the point (1, 1, 1).

14. Find the shortest distance between the diagonals of a rectangular solid and the edges to which it does not meet.

15. Prove that the shortest distance between any two opposite edges of the tetrahedron formed by the planes $y + z = 0, z + x = 0, x + y = 0, x + y + z = a$ is $\frac{2a}{\sqrt{6}}$, and that the three lines of the shortest distance intersect at the point $x = y = z = -a$.

16. If l_1, m_1, n_1 and l_2, m_2, n_2 are the direction cosines, prove that the equations of the planes through the angular bisectors of the lines $\frac{x}{l_1} = \frac{y}{m_1} = \frac{z}{n_1}$;

$\frac{x}{l_2} = \frac{y}{m_2} = \frac{z}{n_2}$ and at right angles to the plane containing them, are

$(l_1 \pm l_2)x + (m_1 \pm m_2)y + (n_1 \pm n_2)z = 0.$

17. The lengths of two opposite edges of a tetrahedron are a, b, their shortest distance is equal to d, where the angle between them is θ, prove that the volume of the tetrahedron is $\frac{abd \sin \theta}{6}$.

18. Prove that the volume of the tetrahedron of which a pair of opposite edges of the lengths r, r' on the lines whose equations are $\frac{x-a}{l} = \frac{y-b}{m} = \frac{z-c}{n}$,

$$\frac{x-a'}{l'} = \frac{y-b'}{m'} = \frac{z-c'}{n'} \text{ is } \frac{1}{6} rr' \begin{vmatrix} a-a' & b-b' & c-c' \\ l & m & n \\ l' & m' & n' \end{vmatrix}.$$

The Sphere

15
CHAPTER

1. Definition

*A **sphere** is the locus of a point which moves such that it remains at a fixed distance from a fixed point in the space.*

The fixed point is called the **centre** of the sphere and the fixed distance is the **radius** of the sphere.

1.1 Equation of a sphere. *To find the equation of a sphere with given centre and radius.*

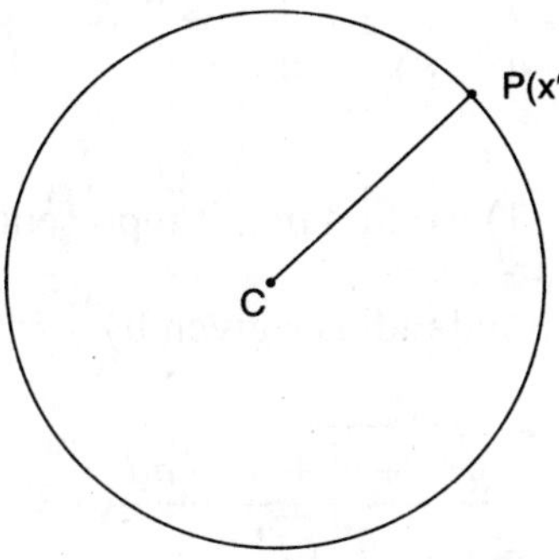

Fig. 15.1

Let $C(\alpha, \beta, \gamma)$ be the centre and r the radius of the sphere. Let $P\ (x', y', z')$ be any point on the locus (sphere). Then, by definition $CP = r$. But

$$CP = \sqrt{(x'-\alpha)^2 + (y'-\beta)^2 + (z'-\gamma)^2}\ .$$

Therefore $$\sqrt{(x'-\alpha)^2 + (y'-\beta)^2 + (z'-\gamma)^2} = r\text{ , i.e.,}$$

$$(x'-\alpha)^2 + (y'-\beta)^2 + (z'-\gamma)^2 = r^2 .$$

Hence, *the equation of the sphere with centre* (α, β, γ) *and radius r is*

$$(x-\alpha)^2 + (y-\beta)^2 + (z-\gamma)^2 = r^2 .$$

1.2 The general form of the equation. The equation of the sphere is

$$(x-\alpha)^2+(y-\beta)^2+(z-\gamma)^2=r^2 \tag{1}$$

which on expanding takes the form

$$x^2+y^2+z^2-2\alpha x-2\beta y-2\gamma z+\alpha^2+\beta^2+\gamma^2-r^2=0.$$

We note that the following about this equation:

(*i*) This is a second degree equation in x, y and z;

(*ii*) The coefficients of x^2, y^2 and z^2 are equal;

(*iii*) There are no terms containing the products yz, zx and xy.

Conversely, we consider the most general equation satisfying these three conditions:

$$ax^2+ay^2+az^2+2ux+2vy+2wz+d=0.\ (a\neq 0) \tag{2}$$

To prove that equation (2) *represents a sphere.*

Equation (2) may be written as

$$\left(x+\frac{u}{a}\right)^2+\left(y+\frac{v}{a}\right)^2+\left(z+\frac{w}{a}\right)^2=\frac{u^2+v^2+w^2-ad}{a^2}.$$

Comparing it with equation (1) we find that it represents a sphere with centre at the point $\left(-\frac{u}{a},-\frac{v}{a},-\frac{w}{a}\right)$ and radius r given by

$$r=\frac{\sqrt{u^2+v^2+w^2-ad}}{a}.$$

Hence, equation (2) always represents a sphere.

There will be no loss of generality if we choose $a = 1$ so that the general equation of the sphere takes the form

$$x^2+y^2+z^2+2ux+2vy+2wz+d=0. \tag{3}$$

This sphere has the centre $(-u, -v, -w)$ and radius equal to $\sqrt{u^2+v^2+w^2-d}$.

Remarks

(*i*) If $u^2+v^2+w^2>d$, equation (3) represents a sphere with centre $(-u, -v, -w)$ and real radius.

(*ii*) If $u^2+v^2+w^2=d$, equation (3) represents a sphere whose centre is $(-u, -v, -w)$ and radius is zero i.e. the sphere coincides with the centre. Such a sphere is called a **point sphere**.

(*iii*) If $u^2 + v^2 + w^2 < d$, equation (3) represents a sphere with centre $(-u, -v, -w)$ and imaginary radius. In this case, there are no real points on the sphere and we call it a **virtual sphere.**

Corollary. The equation of the sphere centred at the origin and of radius a is

$$x^2 + y^2 + z^2 = a^2.$$

1.3 Sphere with given diameter. *To find the equation of the sphere described on the line joining the points* $A(x_1, y_1, z_1)$ *and* $B(x_2, y_2, z_2)$ *as diameter.*

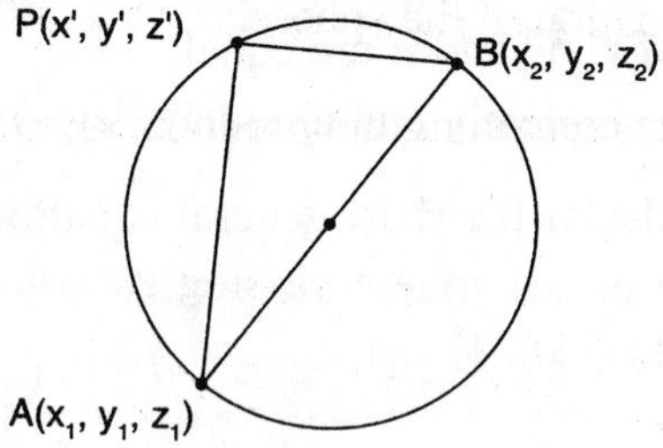

Fig. 15.2

Let $P(x', y', z')$ be any point on the sphere. Then, the direction cosines of AP and BP are proportional to $x' - x_1, y' - y_1, z' - z_1$ and $x' - x_2, y' - y_2, z' - z_2$, respectively. Since $\angle APB = 90°$, we have

$$(x' - x_1)(x' - x_2) + (y' - y_1)(y' - y_2) + (z' - z_1)(z' - z_2) = 0.$$

Hence, the required equation of the sphere is

$$(x - x_1)(x - x_2) + (y - y_1)(y - y_2) + (z - z_1)(z - z_2) = 0.$$

1.4 Sphere through four given points. *To find the equation of the sphere passing through four given points* (x_1, y_1, z_1), (x_2, y_2, z_2), (x_3, y_3, z_3) *and* (x_4, y_4, z_4).

Let the general equation of the sphere be

$$x^2 + y^2 + z^2 + 2ux + 2vy + 2wz + d = 0. \quad (1)$$

Let it pass through the four given points. Then

$$x_1^2 + y_1^2 + z_1^2 + 2ux_1 + 2vy_1 + 2wz_1 + d = 0, \quad (2)$$

$$x_2^2 + y_2^2 + z_2^2 + 2ux_2 + 2vy_2 + 2wz_2 + d = 0, \quad (3)$$

$$x_3^2 + y_3^2 + z_3^2 + 2ux_3 + 2vy_3 + 2wz_3 + d = 0, \quad (4)$$

and $$x_4^2 + y_4^2 + z_4^2 + 2ux_4 + 2vy_4 + 2wz_4 + d = 0. \quad (5)$$

Eliminating u, v, w and d from equations (1), (2), (3), (4) and (5), we get

$$\begin{vmatrix} x^2+y^2+z^2 & x & y & z & 1 \\ x_1^2+y_1^2+z_1^2 & x_1 & y_1 & z_1 & 1 \\ x_2^2+y_2^2+z_2^2 & x_2 & y_2 & z_2 & 1 \\ x_3^2+y_3^2+z_3^2 & x_3 & y_3 & z_3 & 1 \\ x_4^2+y_4^2+z_4^2 & x_4 & y_4 & z_4 & 1 \end{vmatrix} = 0.$$

This is the required equation of the sphere.

Remark. A sphere is completely and uniquely determined by any four points on it.

Ex. Find the equation of the sphere passing through the points (1, –1, –1), (3, 3, 1), (–2, 0, 5) and (–1, 4, 4).

1.5 Examples

Example 1. Find the equation of the sphere passing through the origin and the points $(\alpha, 0, 0)$,$(0, \beta, 0)$ and $(0, 0, \gamma)$.

Solution. Let the equation of the sphere be

$$x^2 + y^2 + z^2 + 2ux + 2vy + 2wz + d = 0. \tag{1}$$

This sphere will pass through the given points if

$$d = 0,\ \alpha^2 + 2u\alpha = 0,\ \beta^2 + 2u\,\beta = 0 \text{ and } \gamma^2 + 2w\gamma = 0$$

$$i.e.,\ d = 0,\ 2u = -\alpha,\ 2v = -\beta \text{ and } 2w = -\gamma$$

Substituting these values of u, v, w in (1), we get

$$x^2 + y^2 + z^2 - \alpha x - \beta y - \gamma z = 0.$$

This is the required equation of the sphere.

Example 2. A sphere of constant radius k passes through the origin and meets the axes in A, B, C. Prove that the centroid of the triangle ABC lies on the sphere $9(x^2 + y^2 + z^2) = 4k^2$.

Solution. Let the distances of the points A, B, C from the origin O be, respectively, α, β, γ. Then the coordinates of A, B, C are, respectively, $(\alpha, 0, 0)$, $(0, \beta, 0)$, $(0, 0, \gamma)$. Therefore, the equation of the sphere $OABC$ is

$$x^2 + y^2 + z^2 - \alpha x - \beta y - \gamma z = 0.$$

The radius of the sphere is $\sqrt{\frac{\alpha^2}{4}+\frac{\beta^2}{4}+\frac{\gamma^2}{4}}$. Therefore

$$\sqrt{\frac{\alpha^2}{4}+\frac{\beta^2}{4}+\frac{\gamma^2}{4}} = k, \textit{ i.e., } \alpha^2+\beta^2+\gamma^2 = 4k^2. \quad (1)$$

Let (x', y', z') be the centroid of the triangle ABC. Then

$$\left.\begin{aligned} x' &= \frac{\alpha+0+0}{3} = \frac{\alpha}{3} \Rightarrow \alpha = 3x' \\ y' &= \frac{0+\beta+0}{3} = \frac{\beta}{3} \Rightarrow \beta = 3y' \\ z' &= \frac{0+0+\gamma}{3} = \frac{\gamma}{3} \Rightarrow \gamma = 3z' \end{aligned}\right\} \quad (2)$$

Eliminating α, β, γ from equations (1) and (2), we get

$$9x'^2 + 9y'^2 + 9z'^2 = 4k^2.$$

Hence, the locus of (x', y', z') is $9(x^2+y^2+z^2) = 4k^2$.

Example 3. A sphere of constant radius k passes through the origin O, and meets the axes in A, B, C. Prove that the locus of the foot of the perpendicular from O to the plane ABC is given by $(x^2+y^2+z^2)^2(x^{-2}+y^{-2}+z^{-2}) = 4k^2$.

Solution. Let the equation of the plane ABC be

$$\frac{x}{\alpha}+\frac{y}{\beta}+\frac{z}{\gamma} = 1. \quad (1)$$

Then, the coordinates of A, B, C are, respectively, $(\alpha, 0, 0)$, $(0, \beta, 0)$, $(0, 0, \gamma)$. Hence, the equation of the sphere $OABC$ is

$$x^2+y^2+z^2-\alpha x-\beta y-\gamma z = 0. \quad (2)$$

Now $\alpha^2+\beta^2+\gamma^2 = 4k^2. \quad (3)$

The direction cosines of the normal to the plane (1) are proportional to $\frac{1}{\alpha}, \frac{1}{\beta}, \frac{1}{\gamma}$. Therefore, the equations of the perpendicular from O to the plane (1) are given by

$$\frac{x}{\frac{1}{\alpha}} = \frac{y}{\frac{1}{\beta}} = \frac{z}{\frac{1}{\gamma}} = r \text{ (say)}. \quad (4)$$

Any point on the line (4) is $\left(\frac{r}{\alpha}, \frac{r}{\beta}, \frac{r}{\gamma}\right)$. This point will lie on the plane (1) if

$$\frac{r}{\alpha^2}+\frac{r}{\beta^2}+\frac{r}{\gamma^2}=1$$

$$\Rightarrow \qquad r=\frac{1}{\frac{1}{\alpha^2}+\frac{1}{\beta^2}+\frac{1}{\gamma^2}}=\frac{1}{\alpha^{-2}+\beta^{-2}+\gamma^{-2}}.$$

Let (x', y', z') be the foot of the perpendicular drawn from O to the plane (1). Then

$$x'=\frac{r}{\alpha}=\frac{\alpha^{-1}}{\alpha^{-2}+\beta^{-2}+\gamma^{-2}},\ y'=\frac{r}{\beta}=\frac{\beta^{-1}}{\alpha^{-2}+\beta^{-2}+\gamma^{-2}},\ z'=\frac{r}{\gamma}=\frac{\gamma^{-1}}{\alpha^{-2}+\beta^{-2}+\gamma^{-2}}.$$

Now

$$\left.\begin{aligned} x'^2+y'^2+z'^2 &= \frac{1}{\alpha^{-2}+\beta^{-2}+\gamma^{-2}} \\ x'^{-2}+y'^{-2}+z'^{-2} &= \frac{\alpha^2+\beta^2+\gamma^2}{(\alpha^{-2}+\beta^{-2}+\gamma^{-2})^{-2}} \end{aligned}\right\}. \qquad (5)$$

Therefore, from equations (3) and (5), we get

$$(x'^2+y'^2+z'^2)^2(x'^{-2}+y'^{-2}+z'^{-2})=4k^2.$$

Hence, the locus of (x', y', z') is $(x^2+y^2+z^2)^2(x^{-2}+y^{-2}+z^{-2})=4k^2$.

Exercises

1. Find the equation of the sphere whose centre is (1, 2, 3) and radius 3.
2. Find the centre and radius of the sphere $x^2+y^2+z^2-6x-8y+10z+1=0$.
3. Find the equation of the sphere whose diameter is the line joining the points (1, –2, 3) and (3, –4, –5).
4. Find the equation of the sphere passing through the point (0, 0, 0), (0, 1, –1), (–1, 2, 0) and (1, 2, 3).
5. Find the equation of the sphere whose centre is (2, –3, 4) and passes through the point (1, 2, 3).
6. Find the equation of the sphere which passes through the origin and the points (2, 1, –1), (1, 5, –4), (–2, 4, –6).
7. Find the equation of the sphere which passes through the points (1, –3, 4), (1, –5, 2), (1, –3, 0) and the centre lies on the plane $x+y+z=0$.

8. Find the equation of the sphere passing through the points (3, 0, 2), (–1, 1, 1), (2, –5, 4) and having its centre on the plane $2x + 3y + 4z = 6$.

9. Find the equation of the sphere passing through the points (0, –2, –4), (2, –1, –1) and having its centre on the line $5y + 2z = 0$, $2x - 3y = 0$.

10. A plane passes through the fixed point (a, b, c) and meets the axes in A, B, C. Prove that the locus of the centre of the sphere $OABC$ is $\frac{a}{x}+\frac{b}{y}+\frac{c}{z}=2$.

11. A sphere of constant radius $2k$, passes through the origin and meets the axes in A, B, C. Prove that the locus of the centroid of the tetrahedron $OABC$ is $x^2 + y^2 + z^2 = k^2$.

2. The Circle

2.1 Plane section of a sphere. *To prove that the section of a sphere by a plane is a circle.*

Let O be the centre of the sphere and P any point on the section of the sphere by the plane.

Draw perpendicular OC to the plane, where C is the foot of the perpendicular. Join the line CP. Then CP is at right angle to OC. Therefore, in right angled triangle OCP, we have $OP^2 = OC^2 + CP^2$, Thus

$$CP = \sqrt{OP^2 - OC^2} = \text{constant}.$$

This shows that the locus of the point P is a circle whose centre is C and radius CP.

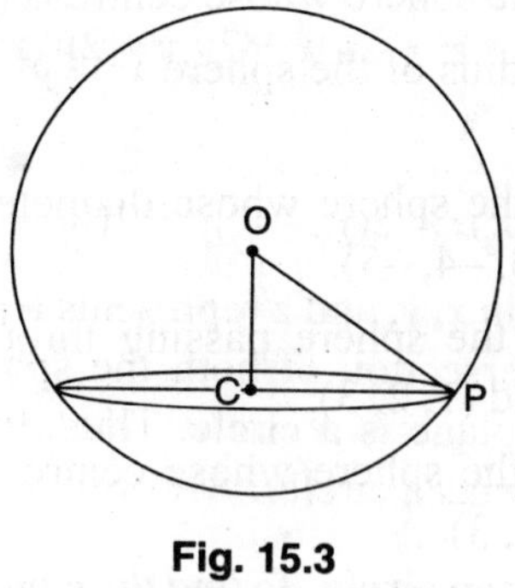

Fig. 15.3

Remarks

(*i*) The centre of the circle is the foot of the perpendicular drawn from the centre of the sphere to the plane.

(*ii*) The radius of the circle is the square root of the quantity which is the difference of the squares of the radius of the sphere and the distance of the plane from the centre of the sphere.

(*iii*) If the radius of the circle is less than the radius of the sphere, the circle is called **small circle**.

(*iv*) If the radius of the circle is equal to the radius of the sphere, the circle is called the **great circle**. In fact, the great circles are obtained as sections of a sphere by the planes which pass through the centre of the sphere.

(*v*) If the radius of the circle is greater than the radius of the sphere, the circle is called **imaginary circle**. In other words, there will be no such real section of a sphere by a plane.

2.2 Equation of a circle. *To find the equation of a circle.*

Since a circle is a plane section of a sphere, the equations of the sphere and the plane together represent a circle. Let the equations of the sphere and the plane be

$$x^2 + y^2 + z^2 + 2ux + 2vy + 2wz + d = 0, \quad (1)$$

and

$$lx + my + nz = p. \quad (2)$$

Then, the equations (1) and (2) represent a circle which may be a small circle, great circle or imaginary circle.

2.3 Intersection of two spheres. *To prove that the curve of intersection of two spheres is a circle.*

Let any two spheres be

$$S_1 \equiv x^2 + y^2 + z^2 + 2u_1x + 2v_1y + 2w_1z + d_1 = 0, \quad (1)$$

and

$$S_2 \equiv x^2 + y^2 + z^2 + 2u_2x + 2v_2y + 2w_2z + d_2 = 0. \quad (2)$$

Now

$$S_1 - S_2 \equiv 2(u_1 - u_2)x + 2(v_1 - v_2)y + 2(w_1 - w_2)z + d_1 - d_2 = 0. \quad (3)$$

Equation (3) being linear in *x*, *y*, and *z* represents a plane and this plane passes through the points of intersection of both the spheres (1) and (2). Also the section of a sphere by a plane is a circle. Thus, the curve of intersection of spheres (1) and (2) represents a circle.

2.4 Spheres through a given circle. *To find the equation of the spheres passing through a given circle.*

Let the given circle be

$$\left.\begin{aligned} x^2 + y^2 + z^2 + 2ux + 2vy + 2wz + d &= 0 \\ lx + my + nz &= p \end{aligned}\right\}. \quad (1)$$

Consider the equation

$$x^2 + y^2 + z^2 + 2ux + 2vy + 2wz + d + \lambda(lx + my + nz - p) = 0. \quad (2)$$

Clearly equation (2) is satisfied by all the points lying on the circle (1) for all values of λ. Thus, equation (2) is the equation of the sphere passing through the circle (1).

Remark. Equation (2), in fact, represents a family of spheres through the circle (1).

Ex. **Prove that the equation $S_1 + \lambda S_2 = 0$ represents a family of spheres passing through the circle of intersection of the spheres $S_1 = 0$ and $S_2 = 0$.**

2.5 Examples

Example 1. Find the centre and radius of the circle

$$x^2 + y^2 + z^2 - 8x + 4y + 8z - 45 = 0, x - 2y + 2z = 3.$$

Solution. The centre of the sphere is (4, –2, –4).

Radius of the sphere $r = \sqrt{16+4+16+45} = 9$.

Distance of the plane from the centre of the sphere is given by

$$d = \frac{-4-4+8+3}{\sqrt{1+4+4}} = 1.$$

Therefore Radius of the circle $= \sqrt{r^2 - d^2} = 4\sqrt{5}$.

Centre of the circle is the foot of the perpendicular drawn from the centre of the sphere to the plane. The equations of the perpendicular are

$$\frac{x-4}{1} = \frac{y+2}{-2} = \frac{z+4}{2} = \lambda \text{ (say)}.$$

Any point on the perpendicular is $(4 + \lambda, -2-2\lambda, -4 + 2\lambda)$. This point will be the centre of the circle if it lies on the plane, i.e., if $\lambda = 1/3$.

Hence, the centre of the circle is $\left(\frac{13}{3}, \frac{-8}{3}, \frac{-10}{3}\right)$.

Example 2. The plane $x/a + y/b + z/c = 1$ meets the axes in the points A, B, C. Find the equations of the circumcircle of the triangle ABC and also find its centre.

Solution. The coordinates of the points A, B, C are, respectively $(a, 0, 0)$, $(0, b, 0)$, $(0, 0, c)$. The equation of the sphere $OABC$ is

$$x^2 + y^2 + z^2 - ax - by - cz = 0. \quad (1)$$

The circumcircle of the triangle ABC is the section of the sphere (1) by the given plane. Thus the equations of the circumcircle are

$$\left.\begin{aligned} x^2+y^2+z^2-ax-by-cz=0 \\ \frac{x}{a}+\frac{y}{b}+\frac{z}{c}=1 \end{aligned}\right\} \qquad (2)$$

Further, the centre of the sphere (1) is $(a/2, b/2, c/2)$. The equations of the line passing through the centre of the sphere and perpendicular to the given plane are

$$\frac{x-a/2}{1/a}=\frac{y-b/2}{1/b}=\frac{z-c/2}{1/c}=r \text{ (say)}.$$

Any point on the line is $\left(\frac{a}{2}+\frac{r}{a}, \frac{b}{2}+\frac{r}{b}, \frac{c}{2}+\frac{r}{c}\right)$. This point will be the centre of the circle (2) if it lies on the given plane, i.e., if

$$\frac{1}{a}\left(\frac{a}{2}+\frac{r}{a}\right)+\frac{1}{b}\left(\frac{b}{2}+\frac{r}{b}\right)+\frac{1}{c}\left(\frac{c}{2}+\frac{r}{c}\right)=1$$

$$\Rightarrow \qquad r=-\frac{1}{2(a^{-2}+b^{-2}+c^{-2})}.$$

Hence, the centre of the circle is

$$\left(\frac{a}{2}-\frac{1/a}{2(a^{-2}+b^{-2}+c^{-2})}, \frac{b}{2}-\frac{1/b}{2(a^{-2}+b^{-2}+c^{-2})}, \frac{c}{2}-\frac{1/c}{(a^{-2}+b^{-2}+c^{-2})}\right)$$

i.e.
$$\left(\frac{a(b^{-2}+c^{-2})}{2(a^{-2}+b^{-2}+c^{-2})}, \frac{b(c^{-2}+a^{-2})}{2(a^{-2}+b^{-2}+c^{-2})}, \frac{c(a^{-2}+b^{-2})}{2(a^{-2}+b^{-2}+c^{-2})}\right)$$

Example 3. Prove that the equation of the sphere which passes through the point (α, β, γ) and the circle $x^2+y^2=a^2$, $z=0$ is $\gamma(x^2+y^2+z^2-a^2)=z(\alpha^2+\beta^2+\gamma^2-a^2)$.

Solution. The equation of the sphere through the given circle is

$$x^2+y^2+z^2-a^2+\lambda z=0. \qquad (1)$$

It will pass through the point (α, β, γ) if $\alpha^2+\beta^2+\gamma^2-a^2+\lambda\gamma=0$, *i.e.*,

$$\lambda=-\frac{\alpha^2+\beta^2+\gamma^2-a^2}{\gamma}.$$

Substituting this value of λ in (1), we get

$$\gamma(x^2+y^2+z^2-a^2)=z(\alpha^2+\beta^2+\gamma^2-a^2).$$

Example 4. Prove that the circles

$$x^2 + y^2 + z^2 - 2x + 3y + 4z - 5 = 0, 5y + 6z + 1 = 0;$$

and

$$x^2 + y^2 + z^2 - 3x - 4y + 5z - 6 = 0, x + 2y - 7z = 0$$

lie on the same sphere and find its equation.

Solution. The equation of the sphere through the first circle is

$$x^2 + y^2 + z^2 - 2x + 3y + 4z - 5 + \lambda(5y + 6z + 1) = 0. \quad (1)$$

The equation of the sphere through the second circle is

$$x^2 + y^2 + z^2 - 3x - 4y + 5z - 6 + \mu(x + 2y - 7z) = 0. \quad (2)$$

The given circles will lie on the same sphere if equations (1) and (2) are identical. Thus, on comparing equations (1) and (2), we get

$$-2 = -3 + \mu, \qquad 3 + 5\lambda = -4 + 2\mu,$$

$$4 + 6\lambda = 5 - 7\mu, \qquad -5 + \lambda = -6.$$

On solving these equations simultaneously, we get $\lambda = -1$, $\mu = 1$.

Hence, the required equation of the sphere is $x^2 + y^2 + z^2 - 2x - 2y - 2z - 6 = 0$.

Exercises

1. Prove that the radius of the circle $x^2 + y^2 + z^2 + x + y + z = 4$, $x + y + z = 0$ is 2.
2. Find the centre of the circle which passes through the points $(-1, 0, 0)$, $(0, 2, 0)$, $(0, 0, 3)$.
3. Find the centre and radius of the circle $x^2 + y^2 + z^2 + 12x - 12y - 16z + 111 = 0$, $2x + 2y + 2z = 17$.
4. Find the equation of the sphere having the circle $x^2 + y^2 + z^2 + 10y - 4z - 8 = 0$, $x + y + z - 3 = 0$ as a great circle.
5. Find the equation of the sphere having the circle $x^2 + y^2 + z^2 = 9$, $x - 2y + 2z = 5$ as a great circle; find its centre and radius.
6. Prove that the plane $x + 2y - z = 4$ cuts the sphere $x^2 + y^2 + z^2 - x + z + 2 = 0$ in a circle of radius unity. Also, find the equation of the sphere which has this circle for one of its great circles.
7. Find the centre and radius of the circle in which the sphere $x^2 + y^2 + z^2 + 2x - 2y - 4z - 19 = 0$ is cut by the plane $x + 2y + 2z + 7 = 0$.
8. Find the equation of the sphere through the circle $x^2 + y^2 + z^2 = 9$, $2x + 3y + 4z = 5$ and the point $(1, 2, 3)$.

9. Find the equation of the sphere through the circle $x^2+y^2+z^2=9$, $x+y-2z=4$ and the point $(1, 2, -1)$.

10. Find the equation of the sphere passing through the circle of intersection of the spheres $x^2+y^2+z^2-2x+4y=11$ and $x^2+y^2+z^2-4x-6z=51$, and the point $(-7, 4, 12)$.

11. Prove that the circles $x^2+y^2+z^2-y+2z=0$, $x-y+z=0$; and $x^2+y^2+z^2+x-3y+z=5$, $2x-y+4z=1$ lie on the same sphere and find its equation.

3. Intersection of a Sphere and a Line

Let

$$x^2+y^2+z^2+2ux+2vy+2wz+d=0, \tag{1}$$

and

$$\frac{x-\alpha}{l}=\frac{y-\beta}{m}=\frac{z-\gamma}{n}=r \tag{2}$$

be the equations of a sphere and a line, respectively.

The coordinates of a point on the line which is at a distance r from (α, β, γ) are $(\alpha+lr, \beta+mr, \gamma+nr)$. If this point be on the sphere, we must have $(\alpha+lr)^2+(\beta+mr)^2+(\gamma+nr)^2+2u(\alpha+lr)+2v(\beta+mr)+2w(\gamma+nr)+d=0$, *i.e.*,

$$r^2(l^2+m^2+n^2)+2r\{l(\alpha+u)+m(\beta+v)+n(\gamma+w)\}$$
$$+\alpha^2+\beta^2+\gamma^2+2u\alpha+2v\beta+2w\gamma+d=0. \tag{3}$$

But this is a quadratic equation in r giving two values of r corresponding to which there are two points common to the sphere and the line. Hence *a line meets a sphere always in two points which may be real and distinct, coincident or imaginary*.

Further, if l, m, n be the actual direction cosines of the line, then equation (3) reduces to

$$r^2+2r\{l(\alpha+u)+m(\beta+v)+n(\gamma+w)\}+\alpha^2+\beta^2+\gamma^2$$
$$+2u\alpha+2v\beta+2w\gamma+d=0. \tag{4}$$

Let r_1 and r_2 be the roots of the equation (4). Then, by theory of equations, we have

$$r_1r_2=\alpha^2+\beta^2+\gamma^2+2u\alpha+2v\beta+2w\gamma+d.$$

We see that the value of r_1r_2 is independent of l, m, n i.e. the direction cosines of the line, and as such the product r_1r_2 is a constant quantity and independent of the line whatsoever we may take through (α, β, γ). We call the magnitude of the product r_1r_2 as the power of the point (α, β, γ) with respect to the sphere. More precisely, *the* **power** *of a point* (α, β, γ) *with respect to the sphere* $x^2 + y^2 + z^2 + 2ux + 2vy + 2wz + d = 0$ *is defined as the quantity* $\alpha^2 + \beta^2 + \gamma^2 + 2u\alpha + 2v\beta + 2w\gamma + d$.

3.1 Tangent line: *A line which meets a sphere in two coincident points is called the* **tangent line** *to the sphere.*

3.2 Tangent plane: *To find the locus of the tangent lines to a sphere at a point on it.*

Let

$$x^2 + y^2 + z^2 + 2ux + 2vy + 2wz + d = 0 \tag{1}$$

be the equation of the sphere and (α, β, γ) be the given point on it.

Equations of any line through the point (α, β, γ) are

$$\frac{x-\alpha}{l} = \frac{y-\beta}{m} = \frac{z-\gamma}{n} = r. \tag{2}$$

Any point on the line (2) at a distance r from (α, β, γ) is $(\alpha + lr, \beta + mr, \gamma + nr)$. If this point be on the sphere, we must have

$$(\alpha + lr)^2 + (\beta + mr)^2 + (\gamma + nr)^2 + 2u(\alpha + lr) + 2v(\beta + mr) + 2w(\gamma + nr) + d = 0$$

i.e. $$r^2(l^2 + m^2 + n^2) + 2r\{l(\alpha + u) + m(\beta + v) + n(\gamma + w)\} + \alpha^2 + \beta^2 + \gamma^2 + 2u\alpha + 2v\beta + 2w\gamma + d = 0. \tag{3}$$

Since the point (α, β, γ) lies on the sphere, we have

$$\alpha^2 + \beta^2 + \gamma^2 + 2u\alpha + 2v\beta + 2w\gamma + d = 0. \tag{4}$$

This shows that one root of equation (3) is zero. Thus, equation (3) reduces to

$$r(l^2 + m^2 + n^2) + 2\{l(\alpha + u) + m(\beta + v) + n(\gamma + w)\} = 0. \tag{5}$$

The line (2) will be tangent line to the sphere (1) if the two points of intersecton of (1) and (2) coincide with the point (α, β, γ). For this the second root of (3), *i.e.*, r in (5) should vanish which requires

$$l(\alpha + u) + m(\beta + v) + n(\gamma + w) = 0. \tag{6}$$

Eliminating l, m, n from (2) and (6), we get

$$(x - \alpha)(\alpha + u) + (y - \beta)(\beta + v) + (z - \gamma)(\gamma + w) = 0$$

$$\Rightarrow \alpha x + \beta y + \gamma z + u(x + \alpha) + v(y + \beta) + w(z + \gamma) + d = 0 \quad \text{(on using (4).)}$$

This equation represents a plane and is the required locus. ▮

The locus of the tangent lines to a sphere at a point on it is called the **tangent plane** *at that point.*

Ex. Prove that the equation of the tangent plane to the sphere $x^2 + y^2 + z^2 = a^2$ at the point (α, β, γ) on it is $\alpha x + \beta y + \gamma z = a^2$.

Working rule. The equation of the tangnet plane at any point (α, β, γ) to a sphere is obtained by writing αx for x^2, βy for y^2, γz for z^2, $x + \alpha$ for $2x$, $y + \beta$ for $2y$ and $z + \gamma$ for $2z$.

3.3 *To prove that the tangent plane to a sphere at any point on it is perpendicular to the radius of the sphere through that point.*

A tangent line at a point to a sphere lies in the tangent plane at that point. Hence, the result follows.

Corollary. The length of the perpendicular from the centre of a sphere to any tangent plane (or tangent line) of it is equal to the radius of the sphere.

Remark. A tangent line to a sphere is the tangent to the circle obtained as section of the sphere by the plane through the tangent line.

3.4 Condition of tangency. *To find the condition when the plane* $lx + my + nz = p$ *becomes a tangent plane to the sphere* $x^2 + y^2 + z^2 + 2ux + 2vy + 2wz + d = 0$.

First Method. Equation of the tangent plane to the given sphere at any point (α, β, γ) on it is $\alpha x + \beta y + \gamma z + u(x + \alpha) + v(y + \beta) + w(z + \gamma) + d = 0$, *i.e.*,

$$(\alpha + u)x + (\beta + v)y + (\gamma + w)z + u\alpha + v\beta + w\gamma + d = 0.$$

The plane $lx + my + nz = p$ is a tangent plane to the given sphere if

$$(\alpha + u)x + (\beta + v)y + (\gamma + w)z + u\alpha + v\beta + w\gamma + d \equiv lx + my + nz - p$$

$$\Rightarrow \quad \frac{\alpha + u}{l} = \frac{\beta + v}{m} = \frac{\gamma + w}{n} = \frac{(u\alpha + v\beta + w\gamma + d)}{-p}$$

$$\Rightarrow \quad \frac{\alpha + u}{l} = \frac{\beta + v}{m} = \frac{\gamma + w}{n} = \frac{-(u\alpha + v\beta + w\gamma + d)}{p}$$

$$= \frac{u(\alpha + u) + v(\beta + v) + w(\gamma + w) - (u\alpha + v\beta + w\gamma + d)}{lu + mv + nw + p}$$

$$= \frac{u^2+v^2+w^2-d}{lu+mv+nw+p}. \qquad (1)$$

Further

$$\frac{\alpha+u}{l} = \frac{\beta+v}{m} = \frac{\gamma+w}{n} = \frac{\sqrt{(\alpha+u)^2+(\beta+v)^2+(\gamma+w)^2}}{\sqrt{l^2+m^2+n^2}}$$

$$= \frac{\sqrt{u^2+v^2+w^2-d}}{\sqrt{l^2+m^2+n^2}} \qquad (2)$$

From (1) and (2), we get

$$\frac{u^2+v^2+w^2-d}{lu+mv+nw+p} = \frac{\sqrt{u^2+v^2+w^2-d}}{\sqrt{l^2+m^2+n^2}}$$

$\Rightarrow$ $(lu + mv + nw + p)^2 = (l^2 + m^2 + n^2)(u^2 + v^2 + w^2 - d)$.

This is the required condition of tangency.

Second Method. The plane $lx + my + nz = p$ will be a tangent plane to the given sphere if perpendicular distance of the plane from the centre of the sphere is equal to the radius of the sphere *i.e.*, if

$$\frac{-lu-mv-nw-p}{\sqrt{l^2+m^2+n^2}} = \sqrt{u^2+v^2+w^2-d}$$

$\Rightarrow$ $(lu + mv + nw + p)^2 = (l^2 + m^2 + n^2)(u^2 + v^2 + w^2 - d)$.

This is the required condition of tangency.

Ex. Prove that the plane $lx + my + nz = p$ touches the sphere $x^2 + y^2 + z^2 = a^2$ if $p = a\sqrt{l^2+m^2+n^2}$.

Remarks.

(*i*) The plane $lx + my + nz = a\sqrt{l^2+m^2+n^2}$ is always tangent plane to the sphere $x^2 + y^2 + z^2 = a^2$ whatever the values of l, m, n may be. In fact, it is a family of tangent planes to the sphere $x^2 + y^2 + z^2 = a^2$.

(*ii*) The point of contact of the tangent plane $lx + my + nz = a\sqrt{l^2+m^2+n^2}$ is $\left(\frac{a^2 l}{p}, \frac{a^2 m}{p}, \frac{a^2 n}{p}\right)$, where $p = a\sqrt{l^2+m^2+n^2}$.

3.5 Examples

Example. Find the equations of the spheres which pass through the circle

$$x^2 + y^2 + z^2 = 5,\ x + 2y + 3z = 3$$

and touch the plane $4x + 3y - 15 = 0$.

Solution. The equation of the sphere passing through the given circle is

$$x^2 + y^2 + z^2 - 5 + \lambda(x + 2y + 3z - 3) = 0. \qquad (1)$$

Centre of (1) is $\left(-\frac{\lambda}{2}, -\lambda, \frac{-3\lambda}{2}\right)$.

Radius of (1) is $\sqrt{\frac{\lambda^2}{4} + \lambda^2 + \frac{9\lambda^2}{4} + 5 + 3\lambda} = \frac{1}{2}\sqrt{14\lambda^2 + 12\lambda + 20}.$

The sphere (1) will touch the plane $4x + 3y - 15 = 0$ if perpendicular distance of the plane from the centre of the sphere is equal to its radius, i.e., if

$$5\lambda^2 - 6\lambda - 8 = 0,\ i.e.,\ \lambda = 2, -4/5.$$

Substituting these values of λ in (1), the required equations of the spheres are

$$x^2 + y^2 + z^2 + 2x + 4y + 6z - 11 = 0,$$

and

$$5(x^2 + y^2 + z^2) - 4x - 8y - 12z - 13 = 0.$$

Exercises

1. If any tangent plane to the sphere $x^2 + y^2 + z^2 = r^2$ makes the intercepts a, b, c on the coordinate axes, prove that $\frac{1}{a^2} + \frac{1}{b^2} + \frac{1}{c^2} = \frac{1}{r^2}$.
2. Find the equations of the spheres through the circle
$$x^2 + y^2 + z^2 = 1,\ 2x + 4y + 5z = 6$$
and touching the plane $z = 0$.
3. Find the equations of the spheres passing through the circle
$$x^2 + y^2 + z^2 - 6x - 2z + 5 = 0,\ y = 0$$
and touching the plane $3y + 4z + 5 = 0$.
4. Prove that the plane $2x - 2y + z + 12 = 0$ touches the sphere
$$x^2 + y^2 + z^2 - 2x - 4y + 2z - 3 = 0;$$
find the coordinates of the point of contact.

5. Find the equations of the tangent planes to the sphere

$$x^2 + y^2 + z^2 - 4x + 2y - 6z + 5 = 0,$$

which are parallel to the plane $2x + y - z = 4$.

6. Find the equations of the spheres which pass through the points (4, 1, 0), (2, –3, 4), (1, 0, 0) and touch the plane $2x + 2y - z = 11$.

7. Find the equations of the tangent planes to the sphere

$$x^2 + y^2 + z^2 + 2x - 4y + 6z - 7 = 0$$

which intersect in the line $6x - 3y - 23 = 0,\ 3x + 2 = 0$.

4. Plane of Contact and Polar Plane

4.1 *To prove that the points of contact of the tangent planes drawn through a point to a sphere lie on a plane.*

Let (α, β, γ) be a given point and

$$x^2 + y^2 + z^2 + 2ux + 2vy + 2wz + d = 0 \qquad (1)$$

a given sphere.

Let (x', y', z') be the point of contact of the tangent plane drawn through the point (α, β, γ) to the sphere. Then, the equation of the tangent plane at (x', y', z') is

$$xx' + yy' + zz' + u(x + x') + v(y + y') + w(z + z') + d = 0.$$

But this plane passes through the point (α, β, γ). Therefore

$$\alpha x' + \beta y' + \gamma z' + u(\alpha + x') + v(\beta + y') + w(\gamma + z') + d = 0.$$

Thus, it is clear that the point (x', y', z') satisfies the equation

$$\alpha x + \beta y + \gamma z + u(x + \alpha) + v(y + \beta) + w(z + \gamma) + d = 0 \qquad (2)$$

which represents a plane.

Corollary. The locus of the points of contact of the tangent planes drawn through a point to a sphere is a circle whose equation is given by

$$\begin{cases} x^2 + y^2 + z^2 + 2ux + 2vy + 2wz + d = 0 \\ \alpha x + \beta y + \gamma z + u(x + \alpha) + v(y + \beta) + w(z + \gamma) + d = 0. \end{cases}$$

Note. In two dimension case, there are only two points of contact of the tangents drawn from a point outside to a circle whereas in case of three dimensional space an infinite number of tangent planes can be drawn through a point outside to a sphere and all the points of contact form a circle.

In analogy with the concept of chord of contact of tangents from a point to a circle in two dimension case, we can call the plane (2) as the plane of contact of tangent planes drawn through the point (α, β, γ) to the sphere (1). More precisely, *the plane through the points of contact of the tangent planes drawn through a point* (α, β, γ) *to a sphere (or a conicoid, in general) is called the* **plane of contact** *of the tangent planes through* (α, β, γ) *to the sphere (or conicoid)*.

Remarks.

(*i*) We note that the equation of the plane of contact of tangent planes drawn through a point (α, β, γ) to a sphere is of the same form as that the equation of the tangent plane at (α, β, γ) to the sphere. The reason is simple in case the point (α, β, γ) lies on the sphere, the points lying on the circle (section of sphere by the plane of contact) coincide with the point (α, β, γ), and in this limiting position the plane of contact becomes the tangent plane to the sphere at the point (α, β, γ).

(*ii*) If the point (α, β, γ) lies outside the sphere, the situation is quite simple and clear geometrically also.

(*iii*) If the point (α, β, γ) lies inside the sphere, the tangent planes drawn through the point (α, β, γ) to the sphere will be imaginary. But the plane of contact still exists and is real since α, β and γ are real. Thus, there is a real plane passing through the imaginary points of contact of the imaginary tangent planes drawn through a point inside the sphere. This is consistent with the result that a plane always meets a sphere in a circle, real or imaginary.

Thus, the plane of contact of the tangent planes drawn through a point (α, β, γ) to a sphere always exists and is real whatever the position of the point (α, β, γ) may be, outside, on or inside the sphere.

4.2 Polar plane

If any line through a fixed point A meets a sphere (or conicoid, in general) in two points P and Q and if R is the harmonic conjugate of A with respect to P and Q (i.e. $1/AP + 1/AQ = 2/AR$*), then the locus of R is a plane which is called the* **polar plane** *of A with respect to the sphere (or conicoid) and the point A is called the* **pole** *of the polar plane.*

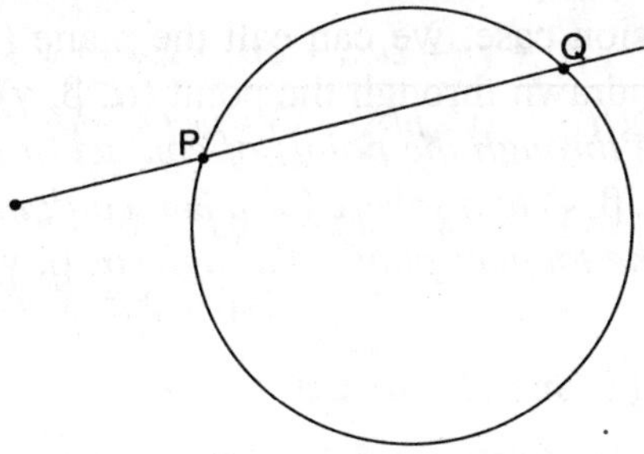

Fig. 15.4

4.3 *To find the equation of the polar plane of the point* (α, β, γ) *with respect to the sphere*

$$x^2 + y^2 + z^2 + 2ux + 2vy + 2wz + d = 0.$$

Let A be the point (α, β, γ). Then, equations of the line through the point A meeting the given sphere in P and Q are

$$\frac{x-\alpha}{l} = \frac{y-\beta}{m} = \frac{z-\gamma}{n} = r, \qquad (1)$$

where l, m, n are the actual direction cosines of the line. Any point on the line (1) is

$$(\alpha + lr, \beta + mr, \gamma + nr).$$

If this point lies on the sphere, we must have

$$(\alpha + lr)^2 + (\beta + mr)^2 + (\gamma + nr)^2 + 2u(\alpha + lr) + 2v(\beta + mr) + 2w(\gamma + nr) + d = 0$$

$$\Rightarrow r^2(l^2 + m^2 + n^2) + 2r\{l(\alpha + u) + m(\beta + v) + n(\gamma + w)\} + \alpha^2 + \beta^2 + \gamma^2 + 2u\alpha + 2v\beta + 2w\gamma + d = 0$$

$$\Rightarrow r^2 + 2r\{l(\alpha + u) + m(\beta + v) + n(\gamma + w)\} + \alpha^2 + \beta^2 + \gamma^2 + 2u\alpha + 2v\beta + 2w\gamma + d = 0.$$

This equation being quadratic has two values of r say r_1 and r_2. Then $AP = r_1$ and $AQ = r_2$. Let R be the point on APQ such that $AR = r$. Then, coordinates of R are $(\alpha + lr, \beta + mr, \gamma + nr)$ and

$$\frac{1}{r_1} + \frac{1}{r_2} = \frac{2}{r}, \text{ i.e. } r = \frac{2r_1r_2}{r_1 + r_2}.$$

But $\quad r_1 + r_2 = -2\{l(\alpha + u) + m(\beta + v) + n(\gamma + w)\},$

and $\quad r_1r_2 = \alpha^2 + \beta^2 + \gamma^2 + 2u\alpha + 2v\beta + 2w\gamma + d.$

Therefore $r = \dfrac{2(\alpha^2+\beta^2+\gamma^2+2u\alpha+2v\beta+2w\gamma+d)}{-2\{l(\alpha+u)+m(\beta+v)+n(\gamma+w)\}}$

$$\Rightarrow \quad lr(\alpha+u)+mr(\beta+v)+nr(\gamma+w) = -(\alpha^2+\beta^2+\gamma^2+2u\alpha+2v\beta+2w\gamma+d). \quad (2)$$

Eliminating l, m, n from (1) and (2), we get

$$(x-\alpha)(\alpha+u)+(y-\beta)(\beta+v)+(z-\gamma)(\gamma+w) = -(\alpha^2+\beta^2+\gamma^2+2u\alpha+2v\beta+2w\gamma+d)$$

$$\Rightarrow \quad \alpha x+\beta y+\gamma z+u(x+\alpha)+v(y+\beta)+w(z+\gamma)+d=0.$$

This is the required equation of the polar plane of the point A with respect to the given sphere.

Ex. Prove that the equation of the polar plane of any point (α, β, γ) with respect to the sphere $x^2+y^2+z^2=a^2$ is $\alpha x+\beta y+\gamma z=a^2$.

4.4 Pole of a plane. *To find the pole of the plane*

$$lx+my+zn=p$$

with respect to the sphere $x^2+y^2+z^2=a^2$.

Let (x_1, y_1, z_1) be the pole of the plane $lx+my+nz=p$ with respect to the sphere $x^2+y^2+z^2=a^2$. Then, this plane must be identical with the polar plane of (x_1, y_1, z_1) with respect to the sphere, i.e., with $xx_1+yy_1+zz_1=a^2$. Therefore

$$\frac{x_1}{l}=\frac{y_1}{m}=\frac{z_1}{n}=\frac{a^2}{p}, \text{ i.e., } x_1=\frac{a^2 l}{p}, y_1=\frac{a^2 m}{p}, z_1=\frac{a^2 n}{p}.$$

Hence, the required pole is $\left(\dfrac{a^2 l}{p}, \dfrac{a^2 m}{p}, \dfrac{a^2 n}{p}\right)$.

4.5 Propositions on poles and polar planes

Proposition I. *The polar plane is perpendicular to the line joining the centre of the sphere to the pole.*

Let $x^2+y^2+z^2=a^2$ be the equation of the sphere and $P(x_1, y_1, z_1)$ the pole. Then, equation of the polar plane is $xx_1+yy_1+zz_1=a^2$.

The direction cosines of the normal to the polar plane are proportional to x_1, y_1, z_1 and the direction cosines of the line joining the centre $O(0, 0, 0)$ to the point $P(x_1, y_1, z_1)$ are proportional to x_1, y_1, z_1. Therefore, the line OP is parallel to the normal to the polar plane. Hence, the result follows.

Proposition II. *The pole and the point of intersection of the polar plane with the join of the centre and the pole are inverse points* with respect to the sphere* $x^2 + y^2 + z^2 = a^2$.

Let C be the centre of the sphere and Q the point of intersection of the polar plane with CP, where $P(x_1, y_1, z_1)$ is the pole. Then, by Proposition I, we note that

$$CQ = \text{perpendicular distance of } C \text{ from the polar plane of } P$$

$$= \frac{a^2}{\sqrt{x_1^2 + y_1^2 + z_1^2}},$$

and $$CP = \sqrt{x_1^2 + y_1^2 + z_1^2}.$$

Therefore $\quad CP \cdot CQ = a^2 = (\text{radius})^2.$

Hence, the points P and Q are the inverse points with respect to the sphere $x^2 + y^2 + z^2 = a^2$.

Proposition III. *If the polar plane of a point P passes through the point Q, then the polar plane of Q passes through P.*

Let (x_1, y_1, z_1) and (x_2, y_2, z_2), respectively, be the coordinates of P and Q. Since the polar plane of the point P passes through the point Q, we have

$$x_1x_2 + y_1y_2 + z_1z_2 = a^2.$$

But this is the condition also that the point (x_1, y_1, z_1) should lie on the plane

$$xx_2 + yy_2 + zz_2 = a^2,$$

which, in fact, is the polar plane of the point Q. This proves the proposition.

Two points such that each lies on the polar plane of the other are called **conjugate points.**

Proposition IV. *If the pole of a plane* L_1, *lies on another plane* L_2, *then the pole of* L_2, *lies on* L_1.

Let $P(x_1, y_1, z_1)$ and $Q(x_2, y_2, z_2)$ be the poles of the planes L_1 and L_2, respectively. In other words, L_1 and L_2 are the polar planes of P and Q, respectively. Since the polar plane of Q passes through P, by Proposition III, the polar plane of P(i.e. the plane L_1) must pass through Q. Hence the proposition follows.

* Two points P and Q are said to be inverse points with respect to a sphere if $CP \cdot CQ = (\text{radius})^2$, C being the centre of the sphere.

Two planes such that each contains the pole the other are called **conjugate planes.**

Proposition V. *The polar planes of all points on a line pass through another line.*

Consider the line

$$\frac{x-\alpha}{l}=\frac{y-\beta}{m}=\frac{z-\gamma}{n}=r\text{ (say)} \tag{1}$$

Any point on the line (1) is $(\alpha + lr, \beta + mr, \gamma + nr)$. The polar plane of this point with respect to the sphere $x^2 + y^2 + z^2 = a^2$ is

$$\alpha x + \beta y + \gamma z - a^2 + r(lx + my + nz) = 0$$

This plane clearly passes through the line of intersection of the planes

$$\left.\begin{aligned}\alpha x+\beta y+\gamma z-a^2&=0\\ lx+my+nz&=0\end{aligned}\right\}. \tag{2}$$

Hence, the polar plane of any point on the line (1) passes through the line (2).

Similarly, we can prove that the polar plane of any point on the line (2) passes through the line (1).

Two lines such that the polar plane of each point of one line passes through the other line are called **polar lines.**

Exercises

1. Prove that the polar plane of any point on the line

$$\frac{x}{2}=\frac{y-1}{3}=\frac{z+3}{4}$$

with respect to the sphere $x^2 + y^2 + z^2 = 1$ passes through the line

$$\frac{2x+3}{13}=\frac{y-1}{-3}=\frac{z}{-1}.$$

2. Prove that the polar line of $\frac{x+3}{1}=\frac{y+1}{2}=\frac{z-2}{3}$ with respect to the sphere $x^2 + y^2 + z^2 = 1$ is the line

$$\frac{x}{-1}=\frac{7y+3}{11}=\frac{7z-2}{-5}.$$

5. Intersection of Two Spheres

Let the equations of any two spheres be

$$S_1 \equiv x^2 + y^2 + z^2 + 2u_1x + 2v_1y + 2w_1z + d_1 = 0, \tag{1}$$

$$S_2 \equiv x^2 + y^2 + z^2 + 2u_2x + 2v_2y + 2w_2z + d_2 = 0. \tag{2}$$

Now, consider the equation $S_1 + \lambda S_2 = 0$, where λ is an arbitrary constant. This equation takes the form of

$$(1 + \lambda)x^2 + (1 + \lambda)y^2 + (1 + \lambda)z^2 + 2(u_1 + \lambda u_2)x + 2(v_1 + \lambda v_2)y + 2(w_1 + \lambda w_2)z + d_1 + \lambda d_2 = 0. \tag{3}$$

The coordinates (x, y, z) which satisfy (1) and (2), also satisfy (3). Therefore, equation (3) represents, in general, a surface (sphere or plane) through the circle of intersection of (1) and (2).

Case I. $\lambda = -1$. Equation (3) becomes a linear equation in x, y and z and as such represents a plane through the circle of intersection of the spheres (1) and (2).

Case II. $\lambda \neq -1$. Equation (3) represents spheres through the circle of intersection of the spheres $S_1 = 0$ and $S_2 = 0$. In this case it gives a family of such spheres since for each value of $\lambda(\neq -1)$ there is a member of the family.

5.1 Angle of intersection of two spheres.

The **angle of intersection** *of any two spheres at a common point is the angle between the tangent planes at that point.*

Two spheres are said to cut **orthogonally** *if their angle of intersection is a right angle.*

5.2 Condition for orthogonal intersection of two spheres. *To find a necessary and sufficient condition for the spheres*

$$x^2 + y^2 + z^2 + 2u_1x + 2v_1y + 2w_1z + d_1 = 0, \tag{1}$$

and

$$x^2 + y^2 + z^2 + 2u_2x + 2v_2y + 2w_2z + d_2 = 0 \tag{2}$$

to intersect orthogonally.

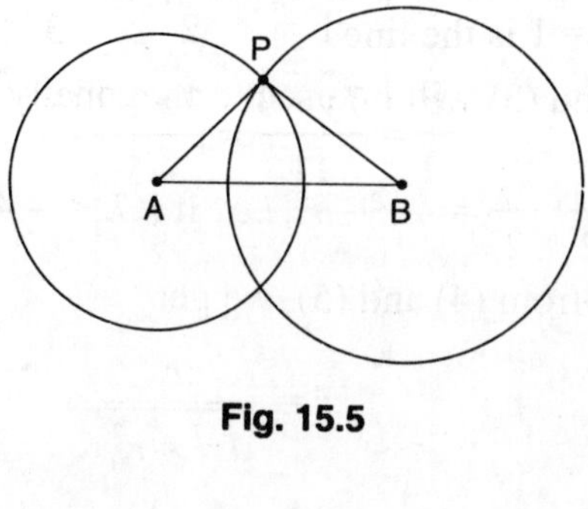

Fig. 15.5

Let A and B be the centres of the two spheres, respectively, and P any point on the circle of intersection. Suppose the two spheres intersect orthogonally. Then, the radii AP and PB are at right angles. Therefore, $AB^2 = AP^2 + PB^2$. But $AB^2 = (u_1 - u_2)^2 + (v_1 - v_2)^2 + (w_1 - w_2)^2$, $AP^2 = u_1^2 + v_1^2 + w_1^2 - d_1$ and $PB^2 = u_2^2 + v_2^2 + w_2^2 - d_2$. Therefore

$$(u_1 - u_2)^2 + (v_1 - v_2)^2 + (w_1 - w_2)^2 = u_1^2 + v_1^2 + w_1^2 - d_1 + u_2^2 + v_2^2 + w_2^2 - d_2$$

$$\Rightarrow \quad 2u_1u_2 + 2v_1v_2 + 2w_1w_2 = d_1 + d_2.$$

This is the required necessary condition for the spheres (1) and (2) to intersect orthogonally. This condition can also be proved to be sufficient by working the same steps backwards.

5.3 Examples

Example 1. Two spheres of radii r_1 and r_2 intersect orthogonally. Prove that the radius of the common circle is $\dfrac{r_1 r_2}{\sqrt{r_1^2 + r_2^2}}$.

Solution. Let the equation of the common circle be

$$x^2 + y^2 = r^2, z = 0. \qquad (1)$$

The equation of the sphere through the circle (1) is $x^2 + y^2 + z^2 - r^2 + \lambda z = 0$, where λ is arbitrary constant. Thus, the two spheres through the circle (1) are given by

$$x^2 + y^2 + z^2 + \lambda_1 z - r^2 = 0, \qquad (2)$$

and

$$x^2 + y^2 + z^2 + \lambda_2 z - r^2 = 0. \qquad (3)$$

Since r_1 and r_2 are the radii of these spheres, we have

$$\left.\begin{aligned} r_1 &= \sqrt{\frac{\lambda_1^2}{4} + r^2} = \frac{1}{2}\sqrt{\lambda_1^2 + 4r^2} \\ \text{and} \quad r_2 &= \sqrt{\frac{\lambda_2^2}{4} + r^2} = \frac{1}{2}\sqrt{\lambda_2^2 + 4r^2} \end{aligned}\right\}. \qquad (4)$$

The spheres (2) and (3) will intersect orthogonally if

$$2 \cdot \frac{\lambda_1}{2} \cdot \frac{\lambda_2}{2} = -r^2 - r^2, \text{ i.e., if } \lambda_1\lambda_2 = -4r^2. \qquad (5)$$

Eliminating λ_1 and λ_2 from (4) and (5), we get

$$r = \frac{r_1 r_2}{\sqrt{r_1^2 + r_2^2}}.$$

Example 2. Two points P and Q are conjugate with respect to a sphere S, prove that the sphere on PQ as diameter cuts S orthogonally.

Solution. Let the equation of the sphere be

$$S \equiv x^2 + y^2 + z^2 - a^2 = 0. \tag{1}$$

Let (x_1, y_1, z_1) and (x_2, y_2, z_2) be the coordinates of the point P and Q, respectively. The polar plane of $P(x_1, y_1, z_1)$ with respect to (1) is

$$xx_1 + yy_1 + zz_1 - a^2 = 0. \tag{2}$$

But the points P and Q are conjugate points with respect to (1). Hence, the polar plane (2) of P will pass through Q.

$$\therefore \quad x_1x_2 + y_1y_2 + z_1z_2 - a^2 = 0. \tag{3}$$

The equation of the sphere on PQ as diameter is

$$(x - x_1)(x - x_2) + (y - y_1)(y - y_2) + (z - z_1)(z - z_2) = 0$$

i.e. $x^2 + y^2 + z^2 - (x_1 + x_2)x - (y_1 + y_2)y - (z_1 + z_2)z + x_1x_2 + y_1y_2 + z_1z_2 = 0.$

This sphere will cut the sphere (1) orthogonally if

$$x_1x_2 + y_1y_2 + z_1z_2 - a^2 = 0,$$

which is true by virtue of (3). Hence the result follows.

Exercises

1. Prove that the two spheres $x^2 + y^2 + z^2 + 6y + 2z + 8 = 0$, and $x^2 + y^2 + z^2 + 6x + 8y + 4z + 20 = 0$ intersect orthogonally.
2. Find the equation of the sphere which passes through the circle
$$x^2 + y^2 + z^2 - 2x + 3y - 4z + 6 = 0,\ 3x - 4y + 5z - 15 = 0$$
and cuts the sphere $x^2 + y^2 + z^2 + 2x + 4y - 6z + 11 = 0$ orthogonally.
3. Prove that every sphere through the circle $x^2 + y^2 - 2ax + r^2 = 0, z = 0$ cuts orthogonally every sphere through the circle $x^2 + z^2 = r^2, y = 0$.
4. Find the equation of the sphere which touches the plane $3x + 2y - z + 2 = 0$ at the point $(1, -2, 1)$ and cuts the sphere $x^2 + y^2 + z^2 - 4x + 6y + 4 = 0$ orthogonally.
5. Prove that a sphere which cuts two spheres $S = 0$ and $S' = 0$ orthogonally also cuts the sphere $\lambda S + \mu S' = 0$ orthogonally.
6. Find the equation of the sphere which cuts orthogonally each of the spheres $x^2 + y^2 + z^2 = a^2 + b^2 + c^2, x^2 + y^2 + z^2 + 2ax = a^2, x^2 + y^2 + z^2 + 2by = b^2, x^2 + y^2 + z^2 + 2cz = c^2$.

6. Radical Plane

6.1 *To find the locus of a point whose powers with respect to two spheres are equal.*

Let the equations of the two spheres be

$$x^2 + y^2 + z^2 + 2u_1x + 2v_1y + 2w_1z + d_1 = 0, \quad (1)$$

and
$$x^2 + y^2 + z^2 + 2u_2x + 2v_2y + 2w_2z + d_2 = 0. \quad (2)$$

Let (x', y', z') be the point such that its powers with respect to the spheres (1) and (2) are equal i.e.

$$x'^2 + y'^2 + z'^2 + 2u_1x' + 2v_1y' + 2w_1z' + d_1 = x'^2 + y'^2 + z'^2 + 2u_2x' + 2v_2y' + 2w_2z' + d_2$$

$$\Rightarrow \quad 2(u_1 - u_2)x' + 2(v_1 - v_2)y' + 2(w_1 - w_2)z' + d_1 - d_2 = 0.$$

Hence, the locus of (x', y', z') is

$$2(u_1 - u_2)x + 2(v_1 - v_2)y + 2(w_1 - w_2)z + d_1 - d_2 = 0.$$

This equation being linear in x, y and z represents a plane. ▌

The locus of a point whose powers with respect to two spheres are equal is called the **radical plane** *of the two spheres.*

Working rule. In order to obtain the radical plane of two spheres, reduce the equations of the spheres to the standard form in such a way that the coefficients of x^2, y^2 and z^2 are unity in the equations of the spheres and then subtract one equation from the other. Thus, the radical plane of two spheres

$$S_1 \equiv x^2 + y^2 + z^2 + 2u_1x + 2v_1y + 2w_1z + d_1 = 0,$$

$$S_2 \equiv x^2 + y^2 + z^2 + 2u_2x + 2v_2y + 2w_2z + d_2 = 0$$

is given by $S_1 - S_2 = 0$.

Note. The radical plane of two spheres passes through the common points, real or imaginary, of the two spheres (compare the equation of the radical plane with that of the plane through the circle of intersection of the two spheres as obtained in § 5). As such, the concept of the radical plane seems to be wider than that of the plane through the circle of intersection or the common tangent plane of the two spheres since if

(*i*) the circle of intersection of two spheres is real, the radical plane coincides with the plane through the circle of intersection;

(*ii*) the circle of intersection of two spheres is a point circle, the radical plane coincides with the tangent plane at the common point;

(*iii*) the circle of intersection of two spheres is imaginary, the radical plane does exist even though there are no real points common to the two spheres.

6.2 Propositions on radical plane

Proposition I. *The radical palne of two spheres is perpendicular to the line joining their centres.*

Let the equations of the two spheres be

$$x^2 + y^2 + z^2 + 2u_1x + 2v_1y + 2w_1z + d_1 = 0, \quad (1)$$

and

$$x^2 + y^2 + z^2 + 2u_2x + 2v_2y + 2w_2z + d_2 = 0. \quad (2)$$

The centers of these spheres are $(-u_1, -v_1, -w_1)$ and $(-u_2, -v_2, -w_2)$, respectively. Therefore, the direction cosines of the line joining them are proportional to

$$u_1 - u_2, v_1 - v_2, w_1 - w_2.$$

Radical plane of the spheres (1) and (2) is

$$2(u_1 - u_2)x + 2(v_1 - v_2)y + 2(w_1 - w_2)z + d_1 - d_2 = 0. \quad (3)$$

The direction cosines of the normal to the radical plane (3) are proportional to

$$u_1 - u_2, v_1 - v_2, w_1 - w_2.$$

Thus, the line joining the centres of the spheres (1) and (2) is parallel to the normal to the radical plane (3). Hence, the proposition follows.

Proposition II. *The radical planes of three spheres taken in pairs pass through a line.*

Let the equations of the spheres be $S_1 = 0$, $S_2 = 0$, $S_3 = 0$. The equations of the radical planes of these spheres taken in pairs are

$$S_1 - S_2 = 0, S_2 - S_3 = 0 \text{ and } S_3 - S_1 = 0.$$

These equations show that the radical planes of the three spheres pass through the line $S_1 = S_2 = S_3$.

Hence the proposition follows. ▌

The line of concurrence of the three radical planes of three spheres taken in pairs is called the **radical line** *of the three spheres.*

Proposition III. *The radical planes of four spheres taken in pairs meet in a point.*

Let the equations of the spheres be

$$S_1 = 0, S_2 = 0, S_3 = 0, S_4 = 0.$$

The equations of the radical planes of these spheres taken in pairs are

$$S_1 - S_2 = 0, S_1 - S_3 = 0, S_1 - S_4 = 0, S_2 - S_3 = 0, S_2 - S_4 = 0, S_3 - S_4 = 0.$$

These equations show that the radical planes of four spheres meet in a point given by

$$S_1 = S_2 = S_3 = S_4.$$

This proves the proposition.

The point of concurrence of the radical planes of four spheres taken in pairs is called the **radical centre** *of the four spheres.*

Exercises

1. Find the equations of the radical planes of the spheres

$$x^2 + y^2 + z^2 + 2x + 2y + 2z + 2 = 0,$$
$$x^2 + y^2 + z^2 + 4y = 0,$$
and
$$x^2 + y^2 + z^2 + 3x - 2y + 8z + 6 = 0.$$

2. Find the equation of the radical line of the spheres

$$(x - 2)^2 + y^2 + z^2 = 1,$$
$$x^2 + (y - 3)^2 + z^2 = 6,$$
and
$$(x + 2)^2 + (y + 1)^2 + (z - 2)^2 = 6.$$

3. Find the locus of a point the powers of which with respect to the spheres

$$x^2 + y^2 + z^2 = 1,$$
$$x^2 + y^2 + z^2 + 2x - 2y + 2z + 1 = 0,$$
and
$$x^2 + y^2 + z^2 - x + 4y - 6z - 2 = 0$$
are equal.

7. Coaxal Spheres

Let

$$S_1 \equiv x^2 + y^2 + z^2 + 2u_1x + 2v_1y + 2w_1z + d_1 = 0,$$
and
$$S_2 \equiv x^2 + y^2 + z^2 + 2u_2x + 2v_2y + 2w_2z + d_2 = 0$$

be the equations of any two spheres.

Now, consider the equation

$$S_1 + \lambda S_2 = 0, \qquad (1)$$

where λ is a constant.

The value $\lambda = -1$ corresponds to the radical plane $S_1 - S_2 = 0$ of the two spheres $S_1 = 0$ and $S_2 = 0$ whereas any other value of λ yields a sphere. Thus, equation (1) represents a system of spheres, except when $\lambda = -1$. Consider now two distinct values λ_1 and λ_2 of λ. Then $S_1+\lambda_1 S_2=0$ and $S_1+\lambda_2 S_2=0$ are the equations of the two spheres of the family (1), and their radical plane is given by

$$\frac{S_1+\lambda_1 S_2}{1+\lambda_1}-\frac{S_1+\lambda_2 S_2}{1+\lambda_2}=0$$

$$\Rightarrow \qquad (\lambda_2-\lambda_1)(S_1-S_2)=0$$

$$\Rightarrow \qquad S_1-S_2=0 \qquad (\because \lambda_1 \neq \lambda_2).$$

Thus, we note that any two distinct spheres of the family (1) have the radical plane $S_1-S_2=0$ which is independent of λ.

A system of spheres is said to be **coaxal** *if every pair of spheres of the system has the same radical plane.*

Thus $S_1+\lambda S_2=0$ represents a coaxal system of spheres.

7.1 *To prove that the centres of all spheres of a coaxal system are collinear and lie on a line perpendicular to the common radical plane.*

We note that the centre of the coaxal system $S_1+\lambda S_2=0$ is

$$\left(-\frac{u_1+\lambda u_2}{1+\lambda}, -\frac{v_1+\lambda v_2}{1+\lambda}, -\frac{w_1+\lambda w_2}{1+\lambda}\right).$$

Clearly, this point lies on the line

$$\frac{x+u_1}{u_1-u_2}=\frac{y+v_1}{v_1-v_2}=\frac{z+w_1}{w_1-w_2}$$

joining the centers of the spheres $S_1=0$ and $S_2=0$. Thus, the centres of all the spheres in the system are collinear. It is trivial to see that the line joining the centres is perpendicular to the common radical plane $S_1-S_2=0$.

7.2 Standard form of the equation to a coaxal system. *To find the equation of a coaxal system of spheres in its standard form.*

Let us choose the line of centres as the *x*-axis and the common radical plane (perpendicular to the line of centres) as *yz*-plane.

Since the centre of each sphere lies on the *x*-axis, the *y*-and *z*-coordinates of it are zero. As such any two spheres of the system may be represented by the equations

$$\left.\begin{array}{l} x^2+y^2+z^2+2ux+d=0 \\ x^2+y^2+z^2+2u_1x+d_1=0 \end{array}\right\}. \qquad (1)$$

The radical plane of these spheres is

$$2(u - u_1)\,x + d - d_1 = 0.$$

But the radical plane is $x = 0$. Therefore $d_1 = d$.

Hence, the equation of any sphere in the system takes the form

$$x^2 + y^2 + z^2 + 2ux + d = 0,$$

where the constant term d is fixed for the whole system, and the coefficient u is a parameter which varies from one sphere to another in the family.

7.3 Limiting points of a coaxal system

The equation $x^2 + y^2 + z^2 + 2ux + d = 0$, which gives any sphere of a coaxal system can be written as

$$(x + u)^2 + y^2 + z^2 = (\sqrt{u^2 - d})^2$$

This equation represents a sphere whose centre is the point $(-u, 0, 0)$ and radius $= \sqrt{u^2 - d}$. If the radius of the sphere is zero *i.e.* $u^2 - d = 0$, the sphere becomes a point sphere. But in this case $u = \pm\sqrt{d}$. Thus, at the particular points $(\sqrt{d}, 0, 0)$ and $(-\sqrt{d}, 0, 0)$, we have point spheres which are in the system.

The points $(\sqrt{d}, 0, 0)$ *and* $(-\sqrt{d}, 0, 0)$ *are called the* **limiting points** *of the coaxal system* $x^2 + y^2 + z^2 + 2ux + d = 0$.

We note that the limiting points are real or imaginary according as d is positive or negative.

Note. The sphere coaxal with two given spheres $S_1 = 0$ and $S_2 = 0$ can be written as $S_1 + \lambda S_2 = 0$ or $S_1 + \lambda L = 0$, where $L = 0$ is the equation of the radical plane of the two given spheres.

7.4 Intersection of spheres of a coaxal system

Let

$$x^2 + y^2 + z^2 + 2u_1 x + d = 0, \qquad (1)$$

and

$$x^2 + y^2 + z^2 + 2u_2 x + d = 0 \qquad (2)$$

be any two members of the coaxal system $x^2 + y^2 + z^2 + 2ux + d = 0$.

The spheres (1) and (2) intersect where

$$2(u_1 - u_2)x = 0, \text{ i.e., } x = 0 \qquad (\because u_1 \neq u_2)$$

Substituting $x = 0$ in (1), we get

$$y^2 + z^2 = -d \text{ i.e., } y^2 + z^2 = (\sqrt{-d})^2.$$

This equation represents a circle in yz-plane, where the radical plane meets a sphere in the system. Also, it is independent of u (the parameter of the system), every sphere of the sytem meets the radical plane in the same circle. ▌

The circle $y^2 + z^2 = (\sqrt{-d})^2$ *is called the* **common circle** *of the coaxal system.*

Now, there are three cases:

Case I. When d is negative, the common circle is real. In this case the system is said to be of **intersecting species.**

Case II. When $d = 0$, the common circle is point circle and hence every sphere of the sytem touches yz-plane (the radical plane) at the origin. In other words any two spheres of the system touch at the origin.

Case III. When d is positive, the common circle is imaginary and the system is said to be of **non-intersecting species.**

Note. The limiting points of a coaxal system are real or imaginary according as the common circle is imaginary or real. Consequently, if the limiting points are real, the system is in the non-intersecting species whereas if the limiting points are imaginary, the system is in the intersecting species. Hence, we see that, except in the second case of § 7.4 which is compromise between the two, there are either limiting points or common circle, of course real, but not both simultaneously.

7.5 Examples

Example 1. Find the equation of the sphere which is coaxal with the spheres

$$x^2 + y^2 + z^2 + 3x - 3y + 2z = 0,$$
$$x^2 + y^2 + z^2 + 2x - y - z + 10 = 0,$$

and passes through the point (0, 1, 2).

Solution. The equation of the sphere which is coaxal with the given spheres is

$$x^2 + y^2 + z^2 + 3x - 3y + 2z + \lambda(x^2 + y^2 + z^2 + 2x - y - z + 10) = 0. \quad (1)$$

This will pass through the point (0, 1, 2) if $\lambda = -\dfrac{1}{2}$.

Substituting this value of λ in (1), we get

$$x^2 + y^2 + z^2 + 4x - 5y + 5z - 10 = 0.$$

Example 2. Find the limiting points of the coaxal system of spheres determined by $x^2 + y^2 + z^2 + 4x - 2y + 2z + 6 = 0$, and $x^2 + y^2 + z^2 + 2x - 4y - 2z + 6 = 0$.

Solution. The radical plane of the given spheres is

$$x + y + 2z = 0.$$

The equation of the coaxal system containing the given spheres is

$$x^2 + y^2 + z^2 + 4x - 2y + 2z + 6 + \lambda(x + y + 2z) = 0$$

i.e. $$x^2 + y^2 + z^2 + (4 + \lambda)x + (-2 + \lambda)y + (2 + 2\lambda)z + 6 = 0. \quad (1)$$

Centre of (1) is $\left(-\dfrac{4+\lambda}{2}, -\dfrac{-2+\lambda}{2}, -\dfrac{2+2\lambda}{2}\right)$.

Radius of (1) is $\sqrt{\left(\dfrac{4+\lambda}{2}\right)^2 + \left(\dfrac{-2+\lambda}{2}\right)^2 + \left(\dfrac{2+2\lambda}{2}\right)^2 - 6}$.

For limiting points of the coaxal system (1), radius of (1) should be zero i.e.

$$\left(\frac{4+\lambda}{2}\right)^2 + \left(\frac{-2+\lambda}{2}\right)^2 + \left(\frac{2+2\lambda}{2}\right)^2 - 6 = 0$$

$$\Rightarrow \qquad \lambda = 0, -2.$$

Substituting these values of λ in the coordinates of the centre of (1), we get $(-2, 1, -1)$ and $(-1, 2, 1)$. These are the required limiting points.

Exercises

1. Find the limiting points of the coaxal system of spheres determined by
$$x^2 + y^2 + z^2 + 3x - 3y + 6 = 0,$$
and $$x^2 + y^2 + z^2 - 6y - 6z + 6 = 0.$$
2. Find the equation of the sphere belonging to the coaxal system of spheres determined by
$$x^2 + y^2 + z^2 - 2ax - 2ay - 2az + 4a^2 = 0, x^2 + y^2 + z^2 - 4ax - 4ay + 4a^2 = 0$$
and which cuts the sphere $x^2 + y^2 + z^2 + 2ax = 0$ orthogonally.
3. Find the equation of the sphere passing through the origin and coaxal with the sphere $x^2 + y^2 + z^2 - 2x - 4y - 6z + 10 = 0$ and the sphere which has the points $(1, 2, -3)$ and $(5, 0, 1)$ as the extremities of a diameter.
4. Prove that the locus of a point whose powers with respect to two given spheres are in a constant ratio is a sphere coaxal with two given spheres except in one particular case.

8. Conjugate System of Coaxal Spheres

A system of spheres through the limiting points, real or imaginary, of a coaxal system is of particular interest.

8.1 *To find the system of spheres through the limiting points* $(\pm\sqrt{d}, 0, 0)$ *of the coaxal system* $x^2 + y^2 + z^2 + 2ux + d = 0$.

Let the sphere

$$x^2 + y^2 + z^2 + 2u'x + 2v'y + 2w'z + d' = 0$$

pass through the limiting point $(\pm\sqrt{d}, 0, 0)$. Then

$$d + 2u'\sqrt{d} + d' = 0 \text{ and } d - 2u'\sqrt{d} + d' = 0$$

$$\Rightarrow \qquad d' = -d \text{ and } u' = 0.$$

Hence, the family of spheres through the limiting points $(\pm\sqrt{d}, 0, 0)$ is given by the equation

$$x^2 + y^2 + z^2 + 2vy + 2wz - d = 0, \tag{1}$$

where v and w (written for v' and w', respectively) are the parameters of the family.

Remark. The spheres of the system $x^2 + y^2 + z^2 + 2vy + 2wz - d = 0$ have a common radical line $y = 0$, $z = 0$, and pass through the fixed points $(\sqrt{d}, 0, 0)$ and $(-\sqrt{d}, 0, 0)$ which are

(*i*) real when $d > 0$, i.e., when the common circle of the system $x^2 + y^2 + z^2 + 2ux + d = 0$ is imaginary and hence the system (1) is of the intersecting species:

(*ii*) imaginary when $d < 0$, i.e. when the common circle of the system $x^2 + y^2 + z^2 + 2ux + d = 0$ is real and hence the system (1) is of the non-intersecting species.

8.2 *To prove that any sphere which cuts any two given spheres of the coaxal system* $x^2 + y^2 + z^2 + 2ux + d = 0$ *orthogonally is sphere of the system*

$$x^2 + y^2 + z^2 + 2vy + 2wz - d = 0.$$

Let the given two spheres of the coaxal system be represented by the equations

$$x^2 + y^2 + z^2 + 2u_1x + d = 0, \tag{1}$$

and

$$x^2 + y^2 + z^2 + 2u_2x + d = 0. \tag{2}$$

Let

$$x^2 + y^2 + z^2 + 2u'x + 2v'y + 2w'z + d' = 0 \tag{3}$$

be a sphere which cuts the spehres (1) and (2) orthogonally. Then, we have

$$2u_1u' - d - d' = 0 \text{ and } 2u_2u' - d - d' = 0$$

$$\Rightarrow \qquad u' = 0, d' = -d.$$

Thus, sphere (3) must be of the form

$$x^2 + y^2 + z^2 + 2v'y + 2w'z - d = 0,$$

which is a member of the family

$$x^2 + y^2 + z^2 + 2vy + 2wz - d = 0.$$

Hence, any sphere orthogonal to the coaxal system

$$x^2 + y^2 + z^2 + 2ux + d = 0,$$

is in the family $x^2 + y^2 + z^2 + 2vy + 2wz - d = 0$, and conversely. ▌

Two systems of spheres such that any sphere of one system cuts any sphere of the other system orthogonally are said to be **conjugate systems** *(or* **orthogonal systems***) of spheres.*

Note. Through each point of the space there is just one member of each system of the conjugate systems of spheres.

Miscellaneous Exercises

1. Find the equation of the sphere passing through the points $(0, 0, 0)$, $(-a, b, c)$, $(a, -b, c)$ and $(a, b, -c)$.
2. Find the equation of the sphere passing through the points (3,0, 2), (–1, 1, 1), (2, –5, 4) and having the centre on the plane $2x + 3y + 4z = 8$.
3. OA, OB, OC are three mutually perpendicular lines through the origin, and their direction cosines are $l_1, m_1, n_1; l_2, m_2, n_2; l_3, m_3, n_3$, respectively. If $OA = a$, $OB = b$, $OC = c$, prove that the equation of the sphere $OABC$ is $x^2 + y^2 + z^2 - x(al_1 + bl_2 + cl_3) - y(am_1 + bm_2 + cn_3) - z(an_1 + bn_2 + cn_3) = 0.$
4. Prove that the equation of the sphere circumscribing the tetrahedron whose faces are $y + z = 0, z + x = 0, x + y = 0$ and $x + y + z = 1$ is $x^2 + y^2 + z^2 - 3(x + y + z) = 0$.
5. Prove that the equation of the sphere circumscribing the tetrahedron whose faces are

$$\frac{y}{b} + \frac{z}{c} = 0, \frac{z}{c} + \frac{x}{a} = 0, \frac{x}{a} + \frac{y}{b} = 0 \text{ and } \frac{x}{a} + \frac{y}{b} + \frac{z}{c} = 1$$

is $x^2 + y^2 + z^2 - (a^2 + b^2 + c^2)\left(\dfrac{x}{a} + \dfrac{y}{b} + \dfrac{z}{c}\right) = 0.$

6. Find the equation of the sphere circumscribing the tetrahedron whose faces are $x = 0, y = 0, z = 0$ and $\frac{x}{a}+\frac{y}{b}+\frac{z}{c}=1$.

7. A point moves so that the sum of the squares of its distances from the six faces of a cube is constant, prove that its locus is a sphere.

8. Find the equation of the sphere having its centre on the plane $4x - 5y - z = 3$, and passing through the circle

$$x^2 + y^2 + z^2 - 2x - 3y + 4z + 8 = 0,\ 3x + 4y - 5z + 3 = 0.$$

9. Find the equation of the sphere through the circle

$$x^2 + y^2 + z^2 + 2x + 3y + 6 = 0,\ x - 2y + 4z - 9 = 0$$

and the centre of the sphere $x^2 + y^2 + z^2 - 2x + 4y - 6z + 5 = 0$.

10. If r is the radius of the circle

$$x^2 + y^2 + z^2 + 2ux + 2vy + 2wz + d = 0,\ lx + my + nz = 0,$$

prove that

$$(r^2 + d)(l^2 + m^2 + n^2) = (mw - nv)^2 + (nu - lw)^2 + (lv - mu)^2.$$

11. Find the conditions that the circles

$$x^2 + y^2 + z^2 + 2ux + 2vy + 2wz + d = 0,\ lx + my + nz = p;$$

and $\quad x^2 + y^2 + z^2 + 2u'x + 2v'y + 2w'z + d' = 0,\ l'x + m'y + n'z = p'$

should lie on the same sphere.

12. Prove that the sphere $x^2 + y^2 + z^2 + 2ux + 2vy + 2wz + d = 0$ cuts the sphere $x^2 + y^2 + z^2 + 2u'x + 2v'y + 2w'z + d' = 0$ in a great circle if $2(uu' + vv' + ww') = 2r'^2 + d + d'$, where r' is the radius of the second sphere.

13. A variable plane parallel to the plane $x/a + y/b + z/c = 0$ and meets the axes in A, B, C, respectively. Prove that the circle ABC lies on the surface

$$yz\left(\frac{b}{c}+\frac{c}{b}\right)+zx\left(\frac{a}{c}+\frac{c}{a}\right)+xy\left(\frac{a}{b}+\frac{b}{a}\right)=0.$$

14. POP' is a variable diameter of the ellipse $x^2/a^2 + y^2/b^2 = 1$, $z = 0$ and a circle is described in the plane $PP'ZZ'$ on PP' as diameter. Prove that as PP' varies, the circle generates the surface

$$(x^2 + y^2 + z^2)\left(\frac{x^2}{a^2}+\frac{y^2}{b^2}\right)=x^2 + y^2.$$

15. A is a point on OX and B on OY, so that the angle OAB is conatant $(= \alpha)$. On AB as diameter a circle is described whose plane is parallel to OZ. Prove that as AB varies the circle generates the conic $2xy - z^2 \sin 2\alpha = 0$.

16. Prove that the lines drawn from the origin touching the sphere

$$x^2 + y^2 + z^2 + 2ux + 2vy + 2wz + d = 0$$

lie on the surface $d(x^2 + y^2 + z^2) = (ux + vy + wz)^2$.

17. Prove that the centres of the spheres which cut both of the spheres

$$x^2 + y^2 + z^2 + 2u_1x + d = 0, x^2 + y^2 + z^2 + 2u_2x + d = 0$$

in great circles, lie on the plane $x + u_1 + u_2 = 0$.

18. Prove that the sum of the squares of the intercepts made by a given sphere on any three mutually perpendicular lines through a fixed point is constant.

19. If d is the distance between the centres of two spheres of radii r_1 and r_2, prove that the angle between them is $\cos^{-1}\left(\frac{r_1^2 + r_2^2 - d^2}{2r_1r_2}\right)$.

20. The equation of the plane ABC is $x/a + y/b + z/c = 1$. Find the equations of the circle passing through A, B and C the points of intersection of the plane with the coordinate axes.

21. Prove that the spheres

$x^2 + y^2 + z^2 = 100$ and $x^2 + y^2 + z^2 - 12x + 4y - 6z + 48 = 0$

touch internally and find the point of contact.

22. Prove that the spheres

$x^2 + y^2 + z^2 = 25$, and $x^2 + y^2 + z^2 - 24x - 40y - 18z + 225 = 0$

touch externally.

23. Prove that the centres of the spheres which touch the lines $y = \pm mx$, $z = \pm c$ lie on the surface $mxy + cz(1 + m^2) = 0$.

24. Prove that the equation of the sphere which lies in the positive octant and touches the coordinate planes can be written in the form

$$x^2 + y^2 + z^2 - 2\lambda(x + y + z) + 2\lambda^2 = 0.$$

25. Prove that the equations of any two spheres can be put in the form

$$x^2 + y^2 + z^2 + 2\lambda_1x + d = 0,$$
$$x^2 + y^2 + z^2 + 2\lambda_2x + d = 0.$$

Hence, prove that the spheres which cut two given spheres along great circles all pass through two fixed points.

26. A varibale sphere passes through the points $(0, 0, \pm c)$ and cuts the lines

$$y = x \tan \alpha, z = c; y = -x \tan \alpha, z = -c$$

in the points P and P'. If PP' is of constant length $2a$, prove that the centre of the sphere lies on the circle $x^2 + y^2 = (a^2 - c^2) \operatorname{cosec}^2 2\alpha, z = 0$.

27. P is a variable point on a given line and A, B, C are the projections on the axes. Prove that the sphere $OABC$ passes through a fixed circle.

28. If two spheres S_1 and S_2 are orthogonal, the polar plane of any point P on S_1 with respect to S_2 passes through the other end of the diameter of S_1 through P.

The Cone

16
CHAPTER

1. Definition

A **cone** *is a surface generated by a variable line which passes through a fixed point and intersects a given curve (or touches a given surface).*

The fixed point is called the **vertex**, the given curve (or the surface) the **guiding curve** and the variable line the **generator**.

Note. There is no loss of generality if we take the guiding curve as a plane curve since any arbitrary plane section of the surface can be taken as guiding curve.

1.1 Equation of a cone. *To find the equation of a cone whose vertex and the guiding curve are given.*

Let the point $A(\alpha, \beta, \gamma)$ be the vertex and the conic

$$ax^2 + 2hxy + by^2 + 2gx + 2fy + c = 0, z = 0 \tag{1}$$

be the guiding curve of the cone.

Let the equations of the generator AP through $A(\alpha, \beta, \gamma)$ be

$$\frac{x-\alpha}{l} = \frac{y-\beta}{m} = \frac{z-\gamma}{n}. \tag{2}$$

This line meets the plane $z = 0$ at the point

$$P\left(\alpha - \frac{l\gamma}{n}, \beta - \frac{m\gamma}{n}, 0\right)$$

If this point lies on the conic (1), then

$$a\left(\alpha - \frac{l\gamma}{n}\right)^2 + 2h\left(\alpha - \frac{l\gamma}{n}\right)\left(\beta - \frac{m\gamma}{n}\right) + b\left(\beta - \frac{m\gamma}{n}\right)^2 +$$

$$2g\left(\alpha - \frac{l\gamma}{n}\right) + 2f\left(\beta - \frac{m\gamma}{n}\right) + c = 0. \tag{3}$$

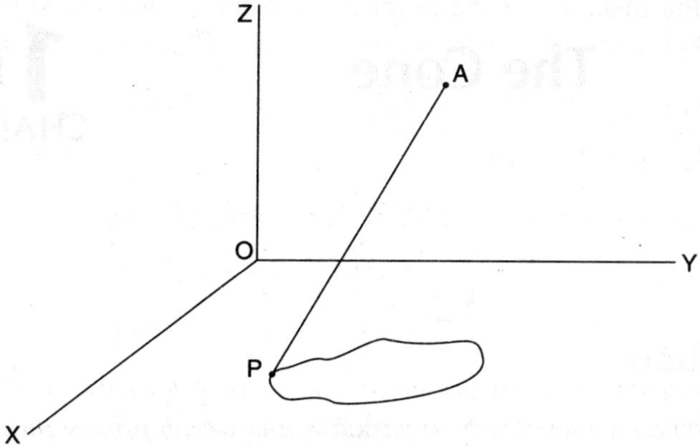

Fig. 16.1

Eliminating l, m, n from (2) and (3), we get

$$a\left(\alpha-\frac{x-\alpha}{z-\gamma}\cdot\gamma\right)^2+2h\left(\alpha-\frac{x-\alpha}{z-\gamma}\cdot\gamma\right)\left(\beta-\frac{x-\beta}{z-\gamma}\cdot\gamma\right)+$$

$$b\left(\beta-\frac{y-\beta}{z-\gamma}\cdot\gamma\right)^2+2g\left(\alpha-\frac{x-\alpha}{z-\gamma}\cdot\gamma\right)+2f\left(\beta-\frac{y-\beta}{z-\gamma}\cdot\gamma\right)+c=0$$

$$\Rightarrow\; a(\alpha z-\gamma x)^2+2h(\alpha z-\gamma x)(\beta z-\gamma y)+b(\beta z-\gamma y)^2+$$

$$2g(\alpha z-\gamma x)(z-\gamma)+2f(\beta z-\gamma z)(z-\gamma)+c(z-\gamma)^2=0. \qquad (4)$$

This is the required equation of the cone.

Note. It appears, from equation (4) that a cone is represented by a second degree equation in x, y and z. However, every second degree equation in x, y and z need not represent a cone.

Ex. Find the equation of the cone whose vertex is (1, 2, 3) and guiding curve is the circle $x^2 + y^2 = 9, z = 0$.

1.2 Equation of a cone with vertex at the origin. *To find the equation of the cone whose vertex is at the origin.*

Let the general equation of second degree in x, y, z *i.e.*

$$ax^2 + by^2 + cz^2 + 2fyz + 2gzx + 2hxy + 2ux + 2vy + 2wz + d = 0 \qquad (1)$$

represent a cone whose vertex is at the origin O. Let $P(x', y', z')$ be any point on the cone. Then, the equations of the generator OP are

$$\frac{x}{x'}=\frac{y}{y'}=\frac{z}{z'}=r \text{ (say)}, \qquad (2)$$

where r is the distance of the point P from the vertex O. Any point on the generator (2) is (rx', ry', rz'). This point will lie on the cone (1) if

$$r^2(ax'^2 + by'^2 + cz'^2 + 2fy'z' + 2gz'x' + 2hx'y') + 2r(ux' + vy' + wz') + d = 0$$

for each value of r. Therefore

$$ax'^2 + by'^2 + cz'^2 + 2fy'z' + 2gz'x' + 2hx'y' = 0, \tag{3}$$

$$ux' + yy' + wz' = 0, \tag{4}$$

$$d = 0. \tag{5}$$

From equation (4), it is clear that $u = v = w = 0$, otherwise the point (x', y', z') will lie on the plane $ux + vy + wz = 0$ which is a contradiction. Hence, the required equation of the cone becomes $ax^2 + by^2 + cz^2 + 2fyz + 2gzx + 2hxy = 0$.

Note 1. The equation of the cone with vertex at the origin is a homogeneous equation of second degree in x, y and z. The converse is also true.

To prove that every homogeneous equation of second degree in x, y and z represents a cone whose vertex is at the origin.

Proof. Let the homogeneous equation of second degree in x, y and z be

$$ax^2 + by^2 + cz^2 + 2fyz + 2gzx + 2hxy = 0. \tag{6}$$

Let $P(x', y', z')$ be any point on the locus represented by (6). Then

$$ax'^2 + by'^2 + cz'^2 + 2fy'z' + 2gz'x' + 2hx'y' = 0. \tag{7}$$

Multiplying equation (7) by r^2, we obtain

$$ar^2x'^2 + br^2y'^2 + cr^2z'^2 + 2fry'rz' + 2grz'rx' + 2hrx'ry' = 0.$$

From this equation it is clear that the point (rx', ry', rz') lies on the locus of (6). But the point (rx', ry', rz') is any point on the line joining the origin O to the point P. Therefore, OP lies completely on the locus of (6). Thus, equation (6) represents a cone whose vertex is at the origin.

Note 2. From above it is clear that a homogeneous equation of second degree

$$ax^2 + by^2 + cz^2 + 2fyz + 2gzx + 2hxy = 0$$

represents a cone whose vertex is at the origin. But, under a special case when the left hand side expression of the equation can be factorised into two real linear factors, it represetns a pair of planes, see Chapter 13, § 10.1. Thus, such a pair of planes (of course intersecting) can be regarded as a cone with vertex to be any point on the line of intersection of the planes.

Corollary. The homogeneous equation of the cone

$$f(x, y, z) \equiv ax^2 + by^2 + cz^2 + 2fyz + 2gzx + 2hxy = 0$$

is satisfied by the direction cosines of its generator.

Proof. Let

$$\frac{x}{l} = \frac{y}{m} = \frac{z}{n}$$

be a generator of the cone $f(x, y, z) = 0$. Then, it is easy to see that

$$f(l, m, n) \equiv al^2 + bm^2 + cn^2 + 2fmn + 2gnl + 2hlm = 0.$$

1.3 Examples

Example 1. Find the equation of the cone whose vertex is the origin and which passes through the curve of intersection of the plane $lx + my + nz = p$ and the surface $ax^2 + by^2 + cz^2 = 1$.

Solution. The equation of the surface is

$$ax^2 + by^2 + cz^2 = 1. \quad (1)$$

The equation of the plane can be written as

$$\frac{lx + my + nz}{p} = 1. \quad (2)$$

Since, the equation of the cone with vertex at the origin is a homogeneous equation of second degree in x, y and z, it is obtained by making equation (1) homogeneous with the help of (2), *i.e.*

$$ax^2 + by^2 + cz^2 = \left(\frac{lx + my + nz}{p}\right)^2.$$

This is the required equation of the cone.

Example 2. Prove that the general equation of the cone of second degree passing through the axes is $fyz + gzx + hxy = 0$.

Solution. The cone passing through the axes will have the vertex at the origin. Let its general equation be

$$ax^2 + by^2 + cz^2 + 2fyz + 2gzx + 2hxy = 0. \quad (1)$$

But it passes through the coordinate axes. Therefore, the direction cosines of the axes 1, 0, 0; 0, 1, 0; 0, 0, 1 will satisfy (1). Thus, we get

$$a = b = c = 0.$$

Hence, the equation of the cone becomes

$$fyz + gzx + hxy = 0.$$

Example 3. The plane $x/a + y/b + z/c = 1$ meets the coordinate axes in A, B, C. Prove that the equation of the cone generated by the lines drawn from O to meet the circle ABC is

$$yz\left(\frac{b}{c}+\frac{c}{b}\right)+zx\left(\frac{c}{a}+\frac{a}{c}\right)+xy\left(\frac{a}{b}+\frac{b}{a}\right)=0.$$

Solution. The plane

$$\frac{x}{a}+\frac{y}{b}+\frac{z}{c}=1 \tag{1}$$

meets the axes in $A(a, 0, 0)$, $B(0, b, 0)$, $C(0, 0, c)$. The equation of the sphere $OABC$ is

$$x^2+y^2+z^2-ax-by-cz=0. \tag{2}$$

The circle ABC is represented by the equations (1) and (2). Thus, the cone generated by the lines drawn through O to meet the circle ABC is obtained by making equation (2) homogeneous with the help of (1), *i.e.,*

$$x^2+y^2+z^2-(ax+by+cz)\left(\frac{x}{a}+\frac{y}{b}+\frac{z}{c}\right)=0$$

$$\Rightarrow \quad yz\left(\frac{b}{c}+\frac{c}{b}\right)+zx\left(\frac{c}{a}+\frac{a}{c}\right)+xy\left(\frac{a}{b}+\frac{b}{a}\right)=0.$$

Example 4. The section of a cone whose guiding curve is the ellipse $x^2/a^2 + y^2/b^2 = 1$, $z = 0$ by the plane $x = 0$, is a rectangular hyperbola. Prove that the locus of the vertex is the surface $\dfrac{x^2}{a^2}+\dfrac{y^2+z^2}{b^2}=1$.

Solution. Let (α, β, γ) be the vertex of the cone. Then, the equations of a generator of the cone are

$$\frac{x-\alpha}{l}=\frac{y-\beta}{m}=\frac{z-\gamma}{n}. \tag{1}$$

This line will meet to the plane $z = 0$ in the point given by $\left(\alpha-\dfrac{l\gamma}{n}, \beta-\dfrac{m\gamma}{n}, 0\right)$. The point $(\alpha - l\gamma/n, \beta - m\gamma/n, 0)$ will lie on the ellipse $x^2/a^2 + y^2/b^2 = 1$, $z = 0$ if

$$\frac{1}{a^2}\left(\alpha-\frac{l\gamma}{n}\right)^2+\frac{1}{b^2}\left(\beta-\frac{m\gamma}{n}\right)^2=1. \tag{2}$$

Eliminating l, m, n from (1) and (2), we get

$$\Rightarrow \quad \frac{1}{a^2}(\alpha z - \gamma x)^2 + \frac{1}{b^2}(\beta z - \gamma y)^2 = (z - \gamma)^2.$$

This is the equation of the cone with vertex at (α, β, γ). The section of this cone by the plane $x = 0$ is given by

$$\frac{\alpha^2 z^2}{a^2} + \frac{1}{b^2}(\beta z - \gamma y)^2 = (z - \gamma)^2$$

$$i.e. \quad \frac{\gamma^2 y^2}{b^2} + \left(\frac{\alpha^2}{a^2} + \frac{\beta^2}{b^2} - 1\right) z^2 - \frac{2\beta\gamma yz}{b^2} + 2\gamma z - \gamma^2 = 0.$$

This section will represent a rectangular hyperbola in yz-plane if coeff. of y^2 + coeff. of $z^2 = 0$, *i.e.*,

$$\frac{\gamma^2}{b^2} + \left(\frac{\alpha^2}{a^2} + \frac{\beta^2}{b^2} - 1\right) = 0.$$

Hence, the locus of (α, β, γ) is $\dfrac{x^2}{a^2} + \dfrac{y^2}{b^2} + \dfrac{z^2}{b^2} = 1.$

Example 5. Prove that a cone of second degree can be found to pass through two sets of rectangular axes through the same origin.

Solution. Let the equation of the cone of second degree having vertex at the origin be

$$ax^2 + by^2 + cz^2 + 2fyz + 2gzx + 2hxy = 0. \quad (1)$$

Let the coordinate axes be one set of rectangular axes through which the cone passes. Then, the equation of the cone passing through the coordinate axes is

$$fyz + gzx + hxy = 0, \quad (2)$$

since the direction cosines 1, 0, 0; 0, 1, 0; 0, 0, 1 of the axes satisfy (1).

Let the second set of rectangular axes through the same origin be OP, OQ, OR having direction cosines $l_1, m_1, n_1; l_2, m_2, n_2; l_3, m_3, n_3$. Then

$$m_1 n_1 + m_2 n_2 + m_3 n_3 = 0, \quad (3)$$

$$n_1 l_1 + n_2 l_2 + n_3 l_3 = 0, \quad (4)$$

$$l_1 m_1 + l_2 m_2 + l_3 m_3 = 0. \quad (5)$$

Let the cone (2) pass through OP and OQ. Then

$$fm_1 n_1 + gn_1 l_1 + hl_1 m_1 = 0 \text{ and } fm_2 n_2 + gn_2 l_2 + hl_2 m_2 = 0.$$

On adding them, we get

$$f(m_1n_1 + m_2n_2) + g(n_1l_1 + n_2l_2) + h(l_1m_1 + l_2m_2) = 0.$$

By using (3), (4) and (5), it becomes

$$fm_3n_3 + gn_3l_3 + hl_3m_3 = 0.$$

This shows that the cone (2) also passes through OR. Hence, the result.

Exercises

1. Find the equation of the cone whose vertex is (α, β, γ) and guiding curve is the parabola $y^2 = 4ax$, $z = 0$.
2. Prove that the line $\frac{x}{l} = \frac{y}{m} = \frac{z}{n}$, where $2l^2 + 3m^2 - 5n^2 = 0$, is a generator of the cone $2x^2 + 3y^2 - 5z^2 = 0$.
3. Prove that the lines drawn through the point (α, β, γ) whose direction cosines satisfy the relation $al^2 + bm^2 + cn^2 = 0$, generate the cone $a(x - \alpha)^2 + b(y - \beta)^2 + c(z - \gamma)^2 = 0$.
4. Find the equations of the cones with vertex at the origin and pass through the curves of intersection given by the equations
   (i) $ax^2 + by^2 = 2z$, $lx + my + nz = p$;
   (ii) $ax^2 + by^2 + cz^2 = 1$, $\alpha x^2 + \beta y^2 = 2z$;
   (iii) $\frac{x^2}{a^2} + \frac{y^2}{b^2} + \frac{z^2}{c^2} = 1, \frac{x^2}{\alpha^2} + \frac{y^2}{\beta^2} = 2z$.
5. A variable plane is parallel to the given plane $\frac{x}{a} + \frac{y}{b} + \frac{z}{c} = 0$, and meets the axes in A, B, C. Prove that the circle ABC lies on the cone
$$yz\left(\frac{b}{c} + \frac{c}{b}\right) + zx\left(\frac{c}{a} + \frac{a}{c}\right) + xy\left(\frac{a}{b} + \frac{b}{a}\right) = 0.$$
6. Find the equation of the cone whose vertex is the origin and base the circle $x = a$, $y^2 + z^2 = b^2$, and prove that the section of the cone by a plane parallel to xy-plane is a hyperbola.
7. Prove that the equation of the cone whose vertex is the origin and base the curve $z = k, f(x, y) = 0$ is $f\left(\frac{xk}{z}, \frac{yk}{z}\right) = 0.$
8. The planes through OX and OY include an angle α; prove that their line of intersection lies on the cone $z^2(x^2 + y^2 + z^2) = x^2y^2 \tan^2\alpha$.

9. Find the equation of the cone which passes through the three coordinate axes as well as the two lines $\frac{x}{1}=\frac{y}{-2}=\frac{z}{3}, \frac{x}{3}=\frac{y}{-1}=\frac{z}{1}$.

10. Find the equation of the cone which passes through the three coordinate axes and the three mutually perpendicular lines

$$\frac{x}{2}=\frac{y}{1}=\frac{z}{-1}, \frac{x}{1}=\frac{y}{3}=\frac{z}{5}, \frac{x}{8}=\frac{y}{-11}=\frac{z}{5}.$$

2. General Second Degree Equation

2.1 Condition that the most general second degree equations in x, y and z represent a cone. *To find a necessary and sufficient condition that the general equation of second degree*

$$ax^2 + by^2 + cz^2 + 2fyz + 2gzx + 2hxy + 2ux + 2vy + 2wz + d = 0 \quad (1)$$

may represent a cone.

In § 1.2, we have proved that the equation of the cone with vertex at the origin is a homogeneous equation of second degree in x, y, z and that the converse is also true. Therefore, equation (1) will represent a cone if and only if there exists a point (α, β, γ) the vertex of the cone such that on transferring the origin to it the equation (1) is reduced to a homogeneous equation in new variables.

Necessity. Let the equation (1) represents a cone with vertex at (α, β, γ). Shifting the origin to the point (α, β, γ) by the transformations $x = X + \alpha$, $y = Y + \beta$, $z = Z + \gamma$, equation (1) reduces to

$$a(X+\alpha)^2 + b(Y+\beta)^2 + c(Z+\gamma)^2 + 2f(Y+\beta)(Z+\gamma) + 2g(Z+\gamma)(X+\alpha)$$
$$+ 2h(X+\alpha)(Y+\beta) + 2u(X+\alpha) + 2v(Y+\beta) + 2w(Z+\gamma) + d = 0$$
$$\Rightarrow \quad aX^2 + bY^2 + cZ^2 + 2fYZ + 2gZX + 2hXY + 2\{(a\alpha + h\beta + g\gamma + u)X$$
$$+ (h\alpha + b\beta + f\gamma + v)Y + (g\alpha + f\beta + c\gamma + w)Z\} + a\alpha^2 + b\beta^2 + c\gamma^2$$
$$+ 2f\beta\gamma + 2g\gamma\alpha + 2h\alpha\beta + 2u\alpha + 2v\beta + 2w\gamma + d = 0. \quad (2)$$

Since, equation (2) represents a cone with vertex at the new origin, it must be a second degree homogeneous equation in X, Y and Z, and consequently the terms of the degree in X, Y, Z and the constant terms must disappear from equation (2) so that we are left only with

$$aX^2 + bY^2 + cZ^2 + 2fYZ + 2gZX + 2hXY = 0.$$

Thus, it is required that the following conditions are satisfied:

$$a\alpha + h\beta + g\gamma + u = 0, \quad (3)$$

$$h\alpha + b\beta + f\gamma + v = 0, \tag{4}$$

$$g\alpha + f\beta + c\gamma + w = 0, \tag{5}$$

and $$a\alpha^2 + b\beta^2 + c\gamma^2 + 2f\beta\gamma + 2g\gamma\alpha + 2h\alpha\beta + 2u\alpha + 2v\beta + 2w\gamma + d = 0. \tag{6}$$

Multiplying equation (3) by α, equation (4) by β, equation (5) by γ and adding, and then subtracting this sum from equation (6), we get

$$u\alpha + v\beta + w\gamma + d = 0. \tag{7}$$

Now, eliminating α, β, γ from equations (3), (4), (5) and (7), we obtain

$$\begin{vmatrix} a & h & g & u \\ h & b & f & v \\ g & f & c & w \\ u & v & w & d \end{vmatrix} = 0. \tag{8}$$

This is the necessary condition for (1) to represent a cone.

Sufficiency. Suppose the condition (8) is satisfied. Then, the equations (3), (4) and (5) could be solved simultaneously to give a set of values of α, β, γ. The point (α, β, γ) is such if the origin is shifted to it, the equation (1) will be reduced to a homogeneous equation in the new system of coordinates and hence will represent a cone.

Note. Let the equation

$$F(x, y, z) \equiv ax^2 + by^2 + cz^2 + 2fyz + 2gzx + 2hxy + 2ux + 2vy + 2wz + d = 0$$

represent a cone. Making this equation homogeneous by introducing a new variable t, we obtain

$$F(x, y, z, t) \equiv ax^2 + by^2 + cz^2 + 2fyz + 2gzx + 2hxy + 2uxt + 2vyt + 2wzt + dt^2 = 0.$$

On differentiating partially, we get

$$\frac{\partial F}{\partial x} = 2(ax + hy + gz + ut)$$

$$\frac{\partial F}{\partial y} = 2(hx + by + fz + vt)$$

$$\frac{\partial F}{\partial z} = 2(gx + fy + cz + wt)$$

$$\frac{\partial F}{\partial t} = 2(ux + vy + wz + dt).$$

Taking $t = 1$, the equations

$$\frac{\partial F}{\partial x} = 0, \frac{\partial F}{\partial y} = 0, \frac{\partial F}{\partial z} = 0, \frac{\partial F}{\partial t} = 0$$

are, in fact, the equations (3), (4), (5) and (7), respectively. Thus, the equation $F(x, y, z) = 0$ represents a cone if and only if the equations

$$\frac{\partial F}{\partial x} = 0, \frac{\partial F}{\partial y} = 0, \frac{\partial F}{\partial z} = 0, \frac{\partial F}{\partial t} = 0$$

are consistent. If these equations are cosistent the coordinates of the vertex can be obtained by solving any three of these equations.

Remark. If the conditions

$$\begin{vmatrix} a & h & g & u \\ h & b & f & v \\ g & f & c & w \\ u & v & w & d \end{vmatrix} = 0 \text{ and } \begin{vmatrix} a & h & g \\ h & b & f \\ g & f & c \end{vmatrix} = 0$$

are satisfied simultaneously, the equation (1) represents a pair of planes.

2.2 Examples

Example. Prove that the equation

$$2x^2 + 2y^2 + 7z^2 - 10yz - 10zx + 2x + 2y + 26z - 17 = 0$$

represents a cone whose vertex is (2, 2, 1).

Solution. Making the equation homogeneous by introducing a new variable t, we get

$$F(x, y, z, t) \equiv 2x^2 + 2y^2 + 7z^2 - 10yz - 10zx + 2xt + 2yt + 26zt - 17t^2 = 0.$$

Now, equating to zero the partial derivatives of F, we get

$$\frac{\partial F}{\partial x} = 4x - 10z + 2t = 0 \text{ i.e. } 2x - 5z + t = 0, \tag{1}$$

$$\frac{\partial F}{\partial y} = 4y - 10z + 2t = 0 \text{ i.e. } 2y - 5z + t = 0, \tag{2}$$

$$\frac{\partial F}{\partial z} = 14z - 10y - 10x + 26t = 0 \text{ i.e. } 5x + 5y - 7z - 13t = 0, \tag{3}$$

$$\frac{\partial F}{\partial t} = 2x + 2y + 26z - 34t = 0 \text{ i.e. } x + y + 13z - 17 = 0. \tag{4}$$

Taking $t = 1$ and solving equations (1), (2) and (3), we get $x = 2, y = 2, z = 1$.

These values of x, y, z also satisfy equation (4). Therefore, the equations (1) – (4) are consistent. Hence, the given equation represents a cone whose vertex is at the point (2, 2, 1).

Exercises

1. Prove that the equation $4x^2 - y^2 + 2z^2 - 3yz + 2xy + 12x - 11y + 6z + 4 = 0$ represents a cone whose vertex is $(-1, -2, -3)$.
2. Prove that the equation
$$x^2 - 2y^2 + 3z^2 + 5yz - 6zx - 4xy + 8x - 19y - 2z - 20 = 0$$
represents a cone whose vertex is $(1, -2, 3)$.
3. Prove that the equation $2y^2 - 8yz - 4zx - 8xy + 6x - 4y - 2z + 5 = 0$ represents a cone whose vertex is $\left(-\frac{7}{6}, \frac{1}{3}, \frac{5}{6}\right)$.
4. Prove that the equation $ax^2 + by^2 + cz^2 + 2ux + 2vy + 2wz + d = 0$ represents a cone if $\frac{u^2}{a} + \frac{v^2}{b} + \frac{w^2}{c} = d$.

3. Intersection of a Cone with a Plane

3.1 *To prove that a cone is cut by a plane through the vertex of the cone in two generating lines.*

Without any loss of generality, let us take the cone with vertex at the origin whose equation is given by

$$f(x, y, z) \equiv ax^2 + by^2 + cz^2 + 2fyz + 2gzx + 2hxy = 0. \tag{1}$$

Let the plane through the vertex of the cone be given by

$$ux + vy + wz = 0. \tag{2}$$

Let the plane (2) cut the cone (1) in the line whose equations are given by

$$\frac{x}{l} = \frac{y}{m} = \frac{z}{n}. \tag{3}$$

Since, the line (3) lies on (1) and (2) both, we have

$$f(l, m, n) \equiv al^2 + bm^2 + cn^2 + 2fmn + 2gnl + 2hlm = 0, \tag{4}$$

and
$$ul + vm + wn = 0. \tag{5}$$

Eliminating n from (4) and (5), we get

$$al^2 + bm^2 + c\left(\frac{ul+vm}{-w}\right)^2 + 2fm\left(\frac{ul+vm}{-w}\right) + 2g\left(\frac{ul+vm}{-w}\right)l + 2hlm = 0$$

$$\Rightarrow \frac{l^2}{m^2}(cu^2 + aw^2 - 2gwu) + \frac{2l}{m}(hw^2 + cuv - fwu - gvw)$$

$$+ bw^2 + cv^2 - 2fvw = 0. \tag{6}$$

This is a quadratic equation in l/m, giving two values of l/m (real and distinct, coincident or imaginary) corresponding to which there are two lines of intersection of the plane and the cone.

Note. Throughout the discussion of this chapter by 'a plane cuts a cone' we shall always mean the plane through the vertex of the cone unless specified otherwise. The section of a cone with an arbitrary plane (not passing through the vertex of the cone) are not the generating lines. In fact, the section would be some conic.

3.2 Angle between two generating lines

First of all we introduce certain notations which we shall be using in what follows.

$$D \equiv \begin{vmatrix} a & h & g \\ h & b & f \\ g & f & c \end{vmatrix}, \quad P^2 \equiv \begin{vmatrix} a & h & g & u \\ h & b & f & v \\ g & f & c & w \\ u & v & w & 0 \end{vmatrix},$$

$$A \equiv \frac{\partial D}{\partial a} = bc - f^2, \; B \equiv \frac{\partial D}{\partial b} = ca - g^2,$$

$$C \equiv \frac{\partial D}{\partial c} = ab - h^2, \; F \equiv \frac{1}{2}\frac{\partial D}{\partial f} = gh - af,$$

$$G \equiv \frac{1}{2}\frac{\partial D}{\partial g} = hf - bg, \; H \equiv \frac{1}{2}\frac{\partial D}{\partial h} = fg - ch.$$

Clearly A, B, C, F, G, H are cofactors of a, b, c, f, g, h, respectively, in the determinant D. One can easily verify that

$$BC - F^2 = aD, \; CA - G^2 = bD, \; AB - H^2 = cD;$$

$$GH - AF = fD, \; HF - BG = gD, \; FG - CH = hD.$$

Also $P^2 = -(Au^2 + Bv^2 + Cw^2 + 2Fvw + 2Gwu + 2Huv)$.

To find the angle between the generating lines in which a plane cuts a cone.

Let the equations of the cone and the plane be

$$f(x, y, z) \equiv ax^2 + by^2 + cz^2 + 2fyz + 2gzx + 2hxy = 0 \quad (1)$$

and $$ux + vy + wz = 0. \quad (2)$$

From equation (6) of § 3.1, it is clear that there are two lines of intersection of (1) and (2). If l_1, m_1, n_1 and l_2, m_2, n_2 are the direction ratios of these lines, then l_1/m_1 and l_2/m_2 will be the roots of the equation (6). By theory of equations, we have

$$\frac{l_1}{m_1} + \frac{l_2}{m_2} = \frac{l_1 m_2 + l_2 m_1}{m_1 m_2} = -\frac{2(hw^2 + cuv - jwu - gvw)}{cu^2 + aw^2 - 2gwu},$$

and $$\frac{l_1}{m_1} \cdot \frac{l_2}{m_2} = \frac{bw^2 + cv^2 - 2fvw}{cu^2 + aw^2 - 2gwu}.$$

These relations can be written as

$$\frac{l_1 l_2}{bw^2 + cv^2 - 2fvw} = \frac{m_1 m_2}{cu^2 + aw^2 - 2gwu}$$

$$= \frac{l_1 m_2 + l_2 m_1}{-2(hw^2 + cuv - fwu - gvw)} = \lambda \text{ (say)}.$$

Now

$$\begin{aligned}(l_1 m_2 - l_2 m_1)^2 &= (l_1 m_2 + l_2 m_1)^2 - 4 l_1 l_2 m_1 m_2 \\ &= 4\lambda^2 (hw^2 + cuv - fwu - gvw)^2 \\ &\quad - 4\lambda^2 (bw^2 + cv^2 - 2fvw) \times (cu^2 + aw^2 - 2gwu) \\ &= 4\lambda^2 \{h^2 w^4 - 2hw^3 (fu + gv) + w^2 (fu + gv)^2 \\ &\quad + 2chuvw^2 - 2cuvw (fu + gv) + c^2 u^2 v^2\} \\ &\quad - 4\lambda^2 \{abw^4 - 2w^3 (afv + bgu) \\ &\quad + w^2 (bcu^2 + cav^2 + 4fguv) - 2cuvw (fu + gv) + c^2 u^2 v^2\} \\ &= 4\lambda^2 w^2 \{(h^2 - ab) w^2 + 2wu (bg - fh) \\ &\quad + 2vw (af - gh) + (f^2 - bc) u^2 + (g^2 - ca) v^2 - 2uv (fg - ch)\} \\ &= -4\lambda^2 w^2 (Au^2 + Bv^2 + Cw^2 + 2Fvw + 2Gwu + 2Huv) \\ &= 4\lambda^2 w^2 P^2.\end{aligned}$$

Similarly $(n_1l_2 - n_2l_1)^2 = 4\lambda^2v^2P^2$, $(m_1n_2 - m_2n_1)^2 = 4\lambda^2u^2P^2$,

and $$\frac{l_1l_2}{bw^2 + cv^2 - 2fvw} = \frac{m_1m_2}{cu^2 + aw^2 - 2gwu} = \frac{n_1n_2}{av^2 + bu^2 - 2huv} = \lambda.$$

Therefore

$$\begin{aligned} l_1l_2 + m_1m_2 + n_1n_2 &= \lambda(bw^2 + cv^2 - 2fvw) \\ &\quad + \lambda(cu^2 + aw^2 - 2gwu) + \lambda(av^2 + bu^2 - 2huv) \\ &= \lambda\{a(v^2 + w^2) + b(w^2 + u^2) + c(u^2 + v^2) - 2fvw - 2gwu - 2huv\} \\ &= \lambda\{(a + b + c)(u^2 + v^2 + w^2) - f(u, v, w)\}, \end{aligned}$$

and $\Sigma(l_1m_2 - l_2m_1)^2 = 4\lambda^2P^2(u^2 + v^2 + w^2)$.

If θ is the angle between the lines of intersection, then

$$\begin{aligned} \tan\theta &= \pm\frac{\sqrt{\Sigma(l_1m_2 - l_2m_1)^2}}{l_1l_2 + m_1m_2 + n_1n_2} \\ &= \pm\frac{2\lambda P\sqrt{u^2 + v^2 + w^2}}{\lambda\{(a+b+c)(u^2+v^2+w^2) - f(u,v,w)\}} \\ &= \pm\frac{2P\sqrt{u^2 + v^2 + w^2}}{(a+b+c)(u^2+v^2+w^2) - f(u,v,w)}. \end{aligned}$$

Corollary 1. The plane $ux + vy + wz = 0$ cuts the cone $f(x, y, z) = 0$ in two coincident generators if and only if $\theta = 0$ *i.e.* $P = 0$.

Corollary 2. The plane $ux + vy + wz = 0$ cuts the cone $f(x, y, z) = 0$ in perpendicular generators if and only if $\theta = 90^\circ$ i.e. $(a + b + c)(u^2 + v^2 + w^2) - f(u, v, w) = 0$.

3.3 Examples

Example 1. Find the equations of the lines in which the plane $3x + 4y + z = 0$ cuts the cone $15x^2 - 32y^2 - 7z^2 = 0$.

Solution. Let

$$\frac{x}{l} = \frac{y}{m} = \frac{z}{n}$$

be one of the lines of intersection of the plane and the cone. Then

$$15l^2 - 32m^2 - 7n^2 = 0, \tag{1}$$

and $$3l + 4m + n = 0. \tag{2}$$

Eliminating n from (1) and (2), we get

$$15l^2 - 32m^2 - 7(-3l - 4m)^2 = 0$$

$$\Rightarrow \qquad \frac{l}{m} = -\frac{3}{2}, -2,$$

From (2), we have $$3\frac{l}{m} + 4 + \frac{n}{m} = 0.$$

If $$\frac{l}{m} = -\frac{3}{2}, \frac{n}{m} = \frac{1}{2}, \text{ then } \frac{l}{-3} = \frac{m}{2} = \frac{n}{1},$$

and if $$\frac{l}{m} = -2, \frac{n}{m} = 2, \text{ then } \frac{l}{2} = \frac{m}{-1} = \frac{n}{-2}.$$

Thus, the required lines of intersection are

$$\frac{x}{-3} = \frac{y}{2} = \frac{z}{1} \text{ and } \frac{x}{2} = \frac{y}{-1} = \frac{z}{-2}.$$

Example 2. Prove that the plane $ax + by + cz = 0$ cuts the cone $yz + zx + xy = 0$ in two perpendicular lines if $\frac{1}{a} + \frac{1}{b} + \frac{1}{c} = 0$.

Solution. Let $\frac{x}{l} = \frac{y}{m} = \frac{z}{n}$ be one of the lines of intersection of the plane and the cone. Then

$$al + bm + cn = 0, \tag{1}$$

and $$mn + nl + lm = 0. \tag{2}$$

Eliminating n from (1) and (2), we get

$$a\frac{l^2}{m^2} + \frac{l}{m}(a + b - c) + b = 0. \tag{3}$$

Let l_1/m_1 and l_2/m_2 be the roots of (3). Then, by theory of equations, we have

$$\frac{l_1}{m_1} \cdot \frac{l_2}{m_2} = \frac{b}{a}.$$

Similarly

$$\frac{m_1 m_2}{n_1 n_2} = \frac{c}{b} \text{ and } \frac{n_1 n_2}{l_1 l_2} = \frac{a}{c}$$

Therefore $$\frac{l_1 l_2}{\frac{1}{a}} = \frac{m_1 m_2}{\frac{1}{b}} = \frac{n_1 n_2}{\frac{1}{c}}.$$

Thus, the lines of intersection

$$\frac{x}{l_1} = \frac{y}{m_1} = \frac{z}{n_1} \text{ and } \frac{x}{l_2} = \frac{y}{m_2} = \frac{z}{n_2}$$

will be at right angles if $l_1 l_2 + m_1 m_2 + n_1 n_2 = 0$, *i.e.*, if

$$\frac{1}{a} + \frac{1}{b} + \frac{1}{c} = 0.$$

Exercises

1. Find the equations of the lines of intersection of the following planes and cones:

   (*i*) $2x + y - z = 0, \quad 4x^2 - y^2 + 3z^2 = 0.$

   (*ii*) $x + 7y - 5z = 0, \quad 3yz + 14zx - 30xy = 0.$

2. Find the angle between the lines of intersection of the plane $10x + 7y - 6z = 0$ and the cone $20x^2 + 7y^2 - 108z^2 = 0$.

3. Prove that the plane $lx + my + nz = 0$ cuts the cone $(b - c)x^2 + (c - a)y^2 + (a - b)z^2 + 2fyz + 2gzx + 2hxy = 0$ in perpendicular lines if

   $$(b - c)l^2 + (c - a)m^2 + (a - b)n^2 + 2fmn + 2gnl + 2hlm = 0.$$

4. Prove that the angle between the lines of intersection of the plane $x + y + z = 0$ with the cone $ayz + bzx + cxy = 0$ is $\frac{\pi}{2}$ if $a + b + c = 0$ and $\frac{\pi}{3}$ if $\frac{1}{a} + \frac{1}{b} + \frac{1}{c} = 0$.

5. Prove that the plane $ax + by + cz = 0$ cuts the cone $yz + zx + xy = 0$ in two lines inclined at an angle

   $$\tan^{-1}\left\{\frac{\sqrt{(a^2 + b^2 + c^2)(a^2 + b^2 + c^2 - 2bc - 2ca - 2ab)}}{bc + ca + ab}\right\}.$$

6. Prove that the lines in which the plane $ux + vy + wz = 0$ cuts the cone $ax^2 + by^2 + cz^2 = 0$ are perpendicular if $(b + c)u^2 + (c + a)v^2 + (a + b)w^2 = 0$.

7. Find the locus of the points from which three mutually perpendicular lines can be drawn to intersect the central conic $ax^2 + by^2 = 1, z = 0$.

8. Find the angle between the lines given by

$$x + y + z = 0, \quad \frac{yz}{b-c} + \frac{zx}{c-a} + \frac{xy}{a-b} = 0.$$

9. Prove that the cone whose vertex is at the origin and passes through the curve of intersection of the sphere $x^2 + y^2 + z^2 = 3a^2$, and any plane at a distance a from the origin, has three mutually perpendicular generators.

4. Intersection of a Cone and a Line

Let

$$ax^2 + by^2 + cz^2 + 2fyz + 2gzx + 2hxy = 0, \tag{1}$$

and

$$\frac{x-\alpha}{l} = \frac{y-\beta}{m} = \frac{z-\gamma}{n} = r \text{ (say)} \tag{2}$$

be the equations of a cone and a line, respectively.

Any point on the line (2) is $(\alpha + lr, \beta + mr, \gamma + nr)$. This point will lie on the cone (1) if

$$a(\alpha + lr)^2 + b(\beta + mr)^2 + c(\gamma + nr)^2 + 2f(\beta + mr)(\gamma + nr)$$
$$+ 2g(\gamma + nr)(\alpha + lr) + 2h(\alpha + lr)(\beta + mr) = 0$$

$$\Rightarrow (al^2 + bm^2 + cn^2 + 2fmn + 2gnl + 2hlm)r^2 + 2\{l(a\alpha + h\beta + g\gamma)$$
$$+ m(h\alpha + b\beta + f\gamma) + n(g\alpha + f\beta + c\gamma)\}r + a\alpha^2 + b\beta^2 + c\gamma^2$$
$$+ 2f\beta\gamma + 2g\gamma\alpha + 2h\alpha\beta = 0.$$

This equation is quadratic in r, hence gives two values of r which may be real and distinct, coincident or imaginary. Accordingly there are two points of intersection of the cone and the line. Thus, a line always meets a cone in two points which may be real and distinct, coincident or imaginary.

4.1 Tangent line. *A line which meets a cone in two coincident points is called the* **tangent line** *to the cone.*

4.2 Tangent plane. *To find the locus of the tangent lines to a cone at a point on it.*

Let

$$ax^2 + by^2 + cz^2 + 2fyz + 2gzx + 2hxy = 0 \tag{1}$$

be the equation of a cone and (α, β, γ) be any point on it. Equations of any line through the point (α, β, γ) are given by

$$\frac{x-\alpha}{l} = \frac{y-\beta}{m} = \frac{z-\gamma}{n} = r \text{ (say)}. \tag{2}$$

Any point on the line (2) is $(\alpha + lr, \beta + mr, \gamma + nr)$. This point will lie on the cone (1) if

$$a(\alpha + lr)^2 + b(\beta + mr)^2 + c(\gamma + nr)^2 + 2f(\beta + mr)(\gamma + nr) + 2g(\gamma + nr)(\alpha + lr) + 2h(\alpha + lr)(\beta + mr) = 0$$

$$\Rightarrow (al^2 + bm^2 + cn^2 + 2fmn + 2gnl + 2hlm)r^2 + 2\{l(a\alpha + h\beta + g\gamma) + m(h\alpha + b\beta + f\gamma) + n(g\alpha + f\beta + c\gamma)\}r + a\alpha^2 + b\beta^2 + c\gamma^2 + 2f\beta\gamma + 2g\gamma\alpha + 2h\alpha\beta = 0. \quad (3)$$

Since (α, β, γ) lies on (1), we have

$$a\alpha^2 + b\beta^2 + c\gamma^2 + 2f\beta\gamma + 2g\gamma\alpha + 2h\alpha\beta = 0. \quad (4)$$

Thus, equation (3) becomes

$$(al^2 + bm^2 + cn^2 + 2fmn + 2gnl + 2hlm)r + 2\{l(a\alpha + h\beta + g\gamma) + m(h\alpha + b\beta + f\gamma) + n(g\alpha + f\beta + c\gamma)\} = 0. \quad (5)$$

This shows that one root of equation (3) is zero. The line (2) will be a tangent line to (1) if both the roots of equation (3) are zero for which

$$l(a\alpha + h\beta + g\gamma) + m(h\alpha + b\beta + f\gamma) + n(g\alpha + f\beta + c\gamma) = 0. \quad (6)$$

Eliminating l, m, n from (2) and (6), we get

$$(x - \alpha)(a\alpha + h\beta + g\gamma) + (y - \beta)(h\alpha + b\beta + f\gamma) + (z - \gamma)(g\alpha + f\beta + c\gamma) = 0$$

$$\Rightarrow (a\alpha + h\beta + g\gamma)x + (h\alpha + b\beta + f\gamma)y + (g\alpha + f\beta + c\gamma)z = a\alpha^2 + b\beta^2 + c\gamma^2 + 2f\beta\gamma + 2g\gamma\alpha + 2h\alpha\beta.$$

On using (4), we get

$$(a\alpha + h\beta + g\gamma)x + (h\alpha + b\beta + f\gamma)y + (g\alpha + f\beta + c\gamma)z = 0.$$

This equation represents a plane and is the required locus.

The locus of the tangent lines to a cone at a point on it is called the **tangent plane** *at that point.*

Remarks

(*i*) The tangent plane at any point of a cone passes through its vertex.

(*ii*) The tangent plane at any point P of a cone touches the cone along the generator through P. Such generator is called the **generator of contact.**

(*iii*) The vertex of the cone through which all the tangent planes pass is called a **singular point** of the surface.

Working rule. In order to write the equation of the tangent plane at any point (α, β, γ) on the cone replace x^2 by αx, y^2 by βy, z^2 by γz, $2yz$ by $\gamma y + \beta z$, $2zx$ by $\alpha z + \gamma x$ and $2xy$ by $\beta x + \alpha y$.

4.3 Condition of tangency. *To find the condition when the plane*

$$ux + vy + wz = 0 \tag{1}$$

becomes a tangent plane to the cone

$$ax^2 + by^2 + cz^2 + 2fyz + 2gzx + 2hxy = 0.$$

Let (α, β, γ) be the point of contact of the plane with the cone. Then, tangnet plane at (α, β, γ) to the cone is

$$(a\alpha + h\beta + g\gamma)x + (h\alpha + b\beta + f\gamma)y + (g\alpha + f\beta + c\gamma)z = 0.$$

The plane (1) will be a tangent plane to the given cone if

$$(a\alpha + h\beta + g\gamma)x + (h\alpha + b\beta + f\gamma)y + (g\alpha + f\beta + c\gamma)z \equiv ux + vy + wz$$

$$\Rightarrow \quad \frac{a\alpha + h\beta + g\gamma}{u} = \frac{h\alpha + b\beta + f\gamma}{v} = \frac{g\alpha + f\beta + c\gamma}{w} = \lambda \text{ (say)}$$

$$\Rightarrow \quad \begin{cases} a\alpha + h\beta + g\gamma - \lambda u = 0, & (2)\\ h\alpha + b\beta + f\gamma - \lambda v = 0, & (3)\\ g\alpha + f\beta + c\gamma - \lambda w = 0. & (4)\end{cases}$$

Since the point (α, β, γ) lies on (1) also, we have

$$u\alpha + v\beta + w\gamma = 0. \tag{5}$$

Eliminating α, β, γ and λ from equations (2), (3), (4), and (5), we obtain

$$\begin{vmatrix} a & h & g & u \\ h & b & f & v \\ g & f & c & w \\ u & v & w & 0 \end{vmatrix} = 0$$

$$\Rightarrow \quad Au^2 + Bv^2 + Cw^2 + 2Fvw + 2Gwu + 2Huv = 0,$$

where A, B, C, F, G, H are defined in § 3.2.

Remark. The condition of tangency is similar to the condition that a plane cuts a cone in two coincident generators (see Cor. 1, § 3.2), which ensures the tangency of the plane.

4.4 Reciprocal cone. *To find the locus of the lines drawn through the vertex of a cone perpendicular to the tangent planes.*

Let the equation of a cone be

$$ax^2 + by^2 + cz^2 + 2fyz + 2gzx + 2hxy = 0. \tag{1}$$

Any tangent plane to the cone (1) is

$$ux + vy + wz = 0, \tag{2}$$

where $$Au^2 + Bv^2 + Cw^2 + 2Fvw + 2Gwu + 2Huv = 0. \tag{3}$$

The normal to the plane (2) passing through the vertex (0, 0, 0) of the cone (1) is

$$\frac{x}{u} = \frac{y}{v} = \frac{z}{w}. \tag{4}$$

Eliminating u, v, w from (3) and (4), we get

$$Ax^2 + By^2 + Cz^2 + 2Fyz + 2Gzx + 2Hxy = 0.$$

This equation represents a cone and is the required locus. ▌

The locus of the lines drawn through the vertex of a cone perpendicular to the tangent planes is called the **reciprocal cone** *of the given cone.*

Further, *to find the reciprocal cone of*

$$Ax^2 + By^2 + Cz^2 + 2Fyz + 2Gzx + 2Hxy = 0. \tag{5}$$

Any tangent plane to (5) is

$$ux + vy + wz = 0, \tag{6}$$

where $$A'u^2 + B'v^2 + C'w^2 + 2F'vw + 2G'wu + 2H'uv = 0. \tag{7}$$

Here $$A' = BC - F^2 = aD,\ B' = CA - G^2 = bD,\ C' = AB - H^2 = cD,$$

$$F' = GH - AF = fD,\ G' = HF - BG = gD,\ H' = FG - CH = hD.$$

Thus, the reciprocal cone of (5) becomes

$$aDx^2 + bDy^2 + cDz^2 + 2fDyz + 2gDzx + 2hDxy = 0$$

$$\Rightarrow \quad ax^2 + by^2 + cz^2 + 2fyz + 2gzx + 2hxy = 0.$$

This is, in fact, the cone (1). Hence, the reciprocal of the reciprocal cone is the cone itself. ▌

Two cones are said to be **reciprocal cones** *if each one is the locus of the normals through the vertex to the tangent planes to the other.*

4.5 Rectangular cone

To find the cone whose generators are the normals to the plane $ux + vy + wz = 0$ which cuts the cone

$$f(x, y, z) \equiv ax^2 + by^2 + cz^2 + 2fyz + 2gzx + 2hxy = 0 \tag{1}$$

in two perpendicular generators.

The plane $ux + vy + wz = 0$ cuts the cone (1) in two perpendicular (rectangular) generators (see § 3.2) if and only if

$$(a + b + c)(u^2 + v^2 + w^2) - f(u, v, w) = 0. \qquad (2)$$

Any normal to the plane $ux + vy + wz = 0$ through the vertex of the cone (1) is

$$\frac{x}{u} = \frac{y}{v} = \frac{z}{w}. \qquad (3)$$

Eliminating u, v, w from (2) and (3), we get

$$(a + b + c)(x^2 + y^2 + z^2) - f(x, y, z) = 0. \qquad (4)$$

This is the required equation of the cone.

Note 1. It may be noted that the cone (4) is not the reciprocal cone of (1). In case the given planes are the tangent planes to (1), then (4) would have been the reciprocal cone.

To find the reciprocal cone of (4).

Equation (4) can be written as

$$(b + c)x^2 + (c + a)y^2 + (a + b)z^2 - 2fyz - 2gzx - 2hxy = 0.$$

Its reciprocal cone is

$$A'x^2 + B'y^2 + C'z^2 + 2F'yz + 2G'zx + 2H'xy = 0,$$

where $A' = (c + a)(a + b) - f^2 = bc - f^2 + a(a + b + c) = A + a\Sigma a,$

$$B' = B + b\Sigma a,\ C' = C + c\Sigma a.$$

$$F' = gh + (b + c)f = gh - af + f(a + b + c) = F + f\Sigma a,$$

$$G' = G + g\Sigma a,\ H' = H + h\Sigma a.$$

Hence, the reciprocal cone of (4) is

$$(A + a\Sigma a)x^2 + (B + b\Sigma a)y^2 + (C + c\Sigma a)z^2 + 2(F + f\Sigma a)yz$$

$$+ 2(G + g\Sigma a)zx + 2(H + h\Sigma a)xy = 0$$

$$Ax^2 + By^2 + Cz^2 + 2Fyz + 2Gzx + 2Hxy + \Sigma a \cdot f(x, y, z) = 0. \qquad (5)$$

Note 2. It is quite clear that a tangent plane to (5) cuts (1) into two perpendicular generators and vice versa. If we wish to work out the possibility of getting a set of three mutually perpendicular generators of the cone (1), we must try to choose a tangent plane to (5) so that its normal is a generator of the given cone (1). But this normal is a generator of the reciprocal cone (4). Further, we note that the generators which are common to (1) and (4) also generate the cone $x^2 + y^2 + z^2 = 0$. On the other hand every generator of the cone $x^2 + y^2 + z^2 = 0$ is orthogonal to itself and lies in its own normal plane. Thus, we are not successful in general to obtain a set of three distinct mutually perpendicular generators. ▌

We shall now be interested in particular to obtain the conditions, if possible, under which we may be successful in obtaining a set of three mutually perpendicular generators.

To find a criteria when a cone has three distinct mutually perpendicular generators.

Let
$$\frac{x}{u} = \frac{y}{v} = \frac{z}{w} \tag{1}$$
be a generator of the cone
$$f(x, y, z) \equiv ax^2 + by^2 + cz^2 + 2fyz + 2gzx + 2hxy = 0. \tag{2}$$
Then
$$f(u, v, w) \equiv au^2 + bv^2 + cw^2 + 2fvw + 2gwu + 2huv = 0. \tag{3}$$

The equation of any plane through the origin perpendicular to (1) is given by
$$ux + vy + wz = 0. \tag{4}$$
The plane (4) will cut the cone (2) is two perpendicular generators if and only if
$$(a + b + c)(u^2 + v^2 + w^2) - f(u, v, w) = 0$$
$$\Leftrightarrow \qquad (a + b + c)(u^2 + v^2 + w^2) = 0; \qquad \text{On using (3).}$$
But
$$u^2 + v^2 + w^2 \neq 0.$$
Therefore
$$a + b + c = 0.$$
This is a necessary and sufficient condition for the cone (2) to have three mutually perpendicular generators, the three generators being given by (1) and the lines of intersection of the plane with the cone.

Note 3. The condition $a + b + c = 0$ for the cone (2) to have a set of three mutually perpendicular generators is independent of the generator given by (1). As such given any generator of the cone (2), we can always find a plane through the vertex (0, 0, 0) perpendicular to the given generator of the cone which cuts the cone into two other perpendicular generators if and only if $a + b + c = 0$, and thus forming a set of three mutually perpendicular generators for each given generator of the cone. Hence, *if $a + b + c = 0$, the cone admits of an infinite number of sets of three mutually perpendicular generators.* In other words *either a cone admits of an infinite number of sets of three mutually perpendicular generators or none.*

A cone having three distinct mutually perpendicular generators is called a **rectangular cone.**

Corollary. The cone $ax^2 + by^2 + cz^2 + 2fyz + 2gzx + 2hxy = 0$ will have three mutually perpendicular tangent planes, if the reciprocal cone

$$Ax^2 + By^2 + Cz^2 + 2Fyz + 2Gzx + 2Hxy = 0$$

has three mutually perpendicular generators for which $A + B + C = 0$, *i.e.*, $f^2 + g^2 + h^2 = bc + ca + ab$.

4.6 Examples

Example 1. Prove that the cones $ax^2 + by^2 + cz^2 = 0$ and $\frac{x^2}{a}+\frac{y^2}{b}+\frac{z^2}{c} = 0$ are reciprocal.

Solution. The reciprocal cone of $ax^2 + by^2 + cz^2 = 0$ is

$$Ax^2 + By^2 + Cz^2 + 2Fyz + 2Gzx + 2Hxy = 0, \tag{1}$$

where $A = bc - f^2 = bc, B = ca - g^2 = ca,$

$$C = ab - h^2 = ab, F = gh - af = 0,$$

$$G = hf - bg = 0, H = fg - ch = 0.$$

Therefore, the equation (1) becomes

$$bcx^2 + cay^2 + abz^2 = 0, \text{ i.e., } \frac{x^2}{a}+\frac{y^2}{b}+\frac{z^2}{c} = 0$$

Further, the reciprocal cone of $ax^2 + by^2 + cz^2 = 0$.

Hence the given cones are reciprocal.

Example 2. Prove that the equation $a^2x^2 + b^2y^2 + c^2z^2 - 2bcyz - 2cazx - 2abxy = 0$ represents a cone which touches the coordinate planes.

Solution. The cone will touch the three coordinate planes (mutually perpendicular tangent planes) if its reciprocal cone has three mutually perpendicular generators.

The reciprocal cone of the given cone is

$$Ax^2 + By^2 + Cz^2 + 2Fyz + 2Gzx + 2Hxy = 0, \tag{1}$$

where $A = b^2c^2 - b^2c^2 = 0, B = c^2a^2 - c^2a^2 = 0, C = a^2b^2 - a^2b^2 = 0,$

$F = a^2bc - (-a^2bc) = 2a^2bc$, $G = b^2ca - (-b^2ca) = 2b^2ca$, $H = c^2ab - (-c^2ab) = 2c^2ab$. Therefore, equation (1) becomes

$$4a^2bcyz + 4b^2cazx + 4c^2abxy = 0, \text{ i.e., } ayz + bzx + cxy = 0. \tag{2}$$

In equation (2), we have

$$\text{coeff. of } x^2 + \text{coeff. of } y^2 + \text{coeff. of } z^2 = 0.$$

Hence, the cone (2) has three mutually perpendicular generators.

Example 3. Prove that the locus of the line of intersection of perpendicular tangent planes to the cone $ax^2 + by^2 + cz^2 = 0$ is the cone

$$a(b + c)x^2 + b(c + a)y^2 + c(a + b)z^2 = 0.$$

Solution. The reciprocal cone of $ax^2 + by^2 + cz^2 = 0$ is

$$bcx^2 + cay^2 + abz^2 = 0. \tag{1}$$

Let the plane $lx + my + nz = 0$ cut the cone (1) along perpendicular generators. Then

$$(bc + ca + ab)(l^2 + m^2 + n^2) = bcl^2 + cam^2 + abn^2$$

$$\Rightarrow \quad a(b + c)l^2 + b(c + a)m^2 + c(a + b)n^2 = 0. \tag{2}$$

The equations of the normal to the plane $lx + my + nz = 0$ through the vertex are given by

$$\frac{x}{l} = \frac{y}{m} = \frac{z}{n}. \tag{3}$$

Eliminating l, m, n from (2) and (3), we get

$$a(b + c)x^2 + b(c + a)y^2 + c(a + b)z^2 = 0.$$

Example 4. Prove that the general equation to a cone which touches the three coordinate planes is $\sqrt{fx} \pm \sqrt{gy} \pm \sqrt{hz} = 0$.

Solution. The general equation of a cone through the coordinate axes is

$$fyz + gzx + hxy = 0.$$

Its reciprocal cone is

$$Ax^2 + By^2 + Cz^2 + 2Fyz + 2Gzx + 2Hxy = 0, \tag{1}$$

where $\quad A = bc - f^2 = -f^2,\ B = ca - g^2 = -g^2,\ C = ab - h^2 = -h^2,$

$$F = gh - af = gh,\ G = hf - bg = hf,\ H = fg - ch = fg.$$

Hence, equation (1) becomes

$$f^2x^2 + g^2y^2 + h^2z^2 - 2ghyz - 2hfzx + 2fgxy = 4fgxy$$

$$\Rightarrow \quad fx + gy - hz = \pm 2\sqrt{fgxy}$$

$$\Rightarrow \quad \sqrt{fx} \pm \sqrt{gh} \pm \sqrt{hz} = 0.$$

Example 5. If $\frac{x}{1}=\frac{y}{2}=\frac{z}{3}$ represents one of the three mutually perpendicular generators of the cone $5yz - 8zx - 3xy = 0$; find the equations of the other two.

Solution. Let the line

$$\frac{x}{l}=\frac{y}{m}=\frac{z}{n} \tag{1}$$

be perpendicular to $\frac{x}{1}=\frac{y}{2}=\frac{z}{3}$. Then

$$l + 2m + 3n = 0. \tag{2}$$

If the line (1) be a generator of the cone $5yz - 8zx - 3xy = 0$, then

$$5mn - 8nl - 3lm = 0. \tag{3}$$

Eliminating n from (2) and (3), we get

$$4\frac{l^2}{m^2}+\frac{l}{m}-5=0, \text{ i.e., } \frac{l}{m}=1, -\frac{5}{4}.$$

Equation (2) can be written as $\frac{l}{m}+2+3\frac{n}{m}=0,$

If $\quad \frac{l}{m}=1, \frac{n}{m}=-1$, then $\frac{l}{1}=\frac{m}{1}=\frac{n}{-1}$.

and if $\quad \frac{l}{m}=-\frac{5}{4}, \frac{n}{m}=-\frac{1}{4}$, then $\frac{l}{5}=\frac{m}{-4}=\frac{n}{1}$.

Thus, the other two mutually perpendicular generators are

$$\frac{x}{1}=\frac{y}{1}=\frac{z}{-1} \text{ and } \frac{x}{5}=\frac{y}{-4}=\frac{z}{1}.$$

Exercises

1. Prove that the perpendiculars drawn from the origin to the tangent planes to the cone $3x^2 + 4y^2 + 5z^2 + 2yz + 4zx + 6xy = 0$ lie on the cone $19x^2 + 11y^2 + 3z^2 + 6yz - 10zx - 26xy = 0$.
2. Prove that the tangent planes to the cone $lyz + mzx + nxy = 0$ are at right angles to the generators of the cone $l^2x^2 + m^2y^2 + n^2z^2 - 2mnyz - 2nlzx - 2lmxy = 0$.

3. Prove that the locus of the points from which three mutually perpendicular planes can be drawn to touch the ellipse $\frac{x^2}{a^2} + \frac{y^2}{b^2} = 1, z = 0$ is the sphere $x^2 + y^2 + z^2 = a^2 + b^2$.

4. Prove that the locus of points from which three mutually perpendicular lines can be drawn to intersect a given circle is a surface of revolution.

5. Find the equation of the cone which touches the three coordinate planes and the planes $x + 2y + 3z = 0, 2x + 3y + 4z = 0$.

6. Find the equation of the cone which touches the coordinate planes and the three mutually perpendicular tangent planes $x - 2y + 3z = 0, 3x + 2y - z = 0, 2x - 5y - 4z = 0$.

5. Enveloping Cone

It has been observed in analytical geometry of two dimensions that two tangents can be drawn from a point to a conic (circle, parabola, ellipse or hyperbola). In analogy with that, an infinite number of tangent lines can be drawn from a point to a conicoid in space. Here, we discuss the case of a sphere in particular. We note that all such tangent lines generate a cone with the given point as vertex, see Fig. 5.2 (for proof see § 5.1). Such a cone is called an **enveloping cone** (or **tangent cone**) since it envelopes all the tangent lines drawn from a point. More precisely, we define

The locus of the tangent lines drawn from a given point to a given surface is called the **enveloping cone** *of the surface.*

The given point is called the vertex of the enveloping cone.

5.1 Equation of the enveloping cone. *To find the equation of the enveloping cone of the sphere* $x^2 + y^2 + z^2 = a^2$ *with vertex at the point* (α, β, γ).

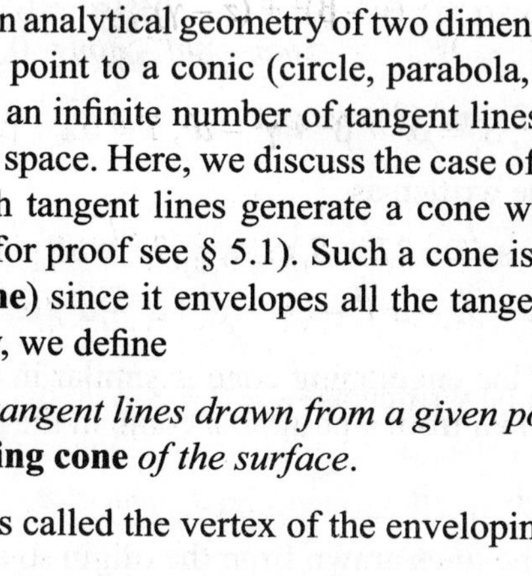

Fig. 16.2

Any line through $P(\alpha, \beta, \gamma)$ is given by

$$\frac{x-\alpha}{l} = \frac{y-\beta}{m} = \frac{z-\gamma}{n} = r \text{ (say).} \qquad (1)$$

Any point on (1) is $(\alpha + lr, \beta + mr, \gamma + nr)$. This point will lie on the given sphere if

$$(\alpha + lr)^2 + (\beta + mr)^2 + (\gamma + nr)^2 = a^2$$

$$\Rightarrow \quad r^2(l^2 + m^2 + n^2) + 2r(\alpha l + \beta m + \gamma n) + \alpha^2 + \beta^2 + \gamma^2 - a^2 = 0. \qquad (2)$$

But this is a quadratic equation in r giving two values of r corresponding to which there are two points common to the sphere and the line (1). Now, if the line (1) be a tangent to the sphere, both the values of r given by equation (2) must be equal. Therefore, we must have

$$(\alpha l + \beta m + \gamma n)^2 = (l^2 + m^2 + n^2)(\alpha^2 + \beta^2 + \gamma^2 - a^2). \qquad (3)$$

The locus of the tangent lines is obtained by eliminating l, m, n from (1) and (3). Thus, the required locus is

$$\{\alpha(x - \alpha) + \beta(y - \beta) + \gamma(z - \gamma)\}^2$$
$$= \{(x - \alpha)^2 + (y - \beta)^2 + (z - \gamma)^2\}(\alpha^2 + \beta^2 + \gamma^2 - a^2). \qquad (4)$$

If we use the notations

$$S \equiv x^2 + y^2 + z^2 - a^2,\ S_1 \equiv \alpha^2 + \beta^2 + \gamma^2 - a^2,\ T \equiv \alpha x + \beta y + \gamma z - a^2,$$

the equation (4) may be written as

$$(T - S_1)^2 = (S - 2T + S_1)S_1, \text{ i.e., } T^2 = SS_1$$

$$\Rightarrow \quad (\alpha x + \beta y + \gamma z - a^2)^2 = (x^2 + y^2 + z^2 - a^2)(\alpha^2 + \beta^2 + \gamma^2 - a^2).$$

Note. The equation of the enveloping cone is similar in appearance to that of the pair of tangents drawn from a point to a conic in the plane.

5.2 Examples

Example. Prove that the lines drawn from the origin so as to touch the sphere $x^2 + y^2 + z^2 + 2ux + 2vy + 2wz + d = 0$, lie on the cone $d(x^2 + y^2 + z^2) = (ux + vy + wz)^2$.

Solution. The lines drawn from the origin to touch the sphere generate the enveloping cone. Hence, the equation of the enveloping cone of the given sphere with vertex at the origin is, $T^2 = SS_1$, *i.e.*,

$$d(x^2 + y^2 + z^2) = (ux + vy + wz)^2.$$

Exercises

1. Find the enveloping cone of the sphere $x^2 + y^2 + z^2 + 2x - 2y = 2$ with its vertex at (1, 1, 1).
2. Find the locus of the tangent lines drawn from the point (α, β, γ) to the sphere $x^2 + y^2 + z^2 + 2ux + 2vy + 2wz + d = 0$.

3. Prove that three mutually perpendicular tangent lines can be drawn to the sphere $x^2 + y^2 + z^2 = a^2$ from any point on the sphere $2(x^2 + y^2 + z^2) = 3a^2$.

6. Right Circular Cone

The surface generated by a line passing through a fixed point and making a constant angle with a fixed line through the fixed point is called the **right circular cone.**

The fixed point is called the **vertex,** the fixed line the axis and the constant angle the **semi-vertical angle** of the cone.

Remarks

(*i*) The section of a right circular cone by a plane perpendicular to its axis is a circle (See Fig. 16.3(a)).

(*ii*) A cone with base curve as a circle is not necessarily a right circular cone (See Fig. 16.3 (b)).

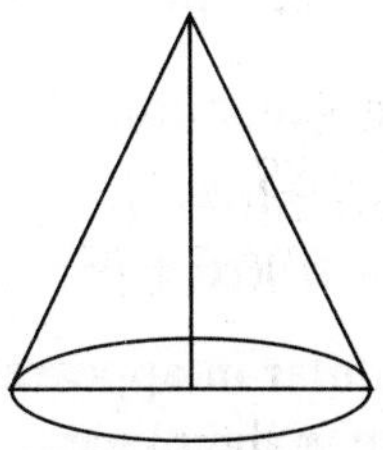

Fig. 16.3(a)

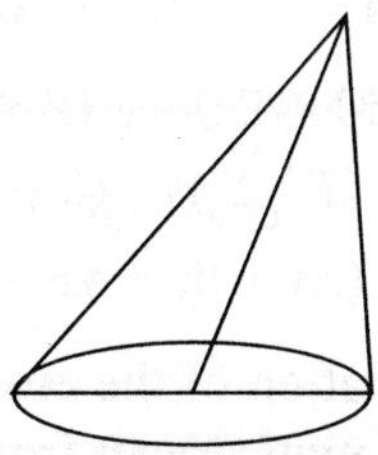

Fig. 16.3(b)

6.1 Equation of a right circular cone. *To find the equation of a right circular cone whose vertex is O(α, β, γ), the semi-vertical angle* is θ *and the axis is*

$$\frac{x-\alpha}{l} = \frac{y-\beta}{m} = \frac{z-\gamma}{n}.$$

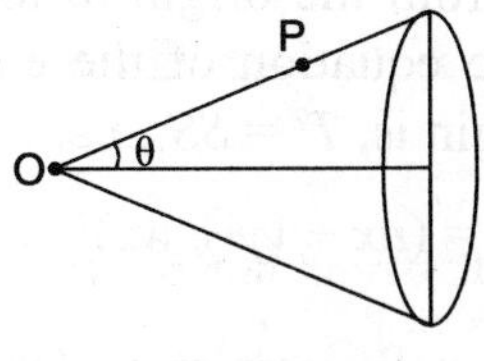

Fig. 16.4

Let $P(x, y, z)$ be any point on the cone. Then, the direction cosines of OP are proportional to $x - \alpha, y - \beta, z - \gamma$. Therefore

$$\cos\theta = \frac{l(x-\alpha)+m(y-\beta)+n(z-\gamma)}{\sqrt{(l^2+m^2+n^2)\{(x-\alpha)^2+(y-\beta)^2+(z-\gamma)^2\}}}$$

$$\Rightarrow \{l(x-\alpha)+m(y-\beta)+n(z-\gamma)\}^2 = (l^2+m^2+n^2)$$
$$\{(x-\alpha)^2+(y-\beta)^2+(z-\gamma)^2\}\cos^2\theta.$$

This is the required equation of the cone. ▌

Corollary 1. The equation of a right circular cone with vertex at the origin is

$$(lx+my+nz)^2 = (l^2+m^2+n^2)(x^2+y^2+z^2)\cos^2\theta.$$

Corollary 2. The equation of a cone with vertex at the origin and axis as z-axis is $x^2+y^2 = z^2\tan^2\theta$.

6.2 Examples

Example 1. If a right circular cone has three mutually perpendicular generators, prove that the semi-vertical angle is $\tan^{-1}\sqrt{2}$.

Solution. The equation of a right circular cone with vertex at the origin, semi-vertical angle θ and axis along z-axis is given by

$$x^2+y^2-z^2\tan^2\theta = 0.$$

This cone will have three mutually perpendicular generators if

$$\text{coeff. of } x^2 + \text{coeff. of } y^2 + \text{coeff. of } z^2 = 0$$

$$\Rightarrow \qquad \theta = \tan^{-1}\sqrt{2}.$$

Example 2. Prove that the semi-vertical angle of a right circular cone which has three mutually perpendicular tangent planes is $\cot^{-1}\sqrt{2}$.

Solution. If a cone has three mutually perpendicular tangent planes, then its reciprocal cone will have three mutually perpendicular generators.

The equation of the cone which has vertex at the origin, semi-vertical angle θ and the axis along z-axis is

$$x^2+y^2-z^2\tan^2\theta = 0.$$

Its reciprocal cone is

$$x^2\tan^2\theta + y^2\tan^2\theta - z^2 = 0.$$

This cone will have three mutually perpendicular generators if

$$\text{coeff. of } x^2 + \text{coeff. of } y^2 + \text{coeff. of } z^2 = 0$$

$$\Rightarrow \qquad \theta = \cot^{-1}\sqrt{2}.$$

Exercises

1. Find the equation of the right circular cone whose vertex is (3, 2, 1), semi-vertical angle 30° and axis the line $\frac{x-3}{4} = \frac{y-2}{1} = \frac{z-1}{3}$.

2. Prove that the equation of the cone generated by rotating the line $\frac{x}{l} = \frac{y}{m} = \frac{z}{n}$ about the line $\frac{x}{a} = \frac{y}{b} = \frac{z}{c}$ as axis is

$$(l^2 + m^2 + n^2)(ax + by + cz)^2 = (al + bm + cn)^2(x^2 + y^2 + z^2).$$

3. Find the equation of the right circular cone with vertex at (1, –2, –1), semi-vertical angle 60° and axis $\frac{x-1}{3} = \frac{y+2}{-4} = \frac{z+1}{5}$. Prove that the plane $3x - 4y + 5z = 56$ cuts it in a circle, find its centre and radius.

4. Find the equation of the cone formed by rotating the line $2x + 3y = 6$, $z = 0$ about y-axis.

5. The axis of the right circular cone, vertex origin, makes equal angles with the coordinate axes. If the equation of the cone be $4(x^2 + y^2 + z^2) + 9(yz + zx + xy) = 0$, prove that the semi-vertical angle is $\cos^{-1}\frac{1}{3\sqrt{3}}$.

6. Find the equation of the right circular cone which passes through the three lines drawn from the origin with direction ratios 1, 2, 2; 2, 1, – 2; 2, – 2, 1.

7. The Cone $ax^2 + by^2 + cz^2 = 0$

The above equation is the particular case of the general equation of the cone whose vertex is at the origin. This equation is also referred to **standard equation** of the cone.

7.1 Certain results

Let the equation of the cone be $ax^2 + by^2 + cz^2 = 0$.

(*i*) The tangent plane at (α, β, γ) is $a\alpha x + b\beta y + c\gamma z = 0$.

(*ii*) The plane $lx + my + nz = 0$ is a tangent plane to the cone if

$$bcl^2 + cam^2 + abn^2 = 0.$$

(*iii*) The angle between the lines of intersection of the plane $ux + vy + wz = 0$ and the cone is

$$\tan^{-1}\frac{2\sqrt{\{-(u^2+v^2+w^2)(bcu^2+cav^2+abw^2)\}}}{a(v^2+w^2)+b(w^2+u^2)+c(u^2+v^2)}.$$

(*iv*) The equation of the pair of tangent planes passing through the line $x/l = y/m = z/n$ is $(ax^2 + by^2 + cz^2)(al^2 + bm^2 + cn^2) = (alx + bmy + cnz)^2$, and the equation of the plane passing through the generators of contact of these planes is $alx + bmy + cnz = 0$.

7.2 Example

Example. Prove that the lines of intersection of pairs of tangent planes to

$ax^2 + by^2 + cz^2 = 0$, which touch along perpendicular generators lie on the cone $a^2(b + c)x^2 + b^2(c + a)y^2 + c^2(a + b)z^2 = 0$.

Solution. Let the equations of the line of intersection of the tangent planes along two perpendicular generators be

$$\frac{x}{l} = \frac{y}{m} = \frac{z}{n}. \qquad (1)$$

The equation of the pair of tangent planes passing through (1) is

$$(ax^2 + by^2 + cz^2)(al^2 + bm^2 + cn^2) = (alx + bmy + cnz)^2, \qquad (2)$$

and the equation of the plane containing the two generators is

$$alx + bmy + cnz = 0. \qquad (3)$$

The plane (3) cuts the cone $ax^2 + by^2 + cz^2 = 0$ in perpendicular lines if

$$(a + b + c)(a^2l^2 + b^2m^2 + c^2n^2) = a(al)^2 + b(bm)^2 + c(cn)^2$$

$$\Rightarrow \qquad a^2(b + c)l^2 + b^2(c + a)m^2 + c^2(a + b)n^2 = 0. \qquad (4)$$

The required locus is obtained by eliminating l, m, n, from (1) and (4), *i.e.*, $a^2(b + c)x^2 + b^2(c + a)y^2 + c^2(a + b)z^2 = 0$.

Exercises

1. Find the equation of the normal plane of the cone $ax^2 + by^2 + cz^2 = 0$ which passes through the generator $\frac{x}{l} = \frac{y}{m} = \frac{z}{n}$.
2. Prove that the plane $lx + my + nz = 0$ cuts the cone $ax^2 + by^2 + cz^2 = 0$ in perpendicular generators if $(b + c)l^2 + (c + a)m^2 + (a + b)n^2 = 0$.
3. Prove that the planes which cut the cone $ax^2 + by^2 + cz^2 = 0$ in perpendicular generators touch the cone $\frac{x^2}{b+c} + \frac{y^2}{c+a} + \frac{z^2}{a+b} = 0$.
4. Prove that the chords of the cone $ax^2 + by^2 + cz^2 = 0$ which are bisected at the point (α, β, γ) lie in the plane $a\alpha(x - \alpha) + b\beta(y - \beta) + c\gamma(z - \gamma) = 0$.

5. Find the equation of the planes through the origin perpendicular to the lines of intersection of the plane $lx + my + nz = 0$ and the cone $ax^2 + by^2 + cz^2 = 0$.

6. Find the equation of the two tangent planes of the cone $ax^2 + by^2 + cz^2 = 0$ which pass through the line $lx + my + nz + 0$, $l'x + m'y + n'z = 0$.

7. Prove that the lines drawn through the origin at right angles to the normal planes of the cone $ax^2 + by^2 + cz^2 = 0$ generate the cone

$$\frac{a(b-c)^2}{x^2} + \frac{b(c-a)^2}{y^2} + \frac{c(a-b)^2}{z^2} = 0.$$

Miscellaneous Exercises

1. Find the equation of the cone whose vertex is (5, 4, 3) and guiding curve the conic $3x^2 + 2y^2 = 6$, $y + z = 0$.

2. Find the equation of the cone above whose vertex is (α, β, γ) and guiding curve the parabola $z^2 = 4ax$, $y = 0$.

3. Find the equation of the cone whose vertex is (1, 2, 3) and guiding curve the circle $x^2 + y^2 + z^2 = 4$, $x + y + z = 1$.

4. A cone having guiding curve the circle $x^2 + y^2 + 2ax + 2by = 0$, $z = 0$, passes through the fixed point $(0, 0, c)$. If the section of the cone by zx-plane is a rectangular hyperbola, prove that the vertex lies on a fixed circle.

5. Prove that the equation of the cone whose vertex is the origin and guiding curve is the circle passing through the points $(a, 0, 0)$, $(0, b, 0)$, $(0, 0, c)$ is
$$a(b^2 + c^2)yz + b(c^2 + a^2)zx + c(a^2 + b^2)xy = 0.$$

6. Two cones with a common vertex pass through the curves $z^2 = 4ax$, $y = 0$ and $z^2 = 4by$, $x = 0$. The plane $z = 0$ meets them in two conics which intersect in four concyclic points. Prove that the vertex lies on the surface
$$z\left(\frac{x}{a} + \frac{y}{b}\right) = 4(x^2 + y^2).$$

7. Prove that the lines of intersection of the plane $lx + my + nz = 0$ and the cone $fyz + gzx + hxy = 0$ are perpendicular if $fmn + gnl + hlm = 0$, and parallel if $\sqrt{fl} + \sqrt{gm} + \sqrt{hn} = 0$.

8. Prove that the equation $\sqrt{fx} + \sqrt{gy} + \sqrt{hz} = 0$ represents a cone which touches the coordinate planes. Find the equation to its reciprocal cone.

9. Find the equation of the cone whose vertex is the origin and guiding curve the circle $y^2 + z^2 = b^2$, $x = a$, and prove that the section of the cone by a plane parallel to xy-plane is a hyperbola.

10. Prove that a cone of second degree can be found to pass through any five concurrent lines, no three of which are coplanar.

11. Find the equation of the cone generated by the lines drawn from the origin to cut the circle through the points (1, 0, 0), (0, 2, 0), (2, 1, 1) and prove that the acute angle between two lines in which the plane $x - 2y = 0$ cuts the cone is $\cos^{-1} 5/14$.

12. Prove that the cone whose vertex is the origin and which passes through the curve of intersection of the surface $2x^2 - y^2 + 2z^2 = 3d^2$ and any plane at a distance d from the origin has three mutually perpendicular generators.

13. A right circular cone passes through the point (1, 1, 1) and has vertex at the point (1, 0, 1). If the axis of the cone is equally inclined to the axes, find the equation of the cone.

14. Find the equations of the right circular cones which contain the three coordinate axes as generators.

15. Find the conditions that the lines of intersection of the plane $lx + my + nz = 0$ and the cones $ax^2 + by^2 + cz^2 = 0, fyz + gzx + hxy = 0$ may be coincident.

16. Prove that the cones $ayz + bzx + cxy = 0$ and $fyz + gzx + hxy = 0$ intersect in the coordinate axes and the line $(bh - cg)x = (cf - ah)y = (ag - bf)z$.

17. Prove that the bisectors of the angles between the lines in which the plane $ux + vy + wz = 0$ cuts the cone $ax^2 + by^2 + cz^2 = 0$ lie on the cone

$$\frac{u(b-c)}{x} + \frac{v(c-a)}{y} + \frac{w(a-b)}{z} = 0.$$

18. Prove that the plane $z = a$ meets any enveloping cone of the sphere $x^2 + y^2 + z^2 = a^2$ in a conic which has a focus at the point $(0, 0, a)$.

The Cylinder

17
CHAPTER

1. Definition

The surface generated by a variable line which remains parallel to a fixed line and intersects a given curve (or touches a given surface) is called the **cylinder**.

The variable line is called the **generator**, the fixed line the **axis** and the given curve (or surface) the **guiding curve**.

1.1 Equation of a cylinder. *To find the equation of the cylinder whose axis and guiding curve are given.*

Let the equation of the guiding curve be the conic

$$ax^2 + 2hxy + by^2 + 2gx + 2fy + c = 0,\ z = 0 \tag{1}$$

and the equations of the axis be

$$\frac{x}{l} = \frac{y}{m} = \frac{z}{n}. \tag{2}$$

Let (α, β, γ) be any point on the cylinder. Then, the equations of a generator are

$$\frac{x-\alpha}{l} = \frac{y-\beta}{m} = \frac{z-\gamma}{n}. \tag{3}$$

This line meets the plane $z = 0$ at the point given by $\left(\alpha - \frac{l\gamma}{n}, \beta - \frac{m\gamma}{n}, 0\right)$.

The generator (3) will intersect the conic (1) if the point $\left(\alpha - \frac{l\gamma}{n}, \beta - \frac{m\gamma}{n}, 0\right)$ lies on (1), *i.e.*,

$$a\left(\alpha - \frac{l\gamma}{n}\right)^2 + 2h\left(\alpha - \frac{l\gamma}{n}\right)\left(\beta - \frac{m\gamma}{n}\right) + b\left(\beta - \frac{m\gamma}{n}\right)^2 +$$

$$2g\left(\alpha - \frac{l\gamma}{n}\right) + 2f\left(\beta - \frac{m\gamma}{n}\right) + c = 0$$

$$\Rightarrow a(n\alpha - l\gamma)^2 + 2h(n\alpha - l\gamma)(n\beta - m\gamma) + b(n\beta - m\gamma)^2$$
$$+ 2gn(n\alpha - l\gamma) + 2fn(n\beta - m\gamma) + cn^2 = 0.$$

Hence, the locus of (α, β, γ) is

$$a(nx - lz)^2 + 2h(nx - lz)(ny - mz) + b(ny - mz)^2$$
$$+ 2gn(nx - lz) + 2fn(ny - mz) + cn^2 = 0.$$

This is the required equation of the cylinder.

Corollary. The equation of the cylinder whose generators are parallel to z-axis and intersect the conic

$$ax^2 + 2hxy + by^2 + 2gx + 2fy + c = 0, z = 0 \text{ is}$$
$$f(x, y) \equiv ax^2 + 2hxy + by^2 + 2gx + 2fy + c = 0.$$

Since, the generators are parallel to z-axis, $l = m = 0$ and $n = 1$, the general equation of the cylinder reduces to the above form.

Remark. It is worth noting that the equation $f(x, y) = 0$ in xy-plane represents a conic whereas the same equation in the space represents a cylinder whose generators are parallel to z-axis.

Note. The equation of the cylinder whose generators are parallel to z-axis and passes through the curve of intersection of the surfaces

$$\phi(x, y, z) = 0 \text{ and } \Psi(x, y, z) = 0$$

is obtained by eliminating z from these equations.

2. Enveloping Cylinder

2.1 *To find the locus of the tangent lines drawn to a sphere and parallel to a given line.*

Let the spehre and the given line be

$$x^2 + y^2 + z^2 = a^2, \tag{1}$$

and

$$\frac{x}{l} = \frac{y}{m} = \frac{z}{n}. \tag{2}$$

Let (α, β, γ) be any point on the locus. Any line through (α, β, γ) parallel to (2) is

$$\frac{x-\alpha}{l} = \frac{y-\beta}{m} = \frac{z-\gamma}{n} = r \text{ (say)}. \tag{3}$$

Any point on (3) is $(\alpha + lr, \beta + mr, \gamma + nr)$. This point will lie on (1) if

$$(\alpha + lr)^2 + (\beta + mr)^2 + (\gamma + nr)^2 = a^2$$

$$\Rightarrow \quad r^2(l^2 + m^2 + n^2) + 2r(\alpha l + \beta m + \gamma n) + \alpha^2 + \beta^2 + \gamma^2 - a^2 = 0. \qquad (4)$$

But this is a quadratic equation in r giving two values of r corresponding to which there are two points common to the sphere (1) and the line (3).

Now, if the line (3) be a tangent to (1), both the values of r given by (4) must be equal. Therefore, we must have

$$(\alpha l + \beta m + \gamma n)^2 = (l^2 + m^2 + n^2)(\alpha^2 + \beta^2 + \gamma^2 - a^2).$$

Hence, the locus of (α, β, γ) is

$$(lx + my + nz)^2 = (l^2 + m^2 + n^2)(x^2 + y^2 + z^2 - a^2).$$

This locus represents a cylinder.

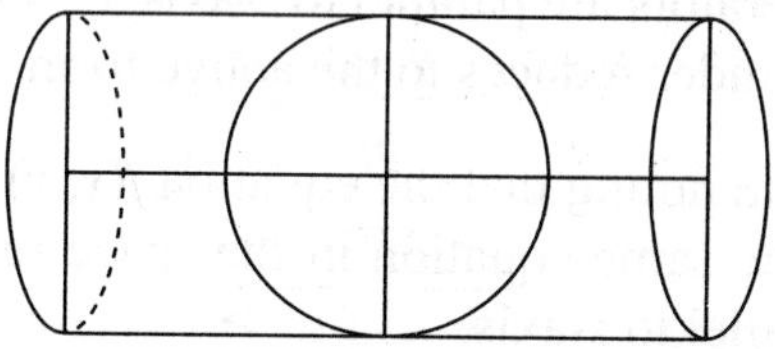

Fig. 17.1

The locus of the tangent lines drawn to a sphere and parallel to a given line is a cylinder and is called the **enveloping cylinder** *of the sphere.*

Note. The plane through the points of contact of the enveloping cylinder and the sphere is $lx + my + nz = 0$.

2.2 Enveloping cylinder as a limiting case of enveloping cone

Any point on the axis

$$\frac{x}{l} = \frac{y}{m} = \frac{z}{n} = r \text{ (say)}$$

of the enveloping cylinder is (lr, mr, nr). Let this point be the vertex of the enveloping cone. Then, its equation is $T^2 = SS_1$ *i.e.*,

$$\left(lx + my + nz - \frac{a^2}{r}\right)^2 = (x^2 + y^2 + z^2 - a^2)\left(l^2 + m^2 + n^2 - \frac{a^2}{r^2}\right).$$

As $r \to \infty$, this equation reduces to

$$(lx + my + nz)^2 = (x^2 + y^2 + z^2 - a^2)(l^2 + m^2 + n^2),$$

which is the equation of the enveloping cylinder.

Thus, *enveloping cylinder may be regarded as a limiting case of the enveloping cone whose vertex is at an infinite distance.*

3. Right Circular Cylinder

A surface generated by a line which intersects a fixed circle (guiding curve) and is perpendicular to the plane of the circle is called the **right circular cylinder**.

The normal to the plane of the circle through its centre is called the **axis** of the cylinder and the radius of the circle the **radius** of the cylinder.

3.1 Equation of right circular cylinder. *To find the equation of the right circular cylinder whose radius is r and axis the line*

$$\frac{x-\alpha}{l}=\frac{y-\beta}{m}=\frac{z-\gamma}{n}.$$

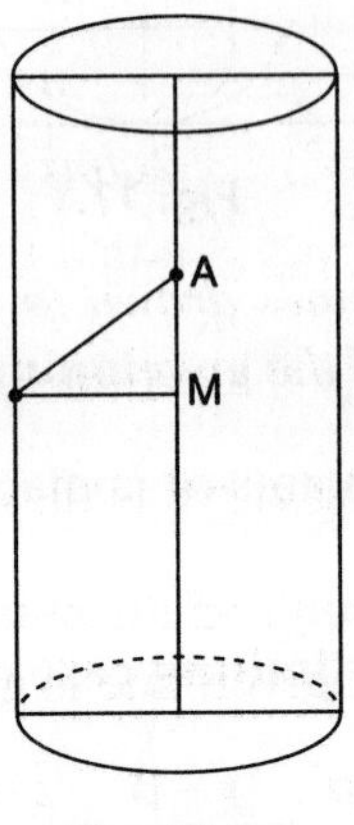

Fig. 17.2

Let $P(x', y', z')$ be any point on the cylinder, $A(\alpha, \beta, \gamma)$ a fixed point on the axis and r the radius. Then $AP^2 - AM^2 = PM^2$, *i.e.,*

$$(x'-\alpha)^2+(y'-\beta)^2+(z'-\gamma)^2-\left\{\frac{l(x'-\alpha)+m(y'-\beta)+n(z'-\gamma)}{\sqrt{l^2+m^2+n^2}}\right\}^2=r^2.$$

Hence, the locus of (x', y', z') is

$$(l^2+m^2+n^2)\{(x-\alpha)^2+(y-\beta)^2+(z-\gamma)^2\}$$
$$-\{l(x-\alpha)+m(y-\beta)+n(z-\gamma)\}^2=r^2(l^2+m^2+n^2).$$

This is the required equation of the right circular cylinder.

Corollary 1. The equation of a right circular cylinder whose axis is $\frac{x}{l} = \frac{y}{m} = \frac{z}{n}$ is $(l^2 + m^2 + n^2)(x^2 + y^2 + z^2 - r^2) = (lx + my + nz)^2$.

Corollary 2. The equation of a right circular cylinder whose axis is z-axis is $x^2 + y^2 = r^2$.

3.2. Examples

Example 1. Find the equation of the cylidner which intersects the curve

$$ax^2 + by^2 + cz^2 = 1,\ lx + my + nz = p,$$

and whose generators are parallel to z-axis.

Solution. Since, the equation of the cylinder whose axis is z-axis is the relation between x and y only, the required equation of the cylinder whose axis is parallel to z-axis is obtained by eliminating z from the given equations, *i.e.*,

$$ax^2 + by^2 + c\left(\frac{lx + my - p}{-n}\right)^2 = 1$$

$$\Rightarrow\ (an^2 + cl^2)x^2 + (bn^2 + cm^2)y^2 + 2clmxy - 2cplx - 2cpmy + cp^2 - n^2 = 0.$$

This is the required equation of the cylinder.

Example 2. Find the equation of the cylinder whose generators are parallel to the line $\frac{x}{1} = \frac{y}{-2} = \frac{z}{3}$ and guiding curve is the ellipse $x^2 + 2y^2 = 1,\ z = 3$.

Solution. Let (α, β, γ) be any point on the cylinder. Then, the generator of the cylinder which is parallel to the given line is given by

$$\frac{x - \alpha}{1} = \frac{y - \beta}{-2} = \frac{z - \gamma}{3}.$$

This generator meets the plane $z = 3$ in the point

$$\left(\frac{(\alpha + 1) - \gamma}{3}, \frac{3(\beta - 2) + 2\gamma}{3}, 3\right)$$

If this point lies on the guiding curve, then

$$3(\alpha^2 + 2\beta^2 + \gamma^2) + 8\beta\gamma - 2\gamma\alpha + 6\alpha - 24\beta - 18\gamma + 24 = 0.$$

Hence, the locus of (α, β, γ) is

$$3(x^2 + 2y^2 + z^2) + 8yz - 2zx + 6x - 24y - 18z + 24 = 0.$$

This is the required equation of the cylinder.

Example 3. Find the equation of the right circular cylinder whose guiding curve is $x^2 + y^2 + z^2 = 9, x - y + z = 3$.

Solution. Since the axis of the cylinder is perpendicular to the plane $x - y + z = 3$, its direction cosines are proportional to $1, -1, 1$. Let (α, β, γ) be any point on the cylinder. Then, the equations of the generator through this point are given by

$$\frac{x-\alpha}{1} = \frac{y-\beta}{-1} = \frac{z-\gamma}{1} = r \text{ (say)}.$$

Any point on the generator is $(\alpha + r, \beta - r, \gamma + r)$. If this point lies on the guiding curve, then

$$\begin{cases} (\alpha+r)^2 + (\beta-r)^2 + (\gamma+r)^2 = 9, \\ (\alpha+r) - (\beta-r) + (\gamma+r) = 3 \end{cases}$$

$$\Rightarrow \begin{cases} \alpha^2 + \beta^2 + \gamma^2 + 2r(\alpha - \beta + \gamma) + 3r^2 = 9, \\ \alpha - \beta + \gamma + 3r = 3. \end{cases}$$

Eliminating r from these equations, we get

$$\alpha^2 + \beta^2 + \gamma^2 + \beta\gamma - \gamma\alpha + \alpha\beta - 9 = 0.$$

Hence, the locus of (α, β, γ) is $x^2 + y^2 + z^2 + yz - zx + xy - 9 = 0$.

Exercises

1. Find the equation of the right circular cylinder of radius 2 whose axis is the line $\frac{x-1}{2} = \frac{y-2}{1} = \frac{z-3}{2}$.
2. Find the equation of the right circular cylinder of radius 3 whose axis is the line $\frac{x-1}{2} = \frac{y-3}{2} = \frac{z-5}{-1}$.
3. Find the equation of the right circular cylinder whose axis is $x = 2y = -z$ and radius is 4. Prove that the area of the section of this cylinder by the plane $z = 0$ is 24π.
4. Find the equation of the right circular cylinder described on the circle through the three points (1, 0, 0), (0, 1, 0) and (0, 0, 1) as the guiding curve.
5. Find the equation of the enveloping cylinder of the sphere $x^2 + y^2 + z^2 - 2x + 4y = 1$ having its generators parallel to the line $x = y = z$. Also, find its guiding curve.

The Central Conicoids

18

CHAPTER

1. Introduction

This chapter deals with central conicoids viz. the conicoids having centre. While discussing various concepts and results we shall be considering the equation of the central conicoid in its standard form,* *i.e.*, $ax^2 + by^2 + cz^2 = 1$.

Throughout the chapter conicoid would mean central conicoid unless specified otherwise.

2. Shapes of the Central Conicoids

Let the equation of the conicoid be $Ax^2 + By^2 + Cz^2 = 1$. This equation can be written in the form

$$\pm\frac{x^2}{a^2} \pm \frac{y^2}{b^2} \pm \frac{z^2}{c^2} = 1.$$

There are four different cases. In the expression on the left hand side of the above, if

(*i*) all the three signs are positive, the surface is called the **ellipsoid**;

(*ii*) one of the signs is negative, the surfac is called the **hyperboloid of one sheet**;

(*iii*) two of the signs are negative, the surface is called the **hyperboloid of two sheets**;

(*iv*) all the three signs are negative, the surface is called the **virtual quadric**.

2.1 The ellipsoid

Consider the surface whose equation is

$$\frac{x^2}{a^2} + \frac{y^2}{b^2} + \frac{z^2}{c^2} = 1.$$

---

* For the reduction of the equation of the central conicoid and conicoids in general to the standard form and other related concepts see Appendix -I.

From the shape of this equation we can immediately draw several conclusions:

(*i*) It is symmetrical with respect to all the three coordinate axes since if (x, y, z) lies on the surface, then so do the points $(-x, y, z)$, $(x, -y, z)$ and $(x, y, -z)$.

(*ii*) The surface is bounded and lies inside the box bounded by the planes $x = \pm a$, $y = \pm b$ and $z = \pm c$ since for any point (x, y, z) on the surface, $|x| \leqslant a$, $|y| \leqslant b$ and $|z| \leqslant c$. Furthermore, the surface is touching the faces of the box at the points $(\pm a, 0, 0)$,$(0, \pm b, 0)$ and $(0, 0, \pm c)$.

(*iii*) The plane section of the surface in the *xy*-plane is the conic given by

$$\frac{x^2}{a^2} + \frac{y^2}{b^2} = 1, z = 0.$$

This, in fact, is an ellipse in the *xy*-plane with *x*-axis and *y*-axis as major axis and minor axis, respectively.

Similarly, the plane sections of the surface in *yz*-plane and *zx*-plane are the ellipses in the respective planes.

(*iv*) The plane sections by the planes $z = k$ (k being an arbitrary but a fixed constant) are the ellipses

$$\frac{x^2}{a^2} + \frac{y^2}{b^2} = \frac{c^2 - k^2}{c^2}, z = k,$$

provided $| k | < c$, and the size of the ellipses decreases as $| k |$ increases. If $| k | = c$, the corresponding planes $z = \pm c$ are tangent to the surface, and if $|k| > c$, there are no real points of intersection of the surface and the planes $z = k$.

Similar conclusions could be drawn for the plane sections by the planes $x = k$ and $y = k$.

It is now possible to draw the sketch of the surface called ellipsoid as shown in Fig. 18.1. In anology with an ellipse in two dimensions, we have a few concepts regarding the ellipsoid:

(*i*) The six points $(\pm a, 0, 0)$, $(0, \pm b, 0)$, $(0, 0, \pm c)$ in which the coordinate axes meets the ellipsoid are called the **vertices** of the ellipsoid.

(*ii*) The three line segments joining the pair of vertices on each coordiante axes are called the **axes** of the ellipsoid.

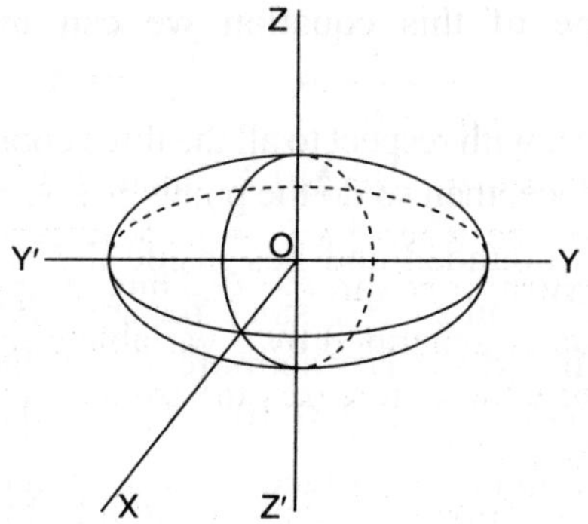

Fig. 18.1

(*iii*) The three axes of the ellipsoid intersect at the point (0, 0, 0) called the **centre** of the ellipsoid.

(*iv*) The line segments joining the centre to the vertices have lengths a, b, c. These lengths are called the **semi-axes** of the ellipsoid.

(*v*) The largest of $2a$, $2b$, $2c$ is called the **major axis**, the next to the largest the **mean axis**, and the smallest the **minor axis** of the ellipsoid.

Note. If $a = b$, the plane section of the surface by the plane $z = k$ is a circle $x^2 + y^2 = a^2\left(1 - \frac{k^2}{c^2}\right)$. The surface (ellipsoid) in this case is generated by revolving an ellipse about minor axis (z-axis) and is called an **ellipsoid of revolution**. If $a = b = c$, the surface is a sphere with centre at the origin.

2.2 The hyperboloid of one sheet

Consider the surface whose equation is $\frac{x^2}{a^2} + \frac{y^2}{b^2} - \frac{z^2}{c^2} = 1$.

We note the following:

(*i*) It is symmetrical with respect to all the three coordinate axes.

(*ii*) The surface is unbounded since each of the coordinates x, y, z can have any values provided only that $\frac{x^2}{a^2} + \frac{y^2}{b^2} \geqslant 1$.

(*iii*) The intersection of the surface with x-axis and y-axis are real meeting at points $(\pm\, a, 0, 0)$ and $(0, \pm\, b, 0)$, respectively, whereas with z-axis is imaginary.

(*iv*) The plane sections by the planes $z = k$, planes parallel to xy-plane, are the ellipses

$$\frac{x^2}{a^2} + \frac{y^2}{b^2} = 1 + \frac{k^2}{c^2}, z = k.$$

These ellipses increase in size as $|k|$ increases, so much so that they become of infinite size when $|k| \to \infty$. However, the smallest ellipse is the one which corresponds to $k = 0$. Thus, it appears that the surface under consideration is generated by a variable ellipse lying in the plane parallel to *xy*-plane which increases in size as it moves farther away on either side of the *xy*-plane.

(*v*) Unlike in the case of ellipsoid, the plane sections by planes parallel to the other coordinate planes are not ellipses. For instance, the plane sections by the planes $x = k$, planes parallel to *yz*-plane, are the hyperbolas

$$\frac{y^2}{b^2} - \frac{z^2}{c^2} = 1 - \frac{k^2}{a^2}, x = k,$$

or the conjugate hyperbolas

$$-\frac{y^2}{b^2} + \frac{z^2}{c^2} = \frac{k^2}{a^2} - 1, x = k$$

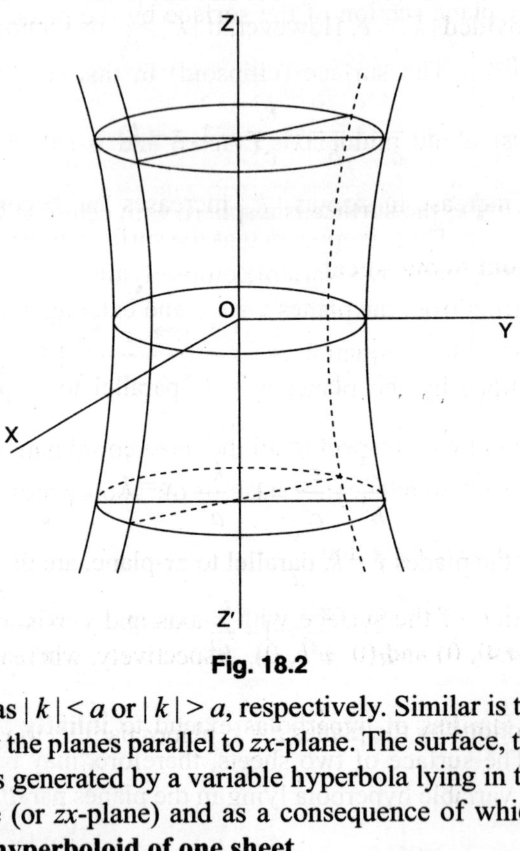

Fig. 18.2

according as $|k| < a$ or $|k| > a$, respectively. Similar is the situation with sections by the planes parallel to *zx*-plane. The surface, therefore, may be regarded as generated by a variable hyperbola lying in the plane parallel to *yz*-plane (or *zx*-plane) and as a consequence of which the surface is called the **hyperboloid of one sheet**.

It is now possible to draw the sketch of the surface as shown in Fig. 18.2.

Note. If $a = b$, the plane section of the surface by the plane $z = k$ is the circle

$x^2 + y^2 = a^2 (1 + k^2/c^2)$. The surface in this case is generated by revolving a hyperbola about its conjugate axis and is called **hyperboloid of revolution**.

2.3 The hyperboloid of two sheets

Consider the surface whose equation is $-\frac{x^2}{a^2} - \frac{y^2}{b^2} + \frac{z^2}{c^2} = 1.$

We note the following:

(*i*) The surface is symmetrical about all the three coordinate axes.

(*ii*) The surface is unbounded since each of coordinates x, y, z can have any values provided only that $|z| \geqslant c$.

(*iii*) The intersection of the surface with x-axis and y-axis are imaginary whereas with z-axis is real meeting at points $(0, 0, \pm c)$.

(*iv*) The xy-plane does not meet the surface nor any plane $z = k$, planes parallel to xy-plane, provided $|k| < c$. However, if $|k| > c$, its section by the plane $z = k$ is the ellipse

$$\frac{x^2}{a^2} + \frac{y^2}{b^2} = \frac{k^2}{c^2} - 1, z = k.$$

These ellipses increase in size as $|k|$ increases and become of infinite size when $|k| \to \infty$. Thus, it appears that the surface under consideration is generated in two parts by a variable ellipse lying in the planes parallel to xy-plane starting from the planes $z = \pm c$ and extending to infinitely on both sides.

(*v*) The plane sections by the planes $x = k$, parallel to yz-plane, are the hyperbolas

$$-\frac{y^2}{b^2} + \frac{z^2}{c^2} = 1 + \frac{k^2}{a^2}, x = k$$

while those by the planes $y = k$, parallel to zx-plane, are the hyperbolas

$$-\frac{x^2}{a^2} + \frac{z^2}{c^2} = 1 + \frac{k^2}{b^2}, y = k.$$

Both of these families of hyperbolas extend to **infinity** as $|k|$ varies from 0 to ∞. The surface of two sheets, therefore, may be regarded as generated by a variable hyperbola lying in the planes parallel to yz-plane (or zx-plane) and as a consequence of which the surface is called the **hyperboloid of two sheets.**

It is now possible to draw the sketch of the surface as shown in Fig. 18.3.

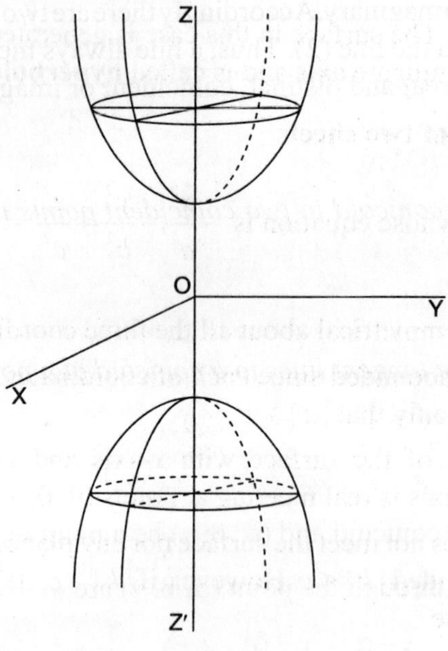

Fig. 18.3.

Note. If $a = b$, the plane section of the surface by the plane $z = k$ is a circle

$$x^2 + y^2 = a^2\left(\frac{k^2}{c^2} - 1\right).$$

The surface in this case is generated by revolving a hyperbola about its conjugate axis and is called **hyperboloid of revolution**.

3. Intersection of a Conicoid and a Line

Let $$ax^2 + by^2 + cz^2 = 1, \tag{1}$$

and $$\frac{x-\alpha}{l} = \frac{y-\beta}{m} = \frac{z-\gamma}{n} = r \text{ (say)} \tag{2}$$

be the equations of a conicoid and a line, respectively.

Any point on the line (2) is

$$(\alpha + lr, \beta + mr, \gamma + nr).$$

This point will lie on the conicoid (1) if

$$a(\alpha + lr)^2 + b(\beta + mr)^2 + c(\gamma + nr)^2 = 1$$

$$\Rightarrow r^2(al^2 + bm^2 + cn^2) + 2r(a\alpha l + b\beta m + c\gamma n) + a\alpha^2 + b\beta^2 + c\gamma^2 - 1 = 0. \quad (3)$$

This equation is quadratic in r giving two values of r which may be real and distinct, coincident or imaginary. Accordingly there are two points of intersection of the conicoid (1) and the line (2). Thus, a line always meets a conicoid in two points which may be real and distinct, coincident or imaginary.

3.1 Tangent line

A line which meets a conicoid in two coincident points is called the **tangent line** *to the conicoid.*

3.2 Tangent plane

To find the locus of the tangent lines to a conicoid at a point on it.

Let

$$ax^2 + by^2 + cz^2 = 1 \quad (1)$$

be the equation of the conicoid and (α, β, γ) be a point on it.

Equations of any line through the point (α, β, γ) are

$$\frac{x-\alpha}{l} = \frac{y-\beta}{m} = \frac{z-\gamma}{n} = r \text{ (say)} \quad (2)$$

Any point on the line (2) is $(\alpha + lr, \beta + mr, \gamma + nr)$. This point will lie on (1) if

$$a(\alpha + lr)^2 + b(\beta + mr)^2 + c(\gamma + nr)^2 = 1$$

$$\Rightarrow r^2(al^2 + bm^2 + cn^2) + 2r(a\alpha l + b\beta m + c\gamma n) + a\alpha^2 + b\beta^2 + c\gamma^2 - 1 = 0. \quad (3)$$

Since the point (α, β, γ) lies on the conicoid, we have

$$a\alpha^2 + b\beta^2 + c\gamma^2 = 1. \quad (4)$$

This shows that one root of equation (3) is zero. Thus, equation (3) reduces to

$$r(al^2 + bm^2 + cn^2) + 2(a\alpha l + b\beta m + c\gamma n) = 0. \quad (5)$$

The line (2) will be tangent line to the conicoid (1) if the two points of intersection of (1) and (2) coincide with the point (α, β, γ). For this the second root of (3) *i.e.* r in (5) should vanish which requires

$$a\alpha l + b\beta m + c\gamma n = 0. \quad (6)$$

Eliminating l, m, n from (2) and (6), we get

$$a\alpha(x - \alpha) + b\beta(y - \beta) + c\gamma(z - \gamma) = 0$$

$$\Rightarrow \quad a\alpha x + b\beta y + c\gamma z = 1, \quad \text{on using (4).}$$

This equation represents a plane and is the required locus. ▮

The locus of the tangent lines to a conicoid at a point on it is called the tangent plane at that point.

Ex. Prove that the equation of the tangent plane to the ellipsoid $x^2/a^2 + y^2/b^2 + z^2/c^2 = 1$ at the point (α, β, γ) on it is $\alpha x/a^2 + \beta y/b^2 + \gamma z/c^2 = 1$.

3.3 Condition of tangency. *To find the condition when the plane*

$$lx + my + nz = p \tag{1}$$

becomes a tangent plane to the conicoid

$$ax^2 + by^2 + cz^2 = 1. \tag{2}$$

Equation of the tangent plane to (2) at any point (α, β, γ) on it is

$$a\alpha x + b\beta y + c\gamma z = 1. \tag{3}$$

The plane (1) is a tangent plane to (2) if

$$a\alpha x + b\beta y + c\gamma z - 1 \equiv lx + my + nz - p$$

$$\Rightarrow \qquad \frac{a\alpha}{l} = \frac{b\beta}{m} = \frac{c\gamma}{n} = \frac{-1}{-p}$$

$$\Rightarrow \qquad \alpha = \frac{l}{ap}, \beta = \frac{m}{bp}, \gamma = \frac{n}{cp}.$$

But (α, β, γ) lies on (2), we have $a\alpha^2 + b\beta^2 + c\gamma^2 = 1$. Therefore

$$a\left(\frac{l}{ap}\right)^2 + b\left(\frac{m}{bp}\right)^2 + c\left(\frac{n}{cp}\right)^2 = 1, \text{ i.e., } \frac{l^2}{a} + \frac{m^2}{b} + \frac{n^2}{c} = p^2$$

$$\Rightarrow \qquad p = \pm\sqrt{\frac{l^2}{a} + \frac{m^2}{b} + \frac{n^2}{c}}.$$

This is the required condition of tangency.

Remarks

(*i*) The plane $lx + my + nz = \sqrt{\frac{l^2}{a} + \frac{m^2}{b} + \frac{n^2}{c}}$ is always tangent plane to the conicoid $ax^2 + by^2 + cz^2 = 1$ whatever the values of l, m, n may be. In fact, it is a family of tangent planes to the conicoid $ax^2 + by^2 + cz^2 = 1$.

(*ii*) The point of contact of the tangent plane $lx + my + nz = \sqrt{\frac{l^2}{a} + \frac{m^2}{b} + \frac{n^2}{c}}$

is $\left(\dfrac{l}{ap}, \dfrac{m}{bp}, \dfrac{n}{cp}\right)$, where $p = \sqrt{\dfrac{l^2}{a} + \dfrac{m^2}{b} + \dfrac{n^2}{c}}$.

Ex. Prove that the plane $lx + my + nz = p$ touches the ellipsoid $x^2/a^2 + y^2/b^2 + z^2/c^2 = 1$, if $p^2 = a^2l^2 + b^2m^2 + c^2n^2$, and the point of contact is $(a^2l/p, b^2m/p, c^2n/p)$.

3.4 Director sphere. *To find the locus of the point of intersection of three mutually perpendicular tangent planes to the conicoid* $ax^2 + by^2 + cz^2 = 1$.

Let the three mutually perpendicular tangent planes to the given conicoid be

$$l_1x + m_1y + n_1z = \sqrt{\frac{l_1^2}{a} + \frac{m_1^2}{b} + \frac{n_1^2}{c}}, \quad (1)$$

$$l_2x + m_2y + n_2z = \sqrt{\frac{l_2^2}{a} + \frac{m_2^2}{b} + \frac{n_2^2}{c}}, \quad (2)$$

$$l_3x + m_3y + n_3z = \sqrt{\frac{l_3^2}{a} + \frac{m_3^2}{b} + \frac{n_3^2}{c}}. \quad (3)$$

Since, l_1, m_1, n_1; l_2, m_2, n_2 and l_3, m_3, n_3 are the direction cosines of the normals to the tangent planes which are mutually perpendicular, we have (See Appendix III)

$$\begin{cases} l_1^2 + l_2^2 + l_3^2 = m_1^2 + m_2^2 + m_3^2 = n_1^2 + n_2^2 + n_3^2 = 1, \\ l_1m_1 + l_2m_2 + l_3m_3 = m_1n_1 + m_2n_2 + m_3n_3 = n_1l_1 + n_2l_2 + n_3l_3 = 0. \end{cases}$$

Hence, on squaring and adding (1), (2) and (3), we get

$$x^2 + y^2 + z^2 = \frac{1}{a} + \frac{1}{b} + \frac{1}{c}.$$

This equation represents a sphere and is the required locus.

The locus of the points of intersection of three mutually perpendicular tangent planes to a conicoid is called the **director sphere** *of the conicoid.*

Ex. Prove that the director sphere of the ellipsoid $x^2/a^2 + y^2/b^2 + z^2/c^2 = 1$ is $x^2 + y^2 + z^2 = a^2 + b^2 + c^2$.

3.5 Examples

Example 1. Prove that the locus of the foot of the central perpendicular on varying tangent planes to the ellipsoid $x^2/a^2 + y^2/b^2 + z^2/c^2 = 1$ is the surface

$$(x^2 + y^2 + z^2)^2 = a^2x^2 + b^2y^2 + c^2z^2.$$

Solution. Any tangent plane to the given ellipsoid is

$$lx + my + nz = \sqrt{a^2l^2 + b^2m^2 + c^2n^2}. \qquad (1)$$

Since, the perpendicular drawn from the centre (origin) to (1) is parallel to the normal to (1), the equations of the central perpendicular are

$$\frac{x}{l} = \frac{y}{m} = \frac{z}{n}. \qquad (2)$$

The locus of the foot of the central perpendicular is obtained by eliminating l, m, n from (1) and (2). Thus, the required locus is

$$x^2 + y^2 + z^2 = \sqrt{a^2x^2 + b^2y^2 + c^2z^2}$$

$$\Rightarrow \qquad (x^2 + y^2 + z^2)^2 = a^2x^2 + b^2y^2 + c^2z^2.$$

Example 2. A tangent plane to the ellipsoid $\frac{x^2}{a^2} + \frac{y^2}{b^2} + \frac{z^2}{c^2} = 1$ meets the coordinate axes in the points A, B, C. Prove that the centroid of the triangle ABC lies on the surface $\frac{a^2}{x^2} + \frac{b^2}{y^2} + \frac{c^2}{z^2} = 9.$

Solution. Any tangent plane to the given ellipsoid is

$$lx + my + nz = p, \qquad (1)$$

where

$$p = \sqrt{a^2l^2 + b^2m^2 + c^2n^2}. \qquad (2)$$

The plane (1) meets the axes in the points $A(p/l, 0, 0)$, $B(0, p/m, 0)$ and $C(0, 0, c/n)$. Let (x', y', z') be the centroid of the triangle ABC. Then

$$\left.\begin{aligned} x' &= \frac{p/l + 0 + 0}{3} = \frac{p}{3l} \Rightarrow l = \frac{p}{3x'} \\ y' &= \frac{0 + p/m + 0}{3} = \frac{p}{3m} \Rightarrow m = \frac{p}{3y'} \\ z' &= \frac{0 + 0 + p/n}{3} = \frac{p}{3m} \Rightarrow n = \frac{p}{3z'} \end{aligned}\right\}. \qquad (3)$$

Eliminating l, m, n from (2) and (3), we get

$$\frac{a^2}{x'^2} + \frac{b^2}{y'^2} + \frac{c^2}{z'^2} = 9.$$

Hence, the locus of (x', y', z') is $\dfrac{a^2}{x^2}+\dfrac{b^2}{y^2}+\dfrac{c^2}{z^2}=9.$

Example 3. Tangent planes are drawn to the ellipsoid

$$\frac{x^2}{a^2}+\frac{y^2}{b^2}+\frac{z^2}{c^2}=1$$

through the point (α, β, γ). Prove that the perpendiculars to them through the origin generate the cone $(\alpha x+\beta y+\gamma z)^2 = a^2x^2+b^2y^2+c^2z^2$.

Solution. Any plane through the point (α, β, γ) is

$$l(x-\alpha)+m(y-\beta)+n(z-\gamma)=0$$

i.e.
$$lx+my+nz=\alpha l+\beta m+\gamma n. \tag{1}$$

This plane will touch the given ellipsoid if

$$(\alpha l+\beta m+\gamma n)^2 = a^2l^2+b^2m^2+c^2n^2. \tag{2}$$

Equations of the perpendicular to the plane (1) passing through the origin are

$$\frac{x}{l}=\frac{y}{m}=\frac{z}{n}. \tag{3}$$

Eliminating l, m, n from (2) and (3), we get

$$(\alpha x+\beta y+\gamma z)^2 = a^2x^2+b^2y^2+c^2z^2.$$

This is the required equation of the cone.

Example 4. If the line of intersection of perpendicular tangent planes to the ellipsoid $\dfrac{x^2}{a^2}+\dfrac{y^2}{b^2}+\dfrac{z^2}{c^2}=1$ passes through the fixed point $(0, 0, k)$. Prove that it lies on the cone $x^2(b^2+c^2-k^2)+y^2(c^2+a^2-k^2)+(z-k)^2(a^2+b^2)=0$.

Solution. The equations of any line through the point $(0, 0, k)$ are

$$\frac{x}{\lambda}=\frac{y}{\mu}=\frac{z-k}{\nu}. \tag{1}$$

The equation of a plane containing the line (1) is

$$lx+my+nz=nk, \tag{2}$$

where
$$l\lambda+m\mu+n\nu=0. \tag{3}$$

The plane (2) will touch the given ellipsoid if

$$a^2l^2+b^2m^2+(c^2-k^2)n^2=0. \tag{4}$$

Eliminating n from (3) and (4), and then dividing by m^2, we get

$$\{a^2\nu^2 + \lambda^2(c^2 - k^2)\}\frac{l^2}{m^2} + 2\lambda\mu(c^2 - k^2)\frac{l}{m} + \{b^2\nu^2 + \mu^2(c^2 - k^2)\} = 0.$$

This equation is quadratic in l/m, it has two roots, say, $\frac{l_1}{m_1}$ and $\frac{l_2}{m_2}$. Therefore, by theory of equations, we have

$$\frac{l_1 l_2}{b^2\nu^2 + \mu^2(c^2 - k^2)} = \frac{m_1 m_2}{a^2\nu^2 + \lambda^2(c^2 k^2)}.$$

Further, eliminating l from (3) and (4), we get

$$\frac{m_1 m_2}{a^2\nu^2 + \lambda^2(c^2 - k^2)} = \frac{n_1 n_2}{a^2\mu^2 + b^2\lambda^2}.$$

Thus, we obtain

$$\frac{l_1 l_2}{b^2\nu^2 + \lambda^2(c^2 - k^2)} = \frac{m_1 m_2}{a^2\nu^2 + \lambda^2(c^2 - k^2)} = \frac{n_1 n_2}{a^2\mu^2 + b^2\lambda^2}. \tag{5}$$

Since the planes through the line (1) are at right angles, we have

$$l_1 l_2 + m_1 m_2 + n_1 n_2 = 0. \tag{6}$$

From equations (5) and (6), we get

$$b^2\nu^2 + \mu^2(c^2 - k^2) + a^2\nu^2 + \lambda^2(c^2 - k^2) + a^2\mu^2 + b^2\lambda^2 = 0$$

$$\Rightarrow \quad \lambda^2(b^2 + c^2 - k^2) + \mu^2(c^2 + a^2 - k^2) + \nu^2(a^2 + b^2) = 0. \tag{7}$$

The cone generated by the line (1) is obtained by eliminating λ, μ, ν from (1) and (7), i.e., $x^2(b^2 + c^2 - k^2) + y^2(c^2 + a^2 - k^2) + (z - k)^2(a^2 + b^2) = 0$.

Exercises

1. Find the equations of the tangent planes to the conicoid $2x^2 - 6y^2 + 3z^2 = 5$ which pass through the line $x + 9y - 3z = 0$, $3x - 3y + 6z - 5 = 0$.
2. Prove that the plane $3x + 12y - 6z - 17 = 0$ touches the conicoid $3x^2 - 6y^2 + 9z^2 + 17 = 0$. Also, find the point of contact.
3. Find the equations of the tangent planes to the conicoid $4x^2 - 5y^2 + 7z^2 + 13 = 0$, which are parallel to the plane $4x + 20y - 21z = 0$.

   Also find their points of contact.
4. Find the equations of the tangent planes to the conicoid $7x^2 + 5y^2 + 3z^2 = 60$ which pass through the line $7x + 10y - 30 = 0$, $5y - 3z = 0$.

5. Prove that the equation of the two tangent planes to the conicoid $ax^2 + by^2 + cz^2 = 1$ which pass through the line $u \equiv lx + my + nz - p = 0$, $u' = l'x + m'y + n'z - p' = 0$ is

$$u^2\left(\frac{l'^2}{a}+\frac{m'^2}{b}+\frac{n'^2}{c}-p'^2\right)-2uu'\left(\frac{ll'}{a}+\frac{mm'}{b}+\frac{nn'}{c}-pp'\right)$$

$$+u'^2\left(\frac{l^2}{a}+\frac{m^2}{b}+\frac{n^2}{c}-p^2\right)=0.$$

6. If $2r$ is the distance between two parallel tangent planes to the ellipsoid $\frac{x^2}{a^2}+\frac{y^2}{b^2}+\frac{z^2}{c^2}=1$, prove that the line through the origin perpendicular to the plane lies on the cone $x^2(a^2 - r^2) + y^2(b^2 - r^2) + z^2(c^2 - r^2) = 0$.

7. Find the tangent planes to the ellipsoid

$$\frac{x^2}{a^2}+\frac{y^2}{b^2}+\frac{z^2}{c^2}=1,$$

which are parallel to the plane $lx + my + nz = 0$.

8. Find the locus of the perpendiculars from the origin to the tangent planes to the ellipsoid

$$\frac{x^2}{a^2}+\frac{y^2}{b^2}+\frac{z^2}{c^2}=1$$

which cuts off from its axes intercepts the sum of whose reciprocals is equal to a constant $1/k$.

9. Prove that the lines drawn through (α, β, γ) and perpendicular to the tangent planes to the ellipsoid

$$\frac{x^2}{a^2}+\frac{y^2}{b^2}+\frac{z^2}{c^2}=1$$

which pass through it generate the cone

$a^2(x-\alpha)^2 + b^2(y-\beta)^2 + c^2(z-\gamma)^2 = \{\alpha(x-\alpha) + \beta(y-\beta) + \gamma(z-\gamma)\}^2$.

10. Prove that the locus of points from which three mutually perpendicular planes can be drawn to touch the ellipse

$$\frac{x^2}{a^2}+\frac{y^2}{b^2}=1,\ z=0$$

is the sphere $x^2 + y^2 + z^2 = a^2 + b^2$.

4. Normal

A line passing through a point P on a surface perpendicular to the tangent plane at P is called the **normal** *to the surface at P.*

4.1 Equation of the normal. *To find the equations of the normal to the conicoid ax2 + by2 + cz2 = 1 at any point (α, β, γ) on it.*

The equation of the tangent plane at (α, β, γ) to the given conicoid is

$$a\alpha x + b\beta y + c\gamma z = 1.$$

The direction cosines of the normal to this plane through the point (α, β, γ) are proportional to $a\alpha$, $b\beta$, $c\gamma$. Hence, the equations of the normal to the conicoid $ax^2 + by^2 + cz^2 = 1$ at (α, β, γ) are

$$\frac{x-\alpha}{a\alpha} = \frac{y-\beta}{b\beta} = \frac{z-\gamma}{c\gamma}.$$

Ex. Prove that the equations of the normal to the ellipsoid

$$\frac{x^2}{a^2} + \frac{y^2}{b^2} + \frac{z^2}{c^2} = 1$$

at any point (α, β, γ) on it are

$$\frac{x-\alpha}{\frac{\alpha}{a^2}} = \frac{y-\beta}{\frac{\beta}{b^2}} = \frac{z-\gamma}{\frac{\gamma}{c^2}}.$$

Note. If p is the length of the perpendicular from the origin to the tangent plane $a\alpha x + b\beta y + c\gamma z = 1$, then $(a\alpha p)^2 + (b\beta p)^2 + (c\gamma p)^2 = 1$. This shows that $a\alpha p$, $b\beta p$, $c\gamma p$ are the actual direction cosines of the normal.

Remark. The equations of the normal to the conicoid $ax^2 + by^2 + cz^2 = 1$ at any point (α, β, γ) on it in terms of its direction cosines are

$$\frac{x-\alpha}{a\alpha p} = \frac{y-\beta}{b\beta p} = \frac{z-\gamma}{c\gamma p},$$

where p is the length of the normal from the origin to the tangent plane $a\alpha x + b\beta y + c\gamma z = 1$.

Ex. Prove that the equations of the normal to the ellipsoid $\frac{x^2}{a^2} + \frac{y^2}{b^2} + \frac{z^2}{c^2} = 1$ at any point (α, β, γ) on it in terms of its direction cosines are

$$\frac{x-\alpha}{\frac{\alpha p}{a^2}} = \frac{y-\beta}{\frac{\beta p}{b^2}} = \frac{z-\gamma}{\frac{\gamma p}{c^2}},$$

where p is the length of the normal from the origin to the tangent plane

$$\frac{\alpha x}{a^2}+\frac{\beta y}{b^2}+\frac{\gamma z}{c^2}=1.$$

4.2 Number of normals from a given point. *To prove that from any point six normals can be drawn to a conicoid.*

Let the equation of conicoid be

$$ax^2+by^2+cz^2=1, \tag{1}$$

and let (f, g, h) be any given point. The equations of the normal at any point (α, β, γ) on (1) are

$$\frac{x-\alpha}{a\alpha}=\frac{y-\beta}{b\beta}=\frac{z-\gamma}{c\gamma}.$$

If it passes through (f, g, h), then

$$\frac{f-\alpha}{a\alpha}=\frac{g-\beta}{b\beta}=\frac{h-\gamma}{c\gamma}=r \text{ (say)}$$

$$\Rightarrow \qquad \alpha=\frac{f}{1+ar}, \beta=\frac{g}{1+br}, \gamma=\frac{h}{1+cr}.$$

But (α, β, γ) lies on (1). Therefore

$$\frac{af^2}{(1+ar)^2}+\frac{bg^2}{(1+br)^2}+\frac{ch^2}{(1+cr)^2}=1. \tag{2}$$

Equation (2) is of sixth degree in r and hence gives six values of r for a given point (f, g, h). Thus, there are six normals that can be drawn from a point (f, g, h) to the conicoid.

Ex. Prove that six normals can be drawn to the ellipsoid $\frac{x^2}{a^2}+\frac{y^2}{b^2}+\frac{z^2}{c^2}=1$ from any given point.

4.3 *To prove that the feet of the normals from (f, g, h) to the conicoid $ax^2+by^2+cz^2=1$ lie on three cylinders which have a common curve of intersection.*

The equations of the normal at any point (α, β, γ) on the given conicoid are

$$\frac{x-\alpha}{a\alpha}=\frac{y-\beta}{b\beta}=\frac{z-\gamma}{c\gamma}. \tag{1}$$

If it passes through (f, g, h), then

$$\frac{f-\alpha}{a\alpha}=\frac{g-\beta}{b\beta}=\frac{h-\gamma}{c\gamma}=r \text{ (say)}. \tag{2}$$

Therefore, the feet of the normals lie on the surface

$$\frac{f-x}{ax}=\frac{g-y}{by}=\frac{h-z}{cz}$$

i.e.

$$\begin{cases} by(f-x)=ax(g-y), \\ cz(g-y)=by(h-z), \\ ax(h-z)=cz(f-x). \end{cases}$$

These are the equations of three cylinders.

Further, from equations (2), we have

$$\alpha=\frac{f}{1+ar},\ \beta=\frac{g}{1+br},\ \gamma=\frac{h}{1+cr}.$$

Hence, the feet of the normals lie on the curve

$$x=\frac{f}{1+ar},\ y=\frac{g}{1+br},\ z=\frac{h}{1+cr},$$

where r is a parameter.

4.4 Cubic curve through the feet of the normals. *To prove that the feet of the six normals from any given point to a conicoid are the points of intersection of the coincoid and a certain cubic curve.*

Let the equation of a conicoid be

$$ax^2+by^2+cz^2=1. \tag{1}$$

The feet of the six normals drawn through the point (f, g, h) to the conicoid (1) lie on it. Also, from § 4.3, it is clear that the feet of the normals lie on the curve

$$x=\frac{f}{1+ar},\ y=\frac{g}{1+br},\ z=\frac{h}{1+cr} \tag{2}$$

where r is a parameter.

Thus, the feet of the six normals drawn through a point (f, g, h) to the conicoid (1) are the points of intersection of the conicoid (1) and the curve (2). Now, we claim that the curve (2) is cubic in r.

The curve (2) meets the plane $Ax + By + Cz + D = 0$, where

$$A\left(\frac{f}{1+ar}\right)+B\left(\frac{g}{1+br}\right)+C\left(\frac{h}{1+cr}\right)+D=0.$$

This equation is cubic in r, hence gives three values of r. This completes the proof.

4.5 Cone through the six normals. *To prove that the six normals drawn from any given point to a conicoid generate a cone of second degree.*

Let $ax^2 + by^2 + cz^2 = 1$ be the equation of the conicoid and (f, g, h) the given point.

The equations of the cubic curve through the feet of the normals are given by

$$x=\frac{f}{1+ar}, y=\frac{g}{1+br}, z=\frac{h}{1+cr}. \tag{1}$$

Any line through the point (f, g, h) is

$$\frac{x-f}{l}=\frac{y-g}{m}=\frac{z-h}{n}. \tag{2}$$

The line (2) will intersect the curve (1), if

$$\frac{af}{l(1+ar)}=\frac{bg}{m(1+br)}=\frac{ch}{n(1+cr)}. \tag{3}$$

Eliminating r from (3), we get

$$\frac{af}{l}(b-c)+\frac{bg}{m}(c-a)+\frac{ch}{n}(a-b)=0. \tag{4}$$

The locus of the line (2) is obtained by eliminating l, m, n from (2) and (4), *i.e.*

$$\frac{af(b-c)}{x-f}+\frac{bg(c-a)}{y-g}+\frac{ch(a-b)}{z-h}=0. \tag{5}$$

This equation represents a cone generated by the lines (2) and with guiding curve (1). These lines, in particular, include all the six normals drawn from (f, g, h) to the conicoid. Hence (5) is the required cone.

4.6 Examples

Example 1. If the normal at a point P on the ellipsoid $\frac{x^2}{a^2}+\frac{y^2}{b^2}+\frac{z^2}{c^2}=1$ meets the coordinate planes in G_1, G_2, G_3 and $PG_1^2+PG_2^2+PG_3^2=k^2$, prove that the point P lies also on the ellipsoid

$$\frac{x^2}{a^4}+\frac{y^2}{b^4}+\frac{z^2}{c^4}=\frac{k^2}{a^4+b^4+c^4}$$

Solution. Let (α, β, γ) be the coordinates of P. Then the equations of the normal at $P(\alpha, \beta, \gamma)$ to the given ellipsoid are

$$\frac{x-\alpha}{\frac{\alpha p}{a^2}}=\frac{y-\beta}{\frac{\beta p}{b^2}}=\frac{z-\gamma}{\frac{\gamma p}{c^2}}=r \text{ (say)}, \tag{1}$$

where
$$p=\frac{1}{\sqrt{\frac{\alpha^2}{a^4}+\frac{\beta^2}{b^4}+\frac{\gamma^2}{c^4}}}. \tag{2}$$

Any point on (1) is $\left(\alpha+\frac{\alpha pr}{a^2}, \beta+\frac{\beta pr}{b^2}, \gamma+\frac{\gamma pr}{c^2}\right)$.

This point will lie on the plane $x = 0$ if

$$\alpha+\frac{\alpha pr}{a^2}=0 \Rightarrow r=PG_1=-\frac{a^2}{p}.$$

Similarly $PG_2=-\frac{b^2}{p}$ and $PG_3=-\frac{c^2}{p}$.

Given that $PG_1^2+PG_2^2+PG_3^2=k^2$. Therefore

$$\frac{1}{p^2}=\frac{k^2}{a^4+b^4+c^4}. \tag{3}$$

From (2) and (3), we get

$$\frac{\alpha^2}{a^4}+\frac{\beta^2}{b^4}+\frac{\gamma^2}{c^4}=\frac{k^2}{a^4+b^4+c^4}.$$

Hence, the locus of (α, β, γ) is

$$\frac{x^2}{a^4}+\frac{y^2}{b^4}+\frac{z^2}{c^4}=\frac{k^2}{a^4+b^4+c^4}.$$

Example 2. Prove that the lines drawn from the origin parallel to the normals to $\frac{x^2}{a^2}+\frac{y^2}{b^2}+\frac{z^2}{c^2}=1$ at the points of intersection with the plane $lx + my + nz = p$, generate the cone

$$p^2(a^2x^2 + b^2y^2 + c^2z^2) = (a^2lx + b^2my + c^2nz)^2.$$

Solution. The equations of the normal at (α, β, γ) on the ellipsoid are given by

$$\frac{x-\alpha}{\frac{\alpha}{a^2}}=\frac{y-\beta}{\frac{\beta}{b^2}}=\frac{z-\gamma}{\frac{\gamma}{c^2}}. \tag{1}$$

The equations of any line passing through origin and parallel to (1) are

$$\frac{x}{\frac{\alpha}{a^2}}=\frac{y}{\frac{\beta}{b^2}}=\frac{z}{\frac{\gamma}{c^2}}. \tag{2}$$

Since (α, β, γ) lies on the given ellipsoid, we have

$$\frac{\alpha^2}{a^2}+\frac{\beta^2}{b^2}+\frac{\gamma^2}{c^2}=1. \tag{3}$$

Let the point (α, β, γ) also lies on the plane $lx + my + nz = p$. Then

$$l\alpha + m\beta + n\gamma = p. \tag{4}$$

From (3) and (4), we get

$$\frac{\alpha^2}{a^2}+\frac{\beta^2}{b^2}+\frac{\gamma^2}{c^2}=\left(\frac{l\alpha+m\beta+n\gamma}{p}\right)^2. \tag{5}$$

Now, eliminating α, β, γ from (2) and (5), we get

$$p^2(a^2x^2 + b^2y^2 + c^2z^2) = (a^2lx + b^2my + c^2nz)^2.$$

Example 3. If the normal at a variable point P on the ellipsoid

$$\frac{x^2}{a^2}+\frac{y^2}{b^2}+\frac{z^2}{c^2}=1$$

meets the plane XOY in A and AQ is drawn parallel to OZ and equal to AP, prove that the locus of Q is given by

$$\frac{x^2}{a^2-c^2}+\frac{y^2}{b^2-c^2}+\frac{z^2}{c^2}=1.$$

Also, find the locus of R if OR is drawn from the centre equal and parallel to AP.

Solution. The equations of the normal at (α, β, γ) on the ellipsoid are given by

$$\frac{x-\alpha}{\frac{\alpha}{a^2}}=\frac{y-\beta}{\frac{\beta}{b^2}}=\frac{z-\gamma}{\frac{\gamma}{c^2}}.$$

This normal meets the plane XOY in A, where $z = 0$ Then

$$\begin{cases} x=\alpha-\dfrac{c^2\alpha}{a^2}, y=\beta-\dfrac{c^2\beta}{b^2}, z=0, \\ AP=-c^2\sqrt{\dfrac{\alpha^2}{a^4}+\dfrac{\beta^2}{b^4}+\dfrac{\gamma^2}{c^4}}, \end{cases}$$

where (x, y, z) are the coordinates of A. Let the coordinates of Q be (x', y', z'). Then

$$x'=\alpha-\frac{c^2\alpha}{a^2}=\frac{\alpha}{a^2}(a^2-c^2), \qquad (1)$$

$$y'=\beta-\frac{c^2\beta}{b^2}=\frac{\beta}{b^2}(b^2-c^2), \qquad (2)$$

$$z'=AP=-c^2\sqrt{\frac{\alpha^2}{a^4}+\frac{\beta^2}{b^4}+\frac{\gamma^2}{c^4}} \qquad (3)$$

Since the point (α, β, γ) lies on the ellipsoid $\frac{x^2}{a^2}+\frac{y^2}{b^2}+\frac{z^2}{c^2}=1$, we have

$$\frac{\alpha^2}{a^2}+\frac{\beta^2}{b^2}+\frac{\gamma^2}{c^2}=1. \qquad (4)$$

Further, equation (3) can be written as

$$\frac{z'^2}{c^2} = c^2\left(\frac{\alpha^2}{a^4} + \frac{\beta^2}{b^4} + \frac{\gamma^2}{c^4}\right)$$

$$\Rightarrow \quad \frac{z'^2}{c^2} = c^2\left(\frac{\alpha^2}{a^4} + \frac{\beta^2}{b^4}\right) + \left(1 - \frac{\alpha^2}{a^2} - \frac{\beta^2}{b^2}\right),$$

$$\Rightarrow \quad \frac{z'^2}{c^2} = 1 - \frac{\alpha^2}{a^4}(a^2 - c^2) - \frac{\beta^2}{b^4}(b^2 - c^2), \text{ on using (4).}$$

$$\Rightarrow \quad \frac{x'^2}{a^2 - c^2} + \frac{y'^2}{b^2 - c^2} + \frac{z'^2}{c^2} = 1, \text{ on using (1) and (2).}$$

Hence, the locus of (x', y', z') is

$$\frac{x^2}{a^2 - c^2} + \frac{y^2}{b^2 - c^2} + \frac{z^2}{c^2} = 1.$$

Further, the equations of OR which is parallel to the normal AP, are

$$\frac{x}{\alpha/a^2} = \frac{y}{\beta/b^2} = \frac{z}{\gamma/c^2} = \frac{\sqrt{x^2 + y^2 + z^2}}{\sqrt{\alpha^2/a^4 + \beta^2/b^4 + \gamma^2/c^4}}$$

$$= \frac{OR}{\sqrt{\alpha^2/a^4 + \beta^2/b^4 + \gamma^2/c^4}}$$

$$= \frac{AP}{\sqrt{\alpha^2/a^4 + \beta^2/b^4 + \gamma^2/c^4}} = -c^2,$$

where (x, y, z) are the coordinates of R. Therefore

$$ax = -\frac{c^2\alpha}{a}, by = -\frac{c^2\beta}{b}, cz = -\frac{c^2\gamma}{c}$$

$$\Rightarrow \quad a^2x^2 + b^2y^2 + c^2z^2 = c^4\left(\frac{\alpha^2}{a^2} + \frac{\beta^2}{b^2} + \frac{\gamma^2}{c^2}\right) = c^4, \qquad \text{on using (4).}$$

Hence, the required locus of R is $a^2x^2 + b^2y^2 + c^2z^2 = c^4$.

Example 4. Prove that the feet of the six normals from (α, β, γ) lie on the curve of intersection of an ellipsoid and the cone

$$\frac{a^2(b^2 - c^2)\alpha}{x} + \frac{b^2(c^2 - a^2)\beta}{y} + \frac{c^2(a^2 - b^2)\gamma}{z} = 0.$$

Solution. Let the equation of the ellipsoid be

$$\frac{x^2}{a^2}+\frac{y^2}{b^2}+\frac{z^2}{c^2}=1. \tag{1}$$

The equations of the normal at (x', y', z') on (1) are

$$\frac{x-x'}{x'/a^2}=\frac{y-y'}{y'/b^2}=\frac{z-z'}{z'/c^2}.$$

If it passes through (α, β, γ), then

$$\frac{\alpha-x'}{x'/a^2}=\frac{\beta-y'}{y'/b^2}=\frac{\gamma-z'}{z'/c^2}=r \text{ (say)}.$$

From these equations, we have

$$r=\frac{a^2\alpha}{x'}-a^2,\ r=\frac{b^2\beta}{y'}-b^2,\ r=\frac{c^2\gamma}{z'}-c^2.$$

Multiplying these equations by b^2-c^2, c^2-a^2 and a^2-b^2, respectively, and on adding, we get

$$0=\frac{a^2(b^2-c^2)\alpha}{x'}+\frac{b^2(c^2-a^2)\beta}{y'}+\frac{c^2(a^2-b^2)\gamma}{z'}.$$

Hence, the locus of (x', y', z') is

$$\frac{a^2(b^2-c^2)\alpha}{x}+\frac{b^2(c^2-a^2)\beta}{y}+\frac{c^2(a^2-b^2)\gamma}{z}=0.$$

Also, the feet of the normals lie on the ellipsoid (1). Thus, the result follows.

Example 5. If P, Q, R; P', Q', R' are the feet of the six normals drawn from a point to the ellipsoid $x^2/a^2+y^2/b^2+z^2/c^2=1$ and the plane PQR is given by $lx+my+nz=p$; prove that the plane $P'Q'R'$ is given by

$$\frac{x}{a^2l}+\frac{y}{b^2m}+\frac{z}{c^2n}+\frac{1}{p}=0.$$

Solution. Let the equation of the plane $P'Q'R'$ be

$$l'x+m'y+n'z=p'. \tag{1}$$

Then, the feet of the six normals lie on the locus

$$(lx+my+nz-p)(l'x+m'y+n'z-p')=0. \tag{2}$$

The coordinates of the feet of the normals drawn from (α, β, γ) are given by $\left(\frac{a^2\alpha}{a^2+r}, \frac{b^2\beta}{b^2+r}, \frac{c^2\gamma}{c^2+r}\right)$, where r is a parameter. Clearly these points will lie on (2). Therefore

$$\left(\frac{la^2\alpha}{a^2+r}+\frac{mb^2\beta}{b^2+r}+\frac{nc^2\gamma}{c^2+r}-p\right)\left(\frac{l'a^2\alpha}{a^2+r}+\frac{m'b^2\beta}{b^2+r}+\frac{n'c^2\gamma}{c^2+r}-p'\right)=0. \quad (3)$$

Also, the feet of the normals lie on the ellipsoid. Therefore, we have

$$\frac{1}{a^2}\left(\frac{a^2\alpha}{a^2+r}\right)^2+\frac{1}{b^2}\left(\frac{b^2\beta}{b^2+r}\right)^2+\frac{1}{c^2}\left(\frac{c^2\gamma}{c^2+r}\right)^2=1. \quad (4)$$

On comparing (3) and (4), we get

$$l'=-\frac{pp'}{a^2l},\ m'=-\frac{pp'}{b^2m},\ n'=-\frac{pp'}{c^2n}.$$

Substituting these values of l', m', n' in (1), we get

$$\frac{x}{a^2l}+\frac{y}{b^2m}+\frac{z}{c^2n}+\frac{1}{p}=0.$$

Exercises

1. Prove that the lines drawn from the origin parallel to the normals to $ax^2 + by^2 + cz^2 = 1$ at the point of intersection with the plane $lx + my + nz = p$ generate the cone

$$p^2\left(\frac{x^2}{a}+\frac{y^2}{b}+\frac{z^2}{c}\right)=\left(\frac{lx}{a}+\frac{my}{b}+\frac{nz}{c}\right)^2.$$

2. If the normal at P to the ellipsoid $x^2/a^2 + y^2/b^2 + z^2/c^2 = 1$ meets the principal planes in G_1, G_2, G_3; prove that $PG_1 : PG_2 : PG_3 = a^2 : b^2 : c^2$.
3. Find the length of the normal chord through P of the ellipsoid

$$\frac{x^2}{a^2}+\frac{y^2}{b^2}+\frac{z^2}{c^2}=1,$$

and prove that if it is equal to $4PG$, where G is the point of intersection of the normal and the xy-plane, P lies on the cone

$$\frac{x^2}{a^6}(2c^2-a^2)+\frac{y^2}{b^6}(2c^2-b^2)+\frac{z^2}{c^4}=0.$$

4. The normals at any point P of the ellipsoid $\frac{x^2}{a^2}+\frac{y^2}{b^2}+\frac{z^2}{c^2}=1$ meets the three principal planes at G_1, G_2, G_3. If a point G is taken on the normal at P such that $3PG = PG_1 + PG_2 + PG_3$, prove that the locus of G is

$$\frac{a^2x^2}{(2a^2-b^2-c^2)^2}+\frac{b^2y^2}{(2b^2-c^2-a^2)^2}+\frac{c^2z^2}{(2c^2-a^2-b^2)^2}=\frac{1}{9}.$$

5. A point Q is taken on the normal at P of the ellipsoid

$$\frac{x^2}{a^2}+\frac{y^2}{b^2}+\frac{z^2}{c^2}=1,$$

such that $PQ = k/p$, where k is a constant and p is the length of the perpendicular from the origin on the tangent plane at P. Prove that the locus of Q is

$$\frac{a^2x^2}{(a^2+k)^2}+\frac{b^2y^2}{(b^2+k)^2}+\frac{c^2z^2}{(c^2+k)^2}=1.$$

6. If the feet of three of the normals from P to the ellipsoid

$$\frac{x^2}{a^2}+\frac{y^2}{b^2}+\frac{z^2}{c^2}=1$$

lie in the plane $\frac{x}{a}+\frac{y}{b}+\frac{z}{c}=1$, prove that the feet of the other three lie in the plane $\frac{x}{a}+\frac{y}{b}+\frac{z}{c}+1=0$, and P lies on the line $a(b^2-c^2)$ $x = b(c^2-a^2)y = c(a^2-b^2)z$.

7. Prove that four normals to the ellipsoid $\frac{x^2}{a^2}+\frac{y^2}{b^2}+\frac{z^2}{c^2}=1$ pass through any point of the curve of intersection of the ellipsoid and the conicoid $x^2(b^2+c^2)+y^2(c^2+a^2)+z^2(a^2+b^2)=b^2c^2+c^2a^2+a^2b^2$.

8. Prove that the normals to the ellipsoid $\frac{x^2}{a^2}+\frac{y^2}{b^2}+\frac{z^2}{c^2}=1$ at all points of intersection with $lyz + mzx + nxy = 0$, intersect the line

$$\frac{a^2x}{l(a^2-b^2)(c^2-a^2)}=\frac{b^2y}{m(b^2-c^2)(a^2-b^2)}=\frac{c^2z}{n(c^2-a^2)(b^2-c^2)}.$$

9. The feet of the six normals from (α, β, γ) to the ellipsoid

$$\frac{x^2}{a^2}+\frac{y^2}{b^2}+\frac{z^2}{c^2}=1$$

are (x_r, y_r, z_r), $r = 1, 2, \ldots, 6$. Prove that

$$a^2\alpha\Sigma\frac{1}{x_r}+b^2\beta\Sigma\frac{1}{y_r}+c^2\gamma\Sigma\frac{1}{z_r}=0.$$

10. Prove that the normals at the points (x_1, y_1, z_1) and (x_2, y_2, z_2) to the ellipsoid $\frac{x^2}{a^2}+\frac{y^2}{b^2}+\frac{z^2}{c^2}=1$, intersect if

$$\frac{(b^2-c^2)x_1}{x_1-x_2}+\frac{(c^2-a^2)y_1}{y_1-y_2}+\frac{(a^2-b^2)z_1}{z_1-z_2}=0,$$

and that if (f, g, h) be their point of intersection, then

$$a^2 f\left(\frac{1}{x_1}-\frac{1}{x_2}\right)+b^2 g\left(\frac{1}{y_1}-\frac{1}{y_2}\right)+c^2 h\left(\frac{1}{z_1}-\frac{1}{z_2}\right)=0.$$

5. Plane of Contact and Polar Plane

5.1 *To prove that the points of contact of the tangent planes drawn through a point to a conicoid lie on a plane.*

Let (α, β, γ) be a given point and

$$ax^2 + by^2 + cz^2 = 1 \tag{1}$$

a given conicoid.

Let (x', y', z') be the point of contact of the tangent plane drawn through the point (α, β, γ) to the conicoid (1). Then the equation of the tangent plane at (x', y', z') is

$$axx' + byy' + czz' = 1.$$

But this plane passes through (α, β, γ). Therefore

$$a\alpha x' + b\beta y' + c\gamma z' = 1.$$

Thus, it is clear that the point (x', y', z') satisfies the equation

$$a\alpha x + b\beta y + c\gamma z = 1. \tag{2}$$

This equation represents a plane.

Corollary. The locus of the points of contact of the tangent planes drawn through a point to a conicoid is a conic whose equation is given by

$$\begin{cases} ax^2+by^2+cz^2=1, \\ a\alpha x+b\beta y+c\gamma z=1. \end{cases}$$ ∎

In analogy with the concept of the chord of contact of tangents drawn from a point to a conic in two dimension case, we can call the plane (2) as the **plane of contact** of the tangent planes drawn through the point (α, β, γ) to the conicoid (1).

Remarks.

(*i*) If the point (α, β, γ) lies inside the conicoid, the tangent planes will be imaginary, but the equation of the plane of contact is real since α, β and γ are real.

(*ii*) If the point (α, β, γ) lies on the conicoid, the plane of contact becomes the tangent plane at that point.

5.2 Polar plane. *To find the equation of the polar plane* of the point* (α, β, γ) *with respect to the conicoid* $ax^2 + by^2 + cz^2 = 1$.

Let *A* be the point (α, β, γ). Then the equations of the line through *A*(α, β, γ) meeting the given conicoid in *P* and *Q* are given by

$$\frac{x-\alpha}{l}=\frac{y-\beta}{m}=\frac{z-\gamma}{n}=r \text{ (say)}. \tag{1}$$

Any point on (1) is $(\alpha + lr, \beta + mr, \gamma + nr)$. If this point lies on the conicoid, we must have

$$(al^2 + bm^2 + cn^2)r^2 + 2(a\alpha l + b\beta m + c\gamma n)r + a\alpha^2 + b\beta^2 + c\gamma^2 - 1 = 0.$$

This equation being quadratic in r has two values, say, r_1 and r_2. Therefore, $AP = r_1$ and $AQ = r_2$. Let R be the point on APQ such that $AR = r$. Then, the coordinates of R are $(\alpha + lr, \beta + mr, \gamma + nr)$ and

$$\frac{1}{r_1}+\frac{1}{r_2}=\frac{2}{r}, \text{ i.e. } r=\frac{2r_1r_2}{r_1+r_2}.$$

But $$r_1+r_2=-\frac{2(a\alpha l+b\beta m+c\gamma n)}{al^2+bm^2+cn^2} \text{ and } r_1r_2=\frac{a\alpha^2+b\beta^2+c\gamma^2-1}{al^2+bm^2+cn^2}.$$

Therefore $$r=-\frac{a\alpha^2+b\beta^2+c\gamma^2-1}{a\alpha l+b\beta m+c\gamma n}$$

*For definition of polar plane and the pole see § 4.2, Chapter 15.

$$\Rightarrow \quad a\alpha lr + b\beta mr + c\gamma nr + a\alpha^2 + b\beta^2 + c\gamma^2 - 1 = 0. \tag{2}$$

Eliminating l, m, n from (1) and (2), we get

$$a\alpha(x-\alpha) + b\beta(y-\beta) + c\gamma(z-\gamma) + a\alpha^2 + b\beta^2 + c\gamma^2 - 1 = 0$$

$$\Rightarrow \quad a\alpha x + b\beta y + c\gamma z = 1.$$

This is the required equation of the polar plane of the point (α, β, γ).

Ex. Prove that the equation of the polar plane of any point (α, β, γ) with respect to the ellipsoid $\frac{x^2}{a^2}+\frac{y^2}{b^2}+\frac{z^2}{c^2}=1$ is $\frac{\alpha x}{a^2}+\frac{\beta y}{b^2}+\frac{\gamma z}{c^2}=1$.

5.3 Pole of a plane. *To find the pole of the plane $lx + my + nz = p$ with respect to the conicoid $ax^2 + by^2 + cz^2 = 1$.*

Let (x_1, y_1, z_1) be the pole of the plane $lx + my + nz = p$ with respect to the conicoid $ax^2 + by^2 + cz^2 = 1$. Then, this plane must be identical with the polar plane of (x_1, y_1, z_1) with respect to the given conicoid, *i.e.*

$$lx + my + nz - p \equiv axx_1 + byy_1 + czz_1 - 1$$

$$\Rightarrow \quad \frac{l}{ax_1} = \frac{m}{by_1} = \frac{n}{cz_1} = \frac{-p}{-1}$$

$$\Rightarrow \quad x_1 = \frac{l}{ap}, y_1 = \frac{m}{bp}, z_1 = \frac{n}{cp}.$$

Hence, the required pole is $\left(\frac{l}{ap}, \frac{m}{bp}, \frac{n}{cp}\right)$.

5.4 Propositions on poles and polar planes

Proposition I. *If the polar plane of a point P passes through the point Q, then the polar plane of Q passes through P.*

Let (x_1, y_1, z_1) and (x_2, y_2, z_2) be the coordinates of P and Q, respectively. Since, the polar plane of the point P passes through the point Q, we have

$$ax_1x_2 + by_1y_2 + cz_1z_2 = 1.$$

But this is the condition also that the point (x_1, y_1, z_1) should lie on the plane

$$axx_2 + byy_2 + czz_2 = 1,$$

which, in fact, is the polar plane of the point Q. This proves the proposition. ▌

Two points such that each lies on the polar plane of the other are called **conjugate points.**

Proposition II. *If the pole of a plane L_1 lies on another plane L_2, then the pole of L_2 lies on L_1.*

Let $P(x_1, y_1, z_1)$ and $Q(x_2, y_2, z_2)$ be the poles of the planes L_1 and L_2, respectively. In other words L_1 and L_2 are the polar planes of P and Q, respectively. Since, the polar plane of Q passes through P, by Proposition I, the polar plane of P (*i.e.* the plane L_1) must pass through Q. Hence, the proposition. ▌

Two planes such that each contains the pole of the other are called **conjugate planes.**

5.5 Polar lines

Consider the line

$$\frac{x-\alpha}{l} = \frac{y-\beta}{m} = \frac{z-\gamma}{n} = r \text{ (say)}. \qquad (1)$$

Any point on the line (1) is $(\alpha + lr, \beta + mr, \gamma + nr)$. The polar plane of this point with respect to the conicoid $ax^2 + by^2 + cz^2 = 1$ is

$$a(\alpha + lr)x + b(\beta + mr)y + c(\gamma + nr)z = 1$$

i.e.
$$a\alpha x + b\beta y + c\gamma z - 1 + r(alx + bmy + cnz) = 0.$$

This plane clearly passes through the line of intersection of the planes

$$\left.\begin{aligned} a\alpha x + b\beta y + c\gamma z - 1 = 0 \\ alx + bmy + cnz = 0 \end{aligned}\right\}. \qquad (2)$$

Hence, the polar plane of any point on the line (1) passes through the line (2). Similarly, we can prove that the polar plane of any point on the line (2) passes through the line (1). ▌

Two lines such that the polar plane of each point of one line passes through the other line are called **polar lines.**

5.6 Examples

Example 1. Prove that the locus of the poles of the tangent planes of the conicoid $ax^2 + by^2 + cz^2 = 1$ with respect to the conicoid $\alpha x^2 + \beta y^2 + \gamma z^2 = 1$, is the conicoid $\dfrac{\alpha^2 x^2}{a} + \dfrac{\beta^2 y^2}{b} + \dfrac{\gamma^2 z^2}{c} = 1.$

Solution. Any tangent plane to the conicoid $ax^2 + by^2 + cz^2 = 1$ is

$$lx + my + nz = p, \qquad (1)$$

where $$p^2 = \frac{l^2}{a} + \frac{m^2}{b} + \frac{n^2}{c}. \tag{2}$$

Let (x', y', z') be the pole of (1). Then, its polar plane with respect to the conicoid $\alpha x^2 + \beta y^2 + \gamma z^2 = 1$ is

$$\alpha xx' + \beta yy' + \gamma zz' = 1. \tag{3}$$

Since (1) and (3) are identical, on comparing, we get

$$\frac{\alpha x'}{l} = \frac{\beta y'}{m} = \frac{\gamma z'}{n} = \frac{1}{p}. \tag{4}$$

Eliminating l, m, n from (2) and (4), we get

$$\frac{\alpha^2 x'^2}{a} + \frac{\beta^2 y'^2}{b} + \frac{\gamma^2 z'^2}{c} = 1.$$

Hence, the locus of (x', y', z') is

$$\frac{\alpha^2 x^2}{a} + \frac{\beta^2 y^2}{b} + \frac{\gamma^2 z^2}{c} = 1.$$

Example 2. Prove that the locus of the poles of the plane $lx + my + nz = p$ with respect to the system of conicoids

$$\frac{x^2}{a^2 + \lambda} + \frac{y^2}{b^2 + \lambda} + \frac{z^2}{c^2 + \lambda} = 1$$

is a line perpendicular to the given plane, where λ is a parameter.

Solution. Let (x', y', z') be the pole of the given plane. Then, its polar plane with respect to the given conicoid is

$$\frac{xx'}{a^2 + \lambda} + \frac{yy'}{b^2 + \lambda} + \frac{zz'}{c^2 + \lambda} = 1.$$

Since this plane is identical with the given plane, we have

$$x' = \frac{l(a^2 + \lambda)}{p}, y' = \frac{m(b^2 + \lambda)}{p}, z' = \frac{n(c^2 + \lambda)}{p}$$

$$\Rightarrow \qquad \frac{x' - \frac{a^2 l}{p}}{l} = \frac{y' - \frac{b^2 m}{p}}{m} = \frac{z' - \frac{c^2 n}{p}}{n} = \frac{\lambda}{p}.$$

Hence, the locus of (x', y', z') is

$$\frac{x-\dfrac{a^2 l}{p}}{l}=\frac{y-\dfrac{b^2 m}{p}}{m}=\frac{z-\dfrac{c^2 n}{p}}{n}.$$

This is a line perpendicular to the given plane.

Example 3. Find the conditions that the lines $\dfrac{x-\alpha}{l}=\dfrac{y-\beta}{m}=\dfrac{z-\gamma}{n}$ and $\dfrac{x-\alpha'}{l'}=\dfrac{y-\beta'}{m'}=\dfrac{z-\gamma'}{n'}$ are polar lines with respect to the conicoid $ax^2 + by^2 + cz^2 = 1$.

Solution. The polar line of the line

$$\frac{x-\alpha}{l}=\frac{y-\beta}{m}=\frac{z-\gamma}{n}$$

is $a\alpha x + b\beta y + c\gamma z - 1 = 0$, $alx + bmy + cnz = 0$. This line will be identical with the line

$$\frac{x-\alpha'}{l'}=\frac{y-\beta'}{m'}=\frac{z-\gamma'}{n'}$$

if
$$\begin{cases} a\alpha\alpha' + b\beta\beta' + c\gamma\gamma' - 1 = 0, \\ a\alpha l' + b\beta m' + c\gamma n' = 0, \\ a\alpha' l + b\beta' m + c\gamma' n = 0, \\ all' + bmm' + cnn' = 0. \end{cases}$$

These are the required conditions.

Exercises

1. Prove that the locus of the poles of the tangent planes of the ellipsoid

$$\frac{x^2}{a^2}+\frac{y^2}{b^2}+\frac{z^2}{c^2}=1$$

with respect to the ellipsoid $\dfrac{x^2}{\alpha^2}+\dfrac{y^2}{\beta^2}+\dfrac{z^2}{\gamma^2}=1$, is the ellipsoid

$$\frac{a^2x^2}{\alpha^4}+\frac{b^2y^2}{\beta^4}+\frac{c^2z^2}{\gamma^4}=1.$$

2. Prove that the polar line of the line $\frac{x-1}{2}=\frac{y-2}{3}=\frac{z-3}{4}$ with respect to the conicoid $x^2-2y^2+3z^2-4=0$ is

$$\frac{x+6}{3}=\frac{y-2}{3}=\frac{z-2}{1}.$$

3. Prove that the polar line of the line PQ with respect to the conicoid $ax^2+by^2+cz^2=1$ is given by $axx_1+byy_1+czz_1-1=0$, $axx_2+byy_2+czz_2-1=0$, where (x_1, y_1, z_1) and (x_2, y_2, z_2) are the coordinates of P and Q, respectively.

4. Prove that any normal to the conicoid

$$\frac{x^2}{pa+q}+\frac{y^2}{pb+q}+\frac{z^2}{pc+q}=1$$

is perpendicular to its polar line with respect to the conicoid

$$\frac{x^2}{a}+\frac{y^2}{b}+\frac{z^2}{c}=1.$$

5. Prove that the locus of the lines drawn through the fixed point (α, β, γ) and at right angles to their polar lines with respect to the conicoid $ax^2+by^2+cz^2=1$, is

$$\Sigma\frac{\alpha}{x-\alpha}\left(\frac{1}{b}-\frac{1}{c}\right)=0.$$

6. Prove that if AB intersects the polar of PQ, then PQ intersects the polar of AB.

7. Prove that the point of intersection of the tangent planes at the points P, Q, R on a conicoid is the pole of the plane PQR with respect to the conicoid.

8. Prove that the lines through (α, β, γ) at right angles to their polars with respect to the surface

$$\frac{x^2}{a+b}+\frac{y^2}{2a}+\frac{z^2}{2b}=1$$

generate the cone $(y-\beta)(\alpha z-\gamma x)+(z-\gamma)(\alpha y-\beta x)=0$.

9. Find the condition that the line $\frac{x-\alpha}{l}=\frac{y-\beta}{m}=\frac{z-\gamma}{n}$ should intersect the polar of the line $\frac{x-\alpha'}{l'}=\frac{y-\beta'}{m'}=\frac{z-\gamma'}{n'}$

with respect to the conicoid $ax^2+by^2+cz^2=1$.

6. Enveloping Cone

The locus of the tangent lines drawn from a given point to a conicoid is a cone and is called a **tangent cone** *or* **enveloping cone.**

The given point is called the **vertex** of the cone.

6.1 Equation of the enveloping cone. *To find the equation of the enveloping cone of a conicoid whose vertex is given.*

Let the equation of the conicoid be

$$ax^2 + by^2 + cz^2 = 1 \tag{1}$$

and (α, β, γ) the vertex of the enveloping cone. The equations of any line through (α, β, γ) are

$$\frac{x-\alpha}{l} = \frac{y-\beta}{m} = \frac{z-\gamma}{n} = r \text{ (say)}. \tag{2}$$

Any point on the line (2) is $(\alpha + lr, \beta + mr, \gamma + nr)$. If this point be on the conicoid (1), we must have

$$(al^2 + bm^2 + cn^2)r^2 + 2(a\alpha l + b\beta m + c\gamma n)r + a\alpha^2 + b\beta^2 + c\gamma^2 - 1 = 0. \tag{3}$$

But this is a quadratic equation in r giving two values of r corresponding to which there are two points common to the conicoid and the line (2).

Now, if the line (2) be a tangent to the conicoid (1), both the values of r given by the equation (3) must be equal. Therefore, we must have

$$(a\alpha l + b\beta m + c\gamma n)^2 = (al^2 + bm^2 + cn^2)(a\alpha^2 + b\beta^2 + c\gamma^2 - 1). \tag{4}$$

The locus of the tangent lines is obtained by eliminating l, m, n from (2) and (4). Thus, the required locus is

$$\{a\alpha(x-\alpha) + b\beta(y-\beta) + c\gamma(z-\gamma)\}^2 = \{a(x-\alpha)^2 + b(y-\beta)^2 + c(z-\gamma)^2\}(a\alpha^2 + b\beta^2 + c\gamma^2 - 1). \tag{5}$$

If we use the notations

$$S \equiv ax^2 + by^2 + cz^2 - 1,$$

$$S_1 \equiv a\alpha^2 + b\beta^2 + c\gamma^2 - 1,$$

$$T \equiv a\alpha x + b\beta y + c\gamma z - 1,$$

the equation (5) may be written as $T^2 = SS_1$, *i.e.,*

$$(a\alpha x + b\beta y + c\gamma z - 1)^2 = (ax^2 + by^2 + cz^2 - 1) \times (a\alpha^2 + b\beta^2 + c\gamma^2 - 1).$$

This is the equation of the enveloping cone.

Ex. Prove that the enveloping cone with its vertex at (α, β, γ) of the ellipsoid $\frac{x^2}{a^2}+\frac{y^2}{b^2}+\frac{z^2}{c^2}=1$ is

$$\left(\frac{\alpha x}{a^2}+\frac{\beta y}{b^2}+\frac{\gamma z}{c^2}-1\right)^2=\left(\frac{x^2}{a^2}+\frac{y^2}{b^2}+\frac{z^2}{c^2}-1\right)\left(\frac{\alpha^2}{a^2}+\frac{\beta^2}{b^2}+\frac{\gamma^2}{c^2}-1\right).$$

6.2 Examples

Example 1. If the section of the enveloping cone of the ellipsoid

$$\frac{x^2}{a^2}+\frac{y^2}{b^2}+\frac{z^2}{c^2}=1,$$

whose vertex is P, by the plane $z=0$ is a rectangular hyperbola; prove that the locus of P is $\frac{x^2+y^2}{a^2+b^2}+\frac{z^2}{c^2}=1$.

Solution. Let (α, β, γ) be the coordinates of the vertex P of the enveloping cone. Then its equation is

$$\left(\frac{\alpha x}{a^2}+\frac{\beta y}{b^2}+\frac{\gamma z}{c^2}-1\right)^2=\left(\frac{x^2}{a^2}+\frac{y^2}{b^2}+\frac{z^2}{c^2}-1\right)\left(\frac{\alpha^2}{a^2}+\frac{\beta^2}{b^2}+\frac{\gamma^2}{c^2}-1\right).$$

The section of this cone by the plane $z=0$ is given by

$$z=0,\ \left(\frac{\alpha x}{a^2}+\frac{\beta y}{b^2}-1\right)^2=\left(\frac{x^2}{a^2}+\frac{y^2}{b^2}-1\right)\left(\frac{\alpha^2}{a^2}+\frac{\beta^2}{b^2}+\frac{\gamma^2}{c^2}-1\right).$$

This section will represent a rectangular hyperbola* if

$$\text{coeff. of } x^2+\text{coeff. of } y^2=0$$

i.e. $$\frac{1}{a^2}\left(\frac{\beta^2}{b^2}+\frac{\gamma^2}{c^2}-1\right)+\frac{1}{b^2}\left(\frac{\alpha^2}{a^2}+\frac{\gamma^2}{c^2}-1\right)=0.$$

Hence, the locus of (α, β, γ) is

$$\frac{1}{a^2}\left(\frac{y^2}{b^2}+\frac{z^2}{c^2}-1\right)+\frac{1}{b^2}\left(\frac{x^2}{a^2}+\frac{z^2}{c^2}-1\right)=0$$

* See P. 280

$$\Rightarrow \quad \frac{x^2+y^2}{a^2+b^2}+\frac{z^2}{c^2}=1.$$

Example 2. Find the locus of a luminous point if the ellipsoid

$$\frac{x^2}{a^2}+\frac{y^2}{b^2}+\frac{z^2}{c^2}=1$$

casts a circular shadow on the plane $z = 0$.

Solution. Let (α, β, γ) be the coordinates of the luminous point. Then, the equation of the enveloping cone of the given ellipsoid with vertex at (α, β, γ) is

$$\left(\frac{\alpha x}{a^2}+\frac{\beta y}{b^2}+\frac{\gamma z}{c^2}-1\right)^2=\left(\frac{x^2}{a^2}+\frac{y^2}{b^2}+\frac{z^2}{c^2}-1\right)\left(\frac{\alpha^2}{a^2}+\frac{\beta^2}{b^2}+\frac{\gamma^2}{c^2}-1\right).$$

The section of this cone by the plane $z = 0$ is given by

$$z=0,\ \left(\frac{\alpha x}{a^2}+\frac{\beta y}{b^2}-1\right)^2=\left(\frac{x^2}{a^2}+\frac{y^2}{b^2}-1\right)\left(\frac{\alpha^2}{a^2}+\frac{\beta^2}{b^2}+\frac{\gamma^2}{c^2}-1\right)$$

This equation will represent a circle if coefficient of x^2 is equal to coefficient of y^2 and coefficient of xy is zero, *i.e,* if

$$\frac{1}{a^2}\left(\frac{\beta^2}{b^2}+\frac{\gamma^2}{c^2}-1\right)=\frac{1}{b^2}\left(\frac{\alpha^2}{a^2}+\frac{\gamma^2}{c^2}-1\right), \quad (1)$$

and $$\alpha\beta = 0. \quad (2)$$

There arise two cases:

Case I. If $\alpha = 0$, then equation (1) becomes

$$\frac{1}{a^2}\left(\frac{\beta^2}{b^2}+\frac{\gamma^2}{c^2}-1\right)=\frac{1}{b^2}\left(\frac{\gamma^2}{c^2}-1\right)$$

Hence, the locus of (α, β, γ) is $x=0,\ \dfrac{y^2}{b^2-a^2}+\dfrac{z^2}{c^2}=1.$

Case II. If $\beta = 0$, then equation (1) becomes

$$\frac{1}{a^2}\left(\frac{\gamma^2}{c^2}-1\right)=\frac{1}{b^2}\left(\frac{\alpha^2}{a^2}+\frac{\gamma^2}{c^2}-1\right)$$

Hence, the locus of (α, β, γ) is $y = 0, \dfrac{x^2}{a^2 - b^2} + \dfrac{z^2}{c^2} = 1.$

Exercises

1. If the section of the enveloping cone of the ellipsoid

$$\frac{x^2}{a^2} + \frac{y^2}{b^2} + \frac{z^2}{c^2} = 1$$

by the plane $z = 0$, is a parabola; prove that the locus of the vertex is $z = \pm c$.

2. Find the locus of points from which three mutually perpendicular tangent lines can be drawn to the conicoid $ax^2 + by^2 + cz^2 = 1$.
3. Through a fixed point $(k, 0, 0)$ pairs of perpendicular tangent lines are drawn to the conicoid $ax^2 + by^2 + cz^2 = 1$. Prove that the plane through any pair touches the cone

$$\frac{(x-k)^2}{(ak^2-1)(b+c)} + \frac{y^2}{c(ak^2-1)-a} + \frac{z^2}{b(ak^2-1)-a} = 0.$$

4. Find the equation of the enveloping cone from the point $(2, -3, 6)$ on the conicoid $x^2 + 2y^2 + z^2 = 2$.
5. Find the focus of the conic in which the plane $z = a$ meets the enveloping cone of the sphere $x^2 + y^2 + z^2 = a^2$.

7. Enveloping Cylinder

The locus of the tangent lines drawn to a conicoid and parallel to a given line is a cylinder and is called **enveloping cylinder** *of the conicoid.*

7.1 Equation of the enveloping cylinder. *To find the equation of the enveloping cylinder of a conicoid whose generators are parallel to a given line.*

Let

$$ax^2 + by^2 + cz^2 = 1, \tag{1}$$

and

$$\frac{x}{l} = \frac{y}{m} = \frac{z}{n} \tag{2}$$

be the equations of a conicoid and a given line, respectively.

Let (x', y', z') be any point on the enveloping cylinder. Then, the equations of the generator through this point are given by

$$\frac{x-x'}{l}=\frac{y-y'}{m}=\frac{z-z'}{n}=r \text{ (say).} \qquad (3)$$

Any point on the line (3) is $(x'+lr, y'+mr, z'+nr)$. This point will lie on the conicoid (1) if $a(x'+lr)^2+b(y'+mr)^2+c(z'+nr)^2=1$, *i.e.*,

$$(al^2+bm^2+cn^2)r^2+2(alx'+bmy'+cnz')r+ax'^2+by'^2+cz'^2-1=0. \qquad (4)$$

But this is a quadratic equation in r giving two values of r corresponding to which there are two points common to the conicoid (1) and the line (3).

Now, if the line (3) be a tangent to the conicoid (1), both the values of r given by equation (4) must be equal. Therefore, we must have

$$(alx'+bmy'+cnz')^2=(al^2+bm^2+cn^2)(ax'^2+by'^2+cz'^2-1).$$

Hence, the locus of (x', y', z') is

$$(alx+bmy+cnz)^2=(al^2+bm^2+cn^2)(ax^2+by^2+cz^2-1).$$

Note. The plane of contact of the enveloping cylinder and the conicoid is $alx+bmy+cnz=0$.

Ex. Prove that the equation of the enveloping cylinder of the ellipsoid $\frac{x^2}{a^2}+\frac{y^2}{b^2}+\frac{z^2}{c^2}=1$ is $\left(\frac{lx}{a^2}+\frac{my}{b^2}+\frac{nz}{c^2}\right)^2=\left(\frac{l^2}{a^2}+\frac{m^2}{b^2}+\frac{n^2}{c^2}\right)\left(\frac{x^2}{a^2}+\frac{y^2}{b^2}+\frac{z^2}{c^2}-1\right)$.

7.2 Enveloping cylinder as a limiting case of enveloping cone

Let the equations of a conicoid and the axis of its enveloping cylinder be, respectively

$$ax^2+by^2+cz^2=1, \qquad (1)$$

$$\frac{x}{l}=\frac{y}{m}=\frac{z}{n}=r \text{ (say).} \qquad (2)$$

Any point on (2) is (lr, mr, nr). Let this point be the vertex of the enveloping cone. Then, the equation of the enveloping cone is given by

$$(alrx+bmry+cnrz-1)^2=(ax^2+by^2+cz^2-1)(al^2r^2+bm^2r^2+cn^2r^2-1)$$

$$\Rightarrow \left(alx+bmy+cnz-\frac{1}{r}\right)^2=(ax^2+by^2+cz^2-1)\left(al^2+bm^2+cn^2-\frac{1}{r^2}\right).$$

As $r\to\infty$, this equation reduces to

$$(alx+bmy+cnz)^2=(ax^2+by^2+cz^2-1)(al^2+bm^2+cn^2),$$

which is the equation of the enveloping cylinder.

7.3 Examples

Example. Prove that the enveloping cylinder of the ellipsoid

$$\frac{x^2}{a^2}+\frac{y^2}{b^2}+\frac{z^2}{c^2}=1,$$

whose generators are parallel to the lines $\frac{x}{0}=\frac{y}{\pm\sqrt{a^2-b^2}}=\frac{z}{c}$ meets the plane $z=0$ in circles.

Solution. Since, the generators are parallel to the given line, the equation of the enveloping cylinder is

$$\left(\frac{0\cdot x}{a^2}+\frac{(\pm\sqrt{a^2-b^2})y}{b^2}+\frac{cz}{c^2}\right)^2=\left(\frac{x^2}{a^2}+\frac{y^2}{b^2}+\frac{z^2}{c^2}-1\right)\left(\frac{a^2-b^2}{b^2}+\frac{c^2}{c^2}\right).$$

This will meet to the plane $z = 0$ in the conic

$$z=0,\ \frac{(a^2-b^2)y^2}{b^4}=\left(\frac{x^2}{a^2}+\frac{y^2}{b^2}-1\right)\left(\frac{a^2}{b^2}\right)$$

$$\Rightarrow \qquad x^2+y^2=a^2,\ z=0$$

which is a circle.

Exercises

1. Enveloping cylinders of the conicoid $ax^2 + by^2 + cz^2 = 1$ meet the plane $z = 0$ in rectangular hyperbolas. Prove that the central perpendiculars to their planes of contact generate the cone $b^2cx^2 + a^2cy^2 + ab(a + b)z^2 = 0$.
2. Prove that the enveloping cylinders of the ellipsoid $\frac{x^2}{a^2}+\frac{y^2}{b^2}+\frac{z^2}{c^2}=1$ with generators parallel to z-axis meet the plane $z = 0$ in parabolas.

8. Diameters and Diametral Planes

A line through the centre of a conicoid is called a **diameter** *of the conicoid.*

A plane through the centre of a conicoid is called a **diametral plane** *of the conicoid.*

8.1 Section with given centre. *To find the locus of the chords of the conicoid $ax^2 + by^2 + cz^2 = 1$ which are bisected at the point* (α, β, γ).

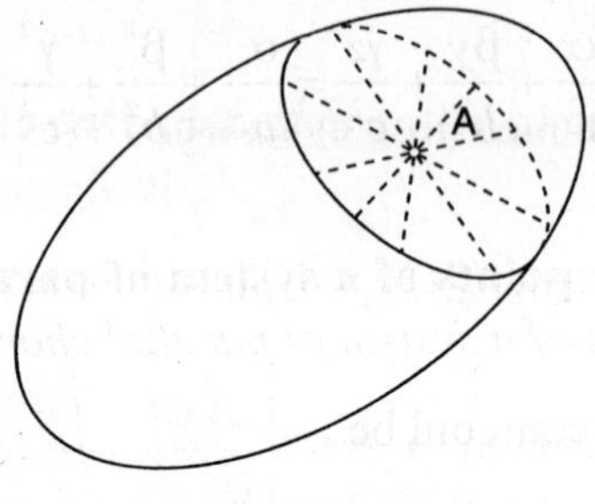

Fig. 18.4

The equations of any chord through $A(\alpha, \beta, \gamma)$ are

$$\frac{x-\alpha}{l}=\frac{y-\beta}{m}=\frac{z-\gamma}{n}=r \text{ (say)}. \qquad (1)$$

Any point on (1) is $(\alpha + lr, \beta + mr, \gamma + nr)$. This point will lie on the given conicoid if $a(\alpha + lr)^2 + b(\beta + mr)^2 + c(\gamma + nr)^2 = 1$, *i.e.*,

$$(al^2 + bm^2 + cn^2)r^2 + 2(a\alpha l + b\beta m + c\gamma n)r + a\alpha^2 + b\beta^2 + c\gamma^2 - 1 = 0. \qquad (2)$$

This equation being quadratic in r, gives two values of r. Since, $A(\alpha, \beta, \gamma)$ is the middle point of the chord, the two values of r given by equation (2) must be equal in magnitude but opposite in sign, *i.e.*, the sum of the roots of equation (2) must be zero. Therefore

$$a\alpha l + b\beta m + c\gamma n = 0. \qquad (3)$$

The required locus is obtained by eliminating l, m, n from (1) and (3) and we get

$$a\alpha(x-\alpha) + b\beta(y-\beta) + c\gamma(z-\gamma) = 0$$

$$\Rightarrow \qquad a\alpha x + b\beta y + c\gamma z = a\alpha^2 + b\beta^2 + c\gamma^2.$$

Hence, the locus of the chords which are bisected at (α, β, γ) is a plane.

Using the notations

$$S \equiv ax^2 + by^2 + cz^2 - 1,$$

$$S_1 \equiv a\alpha^2 + b\beta^2 + c\gamma^2 - 1,$$

$$T \equiv a\alpha x + b\beta y + c\gamma z - 1,$$

the equation of the plane can be written as $T = S_1$.

Clearly, the plane $T = S_1$ meets the conicoid in a conic having centre at (α, β, γ). Thus, the equation $T = S_1$ can be regarded as the equation of the plane in terms of the centre of its section with the conicoid.

Ex. Prove that the locus of the chords of the ellipsoid $\frac{x^2}{a^2}+\frac{y^2}{b^2}+\frac{z^2}{c^2}=1$ which are bisected at the point (α, β, γ) is the plane

$$\frac{\alpha x}{a^2}+\frac{\beta y}{b^2}+\frac{\gamma z}{c^2}=\frac{\alpha^2}{a^2}+\frac{\beta^2}{b^2}+\frac{\gamma^2}{c^2}.$$

8.2 Locus of the middle points of a system of parallel chords. *To find the locus of the middle points of a system of parallel chords of a conicoid.*

Let the equation of a conicoid be

$$ax^2 + by^2 + cz^2 = 1 \tag{1}$$

and let the chords be parallel to the line

$$\frac{x}{l}=\frac{y}{m}=\frac{z}{n}. \tag{2}$$

Let (α, β, γ) be any point on the locus i.e. the middle point of one of the chords parallel to (2). Then, the equations of the chord with middle point (α, β, γ) are given by

$$\frac{x-\alpha}{l}=\frac{y-\beta}{m}=\frac{z-\gamma}{n}. \tag{3}$$

Since, (α, β, γ) is the middle point of (3), from equation (2) of § 8.1, we have

$$a\alpha l + b\beta m + c\gamma n = 0.$$

Hence, the locus of (α, β, γ) is $alx + bmy + cnz = 0$, which is a plane. ▮

We note that the equation (4) represents a plane through the origin, the centre of the conicoid. Therefore, it is diametral plane. On the other hand given any diametral plane, *i.e.*, $Ax + By + Cz = 0$ through the centre of the conicoid, we find that it represents the locus of the middle points of a system of parallel chords having direction cosines proportional to A, B, C.

Therefore, a diametral plane of a central conicoid can alternatively be defined as *the locus of the middle points of a system of parallel chords of the conicoid.*

Note 1. A diametral plane obtained as the locus of the middle points of a system of chords parallel to a line PQ is referred to the diametral plane of the line PQ.

Note 2. Diametral planes which are perpendicular to the chords bisected by them are called **principal planes**. The line of intersection of two principal planes is called a **principal axis**. Thus the planes XOY, YOZ and ZOX are principal planes and $X'OX$, $Y'OY$ and $Z'OZ$ are principal axes of a central conicoid.

8.3 Locus of centres of parallel plane sections. *To find the locus of the centres of parallel plane sections of a conicoid.*

Let the equation of a conicoid be

$$ax^2 + by^2 + cz^2 = 1. \tag{1}$$

Let (x', y', z') be the centre of a plane section of a system of parallel plane sections each of which is parallel to

$$lx + my + nz = p. \tag{2}$$

Then, the equation of the plane section with centre (x', y', z') is

$$axx' + byy' + czz' = ax'^2 + by'^2 + cz'^2. \tag{3}$$

But this plane is parallel to the plane (2). Therefore, on comparing (2) and (3), we get

$$\frac{ax'}{l} = \frac{by'}{m} = \frac{cz'}{n}.$$

Hence, the locus of (x', y', z') is

$$\frac{ax}{l} = \frac{by}{m} = \frac{cz}{n}.$$

These are the equations of a line passing through the centre of the conicoid.

Note. Following the arguments similar to as in § 8.2, the diameter of a conicoid can alternatively be defined as *the locus of the centres of parallel plane sections of a conicoid.*

8.4 Examples

Example 1. Find the equation of the plane which cuts the surface $2x^2 - 3y^2 + 5z^2 = 1$ in a conic whose centre is (2, 1, 3).

Solution. The required equation of the plane is $T = S_1$, *i.e.*,

$$4x - 3y + 15z = 50.$$

Example 2. Prove that the centres of sections of the conicoid $ax^2 + by^2 + cz^2 = 1$ by planes which are at a constant distance p from the origin lie on the surface $(ax^2 + by^2 + cz^2)^2 = p^2(a^2x^2 + b^2y^2 + c^2z^2)$.

Solution. Let (x', y', z') be the centre of the section of the given conicoid. Then equation of the plane section with centre (x', y', z') is

$$axx' + byy' + czz' = ax'^2 + by'^2 + cz'^2. \tag{1}$$

Since, p is the perpendicular distance of the plane (1) from the origin, we have

$$p = \frac{ax'^2 + by'^2 + cz'^2}{\sqrt{a^2x'^2 + b^2y'^2 + c^2z'^2}}.$$

Hence, the locus of (x', y', z') is $(ax^2 + by^2 + cz^2)^2 = p^2(a^2x^2 + b^2y^2 + c^2z^2)$.

Example 3. Find the locus of the middle points of chords of the conicoid $ax^2 + by^2 + cz^2 = 1$ which pass through the point (α, β, γ).

Solution. Let (x', y', z') be the middle point of any chord of the given conicoid. Then, the equation of the plane section with centre (x', y', z') is

$$axx' + byy' + czz' = ax'^2 + by'^2 + cz'^2.$$

This plane will pass through (α, β, γ) if

$$a\alpha x' + b\beta y' + c\gamma z' = ax'^2 + by'^2 + cz'^2.$$

Hence, the locus of (x', y', z') is

$$a\alpha x + b\beta y + c\gamma z = ax^2 + by^2 + cz^2.$$

Example 4. Find the locus of the centres of the sections of the ellipsoid

$$\frac{x^2}{a^2} + \frac{y^2}{b^2} + \frac{z^2}{c^2} = 1$$

which touch the ellipsoid $\dfrac{x^2}{\alpha^2} + \dfrac{y^2}{\beta^2} + \dfrac{z^2}{\gamma^2} = 1.$

Solution. Let (x', y', z') be the centre of a section of the given ellipsoid. Then, its equation is given by

$$\frac{xx'}{a^2} + \frac{yy'}{b^2} + \frac{zz'}{c^2} = \frac{x'^2}{a^2} + \frac{y'^2}{b^2} + \frac{z'^2}{c^2}.$$

This plane will touch the ellipsoid $\dfrac{x^2}{\alpha^2} + \dfrac{y^2}{\beta^2} + \dfrac{z^2}{\gamma^2} = 1$ if

$$\frac{x'^2}{a^2} + \frac{y'^2}{b^2} + \frac{z'^2}{c^2} = \sqrt{\frac{\alpha^2x'^2}{a^4} + \frac{\beta^2y'^2}{b^4} + \frac{\gamma^2z'^2}{c^4}}.$$

Hence, the locus of (x', y', z') is

$$\left(\frac{x^2}{a^2}+\frac{y^2}{b^2}+\frac{z^2}{c^2}\right)^2=\frac{\alpha^2x^2}{a^4}+\frac{\beta^2y^2}{b^4}+\frac{\gamma^2z^2}{c^4}.$$

Exercises

1. Find the equation of the plane which cuts the conicoid $x^2 + 4y^2 - 5z^2 = 1$ in a conic whose centre is the point (2, 3, 4).
2. Find the centre of the conic given by the equations
$$3x^2 + 2y^2 - 15z^2 = 4,\ 2x - 2y - 5z + 5 = 0.$$
3. Prove that the locus of the centres of the sections of the conicoid $ax^2 + by^2 + cz^2 = 1$ which touch the conicoid $\alpha x^2 + \beta y^2 + \gamma z^2 = 1$ is
$$(\alpha x^2 + \beta y^2 + \gamma z^2)^2 = \frac{a^2x^2}{\alpha}+\frac{b^2y^2}{\beta}+\frac{c^2z^2}{\gamma}.$$
4. Prove that the centres of sections of a conicoid which are parallel to a given line lie on a fixed plane.
5. Prove that the centres of sections of a conicoid which pass through a given line lie on a conic.
6. Prove that the middle points of chords of the conicoid $ax^2 + by^2 + cz^2 = 1$ which are parallel to $x = 0$ and touch the sphere $x^2 + y^2 + z^2 = r^2$ lie on the surface $by^2(bx^2 + by^2 + cz^2 - br^2) + cz^2(cx^2 + by^2 + cz^2 - cr^2) = 0$.
7. Prove that the centre of the section of the ellipsoid $\frac{x^2}{a^2}+\frac{y^2}{b^2}+\frac{z^2}{c^2}=1$ by the plane ABC, *i.e.*, $\frac{x}{a}+\frac{y}{b}+\frac{z}{c}=1$ is the centroid of the triangle ABC.

9. Conjugate Diameters and Diametral Planes

In this section, we shall confine our attention to the ellipsoid.

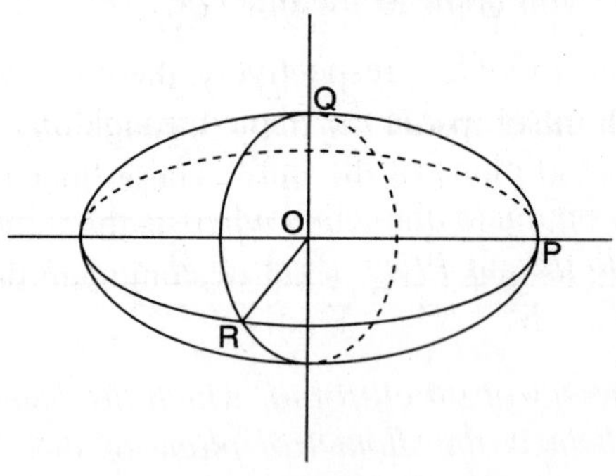

Fig. 18.5

Let $P(x_1, y_1, z_1)$ be any point on the ellipsoid

$$\frac{x^2}{a^2}+\frac{y^2}{b^2}+\frac{z^2}{c^2}=1. \tag{1}$$

Then, the diametral plane of the line OP is

$$\frac{xx_1}{a^2}+\frac{yy_1}{b^2}+\frac{zz_1}{c^2}=0. \tag{2}$$

Let $Q(x_2, y_2, z_2)$ be a point on the plane section of (1) by the plane (2). Since, Q lies on the plane (2), we have

$$\frac{x_1x_2}{a^2}+\frac{y_1y_2}{b^2}+\frac{z_1z_2}{c^2}=0. \tag{3}$$

The condition (3) shows that the point $P(x_1, y_1, z_1)$ lies on the plane

$$\frac{xx_2}{a^2}+\frac{yy_2}{b^2}+\frac{zz_2}{c^2}=0. \tag{4}$$

This, in fact, is the diametral plane of the line OQ. Thus, we conclude that if the diametral plane of the line OP passes through a point Q, then the diametral plane of OQ must passes through P and vice-versa.

Now, suppose that $R(x_3, y_3, z_3)$ be one of the two points where the line common to the diametral planes of OP and OQ meet the ellipsoid (1). Since, R lies on the planes (2) and (4), we have

$$\frac{x_3x_1}{a^2}+\frac{y_3y_1}{b^2}+\frac{z_3z_1}{c^2}=0, \text{ and } \frac{x_2x_3}{a^2}+\frac{y_2y_3}{b^2}+\frac{z_2z_3}{c^2}=0.$$

These are the conditions for the points $P(x_1, y_1, z_1)$ and $Q(x_2, y_2, z_2)$ to lie on the plane

$$\frac{xx_3}{a^2}+\frac{yy_3}{b^2}+\frac{zz_3}{c^2}=0,$$

which, in fact, is the diametral plane of the line OR.

Hence QOR, ROP and POQ are, respectively, the diametral planes of the lines OP, OQ and OR. In other words the plane through any of the two lines OP, OQ, OR is the diametral plane of the third. These lines OP, OQ and OR are said to form a set of conjugate diameters whereas the three corresponding diametral planes QOR, ROP and POQ, a set of conjugate diametral planes. More precisely, we have

The three semi-diameters of an ellipsoid, which are such that the plane containing any two of them is the diametral plane of the third, are called **conjugate semi-diameters** *of the ellipsoid.*

The three diametral planes, which are such that each is the diametral plane of the line of intersection of the other two, are called **conjugate diametral planes** *of the ellipsoid.*

9.1 Relations among the extremities of the conjugate semi-diameters of an ellipsoid.

Let the equations of an ellipsoid be

$$\frac{x^2}{a^2}+\frac{y^2}{b^2}+\frac{z^2}{c^2}=1.$$

Let the extremities of the conjugate semi-diameters of the ellipsoid be $P(x_1, y_1, z_1)$, $Q(x_2, y_2, z_2)$ and $R(x_3, y_3, z_3)$. Since, these points lie on the ellipsoid, we have

$$\left.\begin{aligned} \frac{x_1^2}{a^2}+\frac{y_1^2}{b^2}+\frac{z_1^2}{c^2}=1 \\ \frac{x_2^2}{a^2}+\frac{y_2^2}{b^2}+\frac{z_2^2}{c^2}=1 \\ \frac{x_3^2}{a^2}+\frac{y_3^2}{b^2}+\frac{z_3^2}{c^2}=1 \end{aligned}\right\}. \quad \text{(I)}$$

Also, we have

$$\left.\begin{aligned} \frac{x_1x_2}{a^2}+\frac{y_1y_2}{b^2}+\frac{z_1z_2}{c^2}=0 \\ \frac{x_2x_3}{a^2}+\frac{y_2y_3}{b^2}+\frac{z_2z_3}{c^2}=0 \\ \frac{x_3x_1}{a^2}+\frac{y_3y_1}{b^2}+\frac{z_3z_1}{c^2}=0 \end{aligned}\right\}. \quad \text{(II)}$$

From relations (I) and (II), we conclude that the quantities

$$\frac{x_1}{a}, \frac{y_1}{b}, \frac{z_1}{c}; \frac{x_2}{a}, \frac{y_2}{b}, \frac{z_2}{c}; \frac{x_3}{a}, \frac{y_3}{b}, \frac{z_3}{c}$$

are the direction cosines of three lines which are mutually at right angles.

Also, the quantities

$$\frac{x_1}{a}, \frac{x_2}{a}, \frac{x_3}{a}; \frac{y_1}{b}, \frac{y_2}{b}, \frac{y_3}{b}; \frac{z_1}{c}, \frac{z_2}{c}, \frac{z_3}{c}$$

are the direction cosines of the three mutually perpendicular lines. Hence, we obtain two more sets of relations

$$\left.\begin{aligned} x_1^2 + x_2^2 + x_3^2 &= a^2 \\ y_1^2 + y_2^2 + y_3^2 &= b^2 \\ z_1^2 + z_2^2 + z_3^2 &= c^2 \end{aligned}\right\}. \quad \text{(III)}$$

and

$$\left.\begin{aligned} x_1y_1 + x_2y_2 + x_3y_3 &= 0 \\ y_1z_1 + y_2z_2 + y_3z_3 &= 0 \\ z_1x_1 + z_2x_2 + z_3x_3 &= 0 \end{aligned}\right\}. \quad \text{(IV)}$$

Also

$$\begin{vmatrix} \frac{x_1}{a} & \frac{y_1}{b} & \frac{z_1}{c} \\ \frac{x_2}{a} & \frac{y_2}{b} & \frac{z_2}{c} \\ \frac{x_3}{a} & \frac{y_3}{b} & \frac{z_3}{c} \end{vmatrix} = 1$$

$$\Rightarrow \quad \begin{vmatrix} x_1 & y_1 & z_1 \\ x_2 & y_2 & z_2 \\ x_3 & y_3 & z_3 \end{vmatrix} = abc. \quad \text{(V)}$$

Further

$$\left.\begin{aligned} \frac{x_1}{a} &= \pm\frac{y_2z_3 - y_3z_2}{bc} \\ \frac{y_1}{b} &= \pm\frac{z_2x_3 - z_3x_2}{ca} \\ \frac{z_1}{c} &= \pm\frac{x_2y_3 - x_3y_2}{ab} \text{ etc.} \end{aligned}\right\}. \quad \text{(VI)}$$

Thus, from (I) – (VI) above are a few sets of relations among the extremities of the conjugate semi-diameters of the ellipsoid which are quite useful later on.

9.2 Propositions on conjugate diameters

Proposition I. *The sum of the squares of the lengths of three conjugate semi-diameters of an ellipsoid is constant.*

Let $P(x_1, y_1, z_1)$, $Q(x_2, y_2, z_2)$ and $R(x_3, y_3, z_3)$ be the extremities of the three conjugate semi-diameters of the ellipsoid

$$\frac{x^2}{a^2}+\frac{y^2}{b^2}+\frac{z^2}{c^2}=1.$$

Then

$$OP^2 = x_1^2 + y_1^2 + z_1^2,$$

$$OQ^2 = x_2^2 + y_2^2 + z_2^2,$$

$$OR^2 = x_3^2 + y_3^2 + z_3^2.$$

Therefore $\quad OP^2 + OQ^2 + OR^2 = x_1^2 + x_2^2 + x_3^2 + y_1^2 + y_2^2 + y_3^2 + z_1^2 + z_2^2 + z_3^2$

$$= a^2 + b^2 + c^2 = \text{constant}.$$

Proposition II. *The volume of the parallelopiped formed by three conjugate semi-diameters as coterminous edges is constant.*

The volume of the parallelopiped which has OP, OQ, OR as coterminous edges is

$$6 \times \text{volume of tetrahedron } OPQR = 6 \times \frac{1}{6}\begin{vmatrix} x_1 & y_1 & z_1 \\ x_2 & y_2 & z_2 \\ x_3 & y_3 & z_3 \end{vmatrix} = abc = \text{Const.}$$

Proposition III. *The sum of the squares of the areas of the faces of the parallelopiped which has three conjugate semi-diameters of an ellipsoid as coterminous edge is constant.*

Let A_1, A_2, A_3 be the areas of the triangles QOR, ROP, POQ and l_r, m_r, n_r $(r = 1, 2, 3)$ the direction cosines of the normals to the planes QOR, ROP, POQ. Then, projecting the area A_1 on the plane $x = 0$, we get

$$A_1 l_1 = \frac{y_2 z_3 - y_3 z_2}{2} = \pm\frac{bcx_1}{2a}, \text{ on using (VI) of § 9.1.}$$

Similarly $\quad A_1 m_1 = \pm\dfrac{cay_1}{2b}$ and $A_1 n_1 = \pm\dfrac{abz_1}{2c}$.

On squaring and adding, we get

$$A_1^2 = \frac{1}{4}\left(\frac{b^2c^2x_1^2}{a^2}+\frac{c^2a^2y_1^2}{b^2}+\frac{a^2b^2z_1^2}{c^2}\right).$$

Similarly, on projecting the other areas, we get

$$A_2^2 = \frac{1}{4}\left(\frac{b^2c^2x_2^2}{a^2}+\frac{c^2a^2y_2^2}{b^2}+\frac{a^2b^2z_2^2}{c^2}\right),$$

and $$A_3^2 = \frac{1}{4}\left(\frac{b^2c^2x_3^2}{a^2} + \frac{c^2a^2y_3^2}{b^2} + \frac{a^2b^2z_3^2}{c^2}\right).$$

On adding these, we get

$$A_1^2 + A_2^2 + A_3^2 = \frac{1}{4}(b^2c^2 + c^2a^2 + a^2b^2) = \text{constant}.$$

9.3 Examples

Example 1. If *P*, *Q*, *R* are the extremities of three conjugate semi-diameters of the ellipsoid $\frac{x^2}{a^2} + \frac{y^2}{b^2} + \frac{z^2}{c^2}$, prove that

(*i*) the locus of the pole of the plane *PQR* is the ellipsoid $\frac{x^2}{a^2} + \frac{y^2}{b^2} + \frac{z^2}{c^2} = 3$;

(*ii*) the locus of the foot of the perpendicular from the centre of the ellipsoid on the plane *PQR* is $a^2x^2 + b^2y^2 + c^2z^2 = 3(x^2 + y^2 + z^2)^2$.

Solution. Let (x_1, y_1, z_1), (x_2, y_2, z_2) and (x_3, y_3, z_3) be the coordinates of *P*, *Q* and *R*, respectively. Let the equation of the plane *PQR* be

$$Ax + By + Cz + D = 0. \quad (1)$$

Then $$Ax_1 + By_1 + Cz_1 + D = 0, \quad (2)$$

$$Ax_2 + By_2 + Cz_2 + D = 0, \quad (3)$$

$$Ax_3 + By_3 + Cz_3 + D = 0. \quad (4)$$

Multiplying equation (2) by x_1, (3) by x_2, (4) by x_3 and adding, we get

$$A\Sigma x_1^2 + B\Sigma x_1y_1 + C\Sigma z_1x_1 + D(x_1 + x_2 + x_3) = 0$$

$$\Rightarrow \quad Aa^2 = -D(x_1 + x_2 + x_3)$$

$$\Rightarrow \quad A = -\frac{D}{a^2}(x_1 + x_2 + x_3).$$

Similarly $$B = -\frac{D}{b^2}(y_1 + y_2 + y_3) \text{ and } C = -\frac{D}{c^2}(z_1 + z_2 + z_3).$$

Substituting these values of *A*, *B*, *C* in (1), we get

$$\frac{x}{a^2}(x_1 + x_2 + x_3) + \frac{y}{b^2}(y_1 + y_2 + y_3) + \frac{z}{c^2}(z_1 + z_2 + z_3) = 1. \quad (5)$$

This is the equation of the plane PQR.

(*i*) Let (α, β, γ) be the pole of the plane PQR. Then, its polar plane with respect to the ellipsoid $\frac{x^2}{a^2}+\frac{y^2}{b^2}+\frac{z^2}{c^2}=1$ is $\frac{\alpha x}{a^2}+\frac{\beta y}{b^2}+\frac{\gamma z}{c^2}=1.$ (6)

But (5) and (6) represent the same plane. Therefore, on comparing (5) and (6), we get

$$\frac{\alpha}{a}=\frac{x_1+x_2+x_3}{a}, \frac{\beta}{b}=\frac{y_1+y_2+y_3}{b}, \frac{\gamma}{c}=\frac{z_1+z_2+z_3}{c}.$$

Now

$$\frac{\alpha^2}{a^2}+\frac{\beta^2}{b^2}+\frac{\gamma^2}{c^2}=\frac{x_1^2}{a^2}+\frac{y_1^2}{b^2}+\frac{z_1^2}{c^2}+\frac{x_2^2}{a^2}+\frac{y_2^2}{b^2}+\frac{z_2^2}{c^2}$$

$$+\frac{x_3^2}{a^2}+\frac{y_3^2}{b^2}+\frac{z_3^2}{c^2}+2\Sigma\frac{x_1x_2}{a^2}+2\Sigma\frac{x_2x_3}{a^2}+2\Sigma\frac{x_3x_1}{a^2}. \quad (7)$$

But (x_1, y_1, z_1), (x_2, y_2, z_2) and (x_3, y_3, z_3) lie on the ellipsoid and are extremities of the conjugate semi-diameters. Therefore

$$\frac{x_1^2}{a^2}+\frac{y_1^2}{b^2}+\frac{z_1^2}{c^2}=\frac{x_2^2}{a^2}+\frac{y_2^2}{b^2}+\frac{z_2^2}{c^2}=\frac{x_3^2}{a^2}+\frac{y_3^2}{b^2}+\frac{z_3^2}{c^2}=1,$$

and

$$\Sigma\frac{x_1x_2}{a^2}=\Sigma\frac{x_2x_3}{a^2}=\Sigma\frac{x_3x_1}{a^2}=0.$$

Hence, equation (7) becomes $\frac{\alpha^2}{a^2}+\frac{\beta^2}{b^2}+\frac{\gamma^2}{c^2}=3.$

Thus, the locus of (α, β, γ) is $\frac{x^2}{a^2}+\frac{y^2}{b^2}+\frac{z^2}{c^2}=3.$

(*ii*) Let (α, β, γ) be the foot of the perpendicular drawn from the centre (0, 0, 0) to the plane PQR. Since, (α, β, γ) lies on the plane PQR *i.e.* on (5), we have

$$\frac{\alpha}{a^2}(x_1+x_2+x_3)+\frac{\beta}{b^2}(y_1+y_2+y_3)+\frac{\gamma}{c^2}(z_1+z_2+z_3)=1. \quad (8)$$

Also, the equations of the perpendicular are

$$\frac{x}{\alpha}=\frac{y}{\beta}=\frac{z}{\gamma}. \quad (9)$$

But this perpendicular is parallel to the normal to the plane (5). Therefore, we have

$$\frac{a\alpha}{\frac{x_1+x_2+x_3}{a}} = \frac{b\beta}{\frac{y_1+y_2+y_3}{b}} = \frac{c\gamma}{\frac{z_1+z_2+z_3}{c}} = \frac{\sqrt{a^2\alpha^2+b^2\beta^2+c^2\gamma}}{\sqrt{\Sigma\left(\frac{x_1+x_2+x_3}{a}\right)^2}}$$

$$= \frac{\sqrt{a^2\alpha^2+b^2\beta^2+c^2\gamma^2}}{\sqrt{3}}$$

$$\Rightarrow \qquad \begin{cases} \dfrac{x_1+x_2+x_3}{a^2} = \dfrac{\sqrt{3}\,\alpha}{\sqrt{a^2\alpha^2+b^2\beta^2+c^2\gamma^2}} \\ \dfrac{y_1+y_2+y_3}{b^2} = \dfrac{\sqrt{3}\,\beta}{\sqrt{a^2\alpha^2+b^2\beta^2+c^2\gamma^2}} \\ \dfrac{z_1+z_2+z_3}{c^2} = \dfrac{\sqrt{3}\,\gamma}{\sqrt{a^2\alpha^2+b^2\beta^2+c^2\gamma^2}} \end{cases}$$

Substituting these, values in (8), we get

$$\frac{\sqrt{3}(\alpha^2+\beta^2+\gamma^2)}{\sqrt{a^2\alpha^2+b^2\beta^2+c^2\gamma^2}} = 1.$$

Hence, the locus of (α, β, γ) is $a^2x^2 + b^2y^2 + c^2z^2 = 3(x^2 + y^2 + z^2)^2$.

Example 2. Prove that the locus of the equal conjugate diameters of the ellipsoid $x^2/a^2 + y^2/b^2 + z^2/c^2 = 1$ is

$$\frac{x^2}{a^2}+\frac{y^2}{b^2}+\frac{z^2}{c^2} = \frac{3(x^2+y^2+z^2)}{a^2+b^2+c^2}.$$

Solution. Let OP, OQ, OR be three equal conjugate semi-diameters of the given ellipsoid. Then

$$OP^2 + OQ^2 + OR^2 = a^2 + b^2 + c^2.$$

Also, given that $OP = OQ = OR$, we get

$$OP^2 = \frac{1}{3}(a^2 + b^2 + c^2). \tag{1}$$

Let (x', y', z') be the coordinates of P. Then, the equations of OP are

$$\frac{x}{x'} = \frac{y}{y'} = \frac{z}{z'}. \tag{2}$$

But
$$OP^2 = x'^2 + y'^2 + z'^2, \tag{3}$$

and
$$\frac{x'^2}{a^2} + \frac{y'^2}{b^2} + \frac{z'^2}{c^2} = 1. \tag{4}$$

Therefore, from equations (1), (3) and (4), we get

$$\frac{x'^2}{a^2} + \frac{y'^2}{b^2} + \frac{z'^2}{c^2} = \frac{3(x'^2 + y'^2 + z'^2)}{a^2 + b^2 + c^2}. \tag{5}$$

The locus of the diameter OP is obtained by eliminating x', y', z' from (2) and (5), *i.e.*

$$\frac{x^2}{a^2} + \frac{y^2}{b^2} + \frac{z^2}{c^2} = \frac{3(x^2 + y^2 + z^2)}{a^2 + b^2 + c^2}.$$

Example 3. Prove that the locus of the point of intersection of three tangent planes to the ellipsoid $x^2/a^2 + y^2/b^2 + z^2/c^2 = 1$, which are parallel to the conjugate diametral planes of the ellipsoid $x^2/\alpha^2 + y^2/\beta^2 + z^2/\gamma^2 = 1$ is

$$\frac{x^2}{\alpha^2} + \frac{y^2}{\beta^2} + \frac{z^2}{\gamma^2} = \frac{a^2}{\alpha^2} + \frac{b^2}{\beta^2} + \frac{c^2}{\gamma^2}.$$

Solution. Let (x_r, y_r, z_r), $r = 1, 2, 3$, be the extremities of the conjugate semi-diameters of the ellipsoid $x^2/\alpha^2 + y^2/\beta^2 + z^2/\gamma^2 = 1$. Then, the equations of the conjugate diametral planes are given by

$$\frac{xx_r}{\alpha^2} + \frac{yy_r}{\beta^2} + \frac{zz_r}{\gamma^2} = 0, \; r = 1, 2, 3. \tag{1}$$

The tangent planes to the ellipsoid $x^2/a^2 + y^2/b^2 + z^2/c^2 = 1$ parallel to the conjugate diametral planes (1) are given by

$$\frac{xx_1}{\alpha^2} + \frac{yy_1}{\beta^2} + \frac{zz_1}{\gamma^2} = \sqrt{\frac{a^2x_1^2}{\alpha^4} + \frac{b^2y_1^2}{\beta^4} + \frac{c^2z_1^2}{\gamma^4}}, \tag{2}$$

$$\frac{xx_2}{\alpha^2} + \frac{yy_2}{\beta^2} + \frac{zz_2}{\gamma^2} = \sqrt{\frac{a^2x_2^2}{\alpha^4} + \frac{b^2y_2^2}{\beta^4} + \frac{c^2z_2^2}{\gamma^4}}, \tag{3}$$

$$\frac{xx_3}{\alpha^2} + \frac{yy_3}{\beta^2} + \frac{zz_3}{\gamma^2} = \sqrt{\frac{a^2x_3^2}{\alpha^4} + \frac{b^2y_3^2}{\beta^4} + \frac{c^2z_3^2}{\gamma^4}}. \tag{4}$$

The required locus is obtained by eliminating x_r, y_r, z_r $(r = 1, 2, 3)$ from (2), (3) and (4). Thus on squaring the equations (2), (3), (4) and adding, we get

$$\frac{x^2}{\alpha^4}\Sigma x_1^2 + \frac{y^2}{\beta^4}\Sigma y_1^2 + \frac{z^2}{\gamma^4}\Sigma z_1^2$$
$$+\frac{2xy}{\alpha^2\beta^2}\Sigma x_1 y_1 + \frac{2yz}{\beta^2\gamma^2}\Sigma y_1 z_1 + \frac{2zx}{\gamma^2\alpha^2}\Sigma z_1 x_1$$
$$= \frac{a^2}{\alpha^4}\Sigma x_1^2 + \frac{b^2}{\beta^4}\Sigma y_1^2 + \frac{c^2}{\gamma^4}\Sigma z_1^2$$

$$\Rightarrow \quad \frac{x^2\alpha^2}{\alpha^4} + \frac{y^2\beta^2}{\beta^4} + \frac{z^2\gamma^2}{\gamma^4} = \frac{a^2\alpha^2}{\alpha^4} + \frac{b^2\beta^2}{\beta^4} + \frac{c^2\gamma^2}{\gamma^4}$$

$$\Rightarrow \quad \frac{x^2}{\alpha^2} + \frac{y^2}{\beta^2} + \frac{z^2}{\gamma^2} = \frac{a^2}{\alpha^2} + \frac{b^2}{\beta^2} + \frac{c^2}{\gamma^2}.$$

Exercises

1. Prove that the plane $lx + my + nz = p$ will pass through the extremities of conjugate semi-diameters of the ellipsoid $\frac{x^2}{a^2} + \frac{y^2}{b^2} + \frac{z^2}{c^2} = 1$, if $a^2l^2 + b^2m^2 + c^2n^2 = 3p^2$.

2. Prove that the lines $\frac{x}{1} = \frac{y}{4} = \frac{z}{3}, \frac{x}{4} = \frac{y}{1} = \frac{z}{-9}, \frac{x}{26} = \frac{y}{-28} = \frac{z}{45}$ are three conjugate diameters of the ellipsoid $\frac{x^2}{2} + \frac{y^2}{4} + \frac{z^2}{9} = 1$.

3. If P, Q, R are the extremities of three conjugate semi-diameters of the ellipsoid $\frac{x^2}{a^2} + \frac{y^2}{b^2} + \frac{z^2}{c^2} = 1$, prove that the plane PQR touches the ellipsoid
$$\frac{x^2}{a^2} + \frac{y^2}{b^2} + \frac{z^2}{c^2} = \frac{1}{3}$$
at the centroid of the triangle PQR.

4. Prove that the locus of the centre of the section of the ellipsoid
$$\frac{x^2}{a^2} + \frac{y^2}{b^2} + \frac{z^2}{c^2} = 1$$
by the plane PQR is the ellipsoid
$$\frac{x^2}{a^2} + \frac{y^2}{b^2} + \frac{z^2}{c^2} = \frac{1}{3},$$

where P, Q, R are the extremities of the conjugate semi-diameters of the first ellipsoid. Further prove that this is also the locus of the centroid of the triangle PQR.

5. The enveloping cone from a point P to the ellipsoid $\frac{x^2}{a^2}+\frac{y^2}{b^2}+\frac{z^2}{c^2}=1$ has three generating lines parallel to conjugate diameters of the ellipsoid. Prove that the locus of P is the ellipsoid $\frac{x^2}{a^2}+\frac{y^2}{b^2}+\frac{z^2}{c^2}=\frac{3}{2}$.

6. Prove that if the cone $Ax^2+By^2+Cz^2+2Fyz+2Gzx+2Hxy=0$ has three of its generators along the conjugate diameters of the ellipsoid $\frac{x^2}{a^2}+\frac{y^2}{b^2}+\frac{z^2}{c^2}=1$, then $Aa^2+Bb^2+Cc^2=0$.

7. If OP, OQ, OR are conjugate semi-diameters and p_1, p_2, p_3; π_1, π_2, π_3 are their projections on any two given lines, then prove that $p_1\pi_1+p_2\pi_2+p_3\pi_3$ is constant.

8. If three conjugate diameters of an ellipsoid meet the diameter sphere in P, Q, R; prove that the plane PQR touches the ellipsoid.

9. If λ, μ, ν are the angles between a set of equal conjugate diameters, prove that

$$\cos^2\lambda+\cos^2\mu+\cos^2\nu=\frac{3\Sigma(b^2-c^2)^2}{2(a^2+b^2+c^2)^2}.$$

10. Prove that the plane through a pair of equal conjugate diameters of the ellipsoid $\frac{x^2}{a^2}+\frac{y^2}{b^2}+\frac{z^2}{c^2}=1$ touches the cone

$$\frac{x^2}{a^2(2a^2-b^2-c^2)}+\frac{y^2}{b^2(2b^2-c^2-a^2)}+\frac{z^2}{c^2(2c^2-a^2-b^2)}=0.$$

Miscellaneous Exercises

1. Find the equations of the tangent planes to the conicoid $7x^2-3y^2-z^2+21=0$ which pass through the line $7x-6y+9=0$, $z=3$.

2. Prove that the tangent planes at the extremities of any diameter of a central conicoid are parallel.

3. CP, CQ, CR are three central radii of an ellipsoid which are mutually perpendicular, prove that the plane PQR touches a sphere.

4. Prove that the locus of a point, the sum of whose distances from the points $(a, 0, 0)$, $(-a, 0, 0)$ is constant $(= 2k)$, is the ellipsoid

$$\frac{x^2}{k^2} + \frac{y^2 + z^2}{k^2 - a^2} = 1.$$

5. Tangent planes are drawn to the conicoid $ax^2 + by^2 + cz^2 = 1$ through the point (α, β, γ). Prove that the perpendiculars to them from the origin generate the cone

$$(\alpha x + \beta y + \gamma z)^2 = \frac{x^2}{a} + \frac{y^2}{b} + \frac{z^2}{c}.$$

Further, prove that the reciprocal cone of this cone is the cone

$$(ax^2 + by^2 + cz^2)(a\alpha^2 + b\beta^2 + c\gamma^2 - 1) - (a\alpha x + b\beta y + c\gamma z)^2 = 0,$$

and hence prove that the tangent planes envelope the cone

$$(ax^2 + by^2 + cz^2 - 1)(a\alpha^2 + b\beta^2 + c\gamma^2 - 1) - (a\alpha x + b\beta y + c\gamma z - 1)^2 = 0.$$

6. The tangent planes to an ellipsoid at the points P_1, P_2, P_3, P_4 form a tetrahedron $A_1A_2A_3A_4$, where A_r is the vertex which is not on the tangent plane at P_r. Prove that the planes $A_1A_2P_2$, $A_1A_3P_3$, $A_1A_4P_4$ have a line in common.

7. Prove that if the normals at points P and Q of an ellipsoid, intersect, PQ is at right angles to its polar with respect to the ellipsoid.

8. Prove that the six normals from any point to a central conicoid meet a principal plane in six points which lie on a rectangular hyperbola.

9. Find the locus of the points from which three mutually perpendicular tangent lines can be drawn to the conicoid $ax^2 + by^2 + cz^2 = 1$.

10. Find the locus of the lines drawn through a fixed point (f, g, h) whose polar lines with respect to the conicoids $ax^2 + by^2 + cz^2 = 1$ and $\alpha x^2 + \beta y^2 + \gamma z^2 = 1$ are coplanar.

11. Prove that the planes $l_1x + m_1y + n_1z = p_1$ and $l_2x + m_2y + n_2z = p_2$ are conjugate planes with respect to the conicoid $ax^2 + by^2 + cz^2 = 1$ if

$$\frac{l_1l_2}{a} + \frac{m_1m_2}{b} + \frac{n_1n_2}{c} = p_1p_2.$$

12. If $P, Q, R; P', Q', R'$ are the feet of the six normals from a point to the ellipsoid $\frac{x^2}{a^2} + \frac{y^2}{b^2} + \frac{z^2}{c^2} = 1$ and S, S' are the poles of PQR, $P'Q'R'$, respectively, and O the centre; prove that $SS'^2 - OS^2 - OS'^2 = 2(a^2 + b^2 + c^2)$.

13. Prove that the locus of points on a central conicoid, the normals at which intersect a given diameter is the curve of intersection with a cone having the principal axes of the conicoid as generators.

14. Prove that, in general, two normals to the ellipsoid $\frac{x^2}{a^2}+\frac{y^2}{b^2}+\frac{z^2}{c^2}=1$ lie in a given plane. Determine the coordinates of the two points on the ellipsoid the normals at which lie in the plane $by - cz = \frac{1}{2}(b^2 - c^2)$.

15. The normals at six points on the ellipsoid $\frac{x^2}{a^2}+\frac{y^2}{b^2}+\frac{z^2}{c^2}=1$ meet in the point (f, g, h). Prove that the mean position of the six points is

$$\left(\frac{-f(b^2+c^2-2a^2)a^2}{3(a^2-b^2)(a^2-c^2)}, \frac{-g(c^2+a^2-2b^2)b^2}{3(b^2-c^2)(b^2-a^2)}, \frac{-h(a^2+b^2-2c^2)c^2}{3(c^2-a^2)(c^2-b^2)}\right).$$

16. Prove that the centre of the conic $ax^2 + by^2 + cz^2 = 1$, $lx + my + nz = p$ is the point $\left(\frac{lp}{ap_0^2}, \frac{mp}{bp_0^2}, \frac{np}{cp_0^2}\right)$, where $l^2 + m^2 + n^2 = 1$ and $p_0 = \Sigma\frac{l^2}{a}$.

17. The line joining a point P to the centre of a conicoid passes through the centre of the section of the conicoid by the polar plane of P.

18. Triads of tangent planes at right angles are drawn to the ellipsoid

$$\frac{x^2}{a^2}+\frac{y^2}{b^2}+\frac{z^2}{c^2}=1.$$

Prove that the locus of the centre of the surface by the plane through the point of contact is

$$x^2 + y^2 + z^2 = \left(\frac{x^2}{a^2}+\frac{y^2}{b^2}+\frac{z^2}{c^2}\right)^2 (a^2 + b^2 + c^2).$$

19. Prove that two normals to the ellipsoid $\frac{x^2}{a^2}+\frac{y^2}{b^2}+\frac{z^2}{c^2}=1$ lie in the plane $lx + my + nz = 0$, and the line joining their feet has direction cosines proportional to $a^2(b^2 - c^2)mn$, $b^2(c^2 - a^2)nl$, $c^2(a^2 - b^2)lm$.

Also obtain the coordinates of these points.

20. Prove that the polar of a line AB with respect to a central conicoid is the line of intersection of the planes of contact of the enveloping cone whose vertex is A and the enveloping cylinder whose generators are parallel to AB.

21. Prove that the sum of the squares of the projections of three conjugate semi-diameters on a line is constant. The sum of the squares of the projections on a plane is also constant.

22. Find the locus of the mid-points of chords of the hyperboloid

$$\frac{x^2}{a^2}+\frac{y^2}{b^2}+\frac{z^2}{c^2}=1$$

which are parallel to $x = 0$ and touch the sphere $x^2 + y^2 + z^2 = k^2$.

23. OP, OQ, OR are three radii of an ellipsoid whose centre is O and which are mutually at right angles to one another. Prove that the plane PQR touches a sphere.

24. OP, OQ, OR are conjugate semi-diameters of an ellipsoidn $\frac{x^2}{a^2}+\frac{y^2}{b^2}+\frac{z^2}{c^2}$ $= 1$. At Q and R tangent lines are drawn parallel to OP and p_1, p_2 are their distances from O. The perpendicular from O to the tangent plane at right angles to OP is p. Prove that $p^2 + p_1^2 + p_2^2 = a^2 + b^2 + c^2$.

25. If (x_1, y_1, z_1), (x_2, y_2, z_2), (x_3, y_3, z_3) are the extremities of three conjugate diameters of the ellipsoid $\frac{x^2}{a^2}+\frac{y^2}{b^2}+\frac{z^2}{c^2}=1$, prove that equation of the plane through the points (x_1, x_2, x_3), (y_1, y_2, y_3), (z_1, z_2, z_3) is

$$\left(\frac{x_1}{a^2}+\frac{y_1}{b^2}+\frac{z_1}{c^2}\right)x+\left(\frac{x_2}{a^2}+\frac{y_2}{b^2}+\frac{z_2}{c^2}\right)y+\left(\frac{x_3}{a^2}+\frac{y_3}{b^2}+\frac{z_3}{c^2}\right)z=1,$$

and prove that it touches a fixed sphere.

26. If through a given point (α, β, γ) perpendiculars are drawn to three conjugate diameters of the ellipsoid $\frac{x^2}{a^2}+\frac{y^2}{b^2}+\frac{z^2}{c^2}=1,$

prove that the plane through the feet of the perpendiculars passes through the fixed point

$$\left(\frac{a^2\alpha}{a^2+b^2+c^2}, \frac{b^2\beta}{a^2+b^2+c^2}, \frac{c^2\gamma}{a^2+b^2+c^2}\right).$$

27. OP, OQ, OR are conjugate diameters of the ellipsoid $\frac{x^2}{a^2}+\frac{y^2}{b^2}+\frac{z^2}{c^2}=1,$ prove that the sides of the triangle PQR touch the ellipsoid

$$\frac{x^2}{a^2}+\frac{y^2}{b^2}+\frac{z^2}{c^2}=\frac{1}{2}.$$

28. Prove that the greatest value of the shortest distance between the x-axis and a normal to the ellipsoid

$$\frac{x^2}{a^2}+\frac{y^2}{b^2}+\frac{z^2}{c^2}=1 \text{ is } b-c.$$

29. If two of a set of conjugate diameters of the ellipsoid $\frac{x^2}{a^2}+\frac{y^2}{b^2}+\frac{z^2}{c^2}=1$ lie on the planes

$$\frac{\alpha_1 x}{a^2}+\frac{\beta_1 y}{b^2}+\frac{\gamma_1 z}{c^2}=0 \text{ and } \frac{\alpha_2 x}{a^2}+\frac{\beta_2 y}{b^2}+\frac{\gamma_2 z}{c^2}=0,$$

respectively, prove that the third lies on the cone $\Sigma a^2(\beta_1 z-\gamma_1 y)(\beta_2 z-\gamma_2 y)=0$.

30. P is any point on the ellipsoid $\frac{x^2}{a^2}+\frac{y^2}{b^2}+\frac{z^2}{c^2}=1$, and 2α, 2β are the principal axes of the section of the ellipsoid by the diametral plane of OP. Prove that $OP^2=a^2+b^2+c^2-\alpha^2-\beta^2$ and $\alpha\beta p=abc$,

where p is the perpendicular distance from O to the tangent plane at P.

31. Prove that the projections of the normals to an ellipsoid at P, Q, R the extremities of conjugate diameters in the plane PQR are concurrent.

32. Conjugate diameters of the conicoid $a_1x^2+b_1y^2+c_1z^2=1$ meet the conicoid $a_2x^2+b_2y^2+c_2z^2=1$ in P, Q, R. Prove that the plane PQR touches the conicoid $a_3x^2+b_3y^2+c_3z^2=1$, if

$$\frac{a_3}{a_1}=\frac{b_3}{b_1}=\frac{c_3}{c_1}=\frac{a_2}{a_1}+\frac{b_2}{b_1}+\frac{c_2}{c_1}.$$

33. Prove that two normals to the conicoid $ax^2+by^2+cz^2=1$ lie in the plane $lx+my+nz=p$, and that they are at right angles if $abcp^2\Sigma\{a(b+c)l^2\}=\Sigma\{a^2(b-c)^2m^2n^2\}$.

34. Prove that the locus of a point, the sum of the squares of whose normal distances from the ellipsoid $\frac{x^2}{a^2}+\frac{y^2}{b^2}+\frac{z^2}{c^2}=1$ is constant ($=k^2$), is

$$6(x^2+y^2+z^2)-2\Sigma\left\{\frac{a^2x^2(b^2+c^2-2a^2)}{(a^2-b^2)(c^2-a^2)}\right\}+2a^2+2b^2+2c^2=k^2.$$

The Paraboloid

19
CHAPTER

1. Introduction

This chapter deals with the surfaces given by the equations of the form

$$ax^2 + by^2 = 2z, \; by^2 + cz^2 = 2x \text{ and } cz^2 + ax^2 = 2y.$$

None of the above surfaces possesses a centre* and as such these surfaces are not central conicoids. These surfaces are called **paraboloids**. In what follows we shall be concentrating on the paraboloid given by the equation $ax^2 + by^2 = 2z$ and the other two could be discussed on similar lines.

2. Shapes of the Paraboloids

Let the equation of the paraboloid be $Ax^2 + By^2 = 2z$. This equation can be written in the form

$$\pm\frac{x^2}{a^2} \pm \frac{y^2}{b^2} = 2z.$$

There are two different cases. In the expression of the left hand side of the above, if

(*i*) both the terms are of the same sign, the surface is called **elliptic paraboloid;**

(*ii*) the terms are of the opposite signs, the surface is called **hyperbolic paraboloid.**

2.1 The elliptic paraboloid

Consider the surface whose equation is

$$\frac{x^2}{a^2} + \frac{y^2}{b^2} = 2z.$$

From the shape of this equation we note the following:

(*i*) The surface is symmetrical about *yz*- and *zx*-planes.

*In fact the centre of each of these surfaces lies at infinity. For details see § 4, of Appendix I.

(*ii*) For no negative value of z, the surface has a real point and as such no part of the surface lies below xy-plane.

(*iii*) The xy-plane meets the surface in points where $z = 0$ i.e. $\frac{x^2}{a^2} + \frac{y^2}{b^2} = 0$ *i.e.* in two coincident points at (0, 0, 0). Thus xy-plane touches the surface at the origin.

(*iv*) The yz-plane meets the surface in points where $x = 0$ *i.e.* $y^2 = 2b^2z$, which is an upward parabola in yz-plane. Similarly, the zx-plane meets the surface in an upward parabola $x^2 = 2a^2z$ in that plane.

(*v*) The sections of the surface by the planes $z = k$, planes parallel to xy-plane, are given by

$$\frac{x^2}{a^2} + \frac{y^2}{b^2} = 2k, z = k.$$

These ellipses are real if $k \geqslant 0$. If $k = 0$, the corresponding plane is xy-plane which meets the surface in **point ellipse** $\frac{x^2}{a^2} + \frac{y^2}{b^2} = 0$. Thus xy-plane touches the surface at (0, 0, 0). If $k > 0$, the size of the ellipse increases as k increases beyond limit keeping its centre always on z-axis. Thus, the surface is entirely on the positive side of the xy-plane and extends to infinity. As such the surface can be regarded as generated by the variable ellipse.

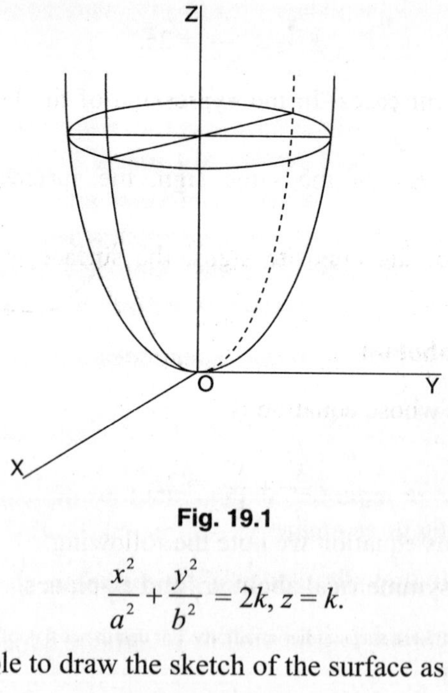

Fig. 19.1

$$\frac{x^2}{a^2} + \frac{y^2}{b^2} = 2k, z = k.$$

It is now possible to draw the sketch of the surface as shown in Fig. 19.1.

Since, the paraboloid is generated by a variable ellipse, it is called **elliptic paraboloid**.

Note. 1. If $a = b$, the plane section of the surface by the plane $z = k$, is a circle $x^2 + y^2 = 2a^2k$. The surface in this case is a **paraboloid of revolution** obtained by revolving the parabola $y^2 = 2a^2z$ about z-axis.

Note 2. The case of the elliptic paraboloid given by the equation $\frac{x^2}{a^2} + \frac{y^2}{b^2} = -2z$ can be dealt with similarly. In this case the surface is entirely lying below the xy-plane.

2.2 The hyperbolic paraboloid

Consider the surface whose equation is $\frac{x^2}{a^2} - \frac{y^2}{b^2} = 2z$.

We note the following:

(*i*) The surface is symmetrical about yz- and zx-planes.

(*ii*) The surface meets the yz-plane in $y^2 = -2b^2z$.
This is a downward parabola with z-axis as its axis.
Further, the surface meets the zx-plane in the upward parabola $x^2 = 2a^2z$ with z-axis as its axis.

(*iii*) The plane sections by the planes $z = k$ $(k \neq 0)$, parallel to the xy-plane, are hyperbolas given by $\frac{x^2}{a^2} - \frac{y^2}{b^2} = 2k$, $z = k$. These hyperbolas are real for all values of k $(\neq 0)$, positive and negative. It will have its transverse axis parallel to x-axis if $k > 0$ and parallel to y-axis if $k < 0$. Let us assume, for definite value, that $k > 0$. The semi-transverse axis, $\sqrt{2k}\, a$, increases as k increases and becomes infinite when k approaches ∞. Thus, the surface extends to infinity above the xy-plane. Similarly, it can be shown, by taking $k < 0$, that the surface extends to infinity below xy-plane.

Further, in case $k = 0$, the hyperbola degenerates into the pair of lines

$$\frac{x}{a} \pm \frac{y}{b} = 0, \; z = 0.$$

Hence, the surface can be regarded as generated by the varibale hyperbola, lying in the plane parallel to xy-plane.

It is now possible to draw the sketch of the surface as shown in Fig. 19.2.

Note 1. Since, the paraboloid is generated by a variable hyperbola, it is called **hyperbolic paraboloid**.

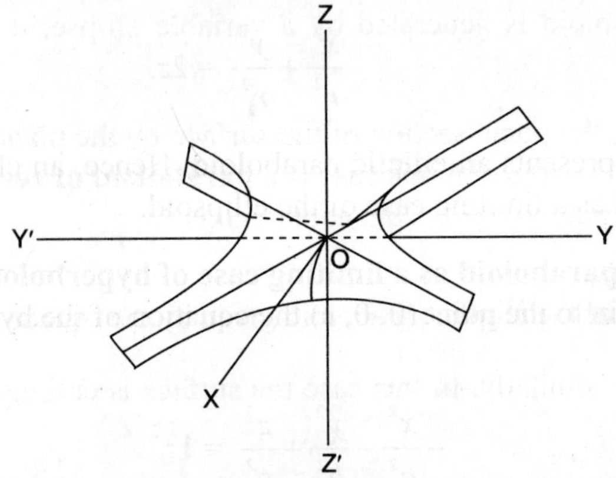

Fig. 19.2

Note 2. The case of the hyperbolic paraboloid given by the equation

$$-\frac{x^2}{a^2}+\frac{y^2}{b^2}=2z,$$

can be dealt with similarly.

3. Paraboloid as Limiting Case of Central Conicoids

We shall now see that by moving the centre of certain central conicoid to a very for off distance, the paraboloids can be obtained as limiting cases of the central conicoids, viz., elliptic paraboloid as a limiting case of an ellipsoid or a hyperboloid of two sheets whereas hyperbolic paraboloid as a limiting case of hyperboloid of one sheet.

3.1 The elliptic paraboloid as a limiting case of an ellipsoid. Shifting the origin to the point $(0, 0, -c)$, the equation of the ellipsoid

$$\frac{x^2}{a^2}+\frac{y^2}{b^2}+\frac{z^2}{c^2}=1$$

takes the form

$$\frac{x^2}{a^2}+\frac{y^2}{b^2}+\frac{z^2}{c^2}=\frac{2z}{c}, \text{ i.e., } \frac{x^2}{\frac{a^2}{c}}+\frac{y^2}{\frac{b^2}{c}}+\frac{z^2}{c}=2z.$$

As $a, b, c \to \infty$, the quantities a^2/c, b^2/c remain finite, say $r_1^2$and r_2^2, respectively, where r_1 and r_2 are finite. Thus in the limiting position the above equation approximates to

$$\frac{x^2}{r_1^2} + \frac{y^2}{r_2^2} = 2z.$$

This equation represents an elliptic paraboloid. Hence, an elliptic paraboloid may be regarded as a limiting case of the ellipsoid.

3.2 The elliptic paraboloid as a limiting case of hyperboloid of two sheets. Shifting the origin to the point (0, 0, *c*) the equation of the hyperboloid of two sheets

$$-\frac{x^2}{a^2} - \frac{y^2}{b^2} + \frac{z^2}{c^2} = 1$$

takes the form

$$\frac{x^2}{a^2} + \frac{y^2}{b^2} - \frac{z^2}{c^2} = \frac{2z}{c}, \text{ i.e., } \frac{x^2}{\frac{a^2}{c}} + \frac{y^2}{\frac{b^2}{c}} - \frac{z^2}{c^2} = 2z.$$

As $a, b, c \to \infty$, the quantities a^2/c, b^2/c remain finite say, r_1^2 and r_2^2, respectively, where r_1 and r_2 are finite. Thus in the limiting position the above equation approximates to

$$\frac{x^2}{r_1^2} + \frac{y^2}{r_2^2} = 2z.$$

This equation represents an elliptic paraboloid. Hence, an elliptic paraboloid may be regarded as a limiting case of the hyperboloid of two sheets.

3.3 The hyperbolic paraboloid as a limiting case of hyperboloid of one sheet. Shifting the origin to the point (0, 0, –*c*) the equation of the hyperboloid of one sheet

$$\frac{x^2}{a^2} - \frac{y^2}{b^2} + \frac{z^2}{c^2} = 1$$

takes the form

$$\frac{x^2}{a^2} - \frac{y^2}{b^2} + \frac{z^2}{c^2} = \frac{2z}{c}, \text{ i.e., } \frac{x^2}{\frac{a^2}{c}} - \frac{y^2}{\frac{b^2}{c}} + \frac{z^2}{c} = 2z$$

As $a, b, c \to \infty$, the quantities a^2/c, b^2/c remain finite, say r_1^2 and r_2^2, respectively, where r_1 and r_2 are finite. Thus, in the limiting position the above equation approximates to

$$\frac{x^2}{r_1^2} - \frac{y^2}{r_2^2} = 2z.$$

This equation represents an hyperbolic paraboloid. Hence, a hyperbolic paraboloid may be regarded as a limiting case of the hyperboloid of one sheet.

4. Some Important Results

Many of the results developed for sphere and central conicoids in the preceding chapters also hold good for the paraboloid. We shall, therefore, only enumerate them and the reader is advised to work out the detailed proofs of them.

Let the equation of the paraboloid be $ax^2 + by^2 = 2cz$.

(*i*) The tangent plane at the point (α, β, γ) is $a\alpha x + b\beta y = c(z + \gamma)$.

(*ii*) The plane $lx + my + nz = p$ is a tangent plane to the paraboloid if

$$\frac{l^2}{a} + \frac{m^2}{b} + \frac{2np}{c} = 0,$$

and then the point of contact will be $\left(-\frac{lc}{an}, -\frac{mc}{bn}, -\frac{p}{n}\right)$.

(*iii*) The equation of the plane of contact of the tangent planes drawn through the point (α, β, γ) to the paraboloid is $a\alpha x + b\beta y = c(z + \gamma)$.

(*iv*) The equation of the polar plane of the point (α, β, γ) with respect to the paraboloid is $a\alpha x + b\beta y = c(z + \gamma)$.

(*v*) The equation of the enveloping cone with vertex at the point (α, β, γ) is $T^2 = SS_1$, where $S \equiv ax^2 + by^2 - 2cz$, $S_1 \equiv a\alpha^2 + b\beta^2 - 2c\gamma$ and $T \equiv a\alpha x + b\beta y - c(z + \gamma)$.

(*vi*) The equation of the enveloping cylinder whose axis is parallel to

$\frac{x}{l} = \frac{y}{m} = \frac{z}{n}$ is $(al^2 + bm^2)(ax^2 + by^2 - 2cz) = (alx + bmy - cn)^2$.

(*vii*) The equation of the section of the paraboloid with centre at (α, β, γ) is

$T = S_1$, *i.e.*, $a\alpha x + b\beta y - cz = a\alpha^2 + b\beta^2 - c\gamma$.

(*viii*) The locus of the middle points of a system of chords parallel to

$\frac{x}{l} = \frac{y}{m} = \frac{z}{n}$ of the paraboloid is $alx + bmy - cn = 0$. This is a plane and is called **diametral plane.**

5. Normal

5.1 Equation of the normal. *To find the equations of the normal to the paraboloid* $ax^2 + by^2 = 2cz$ *at any point* (α, β, γ) *on it.*

The equation of the tangent plane at (α, β, γ) to the given paraboloid is

$$a\alpha x + b\beta y = c(z + \gamma).$$

The direction cosines of the normal to this plane through the point (α, β, γ) are proportional to $a\alpha, b\beta, -c$. Hence, the equations of the normal to the paraboloid $ax^2 + by^2 = 2cz$ at (α, β, γ) are

$$\frac{x-\alpha}{a\alpha} = \frac{y-\beta}{b\beta} = \frac{z-\gamma}{-c}.$$

Ex. Prove that the equations of the normal to the elliptic paraboloid

$$\frac{x^2}{a^2} + \frac{y^2}{b^2} = \frac{2z}{c}$$

at any point (α, β, γ) on it are

$$\frac{x-\alpha}{\alpha/a^2} = \frac{y-\beta}{\beta/b^2} = \frac{z-\gamma}{-1/c}.$$

5.2 Number of normals from a given point. *To prove that from any point five normals can be drawn to a paraboloid.*

Let the equation of a paraboloid be

$$ax^2 + by^2 = 2cz, \tag{1}$$

and let (f, g, h) be any given point. The equations of the normal at any point (α, β, γ) on (1) are

$$\frac{x-\alpha}{a\alpha} = \frac{y-\beta}{b\beta} = \frac{z-\gamma}{-c}.$$

If it passes through (f, g, h), then

$$\frac{f-\alpha}{a\alpha} = \frac{g-\beta}{b\beta} = \frac{h-\gamma}{-c} = r \text{ (say)}$$

$$\Rightarrow \qquad \alpha = \frac{f}{1+ar}, \beta = \frac{g}{1+br}, c = h + cr.$$

But (α, β, γ) lies on (1). Therefore

$$\frac{af^2}{(1+ar)^2} + \frac{bg^2}{(1+br)^2} = 2c(h + cr). \tag{2}$$

Equation (2) is of fifth degree in r and hence gives five values of r for a given point (f, g, h). Thus, there are five normals that can be drawn from a point (f, g, h) to the paraboloid.

Ex. Prove that five normals can be drawn to the elliptic paraboloid

$$\frac{x^2}{a^2}+\frac{y^2}{b^2}=\frac{2z}{c}$$

from any given point.

5.3 *To prove that the feet of the normals from (f, g, h) to the paraboloid $ax^2 + by^2 = 2cz$ lie on the surfaces which have a common curve of intersection.*

The equations of the normal at any point (α, β, γ) on the given paraboloid are

$$\frac{x-\alpha}{a\alpha}=\frac{y-\beta}{b\beta}=\frac{z-\gamma}{-c}. \quad (1)$$

If it passes through (*f*, *g*, *h*), then

$$\frac{f-\alpha}{a\alpha}=\frac{g-\beta}{b\beta}=\frac{h-\gamma}{-c}=r\text{ (say)}$$

$$\Rightarrow \qquad \alpha=\frac{f}{1+ar},\beta=\frac{g}{1+br},\gamma=h+cr.$$

Hence, the feet of the normals lie on the curve

$$x=\frac{f}{1+ar},y=\frac{g}{1+br},z=h+cr,$$

where *r* is a parameter.

5.4 Cubic curve through the feet of the normals. *To prove that the feet of the five normals from any given point to a paraboloid are the points of intersection of the paraboloid and a certain cubic curve.*

Let the equation of a paraboloid be

$$ax^2 + by^2 = 2cz. \quad (1)$$

The feet of the five normals drawn through the point (*f*, *g*, *h*) to the paraboloid (1) lie on it. Also, from § 5.3, it is clear that the feet of the normals lie on the curve

$$x=\frac{f}{1+ar},y=\frac{g}{1+br},z=h+cr, \quad (2)$$

where *r* is a parameter. Thus, the feet of the five normals drawn through a point (*f*, *g*, *h*) to the paraboloid (1) are the points of intersection of the paraboloid (1) and the curve (2). Now we claim that the curve (2) is cubic in *r*.

The curve (2) meets the plane $Ax + By + Cz + D = 0$, where

$$A\left(\frac{f}{1+ar}\right) + B\left(\frac{g}{1+br}\right) + C(h + cr) + D = 0.$$

This equation is cubic in r, hence gives three values of r. This completes the proof.

5.5 Cone through the five normals. *To prove that the five normals drawn from any given point to a paraboloid generate the cone of second degree.*

Let $ax^2 + by^2 = 2cz$ be the equation of the paraboloid and (f, g, h) the given point. The equations of the cubic curve through the feet of the normals are given by

$$x = \frac{f}{1+ar},\ y = \frac{g}{1+br},\ z = h + cr. \tag{1}$$

Any line through the point (f, g, h) is

$$\frac{x-f}{l} = \frac{y-g}{m} = \frac{z-h}{n}. \tag{2}$$

If the line (2) is a normal to the paraboloid at (α, β, γ), then it must be identical with

$$\frac{x-\alpha}{a\alpha} = \frac{y-\beta}{b\beta} = \frac{z-\gamma}{-c}. \tag{3}$$

Therefore, on comparing (2) and (3), we get

$$\frac{l}{a\alpha} = \frac{m}{b\beta} = \frac{n}{-c}.$$

But (α, β, γ) also lies on (1). Therefore

$$\frac{l(1+ar)}{af} = \frac{m(1+br)}{bg} = \frac{n}{-c}$$

$$\Rightarrow \qquad \frac{\frac{1}{a}+r}{\frac{f}{l}} = \frac{\frac{1}{b}+r}{\frac{g}{m}} = -\frac{\frac{1}{c}}{\frac{c}{n}}. \tag{4}$$

Eliminating r from (4), we get

$$\frac{f}{l} - \frac{g}{m} + \left(\frac{1}{a} - \frac{1}{b}\right)\frac{c}{n} = 0. \tag{5}$$

The locus of the line (2) is obtained by eliminating l, m, n from (2) and (4), *i.e.*

$$\frac{f}{x-f}-\frac{g}{y-g}+\frac{c(b-a)}{ab(z-h)}=0. \tag{6}$$

This equation represents a cone generated by the lines (2) and with guiding curve (1), These lines, in particular, include all the five normals drawn from (f, g, h) to the paraboloid. Hence (5) is the required cone.

6. Examples

Example 1. Find the locus of the point of intersection of three mutually perpendicular tangent planes to the paraboloid

$$\frac{x^2}{a^2}+\frac{y^2}{b^2}=\frac{2z}{c}.$$

Solution. Let the three mutually perpendicular tangent planes to the given paraboloid be

$$2cn_1(l_1x+m_1y+n_1z)+(a^2l_1^2+b^2m_1^2)=0, \tag{1}$$

$$2cn_2(l_2x+m_2y+n_2z)+(a^2l_2^2+b^2m_2^2)=0, \tag{2}$$

$$2cn_3(l_3x+m_3y+n_3z)+(a^2l_3^2+b^2m_3^2)=0. \tag{3}$$

Since l_1, m_1, n_1; l_2, m_2, n_2 and l_3, m_3, n_3 are the direction cosines of the normals to the three mutually perpendicular planes, we have

$$\begin{cases}\Sigma l_1^2=\Sigma m_1^2=\Sigma n_1^2=1\\ \Sigma l_1m_1=\Sigma m_1n_1=\Sigma n_1l_1=0.\end{cases}$$

On squaring (1), (2), (3) and adding, we get

$$2cz+a^2+b^2=0.$$

This is the required locus.

Example 2. Prove that the planes $lx+my+p=0$ and $l'x+m'y+p'=0$ are conjugate diametral planes of the paraboloid $ax^2+by^2=2cz$ if

$$\frac{ll'}{a}+\frac{mm'}{b}=0.$$

Solution. Any diametral plane of the given paraboloid is

$$a\lambda x+b\mu y-c\nu=0,$$

which bisects the chords parallel to the line

$$\frac{x}{\lambda} = \frac{y}{\mu} = \frac{z}{\nu}.$$

If the plane $lx + my + p = 0$ is a diametral plane, then

$$lx + my + p \equiv a\lambda x + b\mu y - c\nu$$

$$\Rightarrow \qquad \frac{\lambda}{\frac{l}{a}} = \frac{\mu}{\frac{m}{b}} = \frac{\nu}{\frac{-p}{c}}.$$

Thus, the diametral plane $lx + my + p = 0$ bisects the chords parallel to the line

$$\frac{x}{\frac{l}{a}} = \frac{y}{\frac{m}{b}} = \frac{z}{\frac{-p}{c}}. \qquad (1)$$

The given planes will be conjugate diametral planes if each bisects the chords parallel to the other, *i.e.*, if the plane $l'x + m'y + p' = 0$ is parallel to (1) which gives

$$\frac{ll'}{a} + \frac{mm'}{b} = 0.$$

Exercises

1. Prove that the plane $2x - 4y - z + 3 = 0$ touches the paraboloid $x^2 - 2y^2 = 3z$.
2. Prove that the feet of the normals from the point (α, β, γ) to the paraboloid $x^2 + y^2 = 2az$ lie on the sphere $x^2 + y^2 + z^2 - (a + \gamma)z - \dfrac{(\alpha^2 + \beta^2)y}{2\beta} = 0.$
3. Prove that the normals drawn from the point (α, β, γ) to the paraboloid $\dfrac{x^2}{a^2} + \dfrac{y^2}{b^2} = 2z$ lie on the cone $\dfrac{\alpha}{x-\alpha} - \dfrac{\beta}{y-\beta} + \dfrac{a^2 - b^2}{z-\gamma} = 0.$
4. Prove that, in general, three normals can be drawn from a given point to the paraboloid of revolution $x^2 + y^2 = 2az$, but if the point lies on the surface $27a(x^2 + y^2) + 8(a - z)^2 = 0$, two of the three normals coincide.
5. The perpendicular tangent planes to the paraboloid

$$\frac{x^2}{a} + \frac{y^2}{b} = 2z$$

intersect in a line lying in the plane $x = 0$. Prove that the line touches the parabola $y^2 = (a + b)(2z + a)$, $x = 0$.

6. Find the locus of the points from which three mutually perpendicular tangents can be drawn to the paraboloid $ax^2 + by^2 = 2z$.

7. Prove that the locus of the centres of a system of parallel plane sections of a paraboloid is a diameter.

8. The plane $3x + 4y = 1$ is a diametral plane of the paraboloid $5x^2 + 6y^2 = 2z$. Find the equations of the chord through the point (3, 4, 5) which it bisects.

9. Prove that the feet of the normals from any point to a paraboloid lie on a cubic curve.

10. Prove that the centre of the circle through the feet of the three normals from the point (α, β, γ) to the paraboloid $x^2 + y^2 = 2az$ is $\left(\frac{\alpha}{4}, \frac{\beta}{4}, \frac{\gamma + \alpha}{2}\right)$.

11. Prove that any diametral plane cuts the paraboloid in a parabola, and parallel diametral planes cut it in equal parabolas.

12. Prove that the cubic curve through the feet of the normals drawn to a paraboloid from a given point lies on the cone through the normals themselves.

13. Prove that the perpendicular from a given point $P(\alpha, \beta, \gamma)$ to its polar plane with respect to a paraboloid lies on the cone through the five concurrent normals at P.

14. A variable line intersects the x-axis and the curve given by the equations $x = y$ and $y^2 = cz$, and is parallel to yz-plane. Prove that it generates the paraboloid $xy = cz$.

15. Prove that the equation of the two tangent planes to the paraboloid $ax^2 + by^2 = 2z$ which pass through the line $u \equiv lx + my + nz - p = 0$, $u' \equiv l'x + m'y + n'z - p' = 0$ is

$$u^2\left(\frac{l'^2}{a} + \frac{m'^2}{b} + 2n'p\right) - 2uu'\left(\frac{ll'}{a} + \frac{mm'}{b} + np' + n'p\right)$$
$$+ u'^2\left(\frac{l^2}{a} + \frac{m^2}{b} + 2np\right) = 0.$$

16. If the enveloping cone of a paraboloid has three mutually perpendicular generators, prove that the vertex lies on a paraboloid, and that the polar plane of the vertex touches another paraboloid.

General Equation of Second Degree

APPENDIX II

1. Introduction

The most general equation of second degree in x, y and z is

$$F(x, y, z) \equiv ax^2 + by^2 + cz^2 + 2fyz + 2gzx + 2hxy + 2ux + 2vy + 2wz + d = 0$$

$$\equiv f(x, y, z) + 2ux + 2vy + 2wz + d = 0,$$

where $\quad f(x, y, z) \equiv ax^2 + by^2 + cz^2 + 2fyz + 2gzx + 2hxy.$

Number of constants. The above equation contains ten constants. If we divide the equation by any constant, then there are nine arbitrary constants. Therefore, the general equation of second degree can be made to satisfy nine independent conditions. Conversely, if nine given conditions are satisfied by the arbitrary constants, the equation of the surface can be determined uniquely. The surfaces determined by the second degree equations are called **quadrics** or **conicoids**.

Note. The sphere, cone, pair of planes etc. being second degree equations in x, y and z are particular conicoids.

2. Certain Concepts and Results for General Conicoids

2.1 Intersection of a line and a conicoid. Let a line through a given point $A(\alpha, \beta, \gamma)$ be

$$\frac{x-\alpha}{l} = \frac{y-\beta}{m} = \frac{z-\gamma}{n} = r \text{ (say)}. \tag{1}$$

Any point on this line is $(\alpha + lr, \beta + mr, \gamma + nr)$. This point will lie on the conicoid $F(x, y, z) = 0$, if

$$F(\alpha, \beta, \gamma) + r\left(l\frac{\partial F}{\partial \alpha} + m\frac{\partial F}{\partial \beta} + n\frac{\partial F}{\partial \gamma}\right) + r^2 f(l, m, n) = 0, \tag{2}$$

where $\partial F/\partial\alpha$ is the value of $\partial F/\partial x$ at $x = \alpha$, etc.

Equation (3) being quadratic in r gives two values of r, say, r_1 and r_2. Hence, a line meets a conicoid in two points which may be real and distinct, conicident or imaginary. Now, there arise certain cases:

(*i*) If $F(\alpha, \beta, \gamma) = 0$, one of the values of r is zero and the point (α, β, γ) lies on the conicoid.

(*ii*) If $l\frac{\partial F}{\partial \alpha} = m\frac{\partial F}{\partial \beta} = n\frac{\partial F}{\partial \gamma} = 0$ also, both the values of r are zero, and the line touches the conicoid at A.

(*iii*) If $f(l, m, n) = 0$ also, then (3) becomes an identity and the line lies on the conicoid.

If the conditions (ii) and (iii) are satisfied, we obtain two sets of values for $l : m : n$, and hence through any point on a conicoid two lines can be drawn to lie wholly on the conicoid. These lines are, in fact, parallel to the line in which the plane

$$x\frac{\partial F}{\partial \alpha} + y\frac{\partial F}{\partial \beta} + z\frac{\partial F}{\partial \gamma} = 0$$

cuts the cone $f(x, y, z) = 0$. They may be real and distinct, coincident or imaginary and accordingly the conicoid is an ellipsoid, cone or hyperboloid of one sheet, respectively.

2.2 Tangent line

A line which meets a coincoid at two conicident points is called the **tangent line** *at that point.*

2.3 Tangent plane

If conditions (i) and (ii) are satisfied, the line touches the conicoid at A. Eliminating l, m, n from (1) of § 2. 1 and

$$l\frac{\partial F}{\partial x} + m\frac{\partial F}{\partial y} + n\frac{\partial F}{\partial z} = 0,$$

we get

$$(x - \alpha)\frac{\partial F}{\partial \alpha} + (y - \beta)\frac{\partial F}{\partial \beta} + (z - \gamma)\frac{\partial F}{\partial \gamma} = 0.$$

This is the locus of the tangent lines and represents a plane called the tangent plane. This equation can also be written as

$$x\frac{\partial F}{\partial \alpha} + y\frac{\partial F}{\partial \beta} + z\frac{\partial F}{\partial \gamma} = \alpha\frac{\partial F}{\partial \alpha} + \beta\frac{\partial F}{\partial \beta} + \gamma\frac{\partial F}{\partial \gamma}. \quad (4)$$

If we make the equation $F(x, y, z) = 0$ homogeneous by introducing an auxiliary variable t and taking t equal to unity after differentiation, equation (4) can be expressed as

$$x\frac{\partial F}{\partial \alpha} + y\frac{\partial F}{\partial \beta} + z\frac{\partial F}{\partial \gamma} + t\frac{\partial F}{\partial t} = \alpha\frac{\partial F}{\partial \alpha} + \beta\frac{\partial F}{\partial \beta} + \gamma\frac{\partial F}{\partial \gamma} + t\frac{\partial F}{\partial t}$$

$$= 2F(\alpha, \beta, \gamma, t) = 0, \text{ (by Euler's theorem).}$$

This is the standard equation of the tangent plane.

2.4 The polar plane

If any secant through A meets a conicoid in P and Q and if R is the harmonic conjugate of A with respect to P and Q, the locus of R is the polar plane of A.

If A is (α, β, γ) and the equations of the secant are

$$\frac{x-\alpha}{l} = \frac{y-\beta}{m} = \frac{z-\gamma}{n},$$

then $AP = r_1$ and $AQ = r_2$ are the roots of the equation

$$r^2 f(l, m, n) + r\left(l\frac{\partial F}{\partial \alpha} + m\frac{\partial F}{\partial \beta} + n\frac{\partial F}{\partial \gamma}\right) + F(\alpha, \beta, \gamma) = 0.$$

Let the coordinates of R be (ξ, η, ζ) and $AR = \rho$. Then

$$\rho = \frac{2r_1r_2}{r_1 + r_2} = \frac{2F(\alpha,\beta,\gamma)}{l\frac{\partial F}{\partial \alpha} + m\frac{\partial F}{\partial \beta} + n\frac{\partial F}{\partial \gamma}},$$

and $$\xi - \alpha = l\rho, \eta - \beta = m\rho, \zeta - \gamma = n\rho.$$

Therefore

$$(\xi - \alpha)\frac{\partial F}{\partial \alpha} + (\eta - \beta)\frac{\partial F}{\partial \beta} + (\zeta - \gamma)\frac{\partial F}{\partial \gamma} = -2F(\alpha, \beta, \gamma),$$

and the locus of (ξ, η, ζ) is

$$x\frac{\partial F}{\partial \alpha} + y\frac{\partial F}{\partial \beta} + z\frac{\partial F}{\partial \gamma} + t\frac{\partial F}{\partial t} = \alpha\frac{\partial F}{\partial \alpha} + \beta\frac{\partial F}{\partial \beta} + \gamma\frac{\partial F}{\partial \gamma} - 2F(\alpha, \beta, \gamma, t)$$

$$= 2F(\alpha, \beta, \gamma, t) - 2F(\alpha, \beta, \gamma, t) = 0.$$

This is the equation of the polar plane.

2.5 Locus of chords bisected at a point

Let $$\frac{x-\alpha}{l}=\frac{y-\beta}{m}=\frac{z-\gamma}{n}=r \text{ (say)}, \tag{1}$$

be the equations of a chord and (α, β, γ) its middle point. Any point on (1) is $(\alpha + lr, \beta + mr, \gamma + nr)$. If this point lies on $F(x, y, z) = 0$, then

$$r^2 f(l, m, n) + r\left(l\frac{\partial F}{\partial \alpha}+m\frac{\partial F}{\partial \beta}+n\frac{\partial F}{\partial \gamma}\right) + F(\alpha, \beta, \gamma) = 0. \tag{2}$$

Since (α, β, γ) is the middle point of the chord, we have

$$l\frac{\partial F}{\partial x}+m\frac{\partial F}{\partial y}+n\frac{\partial F}{\partial z} = 0. \tag{3}$$

Eliminating l, m, n from (1) and (3), we get

$$(x-\alpha)\frac{\partial F}{\partial \alpha}+(y-\beta)\frac{\partial F}{\partial \beta}+(z-\gamma)\frac{\partial F}{\partial \gamma} = 0. \tag{4}$$

This equation represents a plane. The section of the conicoid by this plane is a conic whose centre is (α, β, γ).

2.6 The diametral plane

It is clear from § 2.5 that the locus of the middle points of all chords drawn parallel to a fixed line

$$\frac{x}{l}=\frac{y}{m}=\frac{z}{n}$$

is given by

$$l\frac{\partial F}{\partial x}+m\frac{\partial F}{\partial y}+n\frac{\partial F}{\partial z} = 0.$$

This equation represents a plane and is the required diametral plane.

2.7 The principal plane

A diametral plane which is at right angles to the chords which it bisects is called the **principal plane.**

The diametral plane

$$l\frac{\partial F}{\partial x}+m\frac{\partial F}{\partial y}+n\frac{\partial F}{\partial z} = 0$$

i.e., $x(al + hm + gn) + y(hl + bm + fn) + z(gl + fm + cn) + ul + vm + wn = 0$

is at right angles to the line $\frac{x}{l}=\frac{y}{m}=\frac{z}{n}$, if

$$\frac{al+hm+gn}{l}=\frac{hl+bm+fn}{m}=\frac{gl+fm+cn}{n}=\lambda \text{ (say)}.$$

Then
$$\left.\begin{aligned}(a-\lambda)l+hm+gn=0,\\ hl+(b-\lambda)m+fn=0,\\ gl+fm+(c-\lambda)n=0.\end{aligned}\right\} \tag{1}$$

Thus λ is a root of the equation

$$\begin{vmatrix} a-\lambda & h & g \\ h & b-\lambda & f \\ g & f & c-\lambda \end{vmatrix}=0$$

i.e. $\lambda^3-\lambda^2(a+b+c)+\lambda(bc+ca+ab-f^2-g^2-h^2)-D=0$,
where

$$D\equiv\begin{vmatrix} a & h & g \\ h & b & f \\ g & f & c \end{vmatrix}.$$

If $A\equiv\frac{\partial D}{\partial \alpha}$ etc. the above equation may be written as

$$\lambda^3-\lambda^2(a+b+c)+\lambda(A+B+C)-D=0.$$

This equation is called the **discriminating cubic**. By theory of equations it gives three values of λ. For each value of λ there corresponds a set of values for $l:m:n$. Substituting these sets of values in the equation

$$l\frac{\partial F}{\partial x}+m\frac{\partial F}{\partial y}+n\frac{\partial F}{\partial z}=0$$

and using relations in (1), we get

$$\lambda_r(l_rx+m_ry+n_rz)+ul_r+vm_r+wn_r=0,\ (r=1,2,3). \tag{2}$$

These are the equations of the three principal planes of the conicoid.

3. The Discriminating Cubic

The equation of the discriminating cubic is given by

$$\lambda^3 - \lambda^2(a+b+c) + \lambda(A+B+C) - D = 0. \tag{1}$$

3.1 Case of one zero-root. If $D = 0$, one root of equation (1) is zero, then

$$BC - F^2 = aD = 0,\ CA - G^2 = bD = 0,\ AB - H^2 = cD = 0.$$

This shows that A, B, C have the same sign.

If $\lambda = 0$ in equations (1) of § 2.7, we get

$$\frac{l}{G} = \frac{m}{F} = \frac{n}{C} \Rightarrow \frac{1}{\sqrt{A}} = \frac{m}{\sqrt{B}} = \frac{n}{\sqrt{C}}.$$

But A, B, C are not all zero, we obtain a fixed direction and from equation (2) of § 2.7, the equation of the corresponding principal plane is

$$\sqrt{A}x + \sqrt{B}y + \sqrt{C}z + \frac{\sqrt{A}u + \sqrt{B}v + \sqrt{C}w}{0} = 0.$$

If $\sqrt{A}u + \sqrt{B}v + \sqrt{C}w \neq 0$, the plane is at an infinite distance. In this case, the conicoid is a **paraboloid** whose axis is the fixed direction.

Further, if $\sqrt{A}u + \sqrt{B}v + \sqrt{C}w = 0$, the plane is at right angles to the fixed line

$$\frac{x}{\sqrt{A}} = \frac{y}{\sqrt{B}} = \frac{z}{\sqrt{C}}.$$

In this case the conicoid is an **elliptic** or **hyperbolic cylinder** or **pair of intersecting planes** whose axis (line of intersection) is in the fixed direction.

3.2 Case of two zero-roots. If $D = 0$ and $A + B + C = 0$, the two roots of (1) are zero. Also from § 3.1, A, B, C have the same sign. Thus $A = B = C = 0$ and $F = G = H = 0$.

If for two roots $\lambda = 0$, in equations (1) of § 2.7, we observe that all the relations reduce to

$$\sqrt{a}l + \sqrt{b}m + \sqrt{c}n = 0.$$

Therefore, the direction of the normals to two of the three principal planes are indeterminate. In fact, these planes must be at right angles to

$$\sqrt{a}x + \sqrt{b}y + \sqrt{c}z = 0.$$

If $ul + vm + wn \neq 0$, the planes are at an infinite distance. In this case the surface is a **parabolic cylinder** and the axes of the normal sections are parallel to the plane

$$\sqrt{a}x+\sqrt{b}y+\sqrt{c}z = 0.$$

Further, if $ul + vm + wn = 0$, the planes are at a finite distance. In this case the surface is a **pair of planes** parallel to

$$\sqrt{a}x+\sqrt{b}y+\sqrt{c}z = 0.$$

3.3 All the roots of the discriminating cubic are real. The equation of the discriminating cubic may be written as

$$\phi(\lambda) = (\lambda - a)\{(\lambda - b)(\lambda - c) - f^2\} - \{(\lambda - b)g^2 + (\lambda - c)h^2 + 2fgh\} = 0.$$

Let us assume that $a > b > c$. Consider

$$y = \psi(\lambda) \equiv (\lambda - b)(\lambda - c) - f^2$$

when $\lambda = -\infty, c, b, +\infty$, respectively, then

$$\psi(\lambda) = +\infty, f^2, -f^2, +\infty.$$

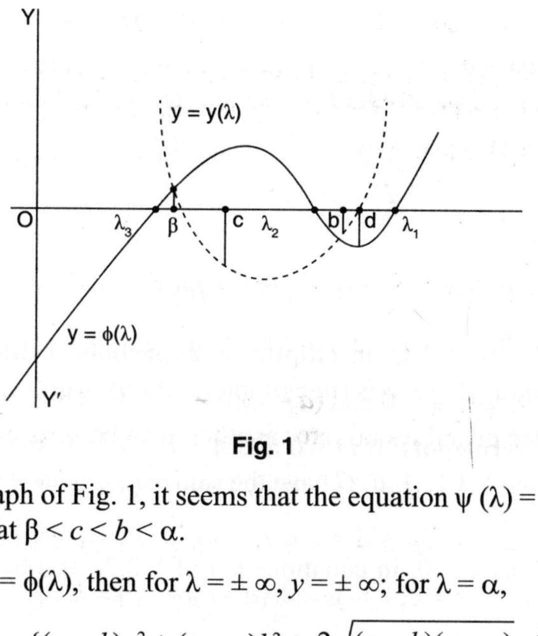

Fig. 1

From the graph of Fig. 1, it seems that the equation $\psi(\lambda) = 0$ has real roots α and β, such that $\beta < c < b < \alpha$.

If we take $y = \phi(\lambda)$, then for $\lambda = \pm\infty$, $y = \pm\infty$; for $\lambda = \alpha$,

$$y = -\{(\alpha - b)g^2 + (\alpha - c)h^2 \pm 2\sqrt{(\alpha - b)(\alpha - c)}\,gh\}$$
$$= -(g\sqrt{\alpha - b} \pm h\sqrt{\alpha - c}\,)^2,$$

where $\sqrt{\alpha - b}$, $\sqrt{\alpha - c}$ are real, and for $\lambda = \beta$,

$$y = (b - \beta)g^2 + (c - \beta)h^2 \pm 2\sqrt{(b - \beta)(c - \beta)}\,gh$$
$$= \left(g\sqrt{(b - \beta)} \pm h\sqrt{(c - \beta)}\,\right)^2,$$

where $\sqrt{(b-\beta)}$, $\sqrt{(c-\beta)}$ are real. Thus, from the graph, we observe that the equation $\phi(\lambda) = 0$ has three real roots, say, $\lambda_1, \lambda_2, \lambda_3$ such that

$$\lambda_3 < \beta < \lambda_2 < \alpha < \lambda_1.$$

If the roots α and β of the equation $\psi(\lambda) = 0$ are real, the above proof fails. In this case $b = c, f = 0$ and then the discriminating cubic becomes

$$(\lambda - b)\{(\lambda - a)(\lambda - b) - g^2 - h^2\} = 0.$$

It is easy to see that the roots of this equation are all real.

3.4 Conditions for equal roots

I. Two roots are equal. If two roots of $\phi(\lambda) = 0$ are equal, then $\dfrac{d\phi}{d\lambda} = 0.$ Therefore

$$(b-\lambda)(c-\lambda) - f^2 + (c-\lambda)(a-\lambda) - g^2 + (a-\lambda)(b-\lambda)h^2 = 0$$

$$\Rightarrow \quad (b-\lambda)(c-\lambda) = f^2, (c-\lambda)(a-\lambda) = g^2, (a-\lambda)(b-\lambda) = h^2. \tag{1}$$

Using the result $F = G = H = 0$ of § 3.2, we get

$$(a-\lambda)f = gh, (b-\lambda)g = hf, (c-\lambda)h = fg. \tag{2}$$

Equations (1) and (2) are same and can be obtained one from the other. These are the sufficient conditions for two equal roots. From equation (2) the value of the repeated root is given by

$$\Rightarrow \quad -\lambda = \frac{F}{f} = \frac{G}{g} = \frac{H}{h}. \tag{3}$$

If either of f, g, h is zero, say, $f = 0$, then from equation (1) $\lambda = b$ or c.

Case I. If $\lambda = b$, then $h = 0$ and $(c-b)(a-b) = g^2$.

Case II. If $\lambda = c$, then $g = 0$ and $(a-c)(b-c) = h^2$. Thus, we observe that if either of the three quantities be zero, another must be zero, and we get

$$\left.\begin{aligned} &\lambda = a, g = h = 0, (b-a)(c-a) = f, \\ \text{or} \quad &\lambda = b, h = f = 0, (c-b)(a-b) = g^2, \\ \text{or} \quad &\lambda = c, f = g = 0, (a-c)(b-c) = h^2. \end{aligned}\right\} \tag{4}$$

The equations (3) and (4) are the conditions for a pair of equal roots and their values in each case. If any of the relations in (4) is satisfied, the surface $F(x, y, z) = 0$ represents a surface of revolution.

II. Three roots are equal. If three roots of the equation $\phi(\lambda) = 0$ are equal, then $\dfrac{d\phi}{d\lambda} = 0$ and $\dfrac{d^2\phi}{d\lambda^2} = 0$, *i.e.*, $(a - \lambda) + (b - \lambda) + (c - \lambda) = 0$.

From above relations (1), we get

$$\{(a-\lambda)+(b-\lambda)+(c-\lambda)\}^2 = (a-\lambda)^2 + (b-\lambda)^2 + (c-\lambda)^2$$
$$+ 2f^2 + 2g^2 + 2h^2,$$
$$\lambda = a = b = c \text{ and } f = g = h = 0.$$

In this case the conicoid represents a sphere.

4. The Centre

Let the origin be the centre of the surface

$$F(x, y, z) \equiv f(x, y, z) + 2ux + 2vy + 2wz + d = 0.$$

If the point (x', y', z') lies on the surface, and the point $(-x', -y', -z')$ also lies on the surface, then

$$\left.\begin{aligned} f(x', y', z') + 2ux' + 2vy' + 2wz' + d &= 0, \\ f(x', y', z') - 2ux' - 2vy' - 2wz' + d &= 0. \end{aligned}\right\} \quad (1)$$

From these equations, we obtain

$$ux' + vy' + wz' = 0$$

for all values of u, v, w. Therefore $u = v = w = 0$.

In general, we assume that the centre of the surface $F(x, y, z) = 0$ be (α, β, γ). Transferring the origin to (α, β, γ), the equation $F(x, y, z) = 0$ becomes

$$F(x + \alpha, y + \beta, z + \gamma) = 0$$

i.e. $$f(x, y, z) + x\frac{\partial F}{\partial \alpha} + y\frac{\partial F}{\partial \beta} + z\frac{\partial F}{\partial \gamma} + F(\alpha, \beta, \gamma) = 0.$$

Thus $$\frac{\partial F}{\partial \alpha} = \frac{\partial F}{\partial \beta} = \frac{\partial F}{\partial \gamma} = 0.$$

On solving these three equations we obtain the values of α, β, γ. Writing these equations in the form

$$a\alpha + h\beta + g\gamma + u = 0,$$
$$h\alpha + b\beta + f\gamma + v = 0,$$
$$g\alpha + f\beta + c\gamma + w = 0,$$

and solving, we get

$$\frac{\alpha}{\begin{vmatrix} h & g & u \\ b & f & v \\ f & c & w \end{vmatrix}} = \frac{-\beta}{\begin{vmatrix} a & g & u \\ h & f & v \\ g & c & w \end{vmatrix}} = \frac{\gamma}{\begin{vmatrix} a & h & u \\ h & b & v \\ g & f & w \end{vmatrix}} = \frac{-1}{\begin{vmatrix} a & h & g \\ h & b & f \\ g & f & c \end{vmatrix}}$$

$$\Rightarrow \quad \alpha = \frac{Au + Hv + Gw}{-D}, \beta = \frac{Hu + Bv + Fw}{-D}, \gamma = \frac{Gu + Fv + Cw}{-D}.$$

Now we discuss the following cases:

I. If $D \neq 0$, the surface has single centre at a finite distance and hence represents an ellipsoid, hyperboloid or cone.

II. If $D = 0, Au + Hv + Gw \neq 0$, the surface has a centre at an infinite distance and hence represents a paraboloid.

III. If $D = 0, Au + Hv + Gw = 0, A \neq 0$, the line of centres of the surface is at a finite distance and hence represents an elliptic or hyperbolic cylinder, pair of intersecting planes.

IV. If A, B, C, D, E, F, G, H are all zero, $fu \neq gv$, the line of centres is at an infinite distance and hence the surface represents a parabolic cylinder.

V. If A, B, C, D, E, F, G, H are all zero, $fu = gv = hw$, central planes are coincident and hence the surface represents a pair of parallel planes.

Change of Axes

APPENDIX III

1. Introduction

Sometimes while solving the problems in coordinate geometry, it is desirable as well as convenient to consider a figure under discussion and to change its equation, either temporarily or permanently, to a different system of axes such that an important set of points assume the coordinates of simple form. In general, we come across with various types of change of axes. We give below two of them; namely, translation of axes and rotation of axes, which are very useful in our considerations.

2. Translation of Axes

To change the origin of a system of coordinate axes without changing the direction of the axes.

Let OX, OY and OZ be the original axes and (x, y, z) be the coordinates of a point P referred to these axes.

Let $O'(\alpha, \beta, \gamma)$ be the new origin and $O'X'$, $O'Y'$ and $O'Z'$ the coordinate axes parallel to the original axes. Let (x', y', z') be the coordinates of the point P with respect to new axes.

Draw PM perpendicular to the plane XOY meeting the plane $X'O'Y'$ in M'. Then $MP = z$ and $M'P = z'$, $MM' = \gamma$. Now

$$z = MP = MM' + M'P = z' + \gamma.$$

Similarly, $x = x' + \alpha$ and $y = y' + \beta$.

Note. The equations $x = x' + \alpha$, $y = y' + \beta$, $z = z' + \gamma$ giving the relationship between the two sets of coordinate axes are called the translation of axes.

Ex. Find the equation of the plane $3x + 4y + 5z + 10 = 0$ referred to the point $(2, 3, 4)$ as origin; directions of the axes remaining the same.

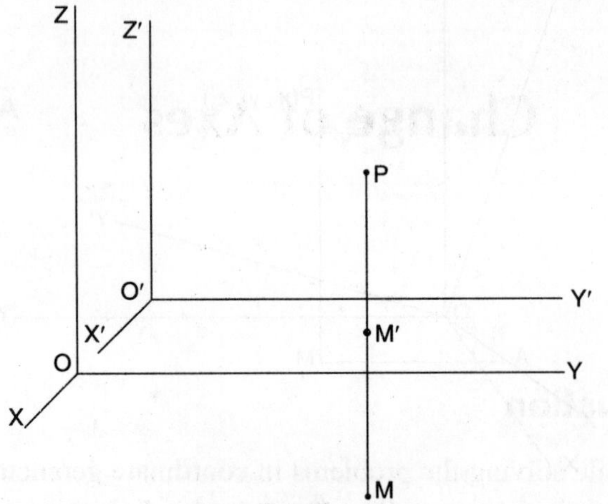

Fig. 1

3. Rotation of Axes

To change the directions of the axes without changing the origin.

Let OX, OY OZ be the original axes and OX', OY', OZ' the new axes. Let l_1, m_1, n_1; l_2, m_2, n_2; l_3, m_3, n_3 be the direction cosines of OX', OY', OZ', respectively, referred to original axes. Then l_1, l_2, l_3; m_1, m_2, m_3; n_1, n_2, n_3 are the direction cosines of OX, OY, OZ, respectively, referred to new axes.

Let (x, y, z) be the coordinates of a point P referred to original axes and (x', y', z') be the coordinates of P referred to new axes.

To express x, y, z in terms of x', y', z'.

Draw PM perpendicular to XOY-plane and MA perpendicular to OX. Then $OA = x$. Now OA is projection of OP on OX. Therefore

$$x = l_1x' + l_2y' + l_3z'.$$

Similarly

$$y = m_1x' + m_2y' + m_3z' \text{ and } z = n_1x' + n_2y' + n_3z'.$$

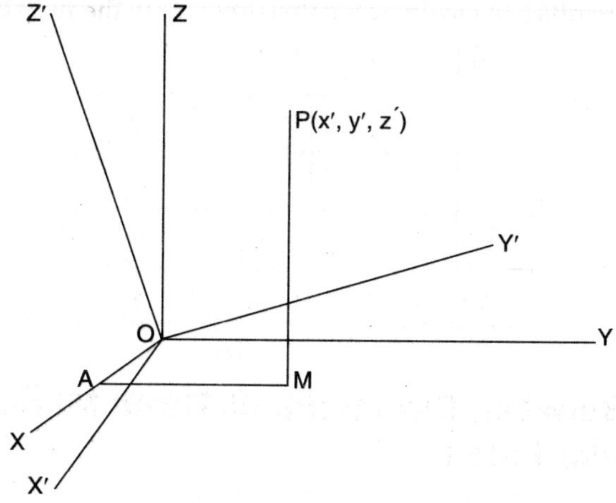

Fig. 2

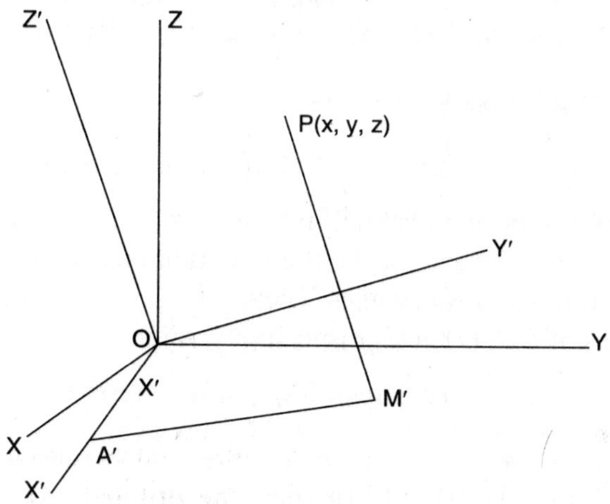

Fig. 3

To express x', y', z' in terms of x, y, z.

Draw PM' perpendicular on $X'OY'$-plane and $M'A'$ perpendicular on OX'. Then $OA' = x'$. Now OA' is the projection of OP on OX'. Therefore

$$x' = l_1x + m_1y + n_1z.$$

Similarly $y' = l_2x + m_2y + n_2z$ and $z' = l_3x + m_3y + n_3z$.

The above results can easily be written down with the help of the following table:

	x	y	z
x'	l_1	m_1	n_1
y'	l_2	m_2	n_2
z'	l_3	m_3	n_3

Relations Between Directions of Three Mutually Perpendicular Lines

Let $OXYZ$ be the given system of coordinate axes. Suppose OA, OB, OC be any three mutually perpendicular lines and their direction cosines, respectively, be l_1, m_1, n_1; l_2, m_2, n_2; l_3, m_3, n_3. Then, these admit the following two sets of relations:

$$\left.\begin{aligned} l_1^2 + m_1^2 + n_1^2 &= 1 \\ l_2^2 + m_2^2 + n_2^2 &= 1 \\ l_3^2 + m_3^2 + n_3^2 &= 1 \end{aligned}\right\} \qquad \text{(I)}$$

$$\left.\begin{aligned} l_1 l_2 + m_1 m_2 + n_1 n_2 &= 0 \\ l_2 l_3 + m_2 m_3 + n_2 n_3 &= 0 \\ l_3 l_1 + m_3 m_1 + n_3 n_1 &= 0 \end{aligned}\right\}. \qquad \text{(II)}$$

Note that l_1, l_2, l_3; m_1, m_2, m_3; n_1, n_2, n_3 are direction cosines of three mutually perpendicular lines OX, OY, OZ (in fact, the original axes of coordinates) referred to new system of coordinate axes $OABC$. Hence, these admit another two sets of relations:

$$\left.\begin{aligned} l_1^2 + l_2^2 + l_3^2 &= 1 \\ m_1^2 + m_2^2 + m_3^2 &= 1 \\ n_1^2 + n_2^2 + n_3^2 &= 1 \end{aligned}\right\} \qquad \text{(III)}$$

$$\left.\begin{aligned} l_1 m_1 + l_2 m_2 + l_3 m_3 &= 0 \\ m_1 n_1 + m_2 n_2 + m_3 n_3 &= 0 \\ n_1 l_1 + n_2 l_2 + n_3 l_3 &= 0 \end{aligned}\right\}. \qquad \text{(IV)}$$

Note. The relations I, II, III, IV are not independent. In fact, the relations III, IV can be algebraically deduced from the relations I, II and vice-versa, without any geometrical considerations at all.

Answers

PART A & B

Part-A

Pages 8 – 9

1. (*i*) $\sqrt{2}$. (*ii*) 11 (*iii*) $2\sqrt{2}$. (*iv*) $3\sqrt{5}$ (*v*) $\sqrt{10}$.

2. 10, 5, $5\sqrt{5}$. **6.** $2\sqrt{2}$.

7. (*i*) $\left(-3, -\frac{1}{9}\right)$. (*ii*) $\left(5, \frac{25}{7}\right)$. (*iii*) (–19, 8).

(*iv*) (11, 3). (*v*) (2, –1).

8. $\left(\frac{9}{5}, \frac{26}{5}\right)$. **9.** (7, 7).

10. (0, 3). **11.** 2:3

12. (*i*) 30. (*ii*) 6. (*iii*) $\frac{1}{2}(a^2 - b^2)$. (*iv*) $a \sin(\alpha - \beta)$.

13. $\frac{11}{2}$. **17.** 7.

Pages 12 – 13

1. (*i*) $\sqrt{34}$ (*ii*) $\sqrt{76}$ (*iii*) $2\sqrt{3}$. (*iv*) $2a\sqrt{5}$.

3. (*i*) $\frac{9}{2} - 2\sqrt{3}$. (*ii*) $\frac{13}{2}\sqrt{3}$. (*iii*) $a^2(8 - 3\sqrt{3})$.

4. (*i*) $r(3\cos\theta + \sin\theta) = 0$. (*ii*) $r = 4$.

(*iii*) $r = 4a \operatorname{cosec}\theta \cot\theta$. (*iv*) $r^2 = a^2 \sin 2\theta$.

5. (*i*) $(x^2 + y^2)^2 = a^2(x^2 - y^2)$. (*ii*) $x^2 + y^2 = 8x$.

(*iii*) $x^2 + y^2 = (4 - x)^2$.

Pages 16 – 17

1. $x - 3y - 1 = 0$. **3.** $x^2 + y^2 = 4$.

5. $x^2 + y^2 - 6x + 2y + 1 = 0$. **6.** $x^2 - 3y^2 = 0$.

7. $x^2 - 3y^2 + 2x - 2y + 2 = 0$. **9.** $x^2 - 6y + 9 = 0$.

10. $\frac{x^2}{16} - \frac{y^2}{9} = 1$.

Pages 27 – 28

1. (*i*) $x + 2y - 6 = 0$. (*ii*) $3x - y + 2 = 0$.

(*iii*) $x + 5y + 13 = 0$. (*iv*) $x + y - 2 = 0$.

2. $x + y - 2 = 0$.

3. (*i*) $x + y + 2 = 0$.

(*ii*) $2x + 3y - 13 = 0$. (*iii*) $x + y - (a + b) = 0$.

(*iv*) $2x - (t_1 + t_2)\, y + 2at_1t_2 = 0$.

(*v*) $\frac{x}{a}\cos\frac{\theta+\phi}{2} + \frac{y}{b}\sin\frac{\theta+\phi}{2} = \cos\frac{\theta-\phi}{2}$.

4. (*i*) $4x + 3y = 12$. (*ii*) $3x - 5y = 15$.

(*iii*) $4x - 3y + 12 = 0$. (*iv*) $x - 4y = 8$. (*v*) $15x - 16y = 10$.

6. $3x - y = 3$.

8. $\frac{4}{3}$; $3x + 4y - 5 = 0$; $4x - 3y - 15 = 0$.

9. $5x + y - 9 = 0$; $4x + 5y - 17 = 0$; $x - 4y + 8 = 0$.

10. $3x + 4y - 2 = 0$. **11.** $3x - y - 24 = 0$.

Page 31

2. (*i*) $(0, 6)$. (*ii*) $(1, 3)$. (*iii*) $(1, -2)$.

3. $2x + y - 8 = 0$.

Page 32

1. (*i*) $\tan^{-1}\frac{1}{7}$. (*ii*) $\tan^{-1}\frac{17}{19}$. (*iii*) $\tan^{-1}31$. (*iv*) $\frac{\pi}{4}$.

2. $y = x + 5$. **3.** $\tan^{-1}\frac{2}{9}, \tan^{-1}\frac{-1}{7}, \tan^{-1} 5$.

Page 35

1. (*i*) $\frac{9}{5}$. (*ii*) $\frac{66}{13}$. (*iii*) $\frac{12}{5}$

2. $\frac{14}{5}$ **3.** $\frac{17}{\sqrt{11}}$.

Page 39

3. (*i*) $= \frac{-2x-3y+1}{\sqrt{13}} = \pm\frac{2x-y+3}{\sqrt{5}}$.

(*ii*) $x + y + 2 = 0, 7x - 7y - 8 = 0$.

4. $2x + 16y + 13 = 0$.

Page 42

1. $8x - 3y - 19 = 0$;

Pages 44 – 45

1. –8. **2.** $\frac{-12}{5}$ **3.** $5x - 12y + 26 = 0$.

4. $3x - 5y + 21 = 0, 5x + 3y + 1 = 0$.

5. $x + y - 9 = 0$.

12. –3. **13.** (4, 3). **14.** $x = y$

Page 52

1. (*i*) $x^2 + y^2 = 4$. (*ii*) $x^2 + 2y^2 + 16 = 0$.

3. $x^2 + 14xy + 49y^2 + 2\sqrt{5}(7x - y) = 0$.

4. $150x^2 + 101y^2 = 875$.

Page 60

1. (*i*) $3x^2 + 8xy - 3y^2 = 0.$ (*ii*) $xy - 4y^2 = 0.$

3. $-3,$ **5.** $(ab' - a'b)^2 + 4(ah' - a'h)(bh' - b'h) = 0.$

7. (*i*) $4x^2 + xy - 4y^2 = 0.$ (*ii*) $x^2 + 42xy - y^2 = 0.$

(*iii*) $7x^2 + 26xy - 7y^2 = 0.$ **9.** -1

Pages 64 – 65

2. (*i*) 2. (*ii*) $\frac{1}{3}$. (*iii*) 28. (*iv*) $1, \frac{5}{3}$.

3. (*i*) $\frac{\pi}{4}$. (*ii*) $\tan^{-1}\frac{3}{5}$. (*iii*) $\frac{\pi}{4}$. **4.** $\lambda = -10, \tan^{-1}\frac{23}{2}$.

Pages 76 – 77

1. (*i*) $x^2 + y^2 - 2x - 4y - 11 = 0.$

(*ii*) $x^2 + y^2 + 6x + 8y + 16 = 0.$

(*iii*) $9x^2 + 9y^2 - 12x - 6y - 22 = 0.$

2. (*i*) $(-2, -4), \sqrt{61}$. (*ii*) $(2, 0), 2\sqrt{5}$. (*iii*) $\left(-1, \frac{5}{6}\right). \frac{\sqrt{61}}{6}$.

3. (*i*) $x^2 + y^2 - 12x - 6y + 40 = 0.$ (*ii*) $x^2 + y^2 + 3x + y - 6 = 0.$

(*iii*) $x^2 + y^2 - 2x + 2y - 11 = 0.$

4. (*i*) $x^2 + y^2 - 19x - 3y = 0.$ (*ii*) $x^2 + y^2 - ax - by = 0.$

(*iii*) $7x^2 + 7y^2 - 27x - 25y + 18 = 0.$

5. $x^2 + y^2 + 14x + 22y - 35 = 0.$

6. $x^2 + y^2 - 4x - 5y = 0.$

7. (*i*) $x^2 + y^2 - 8x - 8y + 16 = 0.$ (*ii*) $x^2 + y^2 + 6x - 2y - 58 = 0.$

8. $x^2 + y^2 - 17x - 19y + 50 = 0.$ **9.** $\pm 5.$

Pages 82 – 83

1. $2x + 3y = 13.$ **5.** $y = x\sqrt{3} \pm 8.$

7. $\frac{lh + mk + n}{\sqrt{l^2 + m^2}} = r.$ **10.** $(-3, 5), (-3, -\frac{1}{3}), (-\frac{1}{2}, 3).$

11. $\lambda = -3$. **12.** $2x + y = 0$. **13.** $n = 0$.

14. $y = (x - a) \tan \alpha$.

Page 87

2. $(g^2 - c)x^2 + (f^2 - c)y^2 + 2fgxy = 0$. **4.** $(2a^2 + 2ab + b^2)^{1/2}$.

Pages 89–90

1. $5x + 18y + 36 = 0$. **2.** $ax - by + a^2 = 0$.

4. $x_1^2 + y_1^2 = 2a^2$. **5.** $x^2 + y^2 = 2a^2$.

Page 94

1. $3x - y = 4$. **2.** $3x + 3y - 7 = 0$.

3. $\left(\frac{5}{3}, \frac{5}{2}\right)$. **4.** $\left(\frac{5}{4}, \frac{1}{2}\right)$.

5. $y^2 + 2ax = a^2$. **8.** $b^2(x^2 + y^2) = (a^2 - cx - dy)^2$.

Page 96

1. $8x - 5y + 21 = 0$. **2.** $\left(\frac{-ln}{l^2 + m^2}, \frac{-mm}{l^2 + m^2}\right)$.

3. $2(x^2 + y^2) - 2cx + c^2 - a^2 = 0$. **5.** Circle.

Page 97

1. $r^2 - r\{r_1 \cos(\theta - \theta_1) + r_2 \cos(\theta - \theta_2)\} + r_1 r_2 \cos(\theta_1 - \theta_2) = 0$.

2. $\left(\frac{1}{2}\sqrt{A^2 + B^2}, \tan^{-1}\frac{B}{A}\right)$.

3. $r^2 - 2ar \operatorname{cosec} \alpha \cos(\theta - \alpha) + a^2 \cot^2 \alpha = 0$.

4. $r \cos(\theta - \theta_1 - \theta_2) = 2a \cos \theta_1 \cos \theta_2$,

$r \cos(\theta - 2\theta_1) = 2a \cos^2 \theta_1$.

5. $2aA + a^2B^2 = 1$.

Pages 98 – 99

1. $x^2 + y^2 - 2x - 2y + 1 = 0.$ **2.** $\left(-\frac{3}{2}, \frac{7}{4}\right); \frac{1}{4}\sqrt{85}.$

5. $\sqrt{l^2 + m^2}\,(x^2 + y^2 - a^2) = c(lx + my + n).$ **6.** Circle.

9. $(x - 2)^2 + \left(y - \frac{3}{2}\right)^2 = 1.$ **14.** $r = a(1 + \cos\theta).$

22. Same circle in both cases.

25. $(ax - by)(ay - bx) + ab(a \pm b)(x \pm y) = 0.$

Pages 103 – 104

2. $3(x^2 + y^2) - 8x + 29y = 0.$ **3.** $3(x^2 + y^2) - 14x + 23y - 15 = 0.$

4. $x^2 + y^2 - 6x - 6y + 9 = 0.$ **5.** $x^2 + y^2 - 16x - 18y - 4 = 0.$

6. $x^2 + y^2 - cx - by + a^2 = 0.$ **7.** $x^2 + y^2 - 5x - 4y + 4 = 0.$

9. Radical axis.

12. $x^2 + y^2 - 2x - 4y + 3 = 0; x^2 + y^2 - 8x - 6y + 17 = 0.$

13. $25(x^2 + y^2) + 49x + 18y - 20 = 0.$

Pages 111 – 112

1. (*i*) $2x + 1 = 0.$ (*ii*) $x + 10y - 2 = 0.$

3. (*i*) (1, 1), (*ii*) (3, 0). (*iii*) (–2, –1).

Pages 119 – 120

1. (*i*) (1, –2), (–2, 1). (*ii*) (0, –3), (–2, –1).

(*iii*) (1, 2), (3, 1). (*iv*) (1, –2), (–1, 2).

2. $x + y + 4 = 0$, (–5, –6).

3. (3, 4), $x^2 + y^2 - 2x + 4y - 115 = 0.$

5. (–1, 1), $\left(\frac{1}{5}, \frac{8}{5}\right)$; $x^2 + y^2 + 102x + 48y - 23 = 0.$

7. $3(x^2 + y^2) - 11x + 3y = 0; 4x - 3y + 3 = 0.$

8. $4(x^2 + y^2) - 3x + 41y = 0.$

9. $x^2 + y^2 + 4y = 0.$

11. $x^2 + y^2 + 2x - 2y + 2 = 0, x^2 + y^2 - 2x + 6y = 0.$

13. (1, 1), (3, 3).

Pages 121 – 123

2. $x^2 + y^2 + x + 2y = 0.$

3. $3x + y = 0; 4(x^2 + y^2) - 9x - 3y + 4 = 0.$

4. $2x + 1 = 0; (1 - 2).$

Pages 130 – 131

1. (*i*) $y^2 - 4y - 8x + 28 = 0.$

(*ii*) $16x^2 - 24xy + 9y^2 - 100x - 50y + 125 = 0.$

(*iii*) $x^2 - 2xy + y^2 - 4y + 6 = 0.$

(*iv*) $16x^2 + 24xy + 9y^2 - 164x + 102y + 589 = 0.$

2. (*i*) $x^2 = -8y$. (*ii*) $y^2 - 6y + 24x - 63 = 0.$

(*iii*) $y^2 + 4y + 28x - 136 = 0.$

3. (*i*) $x^2 + 2x - 2y - 1 = 0.$ (*ii*) $x^2 - 3x - y = 0.$

(*iii*) $y^2 - 3y + x - 4 = 0.$

4. (*i*) Vertex (0, 0), focus $\left(0, -\frac{6}{5}\right)$, directrix $y = \frac{6}{5}$, latus rectum $= \frac{24}{5}$.

(*ii*) Vertex $\left(\frac{1}{2}, 2\right)$, focus (0, 2), directrix $x = 1$, latus rectum = 2.

(*iii*) Vertex (2, –1), focus $\left(2, \frac{1}{4}\right)$, directrix $4y + 9 = 0$, latus rectum = 5.

(*iv*) Vertex $\left(\frac{9}{4}, -1\right)$, focus $\left(\frac{5}{4}, -1\right)$, directrix $4x - 13 = 0$,

latus rectum = 4.

Pages 139 – 140

1. $x - y + a = 0; x + y + a = 0; x + y = 3a; x - y = 3a.$

11. $x(x - h)^2 + y(x - h)(y - k) + a(y - k)^2 = 0.$

12. $y^2 = a(x - 3a).$

16. $27ay^2 = (x - 5a)^2(x - a).$

Page 143

1. $y^2 - 4ax = (a + x)^2.$

3. (*i*) $\frac{y}{x}$ = constant. (*ii*) $y - \lambda x^2 = 2ax$, λ is constant.

Pages 150 – 151

1. (*i*) $y^2 = 2ax.$ (*ii*) $y^2 = 2a(x - a).$

(*iii*) $y^2 = 2a(x - 4a).$

(*iv*) $(y^2 - 2ax + 8a^2)^2 \tan^2 \alpha + 16a^2 (y^2 - 4ax) = 0.$

Pages 158 – 160

3. $y^2 = a(x - 3a).$

8. $y^2 = 2a(x - a); y^2 = a(x - 3a).$

24. $y^2 = 16a(x - 6a).$

Pages 162 – 165

6. $al(2m^2 + l^2) + m^2n = 0.$

Page 173

1. (*i*) $\frac{x^2}{32} + \frac{y^2}{30} = 1.$ (*ii*) $\frac{x^2}{25} + \frac{y^2}{21} = 1.$

(*iii*) $\frac{x^2}{16} + \frac{y^2}{4} = 1.$ (*iv*) $\frac{x^2}{54} + \frac{y^2}{30} = 1.$

(*v*) $13x^2 + 5y^2 = 125.$

2. (*i*) $\frac{\sqrt{5}}{3}$, $(\pm 2\sqrt{5}, 0)$, $x = \pm \frac{18}{\sqrt{5}}$, $\frac{16}{3}$.

(*ii*) $\cos\alpha$, $\left(\pm\frac{\cos^2\alpha}{\sin\alpha}, 0\right)$, $x = \pm \operatorname{cosec}\alpha$, $\sin 2\alpha$.

(*iii*) $\frac{12}{13}$, $(\pm 12, 0)$, $x = \pm \frac{169}{12}$, $\frac{50}{13}$.

(*iv*) $\frac{\sqrt{3}}{2}$, $\left(\pm\frac{3\sqrt{3}}{2}, 0\right)$, $x = \pm 2\sqrt{3}$, $\frac{3}{2}$.

(*v*) $\frac{\sqrt{21}}{5}$, $(0, \pm\sqrt{21})$, $y = \pm \frac{25}{\sqrt{21}}$, $\frac{8}{5}$.

3. $\frac{5x^2}{324} + \frac{(y-3)^2}{36} = 1.$

4. $\frac{(x+2)^2}{40} + \frac{y^2}{54} = 1.$

Pages 181 – 184

1. $y = x\sqrt{3} + \sqrt{3a^2 + b^2}$.

23. $\frac{a^2x^2}{\alpha^4} + \frac{b^2y^2}{\beta^4} = 1.$

Pages 212 – 216

3. $x \pm y \pm \sqrt{a^2 + b^2} = 0.$

20. $\frac{x^2}{a^2} + \frac{y^2}{b^2} = 4.$

Page 223

1. $\frac{x^2}{4} - \frac{y^2}{5} = \frac{4}{9}$.

2. (*i*) $\frac{x^2}{16} - \frac{y^3}{20} = 1.$

(*ii*) $\frac{x^2}{49} - \frac{y^2}{9} = 1.$

(*iii*) $\frac{25x^2}{81} - \frac{y^2}{9} = 1.$

3. (*i*) Foci $(\pm\sqrt{6}, 0)$; Latus rectum $= 2\sqrt{6}$; Eccentricity $= \sqrt{2}$.

(*ii*) Foci $(\pm 5, 0)$; Latus rectum $= 21$; Eccentricity $= \dfrac{5}{2}$.

(*iii*) Foci $\left(\pm\dfrac{\sqrt{21}}{2}, 0\right)$; Latus rectum $= 2\sqrt{7}$;

Eccentricity $= \sqrt{3}$.

4. Foci $(\pm\sqrt{13}, 0)$; Latus rectum $= \dfrac{8}{3}$; Eccentricity $= \dfrac{\sqrt{13}}{3}$.

5. $7x^2 - 9y^2 + 54y - 144 = 0$. 6. $3x^2 - y^2 = 12$.

Pages 229 – 230

4. $4x + 9y = 0$.

5. $(x^2 + y^2)^2 = a^2x^2 - b^2y^2$.

Pages 240 – 241

1. $2xy + 7x - 6y - 21 = 0$; $2xy + 7x - 6y - 24 = 0$.

2. $3x + 2y + 4 = 0$.

3. $12x^2 - 7xy - 12y^2 + 31x + 17y = 0$.

Pages 249 – 250

10. $(xy - c^2)(x^2 + y^2) = c^2xy$.

Pages 267 – 268

1. (*i*) Hyperbola; centre $(-1, 1)$, transverse axis $x + y = 0$; length $2\sqrt{6}$; conjugate axis $x - y + 2 = 0$, length $2\sqrt{2}$.

(*iii*) Parabola; axis $3x - 4y + 7 = 0$; tangent at the vertex $4x + 3y + 2 = 0$; latus rectum 3.

(*iv*) Hyperbola; centre $(-1, 1)$, transverse axis $x - y + 2 = 0$, length $\dfrac{4}{\sqrt{3}}$; conjugate axis $x + y = 0$, length 4.

(*v*) Ellipse; centre $(-2, 3)$; major axis $x + y - 1 = 0$, length $4\sqrt{3}$; minor axis $x - y + 5 = 0$, length 4.

(*vi*) Ellipse; centre (2, –3); major axis $x - y + 3 = 0$, length 6; minor axis $3x - 4y - 18 = 0$, length 4.

(*viii*) Parabola; axis $4x - 3y + 10 = 0$; tangent at the vertex $3x + 4y + 5 = 0$; latus rectum $\frac{1}{5}$.

(*ix*) Ellipse; centre (–1, 1); major axis $2x - y + 3 = 0$, length 4; minor axis $x + 2y - 1 = 0$, length 2; ecc. $\frac{\sqrt{3}}{2}$.

(*x*) Parabola; axis $2x - y - 1 = 0$; tangent at the vertex $x + 2y - 1 = 0$; latus rectum $\frac{4}{\sqrt{5}}$.

2. Parabola; axis $5x - 10y + 13 = 0$; tangent at vertex $10x + 5y + 13 = 0$; latus rectum $\frac{12}{\sqrt{125}}$; vertex $\left(\frac{-39}{25}, \frac{13}{25}\right)$; focus $\left(-\frac{9}{5}, \frac{2}{5}\right)$.

3. Hyperbola; centre (1, 1); transverse axis $x - 3y + 2 = 0$, length $2\sqrt{2}$, conjugate axis $3x + y - 4 = 0$, length $2\sqrt{2}$, ecc. $\sqrt{2}$.

5. $(3x - 4y + 7)(4x + 3y + 1) = 7$; centre (–1, 1); transverse axis $x + 7y - 6 = 0$, length $\frac{2}{5}\sqrt{14}$; conjugate $7x - y + 8 = 0$, length $\frac{2}{5}\sqrt{14}$; asymptotes $3x - 4y + 7 = 0$ and $4x + 3y + 1 = 0$.

6. Hyperbola; centre $\left(0, \frac{1}{2}\right)$; transverse axis $3x + 2y - 1 = 0$, length $\sqrt{5}$; conjugate axis $2x - 3y + \frac{3}{2} = 0$, length $2\sqrt{2}$; foci (1, –1), (–1, 2).

8. Hyperbola; centre (1, 0): transverse axis $x - 2y - 1 = 0$, length 4, conjugate axis $2x + y - 2 = 0$, length 6; foci $\left(1 \pm 2\sqrt{\frac{13}{5}}, \pm\sqrt{\frac{13}{5}}\right)$.

11. $x + y - 1 = 0$; (1, 1).

12. Parabola; axis $3x - 4y + 3 = 0$; tangent at the vertex $4x + 3y + 4 = 0$; latus rectum 1.

Pages 299 – 300

2. $\frac{5}{6}$.

6. If P be the point (h, k) the locus of R is $\frac{h}{x}+\frac{k}{y}=1$.

12. $x^2+xy+y^2+11x+13y+13=0$.

13. $\left(\frac{g-f\cos\omega}{\sin^2\omega}, \frac{f-g\cos\omega}{\sin^2\omega}\right), \frac{\sqrt{f^2+g^2-2fg\cos\omega}}{\sin\omega}$.

Part-B

Page 310

1. 1:2.

2. $2x^2 + 2y^2 + 2z^2 - 4x - 14y + 4z + 109 - k^2 = 0$.

4. $(1, \frac{3}{2}, 2)$.

7. 2:3, 4:5, –7:8.

8. $(-\frac{1}{2}, 2, 0)$.

10. –2:3.

Pages 319–320

2. $\pm\frac{1}{\sqrt{35}}, \pm\frac{5}{\sqrt{35}}, \pm\frac{3}{\sqrt{35}}$.

5. $\pm\frac{1}{\sqrt{3}}, \pm\frac{1}{\sqrt{3}}, \pm\frac{1}{\sqrt{3}}$.

7. $\frac{4}{\sqrt{29}}, \frac{3}{\sqrt{29}}, \frac{-2}{\sqrt{29}}$.

9. $\frac{\pi}{2}, \cos^{-1}\frac{\sqrt{6}}{3}, \cos^{-1}\frac{1}{\sqrt{3}}$.

11. $\cos^{-1}\frac{3}{\sqrt{574}}$.

12. $-\frac{13}{7}$.

13. $\frac{\pi}{2}$.

Pages 321–322

4. $\cos^{-1}\dfrac{20}{3\sqrt{70}}$.

6. $\dfrac{\pi}{3}$.

7. $\cos^{-1}\dfrac{1}{6}$.

8. $\dfrac{4}{3}$.

9. $\cos^{-1}\left(-\dfrac{18\sqrt{2}}{35}\right)$.

15. $\dfrac{2}{\sqrt{30}}, \dfrac{5}{\sqrt{30}}, \dfrac{-1}{\sqrt{30}}$.

16. $-\dfrac{1}{\sqrt{3}}, \dfrac{1}{\sqrt{3}}, \dfrac{1}{\sqrt{3}}$.

Page 330

1. $2x + 3y - z = 14$.

2. $x - 2y - z + 3 = 0$.

3. $7y + 4z - 5 = 0$.

4. $x + 5y - 6z + 18 = 0$.

6. $\lambda = -1$.

Page 332

1. (*i*) $\dfrac{\pi}{3}$; (*ii*) $\dfrac{\pi}{2}$.

3. $3x + 4y - 5z - 9 = 0$.

4. $2x + 2y - 3z + 3 = 0$.

5. $2x - 2y - z = 0$.

6. $3x + 4y - 5z + 4 = 0$.

7. $x + 10y + 7z + 4 = 0$.

8. $y + 4z - 7 = 0, x - z - 1 = 0, 4x + y - 11 = 0$.

Pages 335–336

1. 2.

2. $x^2 + y^2 + z^2 = 9$.

3. $3x^2 + 8y^2 + 35z^2 - 36yz - 24zx + 12xy - 8x - 12y + 24z + 4 = 0$.

Page 337

1. 3, 4.

Page 339

1. $4x + 38y - 62z - 39 = 0$; $22x + 14y + 10z - 195 = 0$.

2. $25x + 17y + 62z - 78 = 0$; $x + 35y - 10z - 156 = 0$.

Page 342

2. $20x + 23y + 26z - 69 = 0$. 3. $2y + z = 2$.

4. $(ac' + a'c)x + (bc' - b'c)y + (dc' - d'c) = 0$.

5. $45x + 15y - 50z + 173 = 0$.

6. $24x - 17v + 15z = 0$.

Pages 350–351

14. $$\frac{abc}{6}\begin{vmatrix} 1 & \cos\alpha & \cos\beta \\ \cos\alpha & 1 & \cos\gamma \\ \cos\beta & \cos\gamma & 1 \end{vmatrix}^{\frac{1}{2}},$$

where $OA = a$, $OB = b$, $OC = c$ are the edges of the tetrahedron $OABC$ and

$\angle AOB = \alpha$, $\angle COA = \beta$, $\angle BOC = \gamma$

Pages 356–357

1. 13.

2. $\left(-\frac{1}{11}, \frac{9}{11}, -\frac{25}{11}\right)$.

3. $(\alpha + lr, \beta + mr, \gamma + nr)$,

where $r = -\dfrac{a\alpha + b\beta + c\gamma + d}{al + bm + cn}$.

5. $\dfrac{x + 1/3}{1} = \dfrac{y + 2/3}{-2} = \dfrac{z}{1}$.

6. $\dfrac{x-2}{2} = \dfrac{y+1}{-3} = \dfrac{z-3}{4}$.

8. $(1, 2, 2)$.

9. $\dfrac{x-1}{-3} = \dfrac{y-2}{5} = \dfrac{z-3}{4}$.

10. 21.

Pages 364–365

2. $29x - 27y - 22z - 85 = 0.$ **3.** $9x - 2y - 5z + 4 = 0.$

5. $21x - 19y + 22z - 125 = 0.$

6. $(2, 1, -3); x + y + z = 0.$ **7.** $(a + a', b + b', c + c').$

8. $x - 2y + z = 0.$ **10.** $x + 2y + 3z - 2 = 0.$

Pages 372–374

1. $3\sqrt{30}$; $4x - 5y - 17z + 79 = 0, 22x - 5y + 19z - 83 = 0.$

2. $14; 117x + 4y - 41z - 490 = 0, 9x - 4y - z - 14 = 0.$

3. $\frac{14}{\sqrt{3}}$; $16x + 11y - z - 18 = 0, 2x + 7y + z - 3 = 0.$

4. 0.

5. $9; \frac{x-1}{1} = \frac{y-3}{-4} = \frac{z+2}{8}.$

6. $\sqrt{35}; \frac{x-4}{1} = \frac{y-2}{3} = \frac{z+3}{-5}.$

7. $6\sqrt{5}.$ **9.** $\frac{13}{5\sqrt{3}}.$

Pages 376–377

1. $\frac{x-16}{1} = \frac{y+7}{-2} = \frac{z}{1}, \frac{x-6}{1} = \frac{y+2}{-2} = \frac{z}{1},$

$\frac{x-36}{1} = \frac{y+22}{-2} = \frac{z}{1}; x - 2y + z + 7 = 0; \frac{25}{\sqrt{6}}.$

2. $\frac{x}{59} = \frac{y}{-5} = \frac{z}{19}.$ **4.** $\left(\frac{24}{5}, -3, -\frac{7}{5}\right).$

Page 380

1. $(x+y)(y+z)=2a(x+z)$.

3. $2x^2+2y^2+2z^2-5yz-5zx-5xy+5ax+5ay+5az-7a^2=0$.

6. $a^2c^2m^2(cy-mzx)^2+b^2c^2(mcx-yz)^2=a^2b^2m^2(c^2-z^2)^2$.

Pages 380–382

1. $\frac{x+1/3}{1}=\frac{y+2/3}{-2}=\frac{z}{1}; x-2y+z=0.$

2. $\frac{1}{2}\sqrt{205}$.

3. $15x+y-7z-53=0$.

5. (9, 13, 15); 14; $9x-4y-z=14$.

7. $\frac{x}{2}=\frac{y}{-1}=\frac{z}{-4};\left(\frac{2}{3},\frac{-1}{3},\frac{-4}{3}\right)$.

11. $(\alpha+\alpha', \beta+\beta', \gamma+\gamma'), \Sigma x(\beta\gamma'-\beta'\gamma)=0$.

13. $\frac{x-1}{0}=\frac{y-1}{1}=\frac{z-1}{3}$.

14. $\frac{ab}{\sqrt{a^2+b^2}}, \frac{bc}{\sqrt{b^2+c^2}}, \frac{ca}{\sqrt{a^2+b^2}}$,

where a, b, c are the edges of the solid.

Pages 388–389

1. $x^2+y^2+z^2-2x-4y-6z+5=0$.

2. (3, 4, –5); 7.

3. $x^2+y^2+z^2-4x+6y+2z-4=0$.

4. $7(x^2+y^2+z^2)-15x-25y-11z=0$.

5. $x^2+y^2+z^2-4x+6y-8z+2=0$.

6. $9(x^2+y^2+z^2)+26x-92y+14z=0$.

7. $x^2 + y^2 + z^2 - 2x + 6y - 4z + 10 = 0$.

8. $x^2 + y^2 + z^2 + 4y - 6z - 1 = 0$.

9. $x^2 + y^2 + z^2 - 6x - 4y + 10z + 12 = 0$.

Pages 393–394

2. $\left(-\frac{13}{98}, \frac{40}{49}, \frac{135}{98}\right)$.

3. (–4, 8, 9); 4.

4. $x^2 + y^2 + z^2 - 4x + 6y - 8z + 4 = 0$.

5. $9(x^2 + y^2 + z^2) - 10x + 20y - 20z - 31 = 0$;

$\left(\frac{5}{9}, \frac{-10}{9}, \frac{10}{9}\right)$; $\frac{2\sqrt{14}}{3}$.

6. $x^2 + y^2 + z^2 - 2x - 2y + 2z + 2 = 0$.

7. $\left(-\frac{7}{3}, -\frac{10}{3}, \frac{-4}{3}\right)$; 3.

8. $3(x^2 + y^2 + z^2) - 2x - 3y - 4z - 22 = 0$.

9. $x^2 + y^2 + z^2 + 3x + 3y - 6z - 21 = 0$.

10. $x^2 + y^2 + z^2 - 6x - 4y - 12z - 91 = 0$.

11. $x^2 + y^2 + z^2 + 3x - 4y + 5z - 6 = 0$.

Pages 398–399

2. $5(x^2 + y^2 + z^2) - 2x - 4y - 5z + 1 = 0$;

$x^2 + y^2 + z^2 - 2x - 4y - 5z + 5 = 0$.

3. $x^2 + y^2 + z^2 - 6x - 4y - 2z + 5 = 0$;

$4(x^2 + y^2 + z^2) - 24x - 11y - 8z + 20 = 0$.

4. (–1, 4, –2).

5. $2x + y - z = \pm 3\sqrt{6}$.

6. $x^2 + y^2 + z^2 - 6x + 2y - 4z + 5 = 0$;

$16(x^2 + y^2 + z^2) - 102x + 50y - 49z + 86 = 0$.

7. $2x - y + 4z - 5 = 0$; $4x - 2y - z - 16 = 0$.

Page 407

2. $5(x^2 + y^2 + z^2) - 13x + 19y - 25z + 45 = 0.$

4. $x^2 + y^2 + z^2 + 7x + 10y - 5z + 12 = 0.$

6. $x^2 + y^2 + z^2 + \frac{b^2 + c^2}{a}x + \frac{c^2 + a^2}{b}y + \frac{a^2 + b^2}{c}z + a^2 + b^2 + c^2 = 0.$

Page 410

1. $x - y + z + 1 = 0; 3x - 6y + 8z + 6 = 0; x - 4y + 6z + 4 = 0.$

2. $\frac{x}{3} = \frac{y}{2} = \frac{z}{7}.$

3. $\frac{x-1}{2} = \frac{y-2}{5} = \frac{z-1}{3}.$

Page 414

1. $(-1, 2, 1)$ and $(-2, 1, -1)$.

2. $x^2 + y^2 + z^2 + 4ax + 4ay - 8az + 4a^2 = 0.$

3. $2(x^2 + y^2 + z^2) - 14x - 3y + 8z = 0.$

Pages 416–419

1. $x^2 + y^2 + z^2 - (a^2 + b^2 + c^2)\left(\frac{x}{a} + \frac{y}{b} + \frac{z}{c}\right) = 0.$

2. $35(x^2 + y^2 + z^2) + 4x + 128y - 238z + 9 = 0.$

6. $x^2 + y^2 + z^2 - ax - by - cz = 0.$

8. $x^2 + y^2 + z^2 + 7x + 9y - 11z + 17 = 0.$

9. $x^2 + y^2 + z^2 + 7y - 8z + 24 = 0.$

11. $\begin{vmatrix} 2(u-u') & l & l' \\ 2(v-v') & m & m' \\ 2(w-w') & n & n' \\ d-d' & -p & -p' \end{vmatrix} = 0.$

20. $x^2 + y^2 + z^2 - ax - by - cz = 0, \ \frac{x}{a} + \frac{y}{b} + \frac{z}{c} = 1.$

21. $\left(\frac{48}{7}, \frac{16}{7}, \frac{24}{7}\right).$

Pages 426–427

1. $(\gamma y - \beta z)^2 = 4a(z - \gamma)(\alpha z - \gamma x)$.

4. (*i*) $p(ax^2 + by^2) = 2z(lx + my + nz)$;

(*ii*) $4z^2(ax^2 + by^2 + cz^2) = (\alpha x^2 + \beta y^2)^2$;

(*iii*) $4z^2\left(\frac{x^2}{a^2} + \frac{y^2}{b^2} + \frac{z^2}{c^2}\right) = \left(\frac{x^2}{\alpha^2} + \frac{y^2}{\beta^2}\right)^2$.

6. $a^2(y^2 + z^2) = b^2x^2$.

9. $3yz + 16zx + 15xy = 0$.

10. $16yz - 33zx - 25xy = 0$.

Pages 435–436

1. (*i*) $\frac{x}{-1} = \frac{y}{4} = \frac{z}{2}; \frac{x}{-1} = \frac{y}{2} = \frac{z}{0}$.

(*ii*) $\frac{x}{1} = \frac{y}{2} = \frac{z}{3}; \frac{x}{3} = \frac{y}{1} = \frac{z}{2}$.

2. $\cos^{-1}\frac{16}{21}$

7. $ax^2 + by^2 + (a + b)z^2 = 1$.

8. $\frac{\pi}{2}$.

Pages 444–445

5. $\sqrt{x} + \sqrt{-6y} + \sqrt{6z} = 0$.

6. $64x^2 + 9y^2 + 25z^2 + 30yz + 80zx - 48xy = 0$.

Pages 446–447

1. $3x^2 - y^2 + 4zx - 10x + 2y - 4z + 6 = 0$.

2. $\{\alpha x + \beta y + \gamma z + u(x + \alpha) + v(y + \beta) + w(z + \gamma) + d\}^2$

$= (x^2 + y^2 + z^2 + 2ux + 2vy + 2wz + d)(\alpha^2 + \beta^2 + \gamma^2 + 2u\alpha + 2v\beta + 2w\gamma + d)$.

Page 449

1. $7x^2 + 37y^2 + 21z^2 - 16xy - 12yz - 48zx + 38x - 88y + 126z - 32 = 0$.

3. $5\{(5y + 4z + 14)^2 + (3z - 5x + 8)^2 + (4x + 3y + 2)^2\} = 75\{(x - 1)^2 + (y + 2)^2 + (z + 1)^2\}$; $(4, -6, 4)$; $5\sqrt{6}$.

6. $8x^2 - 5y^2 - 4z^2 + yz + 5zx + 5xy = 0$.

Pages 450–451

1. $(b - c)mnx + (c - a)nly + (a - b)lmz = 0$.

5. $(bn^2 + cm^2)x^2 + (cl^2 + an^2)y^2 + (am^2 + bl^2)z^2 - 2amnyz - 2bnlzx - 2clmxy = 0$.

6. $$u^2\left(\frac{l'^2}{a}+\frac{m'^2}{b}+\frac{n'^2}{c}\right)-2uu'\left(\frac{ll'}{a}+\frac{mm'}{b}+\frac{nn'}{c}\right)+u'^2\left(\frac{l^2}{a}+\frac{m^2}{b}+\frac{n^2}{c}\right)=0,$$

where $u \equiv lx + my + nz$; $u' \equiv l'x + m'y + n'z$.

Pages 451–452

1. $147x^2 + 87y^2 + 101z^2 + 90yz - 210zx + 210xy + 84y + 84z - 294 = 0$.

2. $(\beta z - \gamma y)^2 = 4a(\beta - y)(\beta x - \alpha y)$.

3. $5x^2 + 3y^2 + z^2 - 6yz - 4zx - 2xy + 6x + 8y + 10z - 26 = 0$.

8. $fyz + gzx + hxy = 0$.

9. $b^2x^2 - a^2y^2 - a^2z^2 = 0$.

13. $yz + zx + xy - x - 2y - z + 1 = 0$.

14. $4z \pm zx \pm xy = 0$.

15. $$\frac{bn^2 + cm^2}{fmn} = \frac{cl^2 + an^2}{gnl} = \frac{am^2 + bl^2}{hlm}.$$

Page 458

1. $5x^2 + 8y^2 + 5z^2 - 4yz - 8zx - 4xy + 22x - 16y - 14z - 10 = 0$.

2. $5x^2 + 5y^2 + 8z^2 + 4yz + 4zx - 8xy - 6x - 42y - 96z + 225 = 0$.

3. $5x^2 + 8y^2 + 5z^2 + 4yz + 8zx - 4xy = 144$.

4. $x^2 + y^2 + z^2 - yz - zx - xy = 1$.

5. $x^2 + y^2 + z^2 - yz - zx - xy - 4x + 5y - z - 2 = 0$.

Pages 470–471

1. $4x + 6y + 3z = 5, 2x - 12y + 9z = 5.$

2. $(-1, 2, 2/3).$

3. $4x + 20y - 21z \pm 13 = 0; (\pm 1, \pm 4, \pm 3).$

4. $7x + 5y + 3z = 30; 14x + 10y + 9z = 50.$

7. $lx + my + nz = \pm\sqrt{a^2l^2 + b^2m^2 + c^2n^2}.$

8. $a^2x^2 + b^2y^2 + c^2z^2 = k^2(x + y + z)^2.$

Pages 488–489

9. $(a\alpha l' + b\beta m' + c\gamma n')(a\alpha' l + b\beta' m + c\gamma' n)$

$= (all' + bmm' + cnn')(a\alpha\alpha' + b\beta\beta' + c\gamma\gamma' - 1).$

Page 493

2. $a(b + c)x^2 + b(c + a)y^2 + c(a + b)z^2 = a + b + c.$

4. $35x^2 + 56y^2 + 19z^2 + 36yz - 12zx + 12xy + 4x - 12y + 12z - 76 = 0.$

5. $(0, 0, a).$

Page 500

1. $x + 6y - 10z + 20 = 0.$

2. $(-2, 3, -1).$

Pages 510–514

1. $7x - 6y - 4z + 21 = 0; 14x - 12y - z + 21 = 0.$

9. $a(b + c)x^2 + b(c + a)y^2 + c(a + b)z^2 = a + b + c.$

10. $\dfrac{(\alpha - a)}{x - f} f(b\gamma - c\beta) + \dfrac{(\beta - b)}{y - g} g(c\alpha - a\gamma) + \dfrac{(\gamma - c)}{z - h} h(a\beta - b\alpha) = 0.$

14. $\left(\pm\dfrac{a}{\sqrt{2}}, \dfrac{b}{2}, \dfrac{c}{2}\right).$

19. $\left(\pm\dfrac{a^2mn(b^2-c^2)}{\sqrt{\Sigma a^2m^2n^2(b^2-c^2)^2}}, \pm\dfrac{b^2nl(c^2-a^2)}{\sqrt{\Sigma a^2m^2n^2(b^2-c^2)^2}},\right.$

$\left.\pm\dfrac{c^2lm(a^2-b^2)}{\sqrt{\Sigma a^2m^2n^2(b^2-c^2)^2}}\right).$

22. $c^4y^2(x^2+y^2-k^2)+b^4z^2(x^2+z^2-k^2)=2b^2c^2y^2z^2.$

Pages 525–526

6. $ab(x^2+y^2)-2(a+b)\,z-1=0.$

8. $\dfrac{x-3}{9}=\dfrac{y-4}{10}=\dfrac{z-5}{15}.$

Index

Subject Index